Turbulent Flows

Aristeu da Silveira Neto

Turbulent Flows

Physical Analysis and Theoretical Modeling

Springer

Aristeu da Silveira Neto
Fluid Mechanics Laboratory (MFLab)
Faculty of Mechanical Engineering
(FEMEC)
Federal University of Uberlândia (UFU)
Uberlândia, Minas Gerais, Brazil

ISBN 978-3-031-89221-9 ISBN 978-3-031-89222-6 (eBook)
https://doi.org/10.1007/978-3-031-89222-6

The original submitted manuscript has been translated into English. The translation was done using artificial intelligence. A subsequent revision was performed by the author(s) to further refine the work and to ensure that the translation is appropriate concerning content and scientific correctness. It may, however, read stylistically different from a conventional translation.

This Springer imprint is published by the registered company Springer Nature Switzerland AG
The registered company address is: Gewerbestrasse 11, 6330 Cham, Switzerland

Foreword

Turbulent flows occur very frequently in a wide variety of geometries, both in natural settings and industrial applications. The solution to a given problem involving such a complex physical phenomenon requires analysis and physical modeling in an appropriate way. This modeling can be done initially with a detailed physical analysis of the problem, which requires a schematic representation, leading to the establishment of suitable simplifying hypotheses, that serve as the basis for the physical and theoretical modeling. Once this is done, one can choose either (1) physical modeling, *i.e.*, setting up the appropriate structure for the desired experiments, or (2) theoretical modeling, *i.e.*, the mathematical representation of the physical reality, using differential mathematical modeling, integral mathematical modeling, as well as algebraic modeling. Within the theoretical modeling, numerical and computational modeling are also required, with which the necessary structure for computational simulation is built. The analyst may choose either of these two methods or undertake two simultaneous lines of work.

The author of the present book aims to contribute to the theoretical modeling of turbulent flows. Based on the existing understanding of the main physical characteristics of turbulent flows, he aims to facilitate an understanding of the mathematical modeling for this flow regime.

The book is organized in the following chapters: 1. Introduction to Turbulent Flows; 2. Mathematical Modeling for Fluid Dynamics; 3. Transition to Turbulence; 4. Homogeneous and Isotropic Turbulence; 5. Equations for Turbulent Flows and the Closure Problem; 6. URANS Methodology and Closure Models; 7. URANS Closure

Models Based on Zero Balance Equation; 8. URANS Closure Models Based on One or More Balance Equations; 9. Sub-Grid Closure Models for LES Methodology and 10. URANS-URANS and LES-URANS Hybrid Closure Models.

Prof. Francisco José de Sousa
Faculty of Mechanical Engineering
(FEMEC)
Federal University of Uberlândia
(UFU)
Uberlândia, Brazil

Prof. Charles Knisely
Bucknell University
Lewisburg, USA

Acknowledgments

First, I thank God, above everything and everyone. In addition, I thank my many advisors who have contributed so much to my development: my parents—advisors for life—Alírio Vicente and Joana Chaves; my Undergraduate Course Completion Work advisor, Prof. Aldo João de Souza—UnB; my master's degree advisor, Prof. Paulo César Phillippi—UFSC; and, my doctoral adviser, Prof. Marcel Lesieur (INPG, Grenoble, France). I thank immensely all the students and researchers at the MFLab-FEMEC-UFU. I thank my family members, in particular my wife, Joveliana, and my son Elias, for the support and incentives given to me. I also thank the mentored students at MFLab and the friends whom I could always count on. I would also like to thank Prof. José Carlos da Silva, Brazilian writer, for reviewing the Portuguese version. Thanks to Prof. Francisco José de Souza (FEMEC-UFU) and Prof. Charles Knisely (Prof. Emeritus, Bucknell University, USA) for reviewing the text in English. I gratefully acknowledge the Brazilian Funding Agencies: CNPq, CAPES and FAPEMIG, and the support of Petrobras. I especially want to thank CNPq for sponsoring the previous edition of this book, Edital Universal 2016, process 401463/2016-8.

Preamble

An italic font was selected to emphasize the most important conceptual expressions regardless of the length of the expressions. Examples of short expressions are *spatial averages*, *fluid filaments*, or *turbulent thermal flux vector*. Samples of longer expressions are *rate of change of a transported information*; *molecular viscous transformation power of an information*; *turbulent diffusive flux of an information* ϕ *over the control surface of a fluid element*, etc.

There was no concern with the number of italicized sentences on a single page. The criterion was to extend the italics until the phrase covered the complete set of ideas exposed at each point in the text. The italic format was not used in table titles and figure legends.

The reviewers of this text, Prof. Francisco José de Sousa and Prof. Charles Knisely, accepted the use of italics in these phrases to assist the reader/student in reading these phrases in blocks, with likely advantages in terms of speed and better comprehension of what is presented.

Expressions that contain names of authorship are presented with initial capital letters and in normal font, as in the following cases: Reynolds Number; Boussinesq Turbulent Viscosity; Van Driest Damping Function; Taylor's Turbulent Structures; MFSim Computational Code; Fourier Pseudo Spectral Computational Code etc.

Paired quotation marks (as in "sample words") are used in the following contexts: (a) to highlight texts by other authors; (b) in cases where the author of this work questions some widely used technical terms found in the fluid dynamics literature with proposed alternatives to them, such as "energy transformation" instead of "energy production"; "viscous transformation" instead of "viscous dissipation" etc.; and (c) in nominal expressions such as: "hairpin"; "horseshoe"; "Robin Wood Effect"; "God does not play dice …," etc. Also, from the author's viewpoint, several new technical terms were introduced, with accompanying justifications. The highlights in some subtitles are denoted using a **bold font**.

Contents

About the Author

Aristeu da Silveira Neto is a professor at the Federal University of Uberlândia (UFU) in Brazil. His academic path began when he entered the University of Brasília (UnB) leading to his subsequent graduation with a bachelor's degree in mechanical engineering in 1983. He then entered the graduate program at the Federal University of Santa Catarina (UFSC) and completed his master's degree in mechanical engineering in 1985. He was employed by Consul S.A. from 1984 to 1985. He then accepted a faculty staff appointment at UFU in 1985. Subsequently, he enrolled at the Polytechnic Institute of Grenoble (INPG—UGA) in France from 1988 to 1991, where he completed his Ph.D. in Engineering. As a doctoral student, he received an award for winning a challenge proposed by a manufacturer of flow measurement systems. He received six other awards during his career as a researcher and professor. After completing his Ph.D., he was hired as a professor of undergraduate and graduate studies in the Faculty of Mechanical Engineering at the Federal University in Uberlândia (FEMEC-UFU). He has mentored more than 80 graduate students and more than 60 undergraduate students (as of 2024) in introductory physical analysis and scientific research as well as final course projects on topics related to fluid mechanics. He taught courses in *Turbulent Flows* at dozens of universities and Brazilian companies. He has held the following positions: Associate Editor of the *Brazilian Journal of Engineering and Mechanical Sciences*; Regional Secretary of the Brazilian Association of Engineering and Mechanical Sciences (ABCM); Reviewer for various technical journals, as well as for scientific and research funding agencies in Brazil. He participated in the organization and proceedings of two meetings of the Brazilian Congress of Engineering and Mechanical Sciences (COBEM) and contributed to in the organization of four proceedings of the Brazilian School of Transition and Turbulence (EPTT).

Notations

Roman Symbols

$A = 26$	Non-dimensional Cebeci Constant
$A^k[W/kg]$	Net advective flux of turbulent kinetic energy across the control surface surrounding a fluid particle
$A^{\varepsilon}[W/(kg.s)]$	Net advective flux of information ε across the surface of the fluid particle
$A^{\omega}[W/(kg.s)]$	Net advective flux of information ω across the control surface of the fluid particle
$A_r[m^2]$	Reference area used in the normalization of drag force and defining the drag coefficient
$A_{ij}[m^2/s^3]$	Net advective flux of transported information, components of the Boussinesq-Reynolds tensor, $\overline{u_i'u_j'}$, across the surface of the fluid particle
$A_{\varepsilon_{\overline{T'T'}}}[^oC^2/s^3]$	Net advective flux of information $\varepsilon_{\overline{T'T'}}$ across the boundary of a fluid particle
$\vec{a}_{rel}[m/s^2]$	Relative acceleration
$A_{\overline{T'T'}}[^oC^2/s]$	Net advective flux of information $\overline{T'T'}$ across the surface of the fluid particle
$A_{\overline{u'T'}}[m\,^oC^2/s^2]$	Advective net flux of information $\overline{u'T'}$, across the surface of the fluid particle

$B[B]$	Any extensive information, for example, mass, linear momentum, energy, etc.
$c = (\omega/\alpha)[m/s]$	Transport speed of a wave
$c[m/s]$	Speed of sound in a material medium
$C_d = F_D/[(\rho U_\infty^2 A_r)/2]$	Drag coefficient
$C_{ij} = \left[\overline{u_i'\bar{u}_j} + \overline{\bar{u}_i u_j'}\right][m^2/s^2]$	Cross tensor
$C_k = 1.4$	Kolmogorov constant
$C_p[J/(kg\,^oC)]$	Thermal capacity of a material medium
C_p	Pressure coefficient
$C_{qj} = \left[\overline{u_i'\bar{T}} + \overline{\bar{u}_j T'}\right][(m/s)\,^oC]$	Cross turbulent thermal flux
$D^{m\varepsilon}[W/(kg.s)]$	Molecular diffusive net flux of information, across the surface of the fluid particle
$D^{t\varepsilon}[W/(kg.s)]$	Turbulent diffusive net flux of information ε across the surface of the fluid particle
$D^{\varepsilon} = (D^{m\varepsilon} + D^{t\varepsilon})[W/(kg.s)]$	Total diffusive fluid flux (molecular and turbulent) of information ε across the surface of the fluid particle
$D^{m\omega}[1/(s^2m)]$	Molecular diffusive flux of ω between a fluid particle and its surroundings
$D^{t\omega}[1/(s^2m)]$	Turbulent diffusive flux of ω between a fluid particle and its surroundings
$D^m_{\varepsilon_{\overline{T'T'}}}[^oC^2/s^2]$	Molecular diffusive flux of information $\varepsilon_{\overline{T'T'}}$ across the boundary of a fluid particle
$D^t_{\varepsilon_{\overline{T'T'}}}[^oC^2/s^2]$	Turbulent diffusive flux of information $\varepsilon_{\overline{T'T'}}$ across the boundary of a fluid particle
$D^m_{ij}[m^2/s^3]$	Molecular diffusive flux of information being transported, that is, the components of the Boussinesq-Reynolds Tensor, $\overline{u_i'u_j'}$, across the surface of the fluid particle
$D^t_{ij}[m^2/s^3]$	Turbulent diffusive flux of information $\overline{u_i'u_j'}$, the components of the Boussinesq-Reynolds Stress Tensor, across the surface of the fluid particle

$D_L[m/s^2]$	Norm of the net flux of linear momentum promoted by turbulence—effect of Leonard and Cross tensors
$D_M[m/s^2]$	Norm of the net flux of linear momentum promoted by molecular diffusion
$D^{km}[W/kg]$	Molecular diffusive net flux of turbulent kinetic energy, across the control surface that surrounds a fluid particle
$D^{kt}[W/kg]$	Turbulent diffusive net flux of turbulent kinetic energy, across the control surface that surrounds a fluid particle
$D_R[m/s^2]$	Norm of the net flux of linear momentum promoted by turbulence—effect of the Boussinesq-Reynolds tensor
$D^m_{\overline{u_i'T'}}[m\,^oC/s^2]$	Molecular diffusive net flux of information $\overline{u_i'T'}$, across the surface of the fluid particle
$D^t_{\overline{u_i'T'}}[m\,^oC/s^2]$	Turbulent diffusive net flux of information $\overline{u_i'T'}$, across the surface of the fluid particle
$D^m_{\overline{T'T'}}[^oC^2/s]$	Molecular diffusive net flux of information $\overline{T'T'}$ across the control surface that surrounds the fluid particle
$D^t_{\overline{T'T'}}[^oC^2/s]$	Turbulent diffusive net flux of information $\overline{T'T'}$ across the control surface that surrounds the fluid particle
$e[J/kg]$	Kolmogorov's Characteristic Specific Kinetic Energy for turbulence (referring to the smallest possible turbulent structures)
$E[J/kg]$	Integral turbulent specific energy (represents the band of turbulent kinetic energy injection into the energy spectrum)
$E(k,t)[(J/kg)/(1/m)]$	Spectral density of specific turbulent kinetic energy, in the spectral space

	of wave number: $E(k,t) \equiv [E_{uu}(k,t) + E_{\vartheta\vartheta}(k,t) + E_{ww}(k,t)]/2$
$E(\vec{x},f)\left[(J/kg)/(1/s)\right]$	Spectral density of specific turbulent kinetic energy, in the spectral frequency space: $E(\vec{x},f) \equiv \left[E_{uu}(\vec{x},f) + E_{\vartheta\vartheta}(\vec{x},f) + E_{ww}(\vec{x},f)\right]/2$
$E_o = \left(g\Delta\rho d_b^2/\sigma\right)$	Dimensionless Eötvos parameter that represents the ratio between gravitational and interfacial tension physical effects
$E_{uu}(\vec{x},f)\left[(J/kg)/(1/s)\right]$	Spectral density of specific turbulent kinetic energy, in the spectral frequency space. It results from the component $u(\vec{x},t)$ of the velocity
$E_{\vartheta\vartheta}(\vec{x},f)\left[(J/kg)/(1/s)\right]$	Spectral density of specific turbulent kinetic energy, in the spectral frequency space. It results from the component $\vartheta(\vec{x},t)$ of the velocity
$E_{ww}(\vec{x},f)\left[(J/kg)/(1/s)\right]$	Spectral density of specific turbulent kinetic energy, in the spectral frequency space. It results from the component $w(\vec{x},t)$ of the velocity
$E_{\phi\phi}(\vec{x},f)$	Spectral density of energy associated with the temporal distribution of an information ϕ
$\vec{f}\left[N/m^3\right]$	Specific force: it can be any type of additional force that appears in Newton's Second Law
$\vec{f}_{ext.field}\left[N/m^3\right]$	Force field, external to a system
$\vec{f}_{ext.surf}\left[N/m^2\right]$	Contact force, external to a system
$f = 1/T = c/\lambda[Hz]$	Frequency
$f(\eta)$	Similarity function
$f(\vec{x},t,w)$	Function that represents a sampling set of an information f, at a position $\vec{x}$ over a time interval, $0 \le t \le T$
$f_c = (2\pi/\Delta t)[1/s]$	Characteristic cut-off frequency of the filtering process
$\hat{f}(\vec{k},t)$	Spatial Fourier Transform
$\left[f(\vec{x},t)\right]$	Ensemble average of an information $f(\vec{x},t,w)$
$\overline{f}(\vec{x},w_i)$	Temporal average of an information $f(\vec{x},t,w)$
$f^{a_n}(\vec{x},t,w)$	Sample of an information $f^{a_n}(\vec{x},t,w)$

$\left[f^{a_1} f^{a_2} \ldots f^{a_n}\right](\vec{x}, t)$	Statistical moment of order n, composed with the information $f^{a_1}, f^{a_2}, \ldots, f^{a_n}(\vec{x}, t, w)$
$F_D[N]$	Drag force
F_1	Hybridization function for the model $k - \omega - SST$
$F_2\left[m^2/s^2\right]$	Second order structure function of velocity
$\vec{g}\left[m/s^2\right]$	Gravitational field
$g(\vec{x})$	Dimensionless function for turbulence intensity, evaluated in the main direction of a turbulent flow
$\vec{g}_{ef}\left[m/s^2\right]$	$\vec{g}_{ef.} = \vec{g} + 0.0005\vec{g} = 1.0005\vec{g}$ effective gravitational field: gravitational field added to the centripetal acceleration, $\vec{\Omega} \times \left(\vec{\Omega} \times \vec{r}\right)$
$G^k\left[W/kg\right]$	Specific power $\left[W/kg\right]$ of transformation of gravitational potential energy into turbulent kinetic energy and vice versa
$G^{\varepsilon}\left[W/(kg.s)\right]$	Process of transforming other information into ε, due to gravitational effects
G^{ω}	Conversion of other information to the information ω, across thermal effects
$G_{ij}\left[m^2/s^3\right]$	Represents the process of transforming gravitational potential energy into components of the Boussinesq-Reynolds tensor, and vice versa
$G_{\overline{u_i'T'}}\left[(m.{}^oC)/s^2\right]$	Represents the transformation of temperature fluctuations into the information $\overline{u'T'}$, due to the gravitational field
$G = (U_\infty\theta/\nu)(\theta/R)^{1/2}$	Görtler's dimensionless parameter that represents the ratio between the physical effects of inertia and diffusive
$Go^2 = k\delta_o Re^2$	Görtler number, where $k[1/m]$ is the wave number, $\delta_o[m]$ is the thickness of a boundary layer and Re is the Reynolds number

$i = \sqrt{-1}$	Complex number
$I(\vec{x}, t)$	Indicator function, used in modeling of two-phase flows
I_1, I_2, I_3	Galilean invariants of the strain rate tensor
$k[W(m^oC)]$	Thermal conductivity
$\vec{k} = 2\pi\left(\frac{1}{\lambda_1}, \frac{1}{\lambda_2}, \frac{1}{\lambda_3}\right)[1/m]$	Wave number vector, composed with the components of the wavelength vector $\vec{\lambda} = (\lambda_1, \lambda_2, \lambda_3)$
$k_c = (2\pi/\Delta x)[1/m]$	Characteristic cutoff wave number of the filtering process
$k_c = (2\pi/\lambda_{max}()[1/m])$	Most amplified wave number in the transition process from a laminar flow regime to a turbulent flow regime
$k \equiv \frac{1}{2}\overline{u_i'u_i'} = \frac{1}{2}\left(\overline{u'u'} + \overline{\vartheta'\vartheta'} + \overline{w'w'}\right)\left[J/kg\right]$	Specific turbulent kinetic energy
$k - \varepsilon$	Closure model for turbulence equations, in the context of the *URANS* methodology
$\ell[m]$	Characteristic length of the turbulent structures
$\ell_I[m]$	Integral length for turbulence (represents the band of turbulent kinetic energy injection in the energy spectrum)
$\ell_m[m]$	Prandtl's Mixing Length, used for the evaluation of turbulent viscosity
$\tilde{\ell}_m[m] = \ell_m/R$	Dimensionless Prandtl's Mixing Length
$\ell_m^+ = \frac{\ell_m}{y_w} = \frac{\ell_m}{\nu/\sqrt{\tau_w/\rho}}$	Prandtl's Mixing Length made dimensionless with wall characteristic quantities
$L_2 = \sqrt{\frac{1}{N}\sum_1^N \lVert\phi_{ni} - \phi_{ri}\rVert^2[\phi]}$	Norm L_2 with which the statistical error between a numerically evaluated information, ϕ_n, and the reference function, ϕ_r is calculated
$L_\infty = max\{\lVert\phi_{ni} - \phi_{ri}\rVert\}[\phi]$	Norm L_∞ with which the maximum error found in the domain of interest is measured, comparing the numerically solution, ϕ_n, of the information ϕ with the reference function, ϕ_r
$L_{ij} = \left[\overline{\bar{u}_i\bar{u}_j} - \bar{u}_i\bar{u}_j\right]\left[m^2/s^2\right]$	Leonard Tensor

$L_{qj} = \left[\overline{\bar{u}_j\bar{T}} - \bar{u}_j\bar{T}\right]\left[m^2/s^2\right]$	Leonard's Turbulent Thermal Flux
$Mo = \left(g\Delta\rho\mu_c^4\right)/\left(\rho_c^2\sigma^3\right)$	Morton's dimensionless parameter represents the ratio between viscous physical effects and interfacial tension
$Mg = -[(d\sigma/dT)](\rho e\Delta T)/(\alpha\nu)$	Marangoni's dimensionless parameter represents the ratio between interfacial tension physical effects and diffusive effects
$p\bar{\bar{I}}\left[N/m^2\right]$	Pressure p and the identity tensor $\bar{\bar{I}}$
$p(r)$	Osculating parabola relative to the function $g(\vec{x}, r)$, for determining the Taylor Characteristic Length λ for turbulence
$P^k\left[W/kg\right]$	Transformation of the kinetic energy, contained in the average flow, to the turbulent kinetic energy spectrum of the turbulent structures that make up the turbulent flow
$P^\varepsilon\left[W/(kg.s)\right]$	Transformation of the information ε, contained in the average flow, into the spectrum of turbulent structures that make up the turbulent flow
$P^\omega\left[W/\left(kg.m^2\right)\right]$	Transformation of the information ω, contained in the average flow, into the spectrum of turbulent structures that make up the turbulent flow
$P_{\overline{T'T'}}\left[{}^oC^2/s\right]$	Transformation of the information $\overline{T'T'}$ contained in the average flow to the same information contained in the entire spectrum of turbulent structures. This transformation is promoted by the gradients of the average temperature
$P^{\bar{T}}_{\varepsilon_{\overline{T'T'}}}\left[{}^oC^2/s^2\right]$	How much of the information $\varepsilon_{\overline{T'T'}}$ is being exchanged between the average temperature field and the thermal variance
$P^{\bar{u}}_{\varepsilon_{\overline{T'T'}}}\left[{}^oC^2/s^2\right]$	How much of the information $\varepsilon_{\overline{T'T'}}$ is being removed from the average speed field
$P_{ij}\left[m^2/s^3\right]$	Represents the transformation of the information object of the balance (Boussinesq-Reynolds Tensor

	Components) from the filtered speed field to the modeled turbulence, not directly resolved
$Pr = \nu/\alpha$	Prandtl number, ratio between the molecular diffusion coefficients of linear momentum and thermal energy
$Pr_{\overline{u'T'}} = \nu_t/\alpha_t$	Turbulent Prandtl number related to the turbulent flux of thermal energy, $\overline{u'T'}$
$P_{\overline{u'T'}}\left[m\,^{o}C/s^2\right]$	Represents the transformation of the information $\overline{u'T'}$, between the average flow and the entire spectrum of turbulent structures that make up the turbulent flow, promoted by the gradients of the average velocity and temperature fields
$\bar{\bar{Q}}$	Transformation tensor of an axis system (x, y, z) into an axis system $\left(x', y', z'\right)$
$Q = \frac{1}{2}\left[\lVert\bar{\bar{\Omega}}\rVert^2 - \lVert\bar{\bar{S}}\rVert^2\right]$	Criterion $Q\left[1/s^2\right]$, serves for an objective visualization of flows. With this criterion, the principle of Euclidean invariance is met
$q_j = \left(\overline{u_jT} - \bar{u}_j\bar{T}\right)[m\,^{o}C/s]$	Global turbulent flux of thermal energy
$\vec{R}[m]$	Absolute position of a non-inertial axis system
$\tilde{\vec{R}}[m]$	Non-inertial position of a non-inertial axis system
$R_1, R_2[m]$	Inner radius and outer radius
$Ra = \left(\rho g \Delta T e^3\right)/(\alpha\nu)$	Rayleigh number, which represents the ratio between gravitational and molecular diffusion effects
Re_D	Reynolds number, based on the characteristic length D
$Re_\tau = (u_\tau R)/\nu$	Reynolds number evaluated with the characteristic velocity of regions close to the walls, $u_\tau[m/s]$
RSM	Closure model of the equations for turbulence (Reynolds Stress Model), in the context of the methodology *URANS*
$\vec{S}[m]$	Absolute position of a fluid particle

$\overline{\overline{S}} \leftrightarrow S_{ij}[1/s]$	$\overline{\overline{S}} = (1/2)\left(\vec{\nabla}\vec{V} + \vec{\nabla}\vec{V}^T\right)$, deformation rate tensor
$\Vert S_{ij}\Vert = \sqrt{2S_{ij}S_{ij}}[1/s]$	Norm of the deformation rate tensor
$\Vert\hat{\overline{S}}_{ij}\Vert = \sqrt{2\hat{\overline{S}}_{ij}\hat{\overline{S}}_{ij}}[1/s]$	Norm of the deformation rate tensor, doubly filtered
$St = f[1/s]/(U/L)[1/s]$	Strouhal number or dimensionless frequency
$t[s]$	Time
$t[s]$	Characteristic time for turbulence (represents the band of turbulent kinetic energy injection in the energy spectrum)
$T_o\left[{}^oC\right]$	Reference temperature
$T_o\left[{}^oCorK\right]$	Temperature
$\bar{T}\left[{}^oCorK\right]$	Filtered Temperature
$\langle T\rangle\left[{}^oCorK\right]$	Averaged Temperature
$T'\left[{}^oCorK\right]$	Temperature fluctuation
$T[s]$	Period
$T(k,t)\left[W.m/kg\right]$	Function with which the net flux of spectral density of turbulent specific kinetic energy, $E(k,t)$, is modeled. This flux occurs between a given wavenumber and the entire band of turbulent structures that make up the spectrum
$T_{ij} = \left(\widehat{\overline{u_i u_j}} - \widehat{\overline{u}}_i\widehat{\overline{u}}_j\right)\left[m^2/s^2\right]$	Turbulent stress tensor, based on two filtering processes. Entity that appears in the sub-grid closure modeling for turbulence
$Ta = r_i L^3\left(\Omega_i^2 - \Omega_o^2\right)/\nu^2$	Dimensionless Taylor parameter that represents the ratio between the physical effects of rotation and diffusivity
$T^k\left[W/kg\right]$	Rate of change of turbulent kinetic energy contained in a fluid particle
$T^\varepsilon\left[W/(kg.s)\right]$	Rate of change of the information ε contained in a fluid particle
$T^\omega[1/s]$	Rate of change of the information ω contained in ae fluid particle
$T_{ij}\left[m^2/s^3\right]$	Rate of change of the components of the Boussinesq-Reynolds tensor, $\overline{u_i' u_i'}$, contained in a differential particle of fluid

$T_{\overline{u'T'}}[m\,^oC/s^2]$	Rate of change of the information $\overline{u_i'T'}$ contained in the fluid particle
$\overline{u'T'}[^oC^2]$	Variance of temperature fluctuations
$T_{\overline{T'T'}}[^oC^2/s]$	Rate of change of the information $\overline{T'T'}$ that is contained in a differential fluid particle
$T_{\varepsilon_{\overline{T'T'}}}[^oC^2/s^2]$	Rate of change of the amount of information $\varepsilon_{\overline{T'T'}}$, contained in the fluid particle
$u[J/kg]$	Specific internal energy of a material particle
$\bar{u}_i(\vec{x})[m/s]$	Filtering of the component i of the velocity vector
$\langle u_i \rangle(\vec{x})[m/s]$	Temporal average of the component i do vetor velocidade
$u_i'(\vec{x}, t)[m/s]$	Fluctuation of component i of the velocity vector
$U[m/s]$	Speed of a free current
$U(y)[m/s]$	Vertical distribution of base velocity, for linear stability analysis
$u_\tau = \sqrt{\tau_w/\rho}[m/s]$	Characteristic velocity of regions close to the wall, based on the calculated viscous stress on the wall, $\tau_w[N/m^2]$
$U_1[m/s]$	Velocity of a free stream of a free shear flow
$U_2[m/s]$	Velocity of a free stream of a free shear flow
$U_\infty\left[\frac{m}{s}\right]$	Speed of a free current of a boundary layer
$\bar{U} = \left[\frac{(U_1+U_2)}{2}\right]\left[\frac{m}{s}\right]$	Average velocity in a free shear flow
$\bar{U} = [(U_1 - U_2)/2][m/s]$	Velocity half-difference in a free shear flow
$U_{ij}(\vec{r}, t) = \langle u_i'(\vec{r}, t)u_j'(\vec{x}+\vec{r}, t)\rangle[m^2/s^2]$	Statistical moment of the velocity components, at two points
$\widehat{U}_{ij}\left(\vec{k}, t\right)\left[(m^2/s^2)/(1/m^3)\right]$	Spectral tensor, based on statistical momentum at two points: $U_{ij} = \langle u_i'(\vec{r}, t)u_j'(\vec{x}+\vec{r}, t)\rangle$
$\langle u'u' \rangle = \overline{u'^2}[m^2/s^2]$	Turbulent intensity relative to the x direction, considering the hypothesis of ergodicity, that is, the ensemble average equal to the temporal average

$\langle u_i' u_j' \rangle \left[m^2/s^2\right]$	Boussinesq-Reynolds tensor in relation to the averaging process
$\overline{u_i' u_j'}\left[m^2/s^2\right]$	Boussinesq-Reynolds tensor in relation to the filtering process
$\langle u_j' T' \rangle [m\,^oC/s]$	Turbulent thermal energy flux vector, resulting from the averaging process
$\overline{u_j' T'}[(m/s)\,^oC]$	Turbulent thermal energy flux, across a given elementary surface resulting from the filtering process
$\vartheta [m/s]$	Characteristic velocity of a flow
$\vartheta_r [m/s]$	Characteristic Velocity of a turbulent structure of size r
$\langle \vartheta_i' \vartheta_j' \rangle = \vartheta'^2 \left[m^2/s^2\right]$	Turbulent intensity relative to the direction y
$\vec{V}_{rel.}[m/s]$	Velocity of a fluid particle relative to a non-inertial observer (x, y, z)
$\vec{\tilde{V}}[m/s]$	Velocity of a fluid particle relative to a non-inertial observer $(\tilde{x}, \tilde{y}, \tilde{z})$
$\vec{V}_{abs.}[m/s]$	Absolute velocity
$\vec{V}_{xyz}[m/s]$	Absolute velocity of the non-inertial axis system (x, y, z), placed at the center of the Earth
$\vec{V}_{\tilde{x}\tilde{y}\tilde{z}}[m/s]$	Absolute velocity of the non-inertial axis system $(\tilde{x}, \tilde{y}, \tilde{z})$, placed in the problem of interest
$\vec{V}_{T\tilde{x}\tilde{y}\tilde{z}}[m/s]$	Velocity relative to the earth, of the non-inertial axis system $(\tilde{x}, \tilde{y}, \tilde{z})$, placed in the problem of interest
$\vec{V}(r_{\Gamma T}, t)[m/s]$	Fluid velocity at the interface between the fluid and the earth
$\vec{V}(r_{\Gamma T\infty}, t)[m/s]$	Fluid velocity at the domain boundary, around the earth and away from it
$\widehat{V}[m/s]$	Characteristic velocity of turbulence, used to compose the model for turbulent viscosity
W	Sample belonging to the sampling space Ω
(x, y, z)	Non-inertial axis system located at the center of the earth
(X, Y, Z)	Absolute axis system
$(\tilde{x}, \tilde{y}, \tilde{z})$	Non-inertial axis system placed in the problem of interest

$x_R[m]$	Reattachment length of a boundary layer
$\langle w'w' \rangle = \langle w'^2 \rangle [m^2/s^2]$	Turbulent intensity relative to the z direction, using the ergodicity hypothesis
$x[m]$	Cartesian coordinate
$y[m]$	Cartesian coordinate
$\tilde{y} = y/R$	Coordinate dimensionless with the characteristic dimension R
$y^+ = \frac{y}{y_w} = \left(y/\frac{\nu}{\sqrt{\tau/\rho}}\right)$	Coordinate normalized with the characteristic dimension $y_w = \frac{\nu}{\sqrt{\tau/\rho}}$
$z[m]$	Cartesian coordinate

Greek Symbols

$\alpha = k/(\rho C_{po})[m^2/s]$	Coefficient of molecular diffusion of thermal energy
$\alpha = V_d/V_t$	Volume fraction—ratio between the volume of the dispersed phase and the total volume
$\alpha_t[m^2/s]$	Concept of turbulent diffusion of thermal energy
$\alpha_{ef} = (\alpha + \alpha_t)[m^2/s]$	Effective thermal diffusion coefficient, which results from the sum of the molecular thermal diffusion coefficient, $\alpha[m^2/s]$, with the turbulent thermal diffusion coefficient, $\alpha_t[m^2/s]$
$\beta[1/^oC]$	Thermal expansion coefficient
$\beta[1/K]$	Thermal expansion coefficient
$\beta[1/s]$	Coefficients
$\beta[B]/[kg]$	Any intensive information, associated with extensive information$B$.
β	Empirical proportionality constant used for the evaluation of Prandtl's mixing length in relation to a characteristic length, $\delta[m]$, of the flow: $\ell_m(x) = \beta\delta(x)$
$\delta[m]$	Boundary layer thickness
$\delta_o[m]$	Initial thickness of a mixing layer or the thickness of a boundary layer
$\delta_{ij} = \begin{cases} 1 if i = j \\ 0 if i \neq j \end{cases}$	Kronecker's tensor
ε_{pqi}	Levi–Civita tensorial operator
$\varepsilon[W/kg]$	Specific power of viscous transformation of turbulent kinetic energy into thermal energy, inside a fluid particle
$\varepsilon_{ij}[m^2/s^3]$	Represents the specific power of viscous transformation of the components of the

	Boussinesq-Reynolds Tensor into other related information
$\varepsilon_{\overline{u_i'T'}}[m^oC/s^2]$	Represents the specific power of molecular viscous transformation of the information $\overline{u_i'T'}$, in another information
$\varepsilon_{\overline{T'T'}}[^oC^2/s]$	Represents the specific power of molecular viscous transformation of the variance of temperature fluctuation, $\overline{T'T'}$, in another information
Γ	Concept of circulation
Γ	Domain limit or boundary
Γ_T	Control surface for a control volume involving the earth
Γ_P	Control surface for a control volume around a specific problem
$\Gamma[\%]$	Percentage of time for which the velocity near a wall remains positive
$\kappa = 0.41$	Von Kármán's Dimensionless Constant
$\xi[m]$	Mean molecular free path
$\lambda[m]$	Wavelength
$\lambda_{max}[m]$	Most amplified wavelength in the transition process from a laminar flow regime to a turbulent regime
$\lambda = \mu_c/\mu_d$	Ratio between the dynamic viscosities of the continuous and dispersed phases of two-phase flows
$\vec{\lambda}(\lambda_1, \lambda_2, \lambda_3)[m]$	Wave length vector
$\lambda[m]$	Taylor's Characteristic Length for turbulence
Λ	Lambda instability of the transition from laminar to turbulent boundary layer regime
$\eta = -p\vec{\nabla} \cdot \vec{V}[W/m^3]$	Specific power of kinetic energy transformation into thermal energy, by the action of the pressure field
$\eta[m]$	Wave displacement
$\eta = \rho_c/\rho_d$	Ratio between the specific masses of the continuous and dispersed phases of two-phase flows
$\eta[m]$	Kolmogorov's Characteristic Length (referring to the smallest possible turbulent structures)
$\eta(y, t)$	Similarity variable
$\mu[N.s/m^2]$	Dynamic viscosity of a fluid
$\mu_c[N.s/m^2]$	Dynamic viscosity of the continuous phase of a two-phase fluid
$\mu_d[N.s/m^2]$	Dynamic viscosity of the dispersed phase of a two-phase fluid
$\nu = (\mu/\rho)[m^2/s]$	Kinematic viscosity or linear momentum diffusion coefficient
$\nu_t[m^2/s]$	Concept of Boussinesq's turbulent kinematic viscosity

$\nu_{ef} = (\nu + \nu_t)\left[m^2/s\right]$	Effective kinematic viscosity, which results from the sum of molecular kinematic viscosity, $\nu\left[m^2/s\right]$, with turbulent kinematic viscosity, $\nu_t\left[m^2/s\right]$
$\mho$	Control volume
$\vec{\Omega}[1/s]$	Angular velocity of a non-inertial axis system (x, y, z)
$\bar{\bar{\Omega}}[1/s]$	Angular velocity of a non-inertial axis system $(\tilde{x}, \tilde{y}, \tilde{z})$
$\Vert\bar{\bar{\Omega}}\Vert = \sqrt{\Omega_{ij}\Omega_{ij}}[1/s]$	Norm of the vorticity tensor, Ω_{ij}
$\vec{\omega} = 2\vec{\Omega}[1/s]$	Vorticity, which, by definition, is given by twice the angular velocity or rotation rate $\vec{\Omega}$
$\omega[1/s]$	Angular frequency
$\omega[1/s]$	Kolmogorov's characteristic vorticity for turbulence (referring to the smallest possible turbulent structures)
$\Pi_{ij}\left[m^2/s^3\right]$	Net flow due to turbulent redistribution of information $\overline{u_i'u_j'}$, components of the Boussinesq-Reynolds stress tensor, across the fluid particle surface
$\Pi_{\overline{u_i'T'}}\left[(m\,{}^oC)/s^2\right]$	Dispersion of the information $\overline{u_i'T'}$, due to the action of turbulence
ψ	Phase constant
$\psi^{\varepsilon}\left[W/(kg.s)\right]$	Viscous transformation process of the information ε into other types of information
$\psi^{\omega}\left[1/s^2\right]$	Viscous transformation of the information ω into other type of information
$\rho[1/s]$	Coefficients
$\rho\left[kg/m^3\right]$	Specific mass of a substance
$\rho'(\vec{x}, t)\left[kg/m^3\right]$	$\rho'(\vec{x}, t) \ll \bar{\rho}$: fluctuation of specific mass, compared to a reference specific mass. This is the basis for the Boussinesq approximation
$\rho_o\left[kg/m^3\right]$	Reference specific mass
$\rho_c\left[kg/m^3\right]$	Specific mass of the continuous phase of a biphasic fluid
$\rho_d\left[kg/m^3\right]$	Specific mass of the dispersed phase of a biphasic fluid
$\sigma[1/s]$	Coefficients
$\bar{\bar{\sigma}}_{abs}\left[N/m^2\right]$	Tensor of viscous stresses and pressure
$\bar{\bar{\sigma}}_{xyz}\left[N/m^2\right]$	Tensor of viscous stresses and pressure, for a non-inertial observer at the center of the earth
$\bar{\bar{\sigma}}_{\tilde{x}\tilde{y}\tilde{z}}\left[N/m^2\right]$	Viscous stress tensor and pressure, for a non-inertial observer in the problem of interest
$\tau_{ij} = \overline{u_i'u_j'}\left[N/m^2\right]$	Boussinesq-Reynolds tensor, composed of the second order statistical moments between velocity fluctuations
$\tau_{ij} = \left[\overline{u_iu_j} - \bar{u}_i\bar{u}_j\right]\left[m^2/s^2\right]$	Global Boussinesq-Reynolds tensor
$\tau[s]$	Kolmogorov's characteristic time for turbulence (referring to the smallest possible turbulent structures)

$\tau_w\left[N/m^2\right]$	Viscous stress calculated on the wall

Other Symbols

$\delta V\left[m^3\right]$	Finite volume element
$\vec{\nabla}[1/m]$	Mathematical operator, $\vec{\nabla} = [\partial(\,)/\partial x, \partial(\,)/\partial y, \partial(\,)/\partial z,]$
$\frac{D()}{Dt} = \frac{\partial()}{\partial t} + \left(\vec{V}\cdot\vec{\nabla}\right)()$	Material derivative operator
$\vec{\nabla}\vec{V}[1/s]$	Velocity vector gradient tensor
$\vec{\nabla}\vec{V}^T[1/s]$	Transposed velocity vector gradient tensor
$\vec{\nabla}\vec{V}\cdot\bar{\bar{\sigma}} = -p\vec{\nabla}\cdot\vec{V} + 2\mu\lVert\bar{\bar{s}}\rVert^2 - (2/3)\mu\left(\vec{\nabla}\cdot\vec{V}\right)^2 = \eta + \phi$	Specific power of kinetic energy transformation into thermal energy—reversible process
$\nabla^2(\,)\left[1/m^2\right]$	Laplacian operator
$\nabla^4(\,)\left[1/m^4\right]$	Double Laplacian or Hamiltonian operator
$\wp_{ij} = \delta_{ij} - k_i k_j/k^2$	Second order tensor, for the projection of any information $\vec{a}$ onto the plane π, to which the wave number vector $\vec{k}$ is normal
$\wp_{ijm}\left(\vec{k},t\right) = k_m\wp_{ij}\left(\vec{k},t\right) + k_j\wp_{im}\left(\vec{k},t\right)$	Third order tensor, for the projection of any third order statistical moment onto the plane π, to which the wave number vector $\vec{k}$ is normal

Acronyms

CFL	Stability control parameter of numerical discretization methods for *ODE*'s and *EDP*'s (Courant-Friedrichs-Lewy)
DNS	Direct Computer Simulation (*Direct Numerical Simulation*)
dof	Degrees of freedom of a dynamical system
$erf\,(\eta) = \frac{2}{\sqrt{\pi}} \int_0^{\eta} e^{-\tilde{\eta}^2} d\tilde{\eta}$	Gaussian Error Function
FEMEC	Faculty of Mechanical Engineering
fps	Images per second (*frames per second*)
IMERSPEC3DP	Computational modeling platform based on the fusion of immersed boundary and Pseudo-Spectral Fourier methods
LES	Large Eddy Simulation
MFLab	Fluid Mechanics Laboratory
ODE	Ordinary differential equation
PDE	Partial differential equation
RANS	Balance equations for linear momentum subjected to Boussinesq-Reynolds averaging (Reynolds Averaged Navier-Stokes Equations)
RFG	Model for random noise (Random Flow Generator)
RK4	Runge-Kutta scheme of order 4
SAS	Scale Adaptative Simulation
SST	Shear Stress Transport
UFU	Federal University of Uberlândia
URANS	Transient balance equations for linear momentum subjected to the Boussinesq-Reynolds average (Unsteady Reynolds Averaged Navier-Stokes Equations)

Chapter 1
Introduction to Turbulent Flows

In this chapter an introduction to the topic of turbulent flows is presented, according to the following sections: Sect. 1.1, an approach to dynamical systems, chaos and turbulence; Sect. 1.2, the characterization of turbulent flows; Sect. 1.3, a physical model for the transformation of energy via viscous process; Sect. 1.4 manifestation of turbulence; and, concluding with Sect. 1.5, exercises for the review and consolidation of the concepts presented in this chapter.

1.1 Dynamical Systems, Chaos and Turbulence

Turbulence is a highly complex natural phenomenon of great practical interest. Its occurrence is not necessarily related only to the movement of fluids. Potentially, any dynamical system with a sufficiently large number of degrees of freedom can meet the criterion for operating in a turbulent regime. Throughout this book, a turbulent regime is characterized by a sufficiently large number of degrees of freedom, among other characteristics. In the turbulent regime of any dynamical system, the values assumed by the quantities that characterize the operation of the system can vary in both time and space, making it difficult to predict the values of these quantities deterministically, since they are strictly dependent upon and highly sensitive to small variations of the initial conditions and boundary conditions of the problem. From this perspective, there are numerous examples where turbulence can be found. It should be noted, however, that unpredictability alone is not enough to characterize a turbulent regime.

The observation of an ant trail between a food source and the entrance to an anthill provides a very illustrative example of a highly complex dynamical system, which very often operates in a chaotic regime. Indeed, even though the starting and ending points of each worker involved in the process are known a priori, their

A. da Silveira Neto, *Turbulent Flows*, https://doi.org/10.1007/978-3-031-89222-6_1

individual trajectory, or even the spatial configuration on the trail over time, is totally unpredictable, even for a few seconds after the instant when the observation begins.

Moreover, a small perturbation injected into the system by an external agent, for example, the introduction of a material obstacle or a small amount of a chemical substance at any point on the track, can transform this globally well-organized dynamical system into true chaos. However, it is interesting to note that, although the basic quantities associated with the system, such as trajectories of individuals, instantaneous flow of food, flux of *informations*, among several others, are unpredictable in time and space, the statistical values of these quantities can be determined with great accuracy. It is also interesting to note, still considering the example of ants, that information can be transmitted from one individual ant to another in various ways: by antennae touching, by diffusion and advection of pheromones, by waves and by radiation.

There are more unknown situations than one can imagine! The state of unrest between the individual ants, when the community feels threatened, or when a major discovery has been made, can put this dynamical system into a regime of turbulence, with all the characteristics that are associated with this word.

The Amazon ecosystem, which impresses us with the luxuriance of the rainforest and the multiplicity of interrelated species, is a typical example of a dynamical system with an almost uncountable degree of freedom. If a particular species, which is part of one of the various food chains of this complex system, is decimated, the balance of the whole is compromised, and the consequences are also unpredictable, providing conditions for the disappearance or uncontrolled proliferation of other species. This type of behavior, containing characteristics of the turbulence regime, originates from the non-linear dynamics involving the various quantities that compose the system.

There are many factors that motivate or force people to move or even migrate from small towns to large urban centers and vice versa. Consider, on the one hand, the advantages found in the big cities, such as substantially greater employment options and more leisure opportunities. On the other hand, life in a small city is generally less tumultuous. Each community may seem more attractive to a certain subset of the population in each community. Unbridled growth in big cities and associated political, economic and social actions adopted by legislators and rulers to regulate this growth often triggers waves of popular dissatisfaction. As a result, social instabilities appear, and people move to other neighborhoods or locations, always in search of better living conditions. Other factors, such as bad weather and prolonged periods of drought, affect the lives of rural communities, inducing entire families to migrate from the rural areas to big cities. A person inserted in this system cannot be seen as an isolated individual, but as part of a whole, where several components interact in a complex way, under the action of imponderable factors. As a result, predicting the future residence of a person inserted into this scenario becomes a very complex task. Nevertheless, the average population growth of a city, which accounts for several factors, has been predicted with relative reliability by government agencies dedicated to demographic studies.

The unpredictable behavior of the stock markets, under the influence of a wide range of different factors, is another typical example of the operation of a turbulent regime. A small disturbance of the political, social, or natural order, injected in a single distant country, can initiate strong economic instabilities, capable of rapid global spreading and, thereby, compromising the equilibrium of the international financial system. In such a situation, solid and well-established companies can go from opulence to bankruptcy in a few hours, as a consequence of our inability to make predictions about the future behavior of a dynamic system such as the complex global stock exchange.

The propagation of an epidemic front of known origin can also easily degenerate and give rise to a chaotic process, due to the important influence exerted by the interactions between individuals and the environment on the mechanisms of proliferation of contagious diseases. In certain situations, no matter how great the efforts undertaken to combat the spread of the disease, successive outbreaks of the epidemic will emerge in places and at times never imagined by health authorities.

Nothing is more chaotic (unpredictable) than traffic in large cities. Changing a tire on the side of a high-speed urban road on a day of heavy traffic can trigger a traffic jam of tens of kilometers, caused by the simple curiosity of drivers passing by, even if no obstruction of the roadway has been caused. Disruptions of greater proportions, however, caused by accidents and construction work on public roads, popular demonstrations, stadium departures, rain and floods, are part of the daily lives of people living in big cities, causing a rapid transition from a situation of relative order to chaos in traffic.

This experience is lived daily by a huge contingent of people who, as they leave their homes to go to work, have learned to find their way around with input from radio stations that use helicopters to transmit continuous information about traffic conditions to drivers along their routes. In a fully interactive process, drivers change their respective routes, using the information provided to them, to avoid the most congested spots. In such a situation, the driver is not able to predict, with a minimum of advance notice, which will be the most effective route to take the driver to a certain destination. Here is an example of nonlinear interaction between drivers and the helicopter traffic spotters. The information sent by helicopter affects the trajectories of the drivers, which in turn affects the aerial observation and resending of updated information.

The examples of dynamical and turbulent behavior that can be drawn from nature and everyday life are both numerous and surprising. Many more pages of this text could be devoted to this kind of approach, providing the reader with an alternative way of looking at the world and the causes of everyday problems. The author's main interest, however, is in the study of turbulence in fluid flows, a subject that has intrigued the minds of philosophers, scientists and engineers over the past centuries. The lack of perfect understanding remains one of the great elusive challenges of modern physics, despite the concerted efforts of some of the best minds in the world.

1.2 Characterization of Turbulent Flows

1.2.1 General Aspects

To define a physical phenomenon precisely, the mechanisms that give rise to it must first be well understood. Despite the remarkable advances in tools for physical and theoretical analyses of turbulent flows, many aspects of the phenomena involved in this flow regime remain quite obscure, even today. These obscure aspects make it difficult to establish a complete and comprehensive definition of turbulence. So far, all attempts in this direction have produced unsatisfactory or controversial definitions. There is still no consensus among the various experts as to how the various fundamental aspects of the phenomenon should be interpreted. However, the marked presence of turbulence in numerous natural phenomena and in technological systems has motivated enormous efforts by researchers dedicated to studying this subject in a vast array of laboratories throughout the world. As a result of the already achieved understanding of turbulence, one can safely establish characteristics associated with the physical nature of turbulent flows. A discussion of this topic is presented in the following Sects. 1.2.2–1.2.9.

1.2.2 Turbulence Is Not Characterized by Vortices

Turbulent flows have often been associated with the presence of vortices within the flow, leading to an incorrect interpretation of their physical nature. In fact, a vortex is a well-organized entity, consisting of a mass of fluid in rotating motion around a central region. From the statistical point of view, the definition of a vortex requires complete invariance of the statistical moments throughout the region it covers. This strict definition of a vortex leads us to the conclusion that we would most likely not find a single vortex in a transitional flow, or even in a fully developed turbulent flow.

With a more flexible definition of a vortex, one might argue that vortices can be identified in both transitional and fully turbulent flows. However, the mere identification of vortices does not guarantee that a given flow can be considered turbulent; the turbulent regime cannot be considered to be a simple superposition of vortices. In this book, the author seeks to give greater consistency to the concept of turbulence; thus, he proposes that turbulence is composed of turbulent structures, which incorporate, above all, other topological forms with different levels of complexity, when compared to the topology of a vortex. Incidentally, the term "eddy", widely used in the English-language turbulence literature, should not always be considered synonymous with "vortex". In most of the situations discussed, "eddy" is synonymous with "turbulent structure". This nomenclature avoids the physical inconsistency that comes from erroneously characterizing turbulence as a superposition of vortices, which does not correspond to physical reality. Characterizing turbulence in this way (i.e., as the superposition of vortices) diminishes all its physical richness in terms of its true composition, which can be better understood based on the concept of turbulent structures. This concept will be discussed and detailed throughout this text.

Images of three flows, characterized by different values of the Reynolds number, are presented in Fig. 1.1. Although these flows are not in the turbulent regime, they are used to illustrate the difficulty in identifying vortices in laminar, transitional and turbulent flows. Figure 1.1a shows a laminar flow, with imminent transition, over a cylinder, with the presence of two counter-rotating recirculating regions, which remain trapped downstream of the immersed solid body. At this point in this text, we will introduce the concept of instability in flows. The concept of instability will be adopted whenever any time variation is identified in the flow. For example, the flows in Fig. 1.1b, c are more unstable than those in Fig. 1.1a, due to the increase in the Reynolds Number, which characterizes the three flows over the same cylinder.

In all three cases, Fig. 1.1a–c, the identification of vortices, according to their characterization presented above, is not possible for any of the three flows. What can be visualized are turbulent structures that become increasingly complex with increasing Reynolds Number, in order from Fig. 1.1a–c.

It is noted that the word "regime" was only used for either the laminar domain or the turbulent domain. It is understood that a regime has its own characteristics, as in the cases of laminar and turbulent regimes. In the transition process, characteristics of both laminar and turbulent regimes are present simultaneously. Therefore, it is not appropriate to speak of a distinct transition regime.

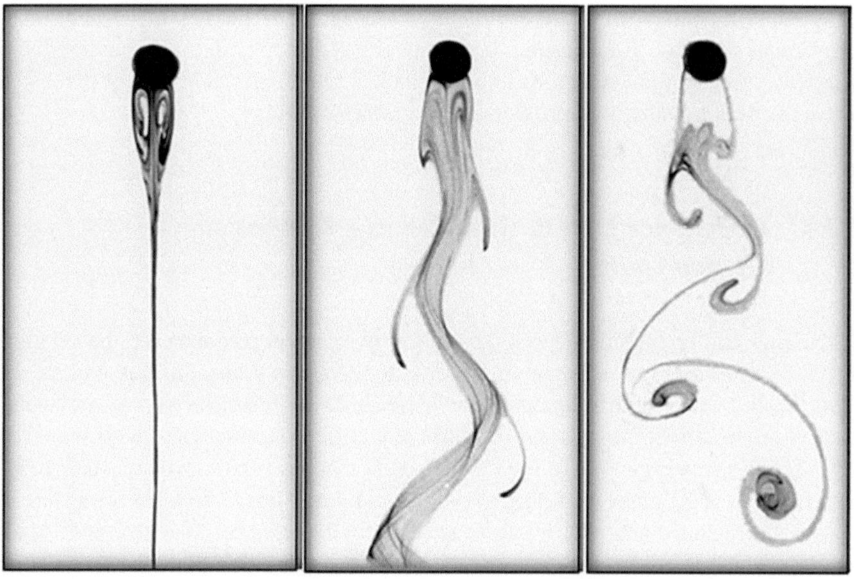

Fig. 1.1 Flows over a circular cylinder: **a** stable laminar flow; **b** flow at the beginning of the transition in laminar unsteady flow; and **c** flow in transition, also laminar unsteady flow. Courtesy of Mansur and Del Rio [5]

1.2.3 Turbulence Is Characterized by Multiple Turbulent Structures

In turbulent flows, turbulent structures present themselves with very diverse topologies, dimensions, wavelengths and periods, and they coexist in space and time, interacting with each other in intense and highly nonlinear ways. This must be understood as a multiplicity of temporal and spatial characteristic quantities. The multiplicity of turbulent structures is one of the fundamental characteristics of turbulent flows.

The dimensions of the largest turbulent structures are of the same order of magnitude as the dimensions of the geometry that gives rise to them and that characterizes the flow. The smaller structures, in turn, take on decreasing magnitudes with increasing values of the Reynolds Number, defined based on the integral length scale of the turbulent flow. This integral length scale represents the most energized turbulent structures of the flow. The relationship between the characteristic size of the largest and smallest turbulent structures and the Reynolds Number will be demonstrated and physically explained in Chap. 5, along with the concepts and a discussion of characteristic magnitudes.

It should be noted that there is a limit to the decrease in magnitude of these small structures; the size of the smallest structures can be evaluated as a multiple of the molecular characteristic length. For this reason, it can be said that the magnitude of these decreasingly smaller structures will approach an asymptotic limit behavior and, consequently, the turbulence becomes independent of the Reynolds Number, as its value increases. In other words, the width of the frequency or wave number spectrum tends towards a constant. If the reader has difficulty in understanding these concepts, it is recommended to wait until Chap. 5, where this topic will be presented in greater detail, which will provide a better understanding.

1.2.4 Turbulence Can Be Modeled as a Continuous Phenomenon

Turbulence can be considered as a continuous phenomenon, since even the smallest turbulent structures are many times larger than the mean molecular free path of the fluid. Disturbances with frequencies or wave numbers characteristic of molecular motion do not affect the spectrum of turbulent structures associated with turbulent flow. The Stokes closure modeling for molecular viscous stresses, relative to Newtonian fluids or some other modeling of non-Newtonian fluids, is suitable for turbulent flows. This means that any flow can be adequately modeled by means of these equations for Mach numbers not greater than 15. Above this value, the mean molecular free path increases to the point of compromising the use of the continuum hypothesis.

1.2.5 Turbulence Can Be Considered Chaos-Deterministic

The system of differential equations through which turbulent flows are modeled can be considered deterministic, in the sense that, given the initial and boundary conditions of the physical domain, it is possible to determine a single solution to this system, which would lead to deterministic and exact predictions of future events. However, to achieve this result, it is necessary that the initial and boundary conditions are known with absolute temporal and spatial accuracy. In practice, however, this is hardly possible. In fact, the determination of the initial and boundary conditions that characterize a given problem always results in a certain degree of inaccuracy. For example, meteorological forecasting of the planetary climate depends on the use of computer models that must be fed with initial and boundary conditions for velocities, temperature, density and so on, which must be measured using sensors and simultaneous data acquisition systems. Of course, in the entire acquisition chain, there are many sources of errors. In addition, other sources of error inherent in the methods of solution of the differential equations that form the mathematical model will always be present.

Physically, the disturbances injected into a flow, by various sources and mechanisms, are amplified by nonlinear physical mechanisms. This process of amplification of disturbances and the promotion of instabilities is modeled through the nonlinear advective terms of the Navier–Stokes equations. Thus, the existence of small errors in the estimation of initial and boundary conditions may imply different results in time and space. This is why one speaks of deterministic turbulent chaos—the equations through which turbulence is modeled are deterministic, but their solutions are not. There is a high sensitivity to initial and boundary conditions. This notion of deterministic chaos was established by Poincaré [8] and transcribed by Lesieur [3].

It is important to note that, equally, the exact repetition of a material experiment is not possible—in contrast to virtual experiments arising from mathematical and computational modeling—since the initial and boundary conditions will never be strictly identical for different realizations. This fact should not, of course, give a false impression about the usefulness of computational and material analysis tools for turbulent flows. In fact, the use of these analysis tools leads to the following practical outcomes, which are independent of unpredictability, namely, understanding physical behavior and obtaining reliable statistical information.

As a quantitative example of this issue of unpredictability, we will use the Lorenz Mathematical Model, through which unpredictable dynamic behavior is represented mathematically. This is a model of a purely mathematical nature, representing no specific physical problem. The system of equations is given as follows:

$$\dot{x} = \sigma(y - x), \tag{1.1}$$

$$\dot{y} = \rho x - y - xz, \tag{1.2}$$

$$\dot{z} = -\beta z + xy. \tag{1.3}$$

The notation $\dot{x}$, $\dot{y}$ and $\dot{z}$ indicates time derivatives. For the present exercise, we fixed $\sigma = 10$, $\beta = 8/3$ and $\rho = 28$, and considered, $x = x(t)$, $y = y(t)$ and $z = z(t)$. Two solutions for this system of equations were computed by using two separate time integrators, one Runge–Kutta and one Gauss–Radal, both of fifth order. Two simulations were performed with the initial conditions of $x(0) = y(0) = z(0) = 0.1$ with the two integrators mentioned above. In this case, the objective was to verify the influence of the integrator on the simulation. Two other simulations were subsequently performed to check the sensitivity of the solutions to the initial conditions. The Runge–Kutta integrator was used for both simulations. The initial conditions were specified as $x(0) = y(0) = z(0) = 0.1$ for one of the simulations. For the second simulation, the initial condition on x was changed ever so slightly to $x(0) = 0.100001$ while keeping $y(0) = z(0) = 0.1$. In other words, the initial conditions on x differ from one simulation to the other by 10^{-5}. For the other two directions, nothing was changed.

Figure 1.2 shows the time distributions of the three dependent variables and the differences between the solutions for the two integration methods, both using the first set of initial conditions. Column (a) shows the time distributions for $x(t)$, $y(t)$ and $z(t)$ determined using the Runge–Kutta time integrator for the system of Eqs. (1.1) through (1.3). Column (b) shows the distributions of these variables determined using the Radal integrator. Column (c) shows the distributions of the differences between the two solutions as a function of time. Note that the two solutions, obtained with both integrators, are similar until a time of $t \approx 15$ [s]. After this time, the solutions are completely different from each other, as can be seen quantitatively in Fig. 1.2c. This makes clear the enormous sensitivity of the solution of this system of ordinary differential equations to the integration method.

In Fig. 1.3, the time distributions of the three dependent variables and the differences between the solutions for the two sets of initial conditions are shown. Column (a) shows the time distributions for $x(t)$, $y(t)$ and $z(t)$, with initial conditions $x(0) = y(0) = z(0) = 0.1$. Column (b) shows the solutions obtained with initial conditions $x(0) = 0.100001$ and $y(0) = z(0) = 0.1$. For both solutions, the Runge–Kutta integrator was used. Column (c) shows the time distributions of the differences between the solutions in columns (a) and (b). The solution for the system of ordinary differential equations is very sensitive to small differences in the initial conditions. A small error in these conditions and the solutions differ completely from one another after a certain amount of time.

An important conclusion is that if a system of first-order ordinary differential equations with such a small degree of complexity already presents such sensitivity to the initial conditions, one can imagine what may happen with the complex mathematical models used for terrestrial climate prediction. It is natural that forecasts must be made daily. After a few days, the errors in the initial and boundary conditions will be amplified, and the predictions will be different from what will really occur!

As an example, the simulation for a temporal mixing layer flow is presented. A computational model based on the Fourier pseudo-spectral method, with a spectral convergence rate, was used. This means that when we solve a problem with this type

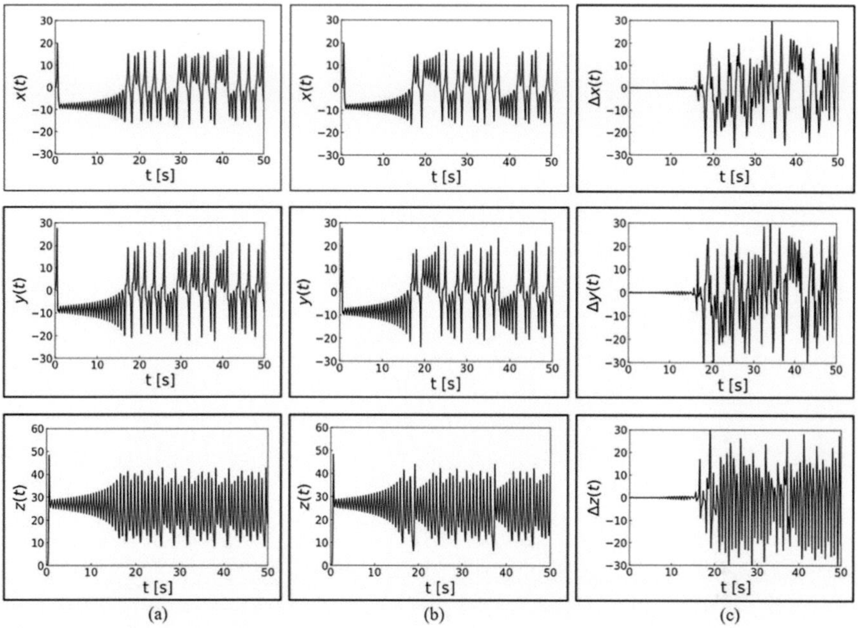

Fig. 1.2 The influence of the integrator used on the evolution of two solutions of the Lorenz Dynamical System: **a** using the Runge–Kutta integrator of fifth order; **b** using the Gauss–Radal integrator also of fifth order; and **c** differences between the solutions obtained with the two integrators. Proprietary data from MFLab

of method, the solution is not very sensitive to the refinement of the computational mesh. With this method, the spatial derivatives are solved numerically with machine error. To model and simulate the flow transition from a steady laminar regime to an unsteady laminar regime, a noise simulator was used to introduce noise into the initial velocity conditions. The seed values used to feed the noise modeler differed by 10^{-5} for the three cases shown in Fig. 1.4. The three rows of figures correspond to the three different initial conditions. The three columns are for three different times. In column (a) with $t = 0.0$ [s] the initial conditions from the vorticity field with injected noise are shown. Visually, they appear identical, but in fact, there are differences in the order of 10^{-5}. In column (b) with $t = 2.0$ [s], the differences in the evolving vorticity fields are already noticeable. The Kelvin–Helmholtz structures are spatially distributed in different ways. In column (c), corresponding to time, $t = 4.0$ [s], the spatial distributions of dynamical structures are completely different. From this example, the high sensitivity of the solution to the differential mathematical model for fluid dynamics has been demonstrated.

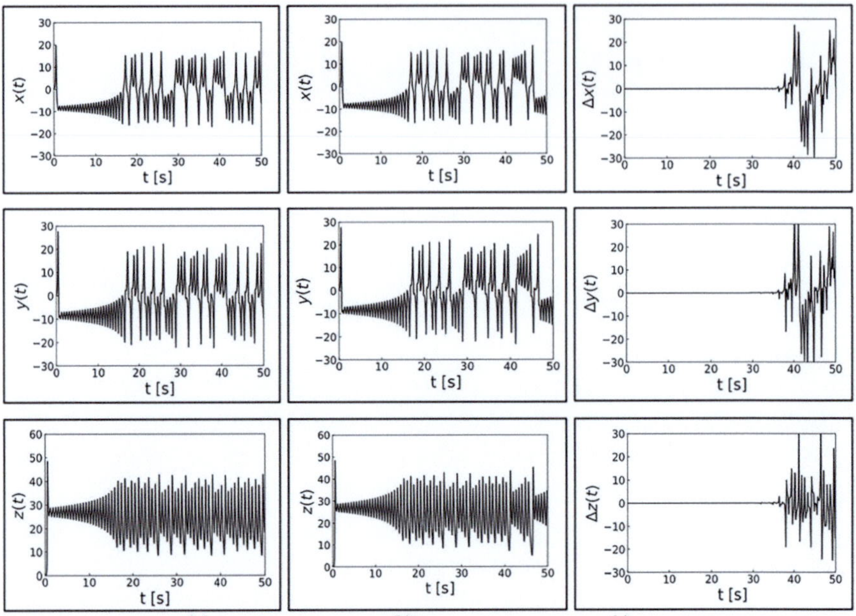

(a) (b) (c)

Fig. 1.3 Influence of initial conditions on the time evolution of the solutions of the Lorenz Dynamical System given by Eqs. (1.1)–(1.3): **a** using initial conditions, $x(0) = y(0) = z(0) = 0.1$ and **b** using initial conditions $x(0) = 0.100001$ and $y(0) = z(0) = 0.1$. Both solutions were achieved with the Runge–Kutta integrator. Proprietary data from MFLab

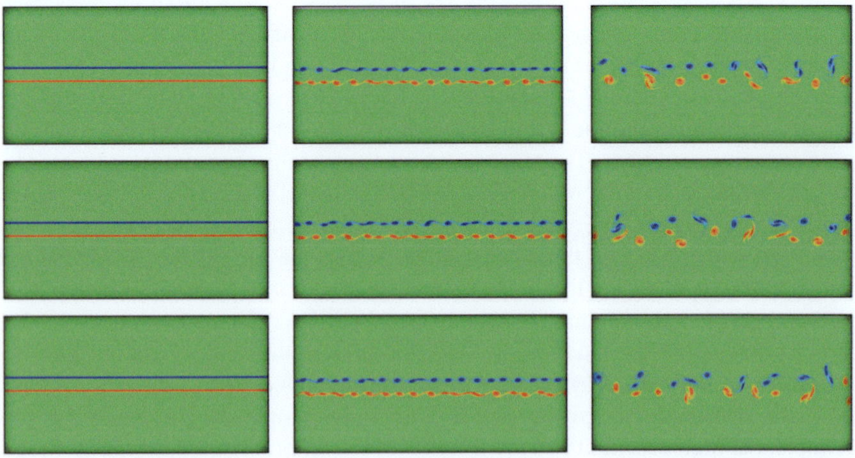

(a) (b) (c)

Fig. 1.4 Unpredictable character of the temporal development of a mixing layer. Three simulations were performed, differing only in the seed value of the random noise simulator: **a** initial conditions $t = 0$ [s]; **b** time $t = 2$ [s]; and **c** time $t = 4$ [s], from Moreira and Silveira-Neto [6]

1.2.6 Turbulence Accelerates the Molecular Diffusion of Transported Physical Quantities

The particularly intense dynamics of turbulent flows are accompanied by strong non-periodic fluid flow fluctuations that result in the intensification of molecular diffusion of physical quantities with associated properties, notably: velocity, pressure, temperature, concentrations of contaminant and chemical species, and density. Increasing turbulence levels are characterized by greater amplitudes of random fluctuations. Consequently, in turbulent flow, there is substantial intensification of diffusion mechanisms and viscous transformation, relative to laminar flow diffusion and viscous transformation of kinetic energy into thermal energy. The molecular diffusion of any flow-related physical information (fluid linear momentum, energy, concentrations of chemical species, etc.) is orders of magnitude greater in the turbulent regime than in the laminar regime. This is because, in the turbulent regime, there are fluctuations in momentum, thermal fluctuations, fluctuations in concentrations of chemical species, or fluctuations of any physical information carried by the flow, which promote numerous intense local gradients, thus making molecular diffusion more intense.

To further clarify this phenomenon, consider a finite parcel of fluid, carrying certain physical information, for example, temperature, within a turbulent flow. As illustrated in Fig. 1.5, the effect of turbulence will be to disperse this parcel of fluid into smaller parts; the smaller turbulent structures will be transported to different regions of space. This process leads to an increase in the number of local gradients and, consequently, leads to an increase in the diffusive transport of the considered physical quantities. Thus, it is concluded that turbulence intensifies the diffusion of physical quantities through the strong advective effects associated with it. The increase in the area over which the exchange of physical quantities occurs is another important factor in enhancing molecular diffusion.

It is observed that the advective mechanism and diffusive mechanism of the transport of physical quantities act simultaneously in a flow. However, one should separate their effects, so as not to take one phenomenon for the other. The diffusion of physical quantities depends on the local gradients of the potential associated with that physical information; it also depends on the diffusion coefficient, which is a physical property of the fluid. On the other hand, advection depends on the kinematics of the flow, which is directly affected by the turbulence itself. Thus, turbulence intensifies and multiplies the gradients of the related potential, in space and time, resulting in an increase in the molecular diffusion process. The term turbulent diffusion must not lead to a false idea of physics. Turbulence itself can only transport quantities by advection. On the other hand, it accelerates and intensifies the purely molecular diffusion process.

In Fig. 1.5, a computational simulation of a spatially developing reactive mixing layer, confined by two parallel horizontal walls, is illustrated. The formation of large coherent Kelvin–Helmholtz-type turbulent structures is observed. A coherent structure, from a statistical point of view, is understood to be a structure that has a

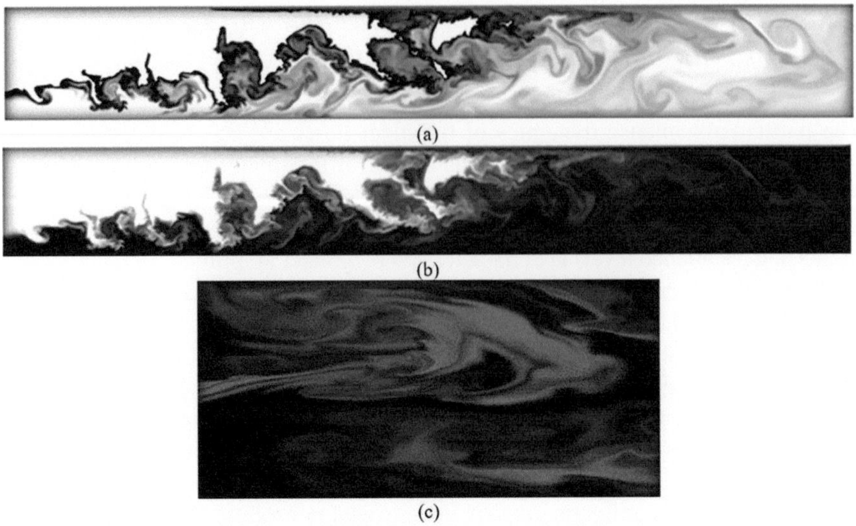

Fig. 1.5 Simulated spatially developing mixing layer confined between two horizontal parallel flat plates: **a** temperature field; **b** chemical reaction rate; and **c** detail of instabilities. Courtesy of Vedovoto [11]

lifetime longer than its characteristic lifetime. This concept will be discussed in more detail in Chap. 5. These structures, in turn, are composed of smaller structures, and this composition process is repeated until the smallest structures have been solved, using a given computational mesh. Figure 1.5a shows the temperature field, and Fig. 1.5b shows the chemical reaction rate. Figure 1.5c presents an amplification of Fig. 1.5b, showing the local concentration gradients. Lighter colors indicate low concentration values, and darker colors stand for higher concentrations. Corrugations in the regions of strong gradients indicate the intensification of these gradients and the increase in the number of material points with high gradients per unit volume.

It is noted that the chemical reaction process depends on the molecular diffusion process, but the reactions should be modeled under the continuum assumption, analogous to Stokes' model for viscous stresses. It is emphasized that the effects of the molecular nature, translated at the continuum level, are all accelerated and intensified by turbulence.

From the standpoint of engineering applications, this characteristic—increased diffusion capacity—is one of the most important characteristics of turbulence, because it implies intensification of internal energy transport, mixing of chemical components, dispersion of pollutants and intensification of chemical reactions.

1.2.7 Turbulence Occurs at High Reynolds Numbers

The transition of a flow to the turbulent regime, as well as its maintenance, both depend on the relative importance of advective and diffusive effects. The advective effects, which are nonlinear, are amplifiers of disturbances that can result in the formation of instabilities. On the other hand, diffusive effects are dampers or inhibitors of instability formation. This is because the greater the relative viscosity effect, the greater the damping of the displacement of a fluid particle in the direction transverse to the flow. The Reynolds Number (Re) is defined here, by Eq. (1.4), as the ratio of advective effects to diffusive effects:

$$\mathrm{Re} = \frac{\text{advective effects (nonlinear)}}{\text{diffusive effects}} \tag{1.4}$$

This dimensionless parameter can also be defined in terms of the physical, geometric and operational properties of the problem, to represent the ratio of inertial effects to viscous effects. Thus, consider a turbulent structure of characteristic size ℓ, transported with a characteristic velocity V, by the flow of a fluid with a kinematic viscosity ν. Through dimensional analysis, it is shown that:

$$\mathrm{Re}_\ell = \frac{V\ell}{\nu} \tag{1.5}$$

A flow can only transition from laminar to turbulent regimes, or remain turbulent, in situations where the nonlinear effects are more important than the viscous effects, which implies Re greater than unity.

1.2.8 Turbulence Is Three-Dimensional and Rotational

Experiments have been performed under a wide variety of conditions and scientists have shown that any turbulent flow is three-dimensional, with the presence of vorticity fluctuations. Physically, vorticity is promoted only in a rotational, three-dimensional flow through a process of stretching of the turbulent structures, a purely three-dimensional mechanism. The stretching mechanisms are presented at the end of Chap. 2. From a mathematical point of view, as will be shown using Eq. (2.104) in Chap. 2, non-zero vorticity promotion can occur only in a three-dimensional, rotational flow. Thus, as will be shown in Chap. 2, a turbulent flow must be both rotational and three-dimensional.

1.2.9 Turbulence Is Transformative

At the molecular level, turbulence accelerates the process of viscous transformation of turbulent kinetic energy resulting in an increase of internal energy. The concept of turbulent kinetic energy transformation is presented in Chap. 5. This transformative process takes place at high frequencies or high wave numbers of turbulence—lower Kolmogorov characteristic times and lengths, where viscous effects predominate over advective effects. It is also known that in the fully developed turbulent regime, the energy injected into the flow to power the large turbulent structures is transferred to all the structures in the turbulence by a nonlinear process of simultaneous interaction between all the structures that form the turbulent spectrum.

This energy transfer process results, overall, in a nonlinear energy transfer process, from the integral-scale structures, the most energized ones, to the smallest structures, also known as Kolmogorov Structures, in which the turbulent kinetic energy is transformed into internal energy by viscous effects. To maintain turbulent flow, therefore, a continuous supply of energy is required, without which the flow will decay to a state of rest.

The transformative characteristic of turbulence is important in identifying flows that cannot be considered turbulent. Internal gravity waves, for example, cannot be turbulent because they do not transform energy.

Figure 1.6a shows a simulated jet in transition to turbulence, developing in a quiescent medium. The flow inside the nozzle is laminar. The energy supply to the developing flow to the right of the nozzle ensures the constant velocity field downstream of the nozzle at the entrance to the domain, as shown schematically in Fig. 1.6a.

This energy supply occurs because at the nozzle outlet, the velocity is held constant over time, modeling the existence of some upstream mechanical system, which ensures the movement of the fluid, for example, a fan. From this point on, the transition begins from the nozzle, which ensures the movement of the fluid, for example, a fan. Also, from this point on, the flow transition begins, from the laminar to the turbulent regime. As the jet propagates through the quiescent medium, it decays under the effect of the viscous transformation of turbulent kinetic energy, coming to rest after a given distance.

Figure 1.6a shows the transition process in the region near the nozzle and the turbulent regime in the region away from the nozzle. In this figure, the decay zone is not visualized. In summary, the jet illustrated in Fig. 1.6a starts at the nozzle exit, develops freely, without interference from walls, transitions from laminar to turbulent, and then enters decay. Its kinetic energy is converted in the axial direction into thermal energy and sound energy until the fluid comes to rest.

Figure 1.6b shows the distribution of the axial component of velocity along the centerline of the jet, showing the decay process, promoted by the expansion of the jet and by the action of viscosity, through which kinetic energy is transformed into thermal energy. Inappropriately, this effect is referred to in the literature as

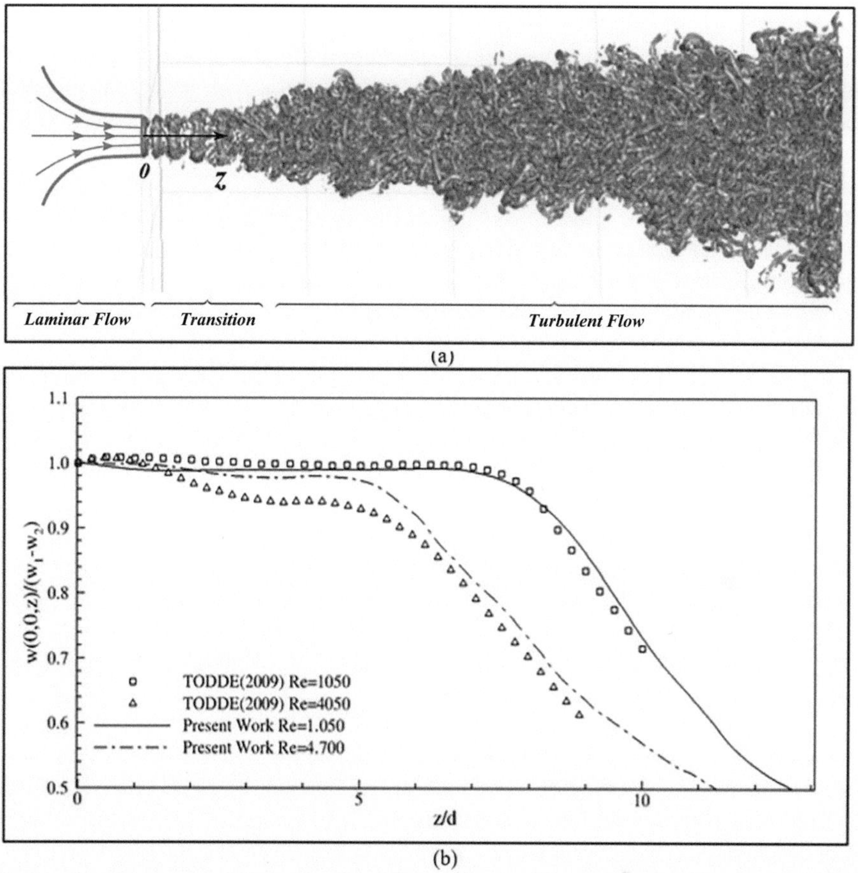

Fig. 1.6 Simulation of a spatially developing jet in a quiescent medium: **a** visualization of the flow using iso-values of the Q criterion; **b** distribution of the axial component of the mean velocity, along the jet centerline, compared with experimental data from Todde et al. [10]. Personal collection, Courtesy of Moreira and Silveira-Neto [6]

"dissipation of kinetic energy". The energy is not dissipated; rather, the energy is transformed!

1.3 Physical Model for the Transformation of Energy via Viscous Process

In this section, a proposed model will be presented, which, in the terminology of this book, is called the viscous transformation process based on the philosophical thought of Lavoisier, who stated: "in nature nothing is destroyed, nothing is created, everything is transformed". A question that remains is: what was Lavoisier referring to when he used the words "everything" and "nothing?" These gaps ("everything" and "nothing") might be filled in the context of this book with the word "physical information" or "physical information", for short, "information". However, information itself is an enormously general word that can represent a multitude of other terms. For physicists and engineers, for example, the term "physical information" devolves to measurable physical quantities (pressure, temperature, speed, etc.) or even those quantities whose existence is inferred through the resources of mathematics (universal cosmic fluid, also called dark matter). In a generic way, word information can be associated with statistical moments of any order, relating to the correlations between any physical information of interest. For example, the average velocity, $\overline{u}\,[\mathrm{m/s}]$, is a statistical moment of the first order. The squared moment, $\overline{u}^2\,\left[\mathrm{J/kg}\right]$, represents the average specific kinetic energy. Rewriting this moment as $\overline{uu}\,\left[(\mathrm{kg\,m/s})(\mathrm{m/s})/\mathrm{kg}\right]$, another interpretation emerges as the average specific advective flux of linear momentum. The pressure, $p\,\left[\mathrm{N/m^2}\right]$, is another physical information of interest. The velocity fluctuation, $u'\,[\mathrm{m/s}]$, when correlated with the velocity fluctuation $\vartheta'\,[\mathrm{m/s}]$, represents a second-order statistical moment, $\overline{u'\vartheta'}\,\left[\mathrm{J/kg}\right]$. This second-order moment is further associated with another "information", or physical information: a component of the Boussinesq–Reynolds Tensor. Still other "information", such as temperature, thermal energy, gravitational potential energy, among many other physical quantities, represents the diversity of information with which engineers and physical scientists must contend.

Substituting the words "physical information" for Lavoisier's "everything" and "nothing", one must conclude that no physical information, in the world of physical sciences, can be created or destroyed (lost), by humans. We can only transform one physical information into another, according to Lavoisier. The author of this book understands, however, that any physical information can also be stored and transported. Thus, it is valid to extend Lavoisier's statement: in nature, nothing is created, nothing is destroyed (lost), everything can be stored, everything can be transported and everything can be transformed. Thus, the "ground" is prepared to proceed with the balancing of any physical information.

The processes of physical information transport are diverse, but in the present text, we highlight:

- advective process, which requires a moving material medium;
- diffusive process which also requires a material medium, as well as potential gradients associated with the physical quantities being transported;
- wave processes, such as thermal radiation. We cannot claim that wave processes can occur in a "vacuum" since, according to the findings of physicists, there is no empty space in the universe.

As for transformation processes, there is also a list of possibilities. For example, consider physical quantities of interest as the different types of energies. Such is the case when

- transforming kinetic energy into thermal energy, by viscous effects;

- transforming electrical energy into thermal energy by the Joule effect;
- transforming chemical potential energy into thermal energy by chemical reactions;
- transforming gravitational potential energy into electrical energy.

We can transform energy; however, we can never generate any kind of energy! Energy generation is an absolute process, while natural or human-induced processes are relative processes.

Due to the importance of viscous transformation processes in laminar flow and especially in turbulent flow, a model for this process is now proposed, based on the continuum hypothesis. The viscous transformation process can be associated with viscous friction between two adjacent fluid layers, as illustrated in Fig. 1.7. In this figure, a material medium is divided in half by a vertical plane, represented by the dashed line. Consider now a molecule model simulating a fluid flow. The fluid is composed of red molecules on the left and blue molecules on the right of the central vertical plane. The red molecules move in a vertical direction with a velocity greater than the velocity of the blue molecules on the right. The smaller red and blue arrows represent the average molecular motion.

The big, red and small, blue arrows represent the motion of the continuous medium when we abstract from the molecules. If the flow is uniform, with no shear, the Brownian motion of the molecules will characterize the thermal energy level of the material medium, as well as the static pressure. When the two parts of the fluid (red

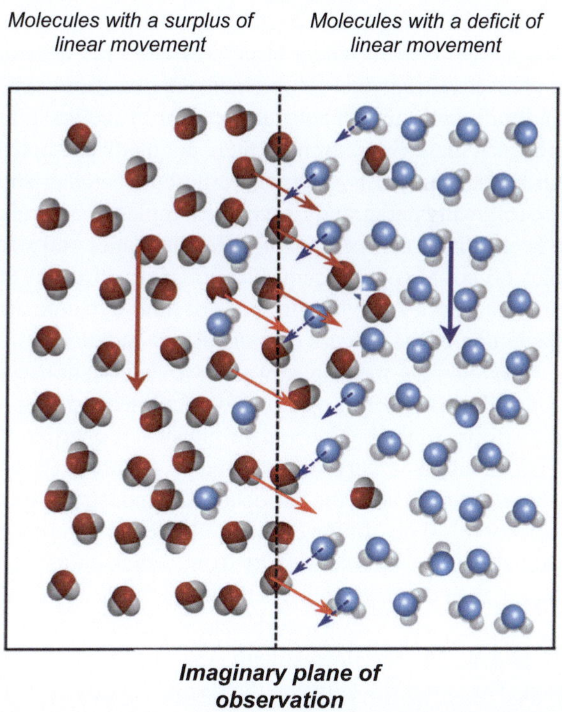

Fig. 1.7 Illustration of molecular flux through an imaginary plane (vertical dashed line) dividing two media moving with different macroscopic velocities

and blue) have different velocities, shear arises, at the macroscopic level, between these two bands of fluid. It also intensifies the energy of the interactions between the red and blue molecules. These hypotheses are illustrated in Fig. 1.7, where the sum of the Brownian motion of the molecules and the macroscopic motion of the fluid leads to an increase in the relative velocities, as indicated in this figure. With this, the viscous stresses (a result of the flux of molecules through the imaginary plane), the pressure (a result of the collisions of the molecules along the imaginary plane), and the thermal energy (a result of the agitation of the molecules) will be altered with the difference in the velocities of the streams. The greater the velocity difference between the two streams, the more intense will be the net linear momentum flux in the dashed plane, increasing the viscous stress. The molecular impacts along the imaginary plane will also be more intense, which increases the static pressure. The molecular shocks throughout the interference band between the two currents (red and blue) will also be more intense, resulting in the intensification of the process of transformation of kinetic energy into thermal energy. The mathematical model for this transformative process will be presented in Chap. 2.

1.4 Manifestation of Turbulence

Observation of the topology of a galaxy provides an intuitive notion of turbulent dynamic behavior. Here, interstellar distances are measured in light-years, and astronomically sized turbulent structures can take thousands of centuries to complete a characteristic cycle. At another extreme, turbulent structures of high frequency and millimeter dimensions are present in the wake left by a small insect in flight. Between these two extremes, there are many examples of natural phenomena where turbulence manifests itself with different characteristic values of time and length. In the following paragraphs, several examples of natural phenomena will be presented, where turbulent structures of different shapes and sizes are observed. These examples do not always present turbulence situations in their fullness. In certain cases, one can identify only the presence of fluid-dynamic instabilities that characterize the beginning of the process of transition to turbulence.

Gas bubbles moving upward within a stationary liquid medium may deform and follow different trajectories—rectilinear, zigzag, or helical—depending on the dimensions of the bubble and the physical properties of the two-phase fluid involved. This situation is very often found in different types of industrial equipment and processes, such as in the petrochemical industry. Experiments and computer simulations of this type of flow have highlighted the presence of turbulent structures inside the bubbles and the wake left by them, which influences the bubble trajectory traveled during their displacement. As the flows inside and outside a bubble become unstable, there can be a transition to turbulence, and bubble trajectories become more complex.

In Fig. 1.8, the downstream wake of an air bubble moving in water is identified. The similarity between the results coming from the two analysis methods, i.e., computational experimentation (Fig. 1.8a) and material experimentation (Fig. 1.8b),

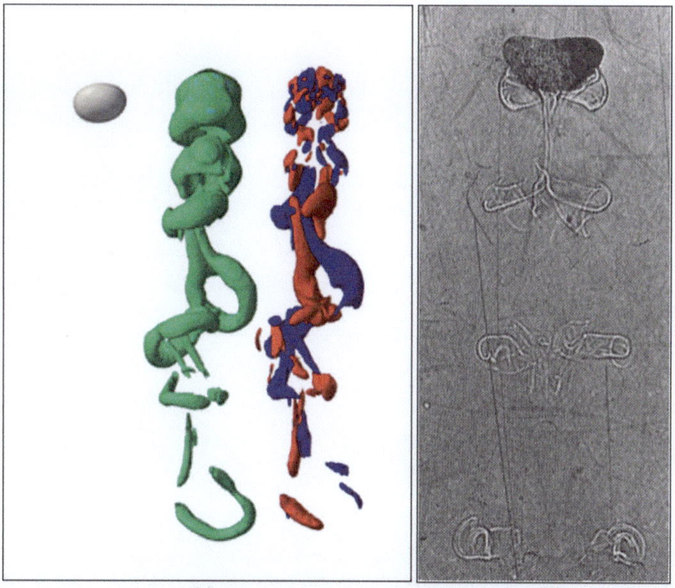

Fig. 1.8 Upward motion of a gas bubble in a liquid medium—turbulent structures with characteristic dimensions of the order of millimeters: **a** computer simulation, courtesy of Pivello et al. [7]; **b** material experiment result, from Clift et al. [1]

is noteworthy. It is expected that by averaging results, the two types of experiments will appear even more similar.

If today the research tools in the field of fluid dynamics allow for results as detailed as those presented in Fig. 1.8, there was a time when visualizing flows in nature was the only tool available to those interested in understanding the different characteristics of fluid motion. The Italian Leonardo da Vinci, in the sixteenth century, left a vast legacy of sketches of great realism, which testify to his thorough observations and reflections on the motions of fluids. In Fig. 1.9, one can see a remarkable similarity between Leonardo da Vinci's drawing, which represents a watercourse discharging into a reservoir, and the photograph of a small waterfall, located in the *Chapada dos Veadeiros*, in the state of Goiás in Brazil. The presence of three-dimensional transient turbulent structures, often found in turbulent flow, is evident in the pictures. The richness of detail contained in da Vinci's painting is impressive, bringing to light a clear notion of the multiplicity of lengths characterizing a turbulent regime.

Today, detailed flux quantities can be extracted by accurate measurements and through three-dimensional and transient flow visualizations, making use of experimental resources in physical laboratories, as well as by modern experimental methods in computational flow facilities. Figure 1.9b shows the result of a photograph taken using conventional equipment that captures images at a rate of about 40 frames per second. The result that one sees shows time-averaged behavior, due to the relatively

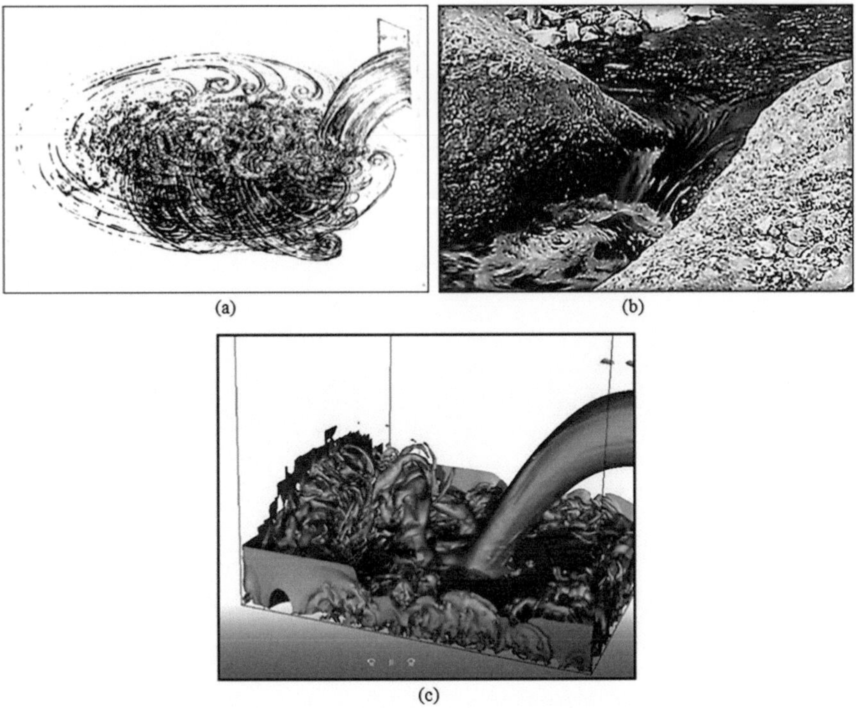

Fig. 1.9 Water movement in cascade—characteristic size of the order of a few meters: **a** drawing by Leonardo da Vinci around the year 1515; **b** photograph of a watercourse in Chapada dos Veadeiros, in the state of Goiás in Brazil, courtesy of Finotti [2]; and **c** computational experiment, developed by Bernardo Alan, at the MFLab of UFU, in 2020

long exposure time. While the diaphragm remains open for a given time, many occurrences with shorter characteristic times have been superimposed on the same image. Therefore, what has been captured by the photographer represents average behavior. Looking at the image in Fig. 1.9a, the level of detail that da Vinci was able to visualize with his naked eyes, or what resulted from his fertile imagination, is impressive. Either or both allowed him to infer a high-frequency physical reality, which is impossible to capture by ordinary people with ordinary eyes. Today, what da Vinci drew so long ago can be observed using high-tech resources, such as cameras operating at thousands of frames per second [fps] or by means of computer simulations of very high spatio-temporal resolutions, as in Fig. 1.9c, the result of a computer simulation performed at MFLab, using the MFSim platform.

When a fluid flows around a solid obstacle, the boundary layers on opposite sides of the obstacle usually detach from the body, forming an inflectional shear region downstream of the immersed body. In this region, instabilities may arise that, when amplified, cause the transition from laminar to turbulent regime. In the flow past cylinders, in particular, alternating vortical flow structures (classically called "von Kármán vortices") may detach from opposing sides of the body, giving rise to the so-called von Kármán wake (or classically called "von Kármán vortex street"). The presence of natural periodic fluid-dynamic instabilities does not in itself imply the existence of a turbulent regime. This is a case illustrated in Fig. 1.10a, where flow visualization in a material experiment shows the formation of a wake with easily identifiable vortical structures. Turbulence, itself, arises for a given value of the Reynolds Number, as the instabilities are transported and subjected to nonlinear effects that lead to three-dimensionalization. This process can be seen in the region near the right side of this same figure. In Fig. 1.10b, the result of a virtual flow visualization experiment (computer modeling and simulation), shows turbulent structures characterizing the transition to turbulence.

The shed von Kármán turbulent structures undergo transition to a turbulent wake in flow past various cylindrical structures in different engineering applications, such as thermal energy exchangers, and can be visualized in physical and computational laboratory experiments with relative ease, such as the visualizations in Fig. 1.10. However, it is in nature that these vortical structures are manifested at characteristic lengths well beyond those of most human-created structures.

Two magnificent images captured by satellites, showing von Kármán wake flows past the Guadeloupe Island and the Canary Islands, can be seen in Fig. 1.11. These

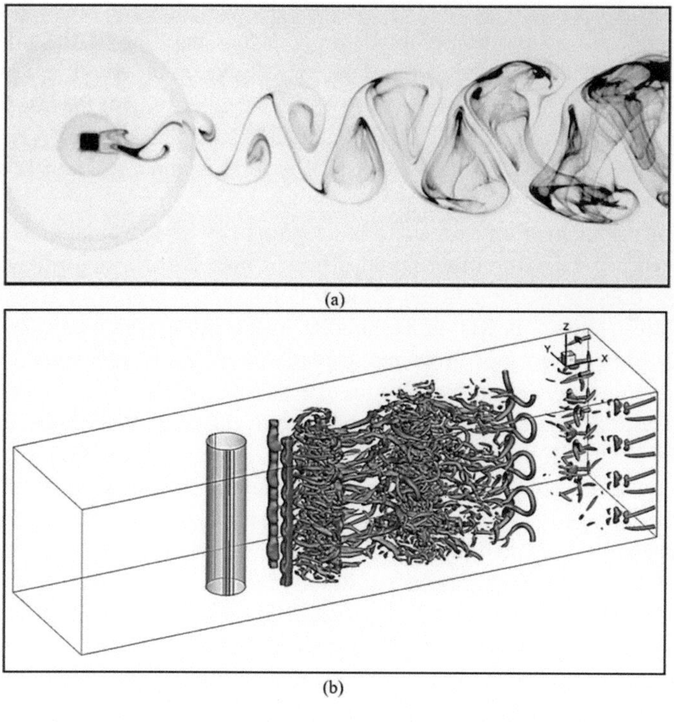

Fig. 1.10 A von Kármán vortex wake formed downstream of a submerged cylinder: **a** with a square cross-section in a water tunnel flow visualization, courtesy of Lindquist [4]; and **b** with a circular cross-section in a computational simulation via finite volumes, courtesy of Vedovoto [11]

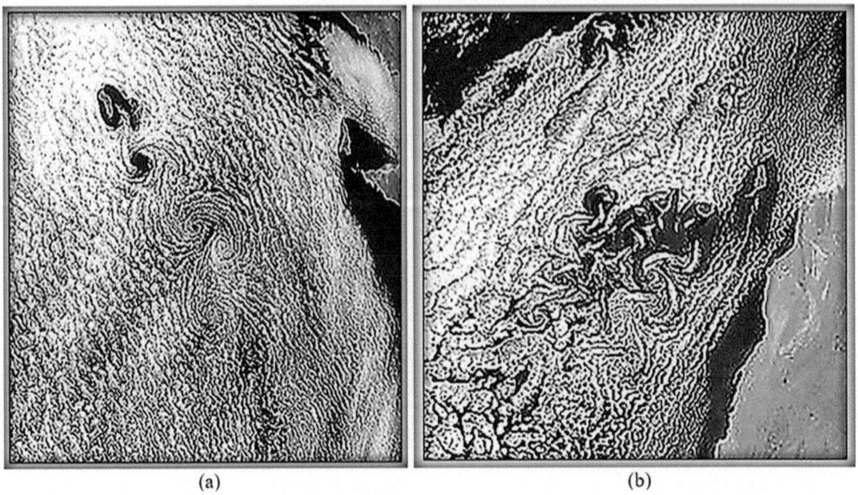

Fig. 1.11 von Kármán wakes observed in nature: flow over **a** Guadeloupe Island, and **b** the Canary Islands

visualizations were made possible by the contrast created by the presence of grainy clouds, organized by the action of the natural convection of the Earth's atmosphere. These images, taken over islands hundreds of kilometers in length, are impressive not only for their beauty but also because of their similarity with the illustrations in Fig. 1.10a, namely, flow past a geometry only a few millimeters in size. Figure 1.11b shows the interaction between the wakes of several islands, giving rise to a very complex flow.

A mixing layer can be understood to be a region between two parallel fluid streams moving in either the same direction at different speeds or in opposite directions. Under these conditions, inflectional velocity fields in the shear layer arise, resulting in instabilities, known as Kelvin–Helmholtz instabilities, which can develop into turbulence. This category of flows includes a wide range of problems of practical interest, either in engineering applications or even for an understanding of the physics associated with flows in transition to turbulence or fully turbulent flows. In fact, the presence of Kelvin–Helmholtz instabilities in flows related to mixing processes in chemical reactors implies an increasing efficiency of such equipment. It can also yield improved performance of thermal energy exchangers. On the other hand, the turbulent wake behind the wing of an aircraft may have this type of fluid-dynamic instability over a wide range of turbulent structures. As a rule, the aircraft wake carries large amounts of turbulent kinetic energy, which directly produces an increase in drag on the aircraft. It is also noteworthy that in chemical reactors and heat exchangers, the presence of these instabilities requires an increase in the pumping power of the fluids.

An example of a temporally developing mixing layer in the terrestrial atmosphere is shown in Fig. 1.12, where, once again, clouds serve as a visualization agent,

Fig. 1.12 Temporally developing mixing layer observed in nature: courtesy of Márcio Mendonça-CTA-IAE-São José dos Campos-Brazil

revealing frequently occurring natural phenomena. Such phenomena are ever-present around us but are invisible to the human eye if one does not have tracers that bring them into focus. The author of this photo witnessed a great coincidence, namely the existence of an elongated cloud superimposed on the inflectional shear region that gave rise to Kelvin–Helmholtz instabilities. Thanks to this coincidence, he was able to visualize and photograph the temporal mixing layer.

With the use of mathematical models, it is possible to represent certain physical realities with great fidelity. Advanced numerical-computational methods, as well as computing machines with high processing and data storage capacity, both used in the solution of mathematical models, have become powerful allies for the understanding of nature and for technological modernization. The data obtained through computer simulations can be visualized through pressure fields, vorticity, or passive scalars, such as temperature and concentrations of chemical species, under conditions such that they do not affect the fluid dynamics. Thus, visualizations like those provided by clouds are obtained using computational algorithms.

When one observes the motion of the earth's atmosphere, or the gaseous mass that makes up the Sun, one sees the presence of turbulence at the most massive characteristic length imaginable. Turbulent structures around these two celestial bodies can be seen in Fig. 1.13. In the case of the earth's atmosphere, turbulence plays a key role in maintaining the climatic conditions of our planet. Turbulent structures, of different characteristics, play the most diverse roles. The turbulent structures of cyclonic or anti-cyclonic type play an important role in the formation or absence of rainfall, depending on the hemisphere where they are located.

Turbulence also has the fundamental role of dispersing and transporting pollutants and contaminants, homogenizing and avoiding environmental conditions that cannot be tolerated by the living beings that inhabit our planet.

An example of turbulence in astronomical dimensions can be seen in Fig. 1.14a, where two turbulent spiral structures can be contemplated. The particles that make

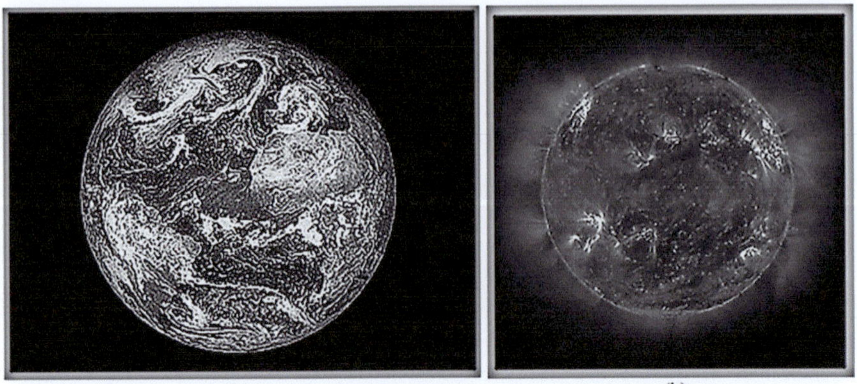

(a) (b)

Fig. 1.13 Turbulent structures in the planetary dimension: **a** on earth; and **b** on the sun

up this "fluid" are celestial bodies as diverse as possible. Visually, one can imagine a moving fluid that tends to organize this matter into structures that move in an apparent pairing process, that is, the winding of two structures that rotate in the same direction. This visualization of a problem in astronomical dimensions is very similar to what is observed in flows in millimeter dimensions in our everyday life.

By way of comparison, one can observe in Fig. 1.14b the pairing of two turbulent structures inside a bubble a few millimeters in diameter. The similarity between the two processes showing approximations of two turbulent structures is striking, because the magnitudes of time and space, which characterize both problems, occupy the extremes of human imagination.

By way of comparison, one can observe in Fig. 1.14b the pairing of two turbulent structures inside a bubble a few millimeters in diameter. The similarity between

(a) (b)

Fig. 1.14 Turbulent structures in the universe in astronomical dimension (**a**); and in millimeter dimension inside an air bubble moving in water (**b**), Silveira-Neto [9] the two approximations of turbulent structures is striking, since the magnitudes of time and space, which characterize both problems, occupy the extremes of human imagination.

Both in nature and in industrial applications, the turbulent regime is the rule, and the laminar regime can be considered the exception. Mathematical modeling must be based on the fundamental principles of mass balances, linear momentum balance, Newton's Second Law, and energy balance. The modeling is the same for all flows, regardless of the magnitude that characterizes the problems. The same differential mathematical model can be employed, from flows in the interiors of bubbles to the astronomical flows of galaxies. It is hoped that with this introduction, the reader will be motivated to study turbulence in fluid flows.

1.5 Exercises

1. Summarize, with your own interpretations, the most important characteristics of the turbulent regime.
2. Find three engineering problems involving flows in the turbulent regime. Justify why they are considered turbulent. It is not enough to use the Reynolds Number. It needs to meet all the characteristics to be considered turbulent.
3. Find three engineering problems that involve laminar flow. Justify why they are considered laminar.
4. Any turbulent regime will also be chaotic. Is the converse true? Discuss!
5. Can the motion of astronomical turbulent structures be modeled with the continuum hypothesis? Look for quantitative information to support your assertions.
6. Discuss and analyze the viscous transformation process in laminar and turbulent flow. Compare the processes in the two regimes.
7. Solve the system of Lorenz ordinary differential equations for different time steps and see if there is dependence of the solutions on this parameter.

References

1. Clift, R., Grace, J.R., Wber, M.E.: Bubbles, Drops and Particles. Academic Press, New York (1978)
2. Finotti, L.: Visualization of Flows Through High-Speed Photography, Scientific Initiation Report. Universidade Federal de Uberlândia, Faculdade de Engenharia Mecânica, Laboratory of Heat and Mass Transfer and Fluid Dynamics, Uberlândia, MG (2001) (in Portuguese)
3. Lesieur, M.: La Turbulence. Presses Universitaires de Grenoble, Grenoble, France (1994) (in French)
4. Lindquist: Experimental study of flows around square and rectangular base cylinders. Master's thesis, UNESP Ilha Solteira (1997) (in Portuguese)

5. Mansur, S.S., Del Rio, E.: Research Technical Report. UNESP-Ilha Solteira, SP (2002) (in Portuguese)
6. Moreira, L.Q., Silveira-Neto: Mathematical modeling of jets in space development using Fourier pseudo-spectral methodology. Ph.D. thesis, PPG Mechanical Engineering, UFU (2011) (in Portuguese)
7. Pivello, M.R., Villar, M.M., Serfaty, R., Roma, A.M., Silveira-Neto, A.: A fully adaptive front tracking method for the simulation of two-phase flows. Int. J. Multiph. Flow 72–82 (2013)
8. Poincaré, H.: Sur le problème de trois corps et les équations de la dynamic. Memoire presenté à L'Académie Royale de Suède (1889)
9. Silveira-Neto, A.: Mathematical modelling and fine numerical simulation of two-phase turbulent flows. Post-doctorate report at CENG-Grenoble-France (1997) (in French)
10. Todde, V., Spazzini, P.G., Sandberg, M.: Experimental analysis of low-Reynolds number free jets evolution along the jet centerline and Reynolds number effects. Exp. Fluids **3**(47), 279–294 (2009)
11. Vedovoto, J.M.: Computational Experimentation. Performed at the Fluid Mechanics Laboratory of the Federal University of Uberlândia (2013) (in Portuguese)

Chapter 2
Mathematical Modeling for Fluid Dynamics

In this chapter, the differential mathematical formulation for fluid dynamics is presented. The only assumptions and restrictions on what is presented are the continuous hypotheses and Newtonian fluids. First, the Reynolds–Leibniz transport theorem is presented in Sect. 2.2, from which balances for mass, linear momentum, total energy, thermic energy, kinetic energy and vorticity were developed and presented in Sect. 2.3. It is an elegant and mathematically formal derivation. It starts from the integral formulation and leads to the differential formulation with the help of the localization theorem. A list of exercises is presented at the end of this chapter.

2.1 Introduction

The mathematical formulation of any physical problem requires hypotheses that allow the problem to be simplified, because physical reality is, as a rule, only partially known by scientists. Another relevant fact is related to the fundamental composition of matter, which is still partially known, leading to simplified models that allow the abstraction of true reality. In this sense, the *continuum hypothesis* is admitted in the present text, by which it is considered that matter is continuous and that, for any decision of modeling, it is assumed that the smallest material fraction to be considered is a fluid particle. It is assumed that the size of a fluid particle is sufficiently large such that all the statistical properties of the material medium are independent of this size. On the other hand, it must be sufficiently small for the properties of differential calculus to be valid, such as the notion of limit, when the dimensions of a material particle tend to "zero".

Any mathematical model must be based on fundamental principles, that is, balance of mass, balance of linear momentum or Newton's Second Law and balance of energy. These principles are part of the laws of Newtonian Classical Mechanics, which were established for closed systems that do not lose or gain mass by advection

A. da Silveira Neto, *Turbulent Flows*, https://doi.org/10.1007/978-3-031-89222-6_2

process during any physical process. The mass of a chemical species is not necessarily conserved, since one species can be transformed into another by chemical reactions. It is important to realize that Newton's Second Law does not represent, forcibly, the conservation of linear momentum. This will only occur for inertial problems and when the vectorial sum of the external forces that are acting on a system is null. Furthermore, for non-inertial systems, the sum of the external forces and the effects of relative acceleration must be null to have conservation of linear momentum. Similarly, the kinetic energy, internal energy, and potential energy of a system are not necessarily conserved, as any of these three forms of energy can be converted into one of the others. For example, kinetic energy can be converted into internal energy by viscous effects, which is an irreversible process. Potential energy can be converted into kinetic energy and vice versa. Therefore, one must be cautious with the use of the word "conservation" when dealing with balances of a given "*information*". A balance does not always imply conservation of the *information* under analysis. The notion of balance requires the concurrent use of the concepts of "closed system" or simply "system" and "open system". The concept of "open system" is also referred to as "control volume". The extension of the laws of classical mechanics to solve problems involving control volumes or open systems, which allow for "advective flux of any *information*", became possible with the establishment of the Reynolds–Leibniz Transport Theorem, which was demonstrated by them in the nineteenth century.

In this chapter, the integral formulation for fluid dynamics, including non-inertial effects, will be presented through the concept of relative acceleration. The Reynolds–Leibniz Transport Theorem will be presented in Sect. 2.2. In Sects. 2.3–2.7, the differential formulation will be presented, with details for the mass balance, for the linear momentum balance, and for the balances of kinetic energy, thermal energy, and total energy. All possible effects, such as Einstein Acceleration, Euler Acceleration and Coriolis Acceleration, will be considered. The "balance equation" for vorticity will also be derived in detail, accompanied by a physical interpretation of all the terms that make up these equations. The concepts of Galilean and Euclidean Invariances associated with different "*information*", such as strain rate, rotation rate, vorticity, among others, will also be presented. This chapter, although seemingly out of place in a book on Turbulent Fluid Flows, is of utmost importance for the development of the other chapters, as it was not found, in a detailed form and focused on the theme of "turbulent flows", in books on Fluid Mechanics. Thus, its detailed presentation in this text is justified.

2.2 Integral Formulation—Reynolds–Leibniz Transport Theorem

The Reynolds–Leibniz Transport Theorem will be presented and discussed here, with focus on deducing the "differential model" for fluid dynamics. The details of its deduction can be found in several Fluid Mechanics books White [12] and Faber [6]. Let B [unit of the *B information*], be a generic "extensive *information*" and its corresponding "intensive *information*" $\beta = \mathrm{d}B/\mathrm{d}m$ [unit of *information B*/unit of mass].

With this theorem, it becomes possible to relate what happens in a system and in a control volume, both coinciding at any arbitrary time. Literally, with this theorem, it is established that the rate of change of *information B* contained in a system must be equal to the rate of change of this same *information*, contained in a control volume $\mho$, added to the net advective flux of this *information*, through the control surface Γ that delimits the control volume $\mho$. Mathematically, we have the following integral–differential equation:

$$\frac{\mathrm{d}B}{\mathrm{d}t}\Big|_{\mathrm{Sist.}} = \frac{\mathrm{d}}{\mathrm{d}t}\int_{\mho} \rho\beta \mathrm{d}\mho + \int_{\Gamma} \rho\beta\left(\vec{V}\cdot \mathrm{d}\vec{\Gamma}\right). \tag{2.1}$$

It is worth justifying, once again, the use of the word *information* (see Sect. 1.3), since, according to current physicists, it can be considered that the universe is a gigantic *reservoir of informations*. Giving some examples of *information*: mass, linear momentum, angular momentum, energies, concentrations, etc. Equation (2.1) is now particularized for Newton's Second Law, which represents a *balance of linear momentum* for *non-inertial systems*, obtaining the following equation:

$$\int_{\mho} \vec{f}_{\mathrm{ext.\,field}} \mathrm{d}\mho + \int_{\mho} \vec{f}_{\mathrm{ext.\,contact}} \mathrm{d}\mho - \int_{\mho} \rho\vec{a}_{\mathrm{rel}} \mathrm{d}\mho = \frac{\mathrm{d}}{\mathrm{d}t}\int_{\mho} \rho\vec{V} \mathrm{d}\mho + \int_{\Gamma} \rho\vec{V}\left(\vec{V}\cdot \mathrm{d}\vec{\Gamma}\right). \tag{2.2}$$

In this equation, $\vec{f}_{\mathrm{ext.\,field}}$ and $\vec{f}_{\mathrm{ext.\,contact}}$ are, respectively, the specific external field force and contact force, both given in [N/m^3], acting on the system, coinciding with the control volume; $\vec{a}_{\mathrm{rel}}$ is the relative acceleration of this fluid particle, relative to any non-inertial axis system. The relative acceleration is given by

$$\vec{a}_{\mathrm{rel}} = \frac{\mathrm{D}^2\vec{R}}{\mathrm{D}t^2} + \frac{\mathrm{D}\vec{\Omega}}{\mathrm{D}t} \times \vec{r} + 2\vec{\Omega} \times \vec{V} + \vec{\Omega} \times \left(\vec{\Omega} \times \vec{r}\right) \tag{2.3}$$

All the quantities used are shown in Fig. 2.1. The vector $\vec{R}$ locates, in an absolute way, the *non-inertial axis system* (x, y, z); the vector $\vec{S}$ locates, in an absolute way, the fluid particle in the *inertial axis system* and the vector $\vec{r}$ locates this same fluid particle in relation to any non-inertial axis system (x, y, z). The vectors $\vec{V}$ and $\vec{\Omega}$

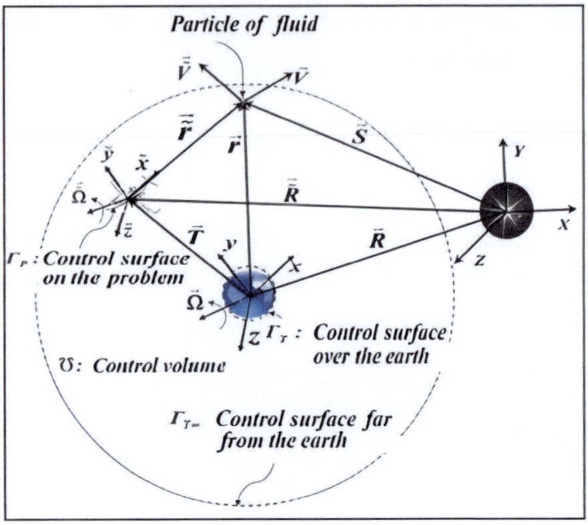

Fig. 2.1 Illustration of the inertial axis system (X, Y, Z) and the non-inertial axis systems (x, y, z) and $(\tilde{x}, \tilde{y}, \tilde{z})$. The associated variables are control volume $\mho$; control surface $\Gamma_{T\infty}$ that isolates the control volume $\mho$ from the rest of the environment; control surface Γ_T that isolates the earth inside the control volume $\mho$; control surface Γ_P that isolates the problem inside the control volume $\mho$

represent, respectively, the translational velocity of the fluid particle in relation to the non-inertial axis system (x, y, z) and the angular velocity of this non-inertial axis system, in relation to the inertial axis system (X, Y, Z). The system $(\tilde{x}, \tilde{y}, \tilde{z})$ is also considered in Fig. 2.1, a second non-inertial axis system. This system is positioned, in an absolute way, by the vector $\vec{\tilde{R}}$. The fluid particle is positioned, in relation to this second non-inertial axis system, by the vector $\vec{\tilde{r}}$. The same particle has a velocity $\vec{\tilde{V}}$, when seen by the observer $(\tilde{x}, \tilde{y}, \tilde{z})$. The axis system $(\tilde{x}, \tilde{y}, \tilde{z})$ has an angular velocity $\vec{\tilde{\Omega}}$.

2.3 Differential Formulation for Fluid Dynamics

The differential formulation represents balances, on a fluid particle, of any *information* that is being transported by a flow, for example, mass, linear momentum, thermal energy, concentrations of chemical species, among other *information.*

2.3.1 Mass Balance

Starting from Eq. (2.1) and for the particular case of $B = m$, where m is the mass contained in a material particle of volume $\delta\mho$, delimited by a control surface $\delta\Gamma$, it is found that $\beta = 1$. As, by definition, a system must be sealed, not allowing advective mass flux, it is found that:

$$\frac{\mathrm{d}}{\mathrm{d}t}\int\limits_{\delta\mho}\rho\mathrm{d}\mho+\int\limits_{\delta\Gamma}\rho\left(\vec{V}\cdot\mathrm{d}\vec{\Gamma}\right)=0. \tag{2.4}$$

Using the Leibniz Theorem and the Stokes Theorem [3] and considering that $\delta\mho$, is a generic control volume, but constant over time, we have:

$$\int\limits_{\delta\mho}\left\{\frac{\partial\rho}{\partial t}+\vec{\nabla}\cdot\left(\rho\vec{V}\right)\right\}\mathrm{d}\mho=0. \tag{2.5}$$

An integral over a generic volume is null only when its integrand is null. This follows from the use of the Localization Theorem, according to Chandrasekharaiah and Debnath [4] and Arpaci and Larsen [1]. Thus, Eq. (2.5) results:

$$\frac{\partial\rho}{\partial t}+\vec{\nabla}\cdot\left(\rho\vec{V}\right)=0. \tag{2.6}$$

This is the continuity equation, which represents the mass balance over an infinitesimal particle of fluid. The first term of this equation, $\vec{\nabla}\cdot\left(\rho\vec{V}\right)$, represents the rate of specific mass variation contained in a fluid particle, $\left[(\mathrm{kg/s})/\mathrm{m}^3\right]$. The term $\vec{\nabla}\cdot\left(\rho\vec{V}\right)$ represents the specific advective net mass flux, $\left[(\mathrm{kg/s})/\mathrm{m}^3\right]$, through the control surface that surrounds the infinitesimal material particle. This is the physical meaning of the divergent operator applied to the mass flux $\left(\rho\vec{V}\right)$. The sum of the rate of mass variation contained in the particle with the advective net mass flux is null. If the flow is compressible, then the mass contained in the fluid particle can vary over time. If the particle gains mass, necessarily, the external environment loses the same amount of mass and vice versa. Thus, when analyzing what is happening in the fluid particle, it makes sense to talk about mass balance, even if its mass does not conserve. It also makes sense to talk about total mass conservation when looking at what happens with the whole, that is, with the material particle and with all the neighborhood that surrounds it.

Next, the balance of linear momentum will be done. This balance requires the use of Newton's Second Law. The balance will be performed for the three-axis systems illustrated in Fig. 2.1.

2.3.2 Linear Momentum Balance for an Absolute Observer Placed at a Distant Point

The observer is illustrated in Fig. 2.1, represented by the axis system (X, Y, Z). Rewriting Eq. (2.2) for a control volume $\mho$ applying Stokes' and Leibniz's Theorems and rearranging the terms, we have:

$$\int_{\mho} \vec{f}_{\text{ext. field}} \mathrm{d}\mho + \int_{\mho} \vec{f}_{\text{ext. contact}} \mathrm{d}\mho - \int_{\mho} \rho \vec{a}_{\text{rel}} = \frac{\mathrm{d}}{\mathrm{d}t} \int_{\mho} \rho \vec{V} \mathrm{d}\mho + \int_{\Gamma} \rho \vec{V} \left(\vec{V} \cdot \mathrm{d}\vec{\Gamma} \right)$$
$$\Rightarrow \int_{\delta\mho} \left\{ \frac{\partial \left(\rho \vec{V} \right)}{\partial t} + \vec{\nabla} \cdot \left(\rho \vec{V} \vec{V} \right) + \rho \vec{a}_{\text{rel}} - \vec{f}_{\text{ext. field}} - \vec{f}_{\text{ext. contact}} \right\} \mathrm{d}\mho$$
$$= \vec{0}. \tag{2.7}$$

The quantity $\mho$ is a generic volume and using the Localization Theorem, it is concluded that the integrand of Eq. (2.7) must be null:

$$\vec{f}_{\text{ext}} = \rho \vec{a}_{\text{rel}} + \frac{\mathrm{D}\left(\rho \vec{V} \right)}{\mathrm{D}t} = \rho \left\{ \vec{a}_{\text{rel}} + \frac{\mathrm{D}\left(\vec{V} \right)}{\mathrm{D}t} \right\}. \tag{2.8}$$

In Eq. (2.8), $\vec{f}_{\text{ext.}} = \vec{f}_{\text{ext. field}} + \vec{f}_{\text{ext. contact}} = \vec{f}_{\text{ext. field}} + \vec{\nabla} \cdot \overline{\overline{\sigma}}$ is the specific force [N/m^3], external to the fluid particle. The tensor $\overline{\overline{\sigma}}$ incorporates the effects of viscous stresses and the pressure field. To establish the second equality of Eq. (2.8), the continuity equation (2.6), was considered. Using the definition of relative acceleration, we have:

$$\vec{a}_{\text{rel}} = \frac{\mathrm{D}\vec{V}_{xyz}}{\mathrm{D}t} + \frac{\mathrm{D}\left(\vec{\Omega} \times \vec{r} \right)}{\mathrm{D}t}. \tag{2.9}$$

Then:

$$\vec{f}_{\text{ext}} = \rho \left\{ \frac{\mathrm{D}\vec{V}_{xyz}}{\mathrm{D}t} + \frac{\mathrm{D}\left(\vec{\Omega} \times \vec{r} \right)}{\mathrm{D}t} + \frac{\mathrm{D}\vec{V}}{\mathrm{D}t} \right\}$$
$$= \rho \frac{\mathrm{D}}{\mathrm{D}t} \left(\vec{V}_{xyz} + \vec{\Omega} \times \vec{r} + \vec{V} \right) = \rho \frac{\mathrm{D}\vec{V}_{\text{abs}}}{\mathrm{D}t}. \tag{2.10}$$

It is known that the external force acting on a fluid particle can be divided into gravitational force, electromagnetic force, interfacial force, pressure force, and viscous forces. Keeping, in this book, only the gravitational, pressure, and viscous forces, and using the tensor $\overline{\overline{\sigma}}$ that represent the effects of pressure and viscosity, together, results in:

$$\rho \frac{\mathrm{D}\vec{V}_{\text{abs}}}{\mathrm{D}t} = \rho \vec{g} + \vec{\nabla} \cdot \overline{\overline{\sigma}}_{\text{abs}}. \tag{2.11}$$

Note that the tensor $\overline{\overline{\sigma}}$, in this point of this deduction, is still unknown and needs to be modeled. When it is modeled, its dependence or not on the observer will be analyzed. For now, the tensor will carry an index relative to the observer. This is the well-known Cauchy Equation.

It is observed in this equation that the rotation of the non-inertial axis systems does not appear in the differential formulation for the absolute velocity of a fluid particle. These effects will be transmitted to the solution of Eq. (2.11), via boundary conditions. The solution to any problem in the vicinity of the Earth would require the joint solution of the complete geophysical flow! The rotation and translation of the Earth must affect the solution of the problem, which requires the imposition of boundary conditions on the surface of the planet. The solution, naturally, requires the boundary conditions of the problem itself. This is what the inertial observer (X, Y, Z) "observes", from a distant celestial body. The solution would provide $\vec{V}_{\text{abs}}$, which is the absolute velocity of a fluid particle seen by the inertial observer. This observer "sees" the Earth rotating and translating, inducing movement of the entire terrestrial atmosphere, and sees the movement of the fluid relative to the problem under analysis (flow around an aircraft, for example).

Considering the establishment of the boundary and initial fluid dynamic conditions, some quantities will be presented again here:

$\vec{V}_{xyz}$: absolute velocity of the non-inertial axis system (x, y, z), placed at the center of the Earth.

$\vec{V}_{\tilde{x}\tilde{y}\tilde{z}}$: absolute velocity of the non-inertial axis system $(\tilde{x}, \tilde{y}, \tilde{z})$, placed at the problem of interest.

$\vec{V}_{T\tilde{x}\tilde{y}\tilde{z}}$: velocity, relative to the Earth, of the non-inertial axis system $(\tilde{x}, \tilde{y}, \tilde{z})$, placed at the problem of interest.

$\vec{V}_{\text{abs}}\left(\vec{S}, t\right)$: absolute velocity of the fluid particle.

$\vec{V}(\vec{r}, t)$: velocity of the fluid particle seen by the observer (x, y, z).

$\vec{\tilde{V}}\left(\vec{\tilde{r}}, t\right)$: velocity of the fluid particle seen by the observer $(\tilde{x}, \tilde{y}, \tilde{z})$.

$\vec{\Omega}$: angular velocity of the axis system (x, y, z).

$\vec{\tilde{\Omega}}$: angular velocity of the axis system $(\tilde{x}, \tilde{y}, \tilde{z})$.

The boundary conditions and the initial condition for Eq. (2.11) are presented below, by which the absolute velocity of the fluid particle is prescribed at all the contours. Boundary conditions:

$$\vec{V}_{\text{abs}}\left(\vec{S}, t\right) = \begin{cases} \vec{V}_{xyz} + \vec{\Omega} \times \vec{r}_{\Gamma T} & \text{if } \vec{S} = \vec{S}_{\Gamma T} \\ \vec{V}_{\tilde{x}\tilde{y}\tilde{z}} + \vec{\tilde{\Omega}} \times \vec{\tilde{r}}_{\Gamma P} & \text{if } \vec{S} = \vec{S}_{\Gamma P} \\ \vec{V}_{xyz} & \text{if } \vec{S} = \vec{S}_{\Gamma T\infty} \end{cases} . \tag{2.12}$$

Initial condition:

$$\vec{V}_{\text{abs}}\left(\vec{S},0\right) = \vec{V}_0\left(\vec{S}\right). \tag{2.13}$$

In the third boundary condition, established with Eq. (2.12), it is considered that the Earth's rotation does not influence the fluid motion in the vicinity of the external boundary $\Gamma_{T\infty}$. Once calculated $\vec{V}_{\text{abs}}$ (fluid particle absolute velocity) the velocity relative to the non-inertial observer, who is at the center of the Earth, can be obtained using the following transformation equation:

$$\vec{V} = \vec{V}_{\text{abs}} - \vec{V}_{xyz} - \vec{\Omega} \times \vec{r}. \tag{2.14}$$

The inertial observer (X, Y, Z) sees the entire Earth's atmosphere in motion, induced by the Earth's rotation and translation and natural convection, and sees the body (airplane) in motion relative to the fluid. The observer (x, y, z) sees the Earth standing still, without rotation or translation, and sees the fluid moving at the velocity $\vec{V}$.

It is also possible to determine the velocity seen by the observer $(\tilde{x}, \tilde{y}, \tilde{z})$, which is given by:

$$\vec{\tilde{V}} = \vec{V}_{\text{abs}} - \vec{\tilde{V}}_{\tilde{x}\tilde{y}\tilde{z}} - \vec{\tilde{\Omega}} \times \vec{\tilde{r}}. \tag{2.15}$$

Again, the inertial observer (X, Y, Z) sees the entire Earth's atmosphere in motion, induced by the Earth's rotation and translation and the natural convection of the Earth's atmosphere, and sees the body in motion relative to the fluid. The observer $(\tilde{x}, \tilde{y}, \tilde{z})$ sees the body, object of analysis, stationary, without rotation or translation, and sees the fluid moving at the velocity $\vec{\tilde{V}}$.

Another way to determine the velocity fields $\vec{V}$ and $\vec{\tilde{V}}$ is by deducing balance equations for these fields. It will be presented in the following topics.

2.3.3 Balance of Linear Momentum for a Non-inertial Observer Placed at the Center of the Earth

This observer is illustrated in Fig. 2.1, represented by the axis system (x, y, z). The total derivative of the absolute velocity, which appears in Eq. (2.11), can be rewritten in terms of the non-inertial velocity $\vec{V}$:

$$\frac{\mathrm{D}\vec{V}_{\mathrm{abs}}}{\mathrm{D}t} = \frac{\mathrm{D}}{\mathrm{D}t}\left(\vec{V} + \frac{\mathrm{D}\vec{R}}{\mathrm{D}t} + \vec{\Omega}\times\vec{r}\right) = \underbrace{\frac{\mathrm{D}\vec{V}}{\mathrm{D}t}}_{\substack{\text{Non-inertial}\\ \text{particle}\\ \text{acceleration}}}$$

$$+ \underbrace{\frac{\mathrm{D}^2\vec{R}}{\mathrm{D}t^2}}_{\substack{\text{Non-inertial}\\ \text{acceleration of}\\ (x,y,z)\text{ axis system}}} + \underbrace{\frac{\mathrm{D}\vec{\Omega}}{\mathrm{D}t}\times\vec{r}}_{\substack{\text{Angular}\\ \text{acceleration}}} + \underbrace{\vec{\Omega}\times\left(\vec{\Omega}\times\vec{r}\right)}_{\substack{\text{Centripetal}\\ \text{acceleration}}} + \underbrace{2\vec{\Omega}\times\vec{V}}_{\substack{\text{Coriolis}\\ \text{acceleration}}} \tag{2.16}$$

Thus, one can rewrite the Cauchy Eq. (2.11), in the following way:

$$\rho\frac{\mathrm{D}\vec{V}}{\mathrm{D}t} = \rho\vec{g} + \vec{\nabla}\overline{\overline{\sigma}}_{xyz} - \rho\left\{\frac{\mathrm{D}^2\vec{R}}{\mathrm{D}t^2} + \frac{\mathrm{D}\vec{\Omega}}{\mathrm{D}t}\times\vec{r} + 2\vec{\Omega}\times\vec{V} + \vec{\Omega}\times\left(\vec{\Omega}\times\vec{r}\right)\right\}. \tag{2.17}$$

In the tensor $\overline{\overline{\sigma}}_{xyz}$, the index indicates that it may depend on the axis system in question. The formulation for the non-inertial velocity, $\vec{V}$, results in an equation with acceleration terms with which relative effects are modeled, which depend on the four components of relative acceleration. The observer who is at the center of the Earth does not perceive these effects. For him, the Earth does not rotate and does not translate. Therefore, the effects of Earth's rotation and translation are computed using these acceleration terms. They appear in the balance equation for the non-inertial velocity, $\vec{V}$ (Eq. 2.17).

Boundary conditions must be specified according to what the observer (x, y, z) perceives in the problem. In this case, he sees the fluid relative motion to the body, $(\tilde{x}, \tilde{y}, \tilde{z})$, which is also in motion. It should be noted that, while a fluid particle was positioned by the vector $\vec{S}$, for an inertial observer, here, it is positioned by the vector $\vec{r}$, for the non-inertial observer, who is at the center of the Earth. These boundary conditions and the initial condition are modeled below:

$$\vec{V}(\vec{r}) = \begin{cases} \vec{0} & \text{if } \vec{r} = \vec{r}_{\Gamma T} \\ \vec{V}_{T\tilde{x}\tilde{y}\tilde{z}} - \vec{\Omega}\times\vec{r}_{\Gamma P} + \vec{\tilde{\Omega}}\times\vec{r}_{\Gamma P} & \text{if } \vec{r} = \vec{r}_{\Gamma P} \\ -\vec{\Omega}\times\vec{r}_{\Gamma T\infty} & \text{if } \vec{r} = \vec{r}_{\Gamma T\infty} \end{cases}, \tag{2.18}$$

$$\vec{V}(\vec{r}, 0) = \vec{V}_0(\vec{r}). \tag{2.19}$$

In Eq. (2.18), it was assumed that far away from the Earth, $\vec{r} = \vec{r}_{\Gamma T\infty}$, the fluid moves only with the Earth, not being influenced by its rotation. Therefore, the non-inertial observer, positioned in the center of the Earth, "observes the fluid" very far from the Earth, moving with a translational velocity equal to $\vec{\Omega} \times \vec{r}_{\Gamma T\infty}$. This observer does not notice the planet's translation and rotation movements. He sees the fluid in movement relative to the body, object of analysis, illustrated in Fig. 2.1.

2.3.4 Balance of Linear Momentum for a Non-inertial Observer Placed at the Problem of Interest

The non-inertial axis system $(\tilde{x}, \tilde{y}, \tilde{z})$ is placed over the control volume that involves the problem under analysis, as illustrated in Fig. 2.2.

Equation (2.11) remains valid. Thus, before modeling the flow seen by the observer $(\tilde{x}, \tilde{y}, \tilde{z})$, the model for the problem seen by the absolute observer (X, Y, Z) is presented:

$$\rho \frac{D\vec{V}_{abs}}{Dt} = \rho \vec{g} + \vec{\nabla} \cdot \overline{\overline{\sigma}}_{abs}. \tag{2.20}$$

Boundary conditions:

$$\vec{V}_{abs}\left(\vec{S}\right) = \begin{cases} \vec{V}_{\tilde{x}\tilde{y}\tilde{z}} + \vec{\tilde{\Omega}} \times \vec{\tilde{r}}_{\Gamma P} & \text{if } \vec{S} = \vec{S}_{\Gamma P} \\ \vec{V}_{abs\Gamma P\infty} & \text{if } \vec{S} = \vec{S}_{\Gamma P\infty} \end{cases}, \tag{2.21}$$

and initial condition:

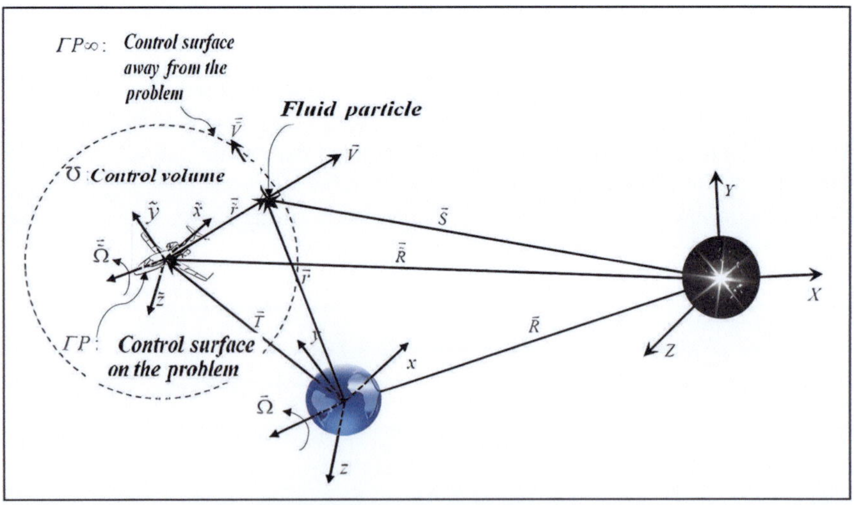

Fig. 2.2 Illustration of the non-inertial problem of interest

$$\vec{V}_{\text{abs}}\left(\vec{S},0\right)=\vec{V}_0\left(\vec{S}\right)\quad\text{for}\quad\vec{S}\in\mho.\tag{2.22}$$

The determination of $\vec{V}_{\text{abs}\Gamma P\infty}$ is extremely complex, as it requires knowledge of the absolute movement of the Earth's atmosphere. Therefore, it becomes very attractive to establish a formulation for the velocity $\vec{\tilde{V}}$, seen by the non-inertial observer, placed on the axis system $(\tilde{x},\tilde{y},\tilde{z})$. Starting from Eq. (2.20), the next model is obtained:

$$\begin{aligned}\rho\frac{\mathrm{D}\vec{\tilde{V}}}{\mathrm{D}t}&=\rho\vec{g}+\vec{\nabla}\cdot\bar{\bar{\sigma}}_{\tilde{x}\tilde{y}\tilde{z}}\\&\quad-\rho\left\{\frac{\mathrm{D}^2\vec{\tilde{R}}}{\mathrm{D}t^2}+\frac{\mathrm{D}\vec{\tilde{\Omega}}}{\mathrm{D}t}\times\vec{\tilde{r}}+2\vec{\tilde{\Omega}}\times\vec{\tilde{V}}+\vec{\tilde{\Omega}}\times\left(\vec{\tilde{\Omega}}\times\vec{\tilde{r}}\right)\right\}\quad\text{or}\\\rho\frac{\mathrm{D}\vec{\tilde{V}}}{\mathrm{D}t}&=\rho\vec{g}+\vec{\nabla}\cdot\bar{\bar{\sigma}}_{\tilde{x}\tilde{y}\tilde{z}}\\&\quad-\rho\left\{\begin{array}{l}\frac{\mathrm{D}^2\vec{R}}{\mathrm{D}t^2}+\frac{\mathrm{D}\vec{V}_{T\tilde{x}\tilde{y}\tilde{z}}}{\mathrm{D}t}+\frac{\mathrm{D}\vec{\Omega}}{\mathrm{D}t}\times\vec{T}+2\vec{\Omega}\times\vec{V}_{T\tilde{x}\tilde{y}\tilde{z}}\\+\vec{\Omega}\times\left(\vec{\Omega}\times\vec{T}\right)+\frac{\mathrm{D}\vec{\tilde{\Omega}}}{\mathrm{D}t}\times\vec{\tilde{r}}+2\vec{\tilde{\Omega}}\times\vec{\tilde{V}}+\vec{\tilde{\Omega}}\times\left(\vec{\tilde{\Omega}}\times\vec{\tilde{r}}\right)\end{array}\right\}\end{aligned}\tag{2.23}$$

The indices appearing in the tensor $\overline{\overline{\sigma}}_{\tilde{x}\tilde{y}\tilde{z}}$ indicate that it refers to the non-inertial observer, placed over the problem. Equation (2.23) is written in two ways. The first shows the absolute acceleration of the axis system $(\tilde{x},\tilde{y},\tilde{z})$, i.e., $\mathrm{D}^2\vec{\tilde{R}}/\mathrm{D}t^2$. In the second form, this acceleration was transformed to have an alternative for its specification, a function of the absolute acceleration of the center of the Earth. To do so, we write the vector $\vec{\tilde{R}}$ in the following form:

$$\vec{\tilde{R}}(t)=\vec{R}(t)+\vec{T}(t),\tag{2.24}$$

which allows us to write the second derivative $\mathrm{D}^2\vec{\tilde{R}}/\mathrm{D}t^2$ as follows:

$$\frac{\mathrm{D}^2\vec{\tilde{R}}}{\mathrm{D}t^2}=\frac{\mathrm{D}^2\vec{R}}{\mathrm{D}t^2}+\frac{\mathrm{D}\vec{V}_{T\tilde{x}\tilde{y}\tilde{z}}}{\mathrm{D}t}+\frac{\mathrm{D}\vec{\Omega}}{\mathrm{D}t}\times\vec{T}+2\vec{\Omega}\times\vec{V}_{T\tilde{x}\tilde{y}\tilde{z}}+\vec{\Omega}\times\left(\vec{\Omega}\times\vec{T}\right).\tag{2.25}$$

This equation represents the velocity and acceleration of the axis system $(\tilde{x},\tilde{y},\tilde{z})$ relative to the center of the Earth with the terms $\vec{V}_{T\tilde{x}\tilde{y}\tilde{z}}$ and $\mathrm{D}\vec{V}_{T\tilde{x}\tilde{y}\tilde{z}}/\mathrm{D}t$. The boundary conditions and initial condition are given below:

$$\vec{\tilde{V}}\left(\vec{\tilde{r}},t\right)=\begin{cases}\vec{0}&\text{if }\vec{\tilde{r}}=\vec{\tilde{r}}_{\Gamma P}\\-\vec{V}_{\tilde{x}\tilde{y}\tilde{z}}+\vec{\tilde{V}}_{\Gamma P\infty}\left(\vec{\tilde{r}}\right)&\text{if }\vec{\tilde{r}}=\vec{\tilde{r}}_{\Gamma P\infty}\end{cases},\tag{2.26}$$

$$\vec{\tilde{V}}\left(\vec{\tilde{r}},0\right)=\vec{\tilde{V}}_0\left(\vec{\tilde{r}}\right)\quad\text{for}\quad\vec{\tilde{r}}\in\mho.\tag{2.27}$$

The boundary conditions for solving Eqs. (2.20) and (2.23) must be specified according to what is seen by the observer, whether inertial or non-inertial. With Eq. (2.20) the absolute velocity field $\vec{V}_{abs}$ is obtained for a fluid particle, located in $\vec{S}$, seen by inertial observer (X, Y, Z). For Eq. (2.23), the non-inertial velocity field $\vec{\tilde{V}}$ is obtained for a fluid particle located in $\vec{\tilde{r}}$, viewed by a non-inertial observer located at $(\tilde{x}, \tilde{y}, \tilde{z})$. The angular velocity, $\vec{\tilde{\Omega}}$, is a characteristic of the non-inertial axis system $(\tilde{x}, \tilde{y}, \tilde{z})$.

2.3.5 *Order of Magnitude Analysis and Appropriate Simplifications*

The models established with Eqs. (2.17) and (2.23), as well as the boundary and initial conditions, are complete. To solve them, the quantities related to the additional terms, arising from relative accelerations, must be established. Depending on the physical, geometric and operational characteristics of the problems to be modeled, it is possible to neglect part or all the terms that make up the relative acceleration. An order-of-magnitude analysis of the terms of these equations allows simplifications to be carried out safely. A demonstration will be presented below for the linear momentum balance equation, for a fluid particle observed from a non-inertial reference frame, placed in the problem of interest, as illustrated in Fig. 2.2. It will be rewritten below to facilitate analysis:

$$\rho \frac{\mathrm{D}\vec{\tilde{V}}}{\mathrm{D}t} = \rho\vec{g} + \vec{\nabla}\cdot\bar{\bar{\sigma}}_{\tilde{x}\tilde{y}\tilde{z}} - \rho\left\{ \boxed{\frac{\mathrm{D}^2\vec{\tilde{R}}}{\mathrm{D}t^2}} + \boxed{\frac{\mathrm{D}\vec{\tilde{\Omega}}}{\mathrm{D}t}\times\vec{\tilde{r}}} + \boxed{2\vec{\tilde{\Omega}}\times\vec{\tilde{V}}} + \boxed{\vec{\tilde{\Omega}}\times\left(\vec{\tilde{\Omega}}\times\vec{\tilde{r}}\right)} \right\}. \tag{2.28}$$

The terms that are inside the rectangles are negligible. To carry out the order of magnitude analysis, it is necessary to define the characteristic quantities of the problem. Considering the case of flows over an aircraft, illustrated in Fig. 2.2, the following characteristic quantities are defined:

- U: characteristic velocity of the problem, that is, the velocity of the aircraft relative to the air, $U \sim 250\,[\mathrm{m/s}]$.
- L: characteristic length of the problem. This length, combined with the characteristic velocity and kinematic viscosity of the fluid, results in the Reynolds Number. This dimensionless parameter represents the relationship between the integral length and the Kolmogorov length. It is natural to adopt L as the wingspan of the aircraft, i.e., $L \sim 80\,[\mathrm{m}]$.

- $T = L/U$: characteristic time of the flow over the aircraft, $T \sim 0.3$ [s].

Once these three characteristic quantities are defined, an order of magnitude analysis can be carried out for each term in Eq. (2.28):

- Rate of change of linear momentum contained in a fluid particle:

$$\left\| \rho \frac{\partial \vec{\tilde{V}}}{\partial t} \right\| \sim \rho \frac{U^2}{L} \sim 10^3 \left[\mathrm{N/m^3}\right].$$

- Net advective flux of linear momentum across the surface of a fluid particle:

$$\left\| \rho \left(\vec{\tilde{V}} \cdot \vec{\tilde{\nabla}} \right) \vec{\tilde{V}} \right\| \sim \rho U^2/L \sim 10^3 \left[\mathrm{N/m^3}\right].$$

- Weight force of the fluid particle: $\|\rho \vec{g}\| \sim 10^1 \left[\mathrm{N/m^3}\right]$. This term will be kept in the equation, as it represents the weight of the fluid particle, which cannot be neglected in view of the other terms.
- Diffusive net flux of linear momentum, due to molecular effects: $\left\| \vec{\tilde{\nabla}} \cdot \overline{\overline{\sigma}}_{\tilde{x}\tilde{y}\tilde{z}}/\rho \right\| \sim U^2/L \sim 10^3 \left[\mathrm{N/m^3}\right]$. This order of magnitude, 10^3, was established considering that this term, for many problems, cannot be ignored compared to other terms.
- Absolute aircraft acceleration: $\left\| \mathrm{D}^2\vec{\tilde{R}}/\mathrm{D}t^2 \right\| \sim \left\| \mathrm{D}^2\vec{R}/\mathrm{D}t^2 \right\| \sim 10^{-13} \left[\mathrm{m/s^2}\right]$. This order of magnitude, 10^{-13}, results from the consideration that the absolute acceleration of the aircraft is of the order of the absolute acceleration of the earth. This data was taken from Kantha and Clayson [10]. Comparing this term to the others, it is concluded that it is negligible.
- Euler or angular acceleration: $\left\| \left(\mathrm{D}\vec{\tilde{\Omega}}/\mathrm{D}t \right) \times \vec{\tilde{r}} \right\| \sim \left\| \frac{\mathrm{D}\vec{\Omega}}{\mathrm{D}t} \right\| \cdot R_T \sim 10^{-21} \left(\mathrm{rad/s^2}\right) 10^6\,\mathrm{m} \sim 10^{-15} \left[\mathrm{m/s^2}\right]$. Comparing this term with the others, we conclude that it is negligible. These data were taken from Kantha and Clayson [10].
- Coriolis acceleration: $\left\| \vec{\tilde{\Omega}} \times \vec{\tilde{V}} \right\| \sim 0\,(\mathrm{rad/s}) \times 250\,(\mathrm{m/s}) \sim 0 \left[\mathrm{m/s^2}\right]$. This result is due to the consideration that the absolute angular velocity of the aircraft is approximately zero. Thus, this term, Coriolis Acceleration, becomes negligible compared to the other terms of Eq. (2.26).
- Centripetal acceleration: $\left\| \vec{\tilde{\Omega}} \times \vec{\tilde{\Omega}} \times \vec{\tilde{r}} \right\| \sim 0 \left[\mathrm{m/s^2}\right]$. This result makes this term negligible compared to the others.

Using these results, Eq. (2.28) can be simplified to:

$$\rho \frac{\mathrm{D}\vec{\tilde{V}}}{\mathrm{D}t} = \rho \vec{g} + \vec{\tilde{\nabla}} \cdot \overline{\overline{\sigma}}_{\tilde{x}\tilde{y}\tilde{z}}, \tag{2.29}$$

$$\vec{\tilde{V}}(\vec{\tilde{r}}) = \begin{cases} \vec{0} & \text{if } \vec{\tilde{r}} = \vec{\tilde{r}}_{\Gamma P} \\ -\vec{V}_{\tilde{x}\tilde{y}\tilde{z}} & \text{if } \vec{\tilde{r}} = \vec{\tilde{r}}_{\Gamma P\infty} \end{cases}$$
$$\vec{\tilde{V}}(\vec{\tilde{r}}, 0) = \vec{\tilde{V}}_0(\vec{\tilde{r}}) \text{ for } \vec{\tilde{r}} \in \mho. \tag{2.30}$$

This simplified model is presented in the literature, without due justification for each simplification adopted. Thus, throughout this text, we sought to deduce the complete models and, subsequently, carry out the simplifications that are usually made. For any other problem to be modeled, its physical, geometric and operational characteristics must be well established, and then an order of magnitude analysis of the terms of the differential equation must be carried out, and decisions must be taken on the appropriate simplifications.

2.3.6 Geophysical Atmospheric Flows

The flows of the Earth's atmosphere are influenced by the Earth's rotation, especially by the Coriolis Acceleration, $2\vec{\Omega} \times \vec{V}$. Consider, for example, an observer positioned at the center of the Earth, as illustrated in Fig. 2.3. We start from the model presented through Eqs. (2.17)–(2.18). To facilitate order-of-magnitude analysis, the model is copied in this section:

$$\rho\frac{D\vec{V}}{Dt} = \rho\vec{g} + \vec{\nabla}\cdot\overline{\overline{\sigma}}_{xyz} - \rho\left\{\frac{D^2\vec{R}}{Dt^2} + \frac{D\vec{\Omega}}{Dt}\times\vec{r} + 2\vec{\Omega}\times\vec{V} + \vec{\Omega}\times\left(\vec{\Omega}\times\vec{r}\right)\right\}. \tag{2.31}$$

The characteristic quantities of geophysical flows seen by an observer placed at the center of the earth are shown below:

- U: characteristic velocity of flow over the Earth, seen by an observer located in the center of the Earth. This will be the transport velocity of atmospheric turbulent structures.
- L: characteristic length, which may be the diameter of the Earth.
- $T = L/U$: characteristic time of geophysical flows.

Once the characteristic quantities are defined, the order of magnitude analysis of the terms of the differential equation above is carried out:

- The term $\frac{D\vec{V}}{Dt}$ has an order of magnitude analyzed below:

$$\left\|\frac{D\vec{V}}{Dt}\right\| \sim \frac{U^2}{L} \sim \frac{(\Omega L)^2}{L} \sim \Omega^2 L \sim \left(7.3\times10^{-5}\right)^2 [\text{rad/s}]^2$$
$$\times\, 10^6\,[\text{m}] \sim 10^{-3}\left[\text{m/s}^2\right];$$

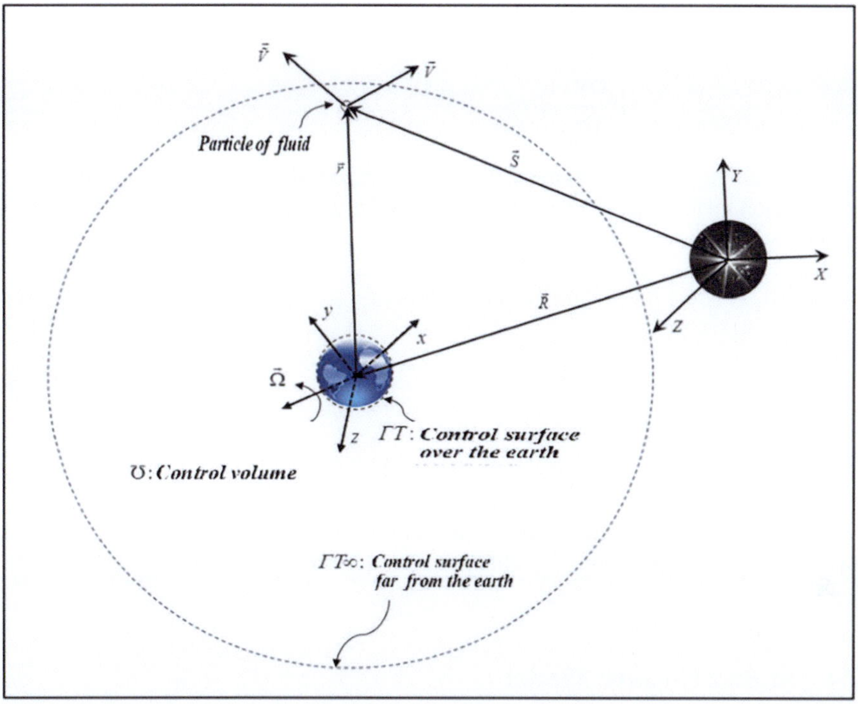

Fig. 2.3 Illustration of the Earth's atmosphere. Non-inertial axis system (x, y, z), placed in the center of the earth. Inertial axis system (X, Y, Z), placed on a distant star

- The term $\|\rho\vec{g}\|$ represents the weight of the particle. Therefore, it must have the same order of magnitude as the others, that is, $\|\rho\vec{g}\| \sim U^2/L$.
- Likewise, the term $\left\|\vec{\nabla}\cdot\overline{\overline{\sigma}}_{xyz}\right\|$ cannot be neglected. Then $\left\|\vec{\nabla}\cdot\overline{\overline{\sigma}}_{xyz}\right\| \sim U^2/L$.
- The absolute acceleration of the Earth is of the order of $\left\|\frac{\mathrm{D}^2\vec{R}}{\mathrm{D}t^2}\right\| \sim 10^{-13}\left[\mathrm{m/s^2}\right]$, as presented by Kantha and Clayson [10]. So, this term can be neglected.
- The term $\left\|\frac{\mathrm{D}\vec{\Omega}}{\mathrm{D}t}\times\vec{r}\right\| \sim \left\|\frac{\mathrm{D}\vec{\Omega}}{\mathrm{D}t}\right\|\cdot R_T \sim 10^{-21}\left(\mathrm{rad/s^2}\right)\times 6.4\times 10^7\,\mathrm{m} \sim 10^{-14}\left[\mathrm{m/s^2}\right]$, according to Kantha and Clayson [10]. So, it can be neglected.
- The term $2\vec{\Omega}\times\vec{V} \sim 2\Omega U \sim 10^{-3}\left[\mathrm{rad/s^2}\right]$, according to Kantha and Clayson [10]. Therefore, for atmospheric flows, this term cannot be neglected.
- Centripetal acceleration $\vec{\Omega}\times\left(\vec{\Omega}\times\vec{r}\right)$ can be estimated as follows: $\vec{\Omega}\times\left(\vec{\Omega}\times\vec{r}\right) \sim \left(7.3\times 10^{-5}\right)^2\times 6.4\times 10^6 \sim 5\times 10^{-3} \sim 5\times 10^{-4}g$. This term can be incorporated into the gravitational field, giving rise to a modified or effective gravitational field, $g_{\mathrm{ef}} = g + 0.0005g = 1.0005g$. It is observed that the influence of this term is small, and it may or may not be considered. Normally, when measuring the gravitational field on the earth's surface, this effect is already incorporated into the measured value.

Therefore, the model used for this type of application is given by:

$$\rho \frac{\mathrm{D}\vec{V}}{\mathrm{D}t} = \rho \vec{g}_{\mathrm{ef}} + \vec{\nabla} \cdot \overline{\overline{\sigma}}_{xyz} - 2\rho\vec{\Omega} \times \vec{V}. \tag{2.32}$$

Emphasizing, with this analysis of order of magnitude, it is shown that the relative acceleration term, $2\rho\vec{\Omega} \times \vec{V}$, is of the same order as the other terms. The following boundary conditions are established:

$$\vec{V}(\vec{r}_{\Gamma T}, t) = \vec{0}, \tag{2.33}$$

$$\vec{V}(\vec{r}_{\Gamma T\infty}, t) = \vec{V}_{\Gamma T\infty}, \tag{2.34}$$

$$\vec{V}(\vec{r}, 0) = \vec{V}_0(\vec{r}), \quad \text{for } \vec{r} \in \mho, \tag{2.35}$$

where $\vec{V}_{\Gamma T\infty}$ is the velocity of the fluid, seen by (x, y, z), on a control surface far away from the Earth. Figure 2.3 illustrates this physical problem.

2.4 Stokes Closure Model

In all the models presented and discussed in the previous sections, in the different ways of writing the Cauchy Equations, the tensors $\overline{\overline{\sigma}}_{\mathrm{abs}}$, $\overline{\overline{\sigma}}_{xyz}$ and $\overline{\overline{\sigma}}_{\bar{x}\bar{y}\bar{z}}$ are open. Closures for them must still be established. Stokes proposed, for Newtonian fluids, the following closure model:

$$\overline{\overline{\sigma}}(\vec{r}, t) = \mu\left(\vec{\nabla}\vec{V} + \vec{\nabla}\vec{V}^T\right) - \frac{2}{3}\mu\left(\vec{\nabla} \cdot \vec{V}\right)\overline{\overline{I}} - p\overline{\overline{I}}. \tag{2.36}$$

In Eqs. (2.11), (2.17) and (2.23), the tensor $\overline{\overline{\sigma}}$ appears dependent on the observer, i.e., $\overline{\overline{\sigma}}_{\mathrm{abs}}$, $\overline{\overline{\sigma}}_{xyz}$ and $\overline{\overline{\sigma}}_{\bar{x}\bar{y}\bar{z}}$. As can be seen in Eq. (2.36), this tensor was modeled as a function of the strain rate tensor, $\overline{\overline{S}} = (1/2)\left(\vec{\nabla}\vec{V} + \vec{\nabla}\vec{V}^T\right)$, of the tensor $p\overline{\overline{I}}$ and the tensor $\left(\vec{\nabla} \cdot \vec{V}\right)\overline{\overline{I}}$. It will be presented, at the end of this chapter, that the strain rate tensor ($\overline{\overline{S}}$) is an objective *information*, that is, it is Euclidean invariant, therefore not depending on the observer, even if the observer is non-inertial. The tensors $p\overline{\overline{I}}$ and $\left(\vec{\nabla} \cdot \vec{V}\right)\overline{\overline{I}}$ are products of scalars by the identity tensor and are therefore also objective quantities. Therefore, it is established that, for Newtonian fluids, the tensor $\overline{\overline{\sigma}}$, modeled with Eq. (2.36), is an objective *information*, not depending on the observer. Therefore, the indices of this tensor are no longer necessary and will no longer be used. It should be noted that the same cannot be established for non-Newtonian fluids, for which viscosity is also a function of the strain rate tensor

$\overline{\overline{S}}$. This will lead to the fact that the tensor $\overline{\overline{\sigma}}$, for non-Newtonian fluids, will be a non-objective *information* and will depend on the observer.

All open models presented in this text involving tensors $\overline{\overline{\sigma}}_{\text{abs}}$, $\overline{\overline{\sigma}}_{xyz}$ and $\overline{\overline{\sigma}}_{\bar{x}\bar{y}\bar{z}}$, can be closed with the Stokes Model. For example, flows for which relative acceleration can be neglected take the form below:

$$\frac{\partial \rho}{\partial t} + \vec{\nabla} \cdot \left(\rho \vec{V}\right) = 0, \tag{2.37}$$

$$\frac{\partial\left(\rho \vec{V}\right)}{\partial t} + \vec{\nabla} \cdot \left(\rho \vec{V} \vec{V}\right) = -\vec{\nabla} p + \rho \vec{g} + \vec{\nabla} \cdot \left[\mu\left(\vec{\nabla} \vec{V} + \vec{\nabla} \vec{V}^T\right) - \frac{2}{3}\mu\left(\vec{\nabla} \cdot \vec{V}\right)\overline{\overline{I}}\right]. \tag{2.38}$$

Flows with thermal effects require another equation, which must be derived, making an internal energy balance.

2.5 Internal Energy Balance Equation

The internal energy balance equation is derived by particularizing the Reynolds–Leibniz Transport Theorem, Eq. (2.1), for this *information* (internal energy), here denoted by U. Thus, the *information* B is now particularized to the internal energy, U [J], and the *information* β is particularized to the internal energy per unit mass, $u\left[\text{J/kg}\right]$. The exchange of internal energy between a system and its surroundings, on the left side of Eq. (2.1), can only occur through processes called work and heat. With the right side of this equation, the rate of variation of the internal energy contained in the control volume, coincident with the system, at any time, is modeled. The net advective flux of internal energy through the control surface that delimits the control volume is also modeled. After applying the divergence theorem to convert surface integrals into volume integrals, it results in a zero integral. As the integration volume is generic, this implies that, according to the localization theorem, the integrand must be zero. The resulting differential equation is presented below:

$$\rho\left[\frac{\partial u}{\partial t} + \left(\vec{V} \cdot \vec{\nabla}\right) u\right] = \vec{\nabla} \vec{V} \cdot \overline{\overline{\sigma}} + \vec{\nabla} \cdot \vec{q}. \tag{2.39}$$

The diffusive flux vector of internal energy, $\vec{q}$, is modeled by Fourier's Law. The term deformation work can be broken down as follows:

$$\vec{\nabla} \vec{V} \cdot \overline{\overline{\sigma}} = -p \vec{\nabla} \cdot \vec{V} + 2\mu \left\|\overline{\overline{S}}\right\|^2 - \frac{2}{3}\mu\left(\vec{\nabla} \cdot \vec{V}\right)^2 = \eta + \phi, \tag{2.40}$$

being

$$\eta = -p\vec{\nabla} \cdot \vec{V}, \tag{2.41}$$

$$\phi = 2\mu \left\| \overline{\overline{S}} \right\|^2 - \frac{2}{3}\mu \left(\vec{\nabla} \cdot \vec{V} \right)^2. \tag{2.42}$$

With the function η, Eq. (2.41), the process of heating or cooling the fluid is modeled, due to the compression or expansion of fluid particles. This is a reversible process of energy transformation, which partly justifies the accumulation of internal energy in the material particle, the object of the balance represented through Eq. (2.39). This term can contribute to a positive value, if the process is compression, or a negative value, if the process is expansion of the fluid particle.

With the function ϕ, Eq. (2.42), called, in this text, specific power of viscous transformation [W/m^3], the irreversible transformation of kinetic energy into thermal energy is modeled. This is a viscous transformation process, which is relevant in flows that have strong viscous effects. These effects become more important as velocity gradients or fluid viscosity intensify. Re-entry problems in the Earth's atmosphere are examples for which viscous effects are very important, due to strong velocity gradients. In shaft-bearing systems, viscous effects are also important due to high viscosity values and high-velocity gradients. The specific power function of viscous transformation depends on the strain rate:

$$\overline{\overline{S}} = \frac{1}{2}\left(\vec{\nabla}\vec{V} + \vec{\nabla}\vec{V}^T \right). \tag{2.43}$$

Considering incompressible flows, Eq. (2.42) is rewritten in the simplified form below:

$$\phi = 2\mu \left\| \overline{\overline{S}} \right\|^2, \tag{2.44}$$

or, in Cartesian coordinates:

$$\begin{aligned} \phi = \mu \Bigg\{ 2\left[\left(\frac{\partial u}{\partial x} \right)^2 + \left(\frac{\partial v}{\partial y} \right)^2 + \left(\frac{\partial w}{\partial z} \right)^2 \right] + \left(\frac{\partial u}{\partial y} + \frac{\partial v}{\partial x} \right)^2 \\ + \left(\frac{\partial u}{\partial z} + \frac{\partial w}{\partial x} \right)^2 + \left(\frac{\partial v}{\partial z} + \frac{\partial w}{\partial y} \right)^2 \Bigg\}. \end{aligned} \tag{2.45}$$

For problems in which variations in internal energy are limited to variations in thermal energy, internal energy can be rewritten as a function of temperature and thermal capacity: $u = C_p T$. Thus, the thermal energy balance equation is obtained:

$$\rho C_p \left[\frac{\partial T}{\partial t} + \left(\vec{V} \cdot \vec{\nabla} \right) T \right] = \vec{\nabla} \cdot \left(k \vec{\nabla} T \right) + \phi. \tag{2.46}$$

Equation (2.46) is written in non-divergent form. It can also be written in divergent form. In this text, the nomenclature conservative form was replaced by the divergent form, and the non-conservative form was replaced by the non-divergent form. This care is because thermal energy is not necessarily conserved! The divergent form is presented with Eq. (2.47):

$$\frac{\partial\left(\rho C_{\mathrm{p}} T\right)}{\partial t}+\vec{\nabla} \cdot\left(\rho C_{\mathrm{p}} \vec{V} T\right)=\vec{\nabla} \cdot\left(k \vec{\nabla} T\right)+\phi. \tag{2.47}$$

Equation (2.37) was used to rewrite Eq. (2.46) as Eq. (2.47). It was also considered that the material derivative of the thermal capacity is zero, $\left(\mathrm{D} C_{\mathrm{p}}/\mathrm{D}t\right)=0$, over a trajectory of a fluid particle.

Once the thermal energy balance equation has been deduced, we have the complete differential model for flows with thermal effects of Newtonian fluids. The name used in this text for the equations that make up this model is as follows: mass balance equation, also known as continuity Eq. (2.37); Navier–Stokes Equation or linear momentum balance, Eq. (2.38), and finally thermal energy balance Eq. (2.47). The mass balance equation does not always represent the principle of mass conservation. If the balance is made over the total mass, we can speak of conservation. If it is the mass balance of a chemical component, the mass of that component is not necessarily conserved. It may be converted into other chemical components. The thermal energy equation does not necessarily represent a conservation principle, since, as a rule, the energy transformation process is not null. This means that one *information* can be converted into another, which implies that this *information*—thermal energy—is not conservative.

These equations, plus equations of state, constitute a closed mathematical model that represents any Newtonian fluid flow, for values of Mach Number lower than 15. This model is valid for laminar flows, flows in transition to turbulence and fully turbulent flows. In fact, it will be demonstrated in Chap. 5 that the smallest structures characteristic of turbulence are much larger than the molecular mean free path. The particularization of this model, for a given problem with specific geometric, physical and operational characteristics, is done by establishing the boundary and initial conditions representative of the problem under analysis.

The physical interpretation of all the terms that make up this differential model is important in the context of the topic of turbulent flows. The terms in the first parenthesis of Eq. (2.36) represent the effects of viscous, shear and normal stresses on a fluid particle, exerted or suffered by the environment that surrounds it. The second parenthesis represents the tensions due to the expansion or contraction of a particle, and the third term represents the effects of the pressure field that acts on this same particle.

Equation (2.37) represents the mass balance at the differential level. The first term of this equation represents the rate of change of mass, per unit volume, contained in a differential fluid particle. The second term represents the net advective flux of mass across the control surface surrounding the differential fluid particle. The divergent operator is a differential way of representing the net flux of *information* that is the object of the balance.

Equation (2.38) represents the linear momentum balance or Newton's Second Law, applied to a differential fluid particle. Thus, this equation can be interpreted in two equivalent ways: as Newton's second law or as a linear momentum balance equation.

From the first perspective, the left side of Eq. (2.38), these two terms represent the density of a fluid particle multiplied by its total acceleration. The right side represents the external forces acting on the particle, that is, the sum of pressure, viscous and gravitational forces.

From the second perspective, that is, looking at Eq. (2.38) as a balance equation, we must interpret terms such as accumulation rate, transport terms and transformation terms. Thus, the first term, $\left[\partial\left(\rho\vec{V}\right)/\partial t\right]\left[(\mathrm{kg\,m/s})/\mathrm{m}^3/\mathrm{s}\right]$, the left side of this equation represents the rate of change of the specific linear momentum contained in a fluid particle. The second term on the left side of this equation, $\vec{\nabla}\cdot\left[\rho\vec{V}\vec{V}\right]\left[(\mathrm{kg\,m/s})/\mathrm{m}^3/\mathrm{s}\right]$, is the advective net flux of specific linear momentum through the control surface surrounding the particle. This term can also be interpreted, for the turbulent regime, as the nonlinear process of transfer of linear momentum between the turbulent structures. The first term on the right side of the equation, $\vec{\nabla}p\left[\mathrm{N/m}^3\right] = \vec{\nabla}p\left[(\mathrm{kg\,m/s})/\mathrm{m}^3/\mathrm{s}\right]$, represents the specific diffusive net flux of linear momentum through the control surface, promoted by the pressure gradient. The second term is the external gravitational force or the specific weight of the fluid particle, which represents the process of converting gravitational effects into linear momentum, $\rho\vec{g}\left[\mathrm{N/m}^3\right] = \rho\vec{g}\left[(\mathrm{kg\,m/s})/\mathrm{m}^3/\mathrm{s}\right]$. The last term, given by the divergence of the product of the viscosity coefficient by the gradient of the velocity vector, added to its transpose, represents the specific diffusive liquid flux of linear momentum, through the control surface, promoted by velocity gradients. The divergence of the last term within the bracket represents the specific net diffusive flux of linear momentum through the control surface, promoted by particle expansion or contraction effects.

The first term on the left side of Eq. (2.47) represents the rate of change of the specific thermal energy contained in the fluid particle. The second term represents the net advective flux of specific thermal energy across the particle surface. The first term on the right side of this equation represents the net diffusive flux of specific thermal energy. The last term represents the specific power of viscous transformation of specific kinetic energy into specific thermal energy. For flows in a statistically equilibrium regime, all energy injected into the flow must be transformed by viscous effects.

It will be demonstrated in chapter [5] that, for $M < 15$, the smallest turbulent structures are larger than the mean molecular free path, leading to the conclusion that disturbances with wave numbers (or frequencies) characteristic of molecular activity have a very low probability of giving rise to fluid dynamic instabilities within the spectrum of turbulent structures. It will also be seen that the differential modeling presented is sufficient for the solution of any Newtonian fluid flow, in any regime, be it laminar, transitional or turbulent. The biggest challenge in this research and application domain—transitional flows and turbulent flows—lies in resolving this differential model to cover the entire spectrum of turbulent structures, from the largest to the Kolmogorov Structures, which characterize turbulent flows.

2.6 Specific Differential Models

The differential equations presented can be used to model any type of flow, respecting the hypotheses established for the deduction. They will be rewritten for specific flow groups, for which new simplifying assumptions will be made. These simplified equations will be presented below.

2.6.1 Incompressible Flows Considering Variable Physical Properties

Using the hypothesis of incompressible flows, but maintaining the physical properties k, ρ, C_p and μ variables in time and space, the following simplified differential model is obtained:

$$\vec{\nabla} \cdot \vec{V} = 0, \tag{2.48}$$

$$\frac{\partial\left(\rho\vec{V}\right)}{\partial t} + \vec{\nabla} \cdot \left(\rho\vec{V}\vec{V}\right) = -\vec{\nabla}p + \rho\vec{g} + \vec{\nabla} \cdot \left[\mu\left(\vec{\nabla}\vec{V} + \vec{\nabla}\vec{V}^T\right)\right], \tag{2.49}$$

$$\frac{\partial(\rho C_\mathrm{p} T)}{\partial t} + \vec{\nabla} \cdot \left(\rho C_\mathrm{p} T\vec{V}\right) = \vec{\nabla} \cdot \left(k\vec{\nabla}T\right) + \phi. \tag{2.50}$$

This model is the starting point for the analysis of turbulent flows, using modeling for the problem of closing the filtered equations for turbulence. It will be seen that this modeling requires the calculation of diffusion coefficients for turbulence, which are variable in time and space. Thus, even if the physical properties μ and k are constant, it is convenient to have the differential equations written in the form of Eqs. (2.49) and (2.50), where μ and k appear within the divergent operator.

2.6.2 Incompressible Flows Considering Constant Physical Properties

In this case, the properties μ and k can leave divergent operators and, considering that $\rho = \rho_o$ and $C_p = C_{po}$ are constants, the equations are divided by these constants, obtaining the following simplified equations:

$$\vec{\nabla} \cdot \vec{V} = 0, \tag{2.51}$$

$$\frac{\partial \vec{V}}{\partial t} + \vec{\nabla} \cdot \left(\vec{V}\vec{V} \right) = -\frac{1}{\rho_o} \vec{\nabla} p + \vec{g} + \nu \nabla^2 \vec{V}, \tag{2.52}$$

$$\frac{\partial T}{\partial t} + \vec{\nabla} \cdot \left(\vec{V} T \right) = \alpha \nabla^2 T + \phi / \left(\rho_o C_{po} \right). \tag{2.53}$$

The constant $\alpha = k / \left(\rho_o C_{po} \right)$ is the thermal diffusion coefficient, with which the diffusive transport of thermal energy is modeled. In these equations, the dimensionless parameters appear: Reynolds Number, Mach Number and Peclet Number. When the Mach Number assumes values below $M < 0.34$, flows can be considered incompressible, even if the fluids are compressible. In this sense, it is considered more appropriate to speak about a formulation for incompressible flows and not a formulation for incompressible fluids.

2.6.3 Incompressible Flows with Variable Density

This group of flows can be subdivided into three subgroups: (1) natural convection-type flows, which are promoted only by gravitational effects; (2) forced convection-type flows, which are promoted only by mechanical effects and (3) mixed convection-type flows, which are promoted simultaneously by the two preceding effects, gravitational and mechanical effects.

Gravitational effects manifest themselves due to variations in the density field, which promotes an imbalance between the buoyant and weight forces acting on a fluid particle. These variations in density occur due to two reasons: variations in the temperature field or in multiphase fluid flows. Both effects can also occur simultaneously: multiphase flows with thermal effects.

Flows promoted by variations in density are known as natural convection. Normally, in literature, natural convection is associated only when there are thermal effects. In the present text, natural convection is any flow promoted by gravitational effects, including isothermal and non-isothermal multiphase flows.

2.6.3.1 Single-Phase Incompressible Flows with Thermal Effects

These flows are modeled using the classic Oberbeck–Boussinesq Approximation, under which the density is considered constant in all terms of the balance equations, except for the weight term of the fluid particle. This approximation is acceptable for flows in which the space–time variations of the density are much smaller than the average value of this thermodynamic property, that is, $\rho'(\vec{x}, t) \ll \overline{\rho}$. Thus, the simplified differential model, established from the general model, is as follows:

$$\vec{\nabla} \cdot \vec{V} = 0, \tag{2.54}$$

$$\frac{\partial \vec{V}}{\partial t} + \vec{\nabla} \cdot \left(\vec{V}\vec{V}\right) = -\frac{1}{\rho_o}\vec{\nabla}p - \beta(T - T_o)\vec{g} + \nu\nabla^2\vec{V}, \tag{2.55}$$

$$\frac{\partial T}{\partial t} + \vec{\nabla} \cdot \left(\vec{V}T\right) = \alpha\nabla^2 T + \phi/\left(\rho_o C_{po}\right). \tag{2.56}$$

The term $\beta(T - T_o)\vec{g}$, which appears in the Navier–Stokes Equations, represents the combined effect of the weight force and the buoyancy force acting on the fluid particle. A reference state of the problem is represented by the properties ρ_o and T_o. They can be, for example, the average values of the density and temperature within the domain of interest, the object of the problem under analysis. Each problem must be carefully analyzed to establish these reference properties. It should be borne in mind that, for turbulence modeling, it is convenient that the diffusive terms of these equations are rewritten so that the effective diffusion coefficients, which include the effects of turbulence, are considered variables.

2.6.3.2 Isothermal Two-Phase Incompressible Flows

For flows whose relative movements between phases are promoted only by gravitational effects, that is, without mechanical effects, these flows are also classified as natural convection. For this group of flows, the density $\rho(\vec{x}, t)$ and dynamic viscosity $\mu(\vec{x}, t)$ vary in time and in space, due to the displacement of the phases that make up the flowing fluid. However, these physical properties are constant in each phase. Under these hypotheses, the model for this group of flows is presented below:

$$\vec{\nabla} \cdot \vec{V} = 0, \tag{2.57}$$

$$\frac{\partial\left(\rho\vec{V}\right)}{\partial t} + \vec{\nabla} \cdot \left(\rho\vec{V}\vec{V}\right) = -\vec{\nabla}p + \rho\vec{g} + \vec{\nabla} \cdot \left[\mu\left(\vec{\nabla}\vec{V} + \vec{\nabla}\vec{V}^T\right)\right] + \vec{f}(\vec{x}, t). \tag{2.58}$$

Equation (2.57) is derived from Eq. (2.37), using the fact that the density $\rho(\vec{x}, t)$ be a constant over any stream line, that is, $D\rho/Dt = 0$. The additional force term, $\vec{f}(\vec{x}, t)$, which appears in Eq. (2.58), represents the interfacial force field, which appears between immiscible fluids. The physical properties of the fluid, $\rho(\vec{x}, t)$ and $\mu(\vec{x}, t)$, must be specified as a function of time and space, with a view to solving the system of Eqs. (2.57) and (2.58). To this purpose, an indicator function, $0 \leq I(\vec{x}, t) \leq 1$, where $I = 0$ in the continuous phase and $I = 1$ in the dispersed phase. The function $I(\vec{x}, t)$ must be used to establish the two fields, using the following equations:

$$\rho(\vec{x}, t) = I(\vec{x}, t)\rho_\mathrm{d} + (1 - I(\vec{x}, t))\rho_\mathrm{c}, \tag{2.59}$$

$$\mu(\vec{x}, t) = I(\vec{x}, t)\mu_\mathrm{d} + (1 - I(\vec{x}, t))\mu_\mathrm{c}. \tag{2.60}$$

The physical properties ρ_c, ρ_d, μ_c and μ_d are relative to the continuous and dispersed phases, respectively. For isothermal two-phase flows, these properties are considered constant over the trajectory of a fluid particle; that is, the material derivatives of these properties are zero when a fluid particle is followed. In mathematical terms, $\mathrm{D}\rho/\mathrm{D}t = 0$ and $\mathrm{D}\mu/\mathrm{D}t = 0$.

The terms $\vec{\nabla} p$ and $\rho\vec{g}$ are used to model the effects of buoyancy and fluid particle weight. Buoyancy is promoted by the fluid-static pressure gradient. Thus, the use of Eq. (2.58) requires care about boundary conditions, which must be applied so that the fluid-static pressure field is correctly recovered with the solution of the balance equations.

If it is of interest, in accordance with the problem to be analyzed, it is worth paying attention to the following: when using, for example, periodic boundary conditions in the direction of the gravitational field vector, the fluid-static pressure field cannot be recovered with the solution of Eq. (2.49). In this case, it is convenient to rewrite this equation in an alternative form. Knowing that the degree of fluid-static pressure must be increased by the two-phase fluid, it can be expressed as follows: $\vec{\nabla} p_\mathrm{fe} = \rho_\mathrm{o}\vec{g}$ where $\rho_0 = \alpha\rho_\mathrm{d} + (1 - \alpha)\rho_\mathrm{c}$, being $\alpha = V_\mathrm{d}/V_\mathrm{t}$ the volume fraction and $V_\mathrm{t} = V_\mathrm{c} + V_\mathrm{d}$ the total volume of fluid, where V_d is the volume occupied by the dispersed phase and V_c is the volume of the continuous phase. Decomposing the pressure field into a fluid-static pressure parcel and a fluid-dynamic pressure parcel, then the system of Eqs. (2.57) and (2.58) can be rewritten:

$$\vec{\nabla} \cdot \vec{V} = 0, \tag{2.61}$$

$$\frac{\partial\left(\rho\vec{V}\right)}{\partial t} + \vec{\nabla} \cdot \left(\rho\vec{V}\vec{V}\right) = -\vec{\nabla} p_\mathrm{fd} + (\rho - \rho_\mathrm{o})\vec{g} + \vec{\nabla} \cdot \left[\mu\left(\vec{\nabla}\vec{V} + \vec{\nabla}\vec{V}^T\right)\right]. \tag{2.62}$$

This model can be used with periodic boundary conditions, even for the fluid-dynamic pressure field.

2.6.3.3 Non-isothermal Two-Phase Incompressible Flows

For this group of flows, the physical properties vary for two reasons: due to space–time variation, when the phases move, and due to the variation in the physical properties of each phase, due to the dependence of these properties on the temperature field. For this class of flows, variations in density can be modeled based on a mixed procedure, including what is done for natural thermal convection (Oberbeck–Boussinesq Approximation), Sect. 2.6.3.1 and what is done for isothermal two-phase flows, Sect. 2.3.2. The differential model for two-phase flows with thermal energy transfer is presented below, where, using the Oberbeck–Boussinesq Approximation, the continuity equation is written for incompressible flows. On the other hand, the density, viscosity, thermal capacity and thermal conductivity are variable when passing from one phase to another. Therefore, the Continuity, Navier–Stokes and Energy Equations must be written to take these facts into account:

$$\vec{\nabla} \cdot \vec{V} = 0, \tag{2.63}$$

$$\frac{\partial\left(\rho\vec{V}\right)}{\partial t} + \vec{\nabla} \cdot \left(\rho\vec{V}\vec{V}\right) = -\vec{\nabla}p_{\text{fd}} + (\rho - \rho_{\text{o}})\vec{g} + \vec{\nabla} \cdot \left[\mu\left(\vec{\nabla}\vec{V} + \vec{\nabla}\vec{V}^{T}\right)\right], \tag{2.64}$$

$$\frac{\partial\left(\rho C_{\text{p}}T\right)}{\partial t} + \vec{\nabla} \cdot \left(\rho C_{\text{p}}\vec{V}T\right) = \vec{\nabla} \cdot \left(k\vec{\nabla}T\right) + \phi. \tag{2.65}$$

The Energy Eq. (2.65) is written in divergent form, in the sense that the density and thermal capacity are within the differential operators. This requires the Continuity Eq. (2.63), as well as the equation $\text{D}C_{\text{p}}/\text{D}t = 0$. This last condition is not strictly met, since this property changes with temperature. Another option, depending on the numerical method to be used, is to write the energy equation with C_{p} outside of derivative operators.

The modeling of physical properties must consider the variations caused by phases and temperature variations:

$$\rho(\vec{x}, t) = I(\vec{x}, t)\rho_{\text{d}}(T(\vec{x}, t)) + (1 - I(\vec{x}, t))\rho_{\text{c}}(T(\vec{x}, t)), \tag{2.66}$$

$$\mu(\vec{x}, t) = I(\vec{x}, t)\mu_{\text{d}}(T(\vec{x}, t)) + (1 - I(\vec{x}, t))\mu_{\text{c}}(T(\vec{x}, t)), \tag{2.67}$$

$$C_{\text{p}}(\vec{x}, t) = I(\vec{x}, t)C_{\text{pd}}(T(\vec{x}, t)) + (1 - I(\vec{x}, t))C_{\text{pc}}(T(\vec{x}, t)), \tag{2.68}$$

$$k(\vec{x}, t) = I(\vec{x}, t)k_{\text{d}}(T(\vec{x}, t)) + (1 - I(\vec{x}, t))k_{\text{c}}(T(\vec{x}, t)). \tag{2.69}$$

The physical properties of the continuous and dispersed phases depend on temperature and can be calculated using equations of state. The reference density, which appears in Eq. (2.64), can be evaluated using the following equation:

$$\rho_{\mathrm{o}} = \frac{1}{V}\int_{V}\left[\alpha\rho_{\mathrm{d}}(T(\vec{x},t)) + (1-\alpha)\rho_{\mathrm{c}}(T(\vec{x},t))\right]\mathrm{d}V. \tag{2.70}$$

The *information* α, which appears in Eq. (2.70), is the volume fraction, given by the ratio between the volume of the dispersed phase and the total volume.

2.7 Kinetic Energy and Total Energy Equations

2.7.1 Kinetic Energy Equation

The specific kinetic energy (per unit mass) is defined as $e = \vec{V}\cdot\vec{V}/2\,[\mathrm{J/kg}]$. The balance equation for this variable can be deduced starting from Eq. (2.17), neglecting the effects of relative acceleration and taking the scalar product of the velocity vector with the terms of that equation:

$$\vec{V}\cdot\frac{\mathrm{D}\left(\rho\vec{V}\right)}{\mathrm{D}t} = \rho\vec{V}\cdot\vec{g} + \vec{V}\cdot\left(\vec{\nabla}\cdot\overline{\overline{\sigma}}\right). \tag{2.71}$$

But,

$$\vec{V}\cdot\frac{\mathrm{D}\left(\rho\vec{V}\right)}{\mathrm{D}t} = \frac{1}{2}\frac{\mathrm{D}\left(\rho\vec{V}\cdot\vec{V}\right)}{\mathrm{D}t} = \frac{\mathrm{D}(\rho e)}{\mathrm{D}t}, \tag{2.72}$$

$$\vec{V}\cdot\left(\vec{\nabla}\cdot\overline{\overline{\sigma}}\right) = \vec{\nabla}\cdot\left(\vec{V}\cdot\overline{\overline{\sigma}}\right) - \vec{\nabla}\vec{V}\cdot\overline{\overline{\sigma}}. \tag{2.73}$$

Therefore,

$$\rho\left[\frac{\partial e}{\partial t} + \left(\vec{V}\cdot\vec{\nabla}\right)e\right] = \rho\vec{V}\cdot\vec{g} + \vec{\nabla}\cdot\left(\overline{\overline{\sigma}}\cdot\vec{V}\right) - \vec{\nabla}\vec{V}\cdot\overline{\overline{\sigma}}. \tag{2.74}$$

This is the balance equation for the kinetic energy of a fluid particle. The first term on the left is used to model the rate of change of the kinetic energy contained in the fluid particle. With the second term, the net advective flux of kinetic energy through the particle surface is modeled. The first term on the right represents the conversion of potential energy into kinetic energy and vice versa, which is a reversible process. The second term on the right represents the net exchange of thermal energy between the particle and its surroundings, in the form of deformation work, promoted by the pressure field and viscous stresses, this term being also reversible. The third term is used to represent two types of energy transformations. For a better understanding, this term is broken down into two parts, according to Eqs. (2.40), (2.41) and (2.42).

This term represents the irreversible transformation of kinetic energy into internal energy, called in this text the specific power of viscous transformation.

2.7.2 *Total Energy Equation*

Adding the specific internal energy equation, $u\left[\mathrm{J/kg}\right]$, Eq. (2.74), to Eq. (2.39), results in the following equation:

$$\rho\left[\frac{\partial(e+u)}{\partial t}+\left(\vec{V}\cdot\vec{\nabla}\right)(e+u)\right]=\rho\vec{V}\cdot\vec{g}+\vec{\nabla}\cdot\left(\overline{\overline{\sigma}}\cdot\vec{V}\right)-\vec{\nabla}\cdot\vec{q}. \tag{2.75}$$

This is a balance equation for the total energy of a fluid particle, with the potential energy being considered constant. The first term on the left side of Eq. (2.75) serves to model the rate of change of the total energy contained in the fluid particle. With the second term, the net advective flux of total energy through the particle surface is represented. The first term on the right side, the gravitational term, serves to model the conversion of potential energy into kinetic energy and vice versa, this being a reversible process. With the second term on the right side, the net flux of thermal energy, in the form of deformation work, through the surface of the particle, promoted by the pressure and viscous stress fields, is modeled. The last term is representative of the net diffusive flux of thermal energy (one of the components of internal energy), which results from the thermal interaction between the fluid particle and its surroundings. All these processes serve to justify the rate of change in the total energy contained in the fluid particle through the interaction of the particle with its surroundings. The portion of irreversible conversion of kinetic energy into internal energy does not appear in the equation under study.

This equation, once used to represent the conservative interaction between a fluid particle and its surroundings, can be taken as a total energy conservation equation. But, to do so, the balance between the particle and its surroundings must be incorporated. Even the total energy contained in a particle cannot be considered as conservative *information*. It is preserved only when the interaction between the particle and its surroundings is observed.

2.7.3 *Comparison Between the Kinetic, Internal and Total Energy Equations*

Analyzing Eqs. (2.39), (2.74) and (2.75), it can be observed that in the internal energy balance equation, Eq. (2.39), the term $\vec{\nabla}\vec{V}\cdot\overline{\overline{\tau}}$ serves to model the specific power of viscous transformation of kinetic energy into internal energy. This is an irreversible process and, consequently, this term is strictly positive in the internal energy balance equation. Thus, it is possible to add internal energy to a fluid particle without this same amount being subtracted from the environment that surrounds it. This leads to the conclusion that the internal energy equation does not represent a principle of conservation.

It is observed that in the kinetic energy balance equation, Eq. (2.74), the specific power of viscous transformation appears with a negative sign. With this term, the conversion of kinetic energy into internal energy is modeled. Thus, this portion of kinetic energy, which is subtracted from the fluid particle, is not transferred to its surroundings in the form of kinetic energy. Therefore, kinetic energy cannot be conserved either. In the total energy equation, Eq. (2.75), this term disappears, since the total energy balance equation is obtained by adding the balance equations for internal energy and kinetic energy. Thus, kinetic and internal energy are not conserved, since one type of energy can be converted into another. The total energy of the fluid particle and its surroundings is conserved, since no term appears to model transformative processes in its balance equation. The flows of different types of energy between the particle and its surroundings characterize, when added together, total energy flux. Thus, the amount of total energy that the particle loses, or gains, must be equal to the total energy that the surroundings gain or lose. This characterizes the conservation of total energy when observing the global balance in interaction processes between the particle and its surroundings.

2.8 Strain Rate, Angular Velocity, Vorticity and Vorticity Balance Equation

2.8.1 Concepts of Strain Rate, Angular Velocity and Vorticity

The concepts of strain rate, angular velocity and vorticity are related to each other and will be presented here without further details, as they can be found in Fluid Mechanics books, White [12], Fox and Mac Donald [5], Çengel and Cimbala [3]. The rotation rate, or angular velocity, is given, in Cartesian coordinates, by:

$$\vec{\Omega} = \frac{\mathrm{d}\vec{\Theta}}{\mathrm{d}t} = \left(\frac{\mathrm{d}\Theta_x}{\mathrm{d}t}, \frac{\mathrm{d}\Theta_y}{\mathrm{d}t}, \frac{\mathrm{d}\Theta_z}{\mathrm{d}t}\right) = \left[\frac{1}{2}\left(\frac{\partial w}{\partial y} - \frac{\partial v}{\partial z}\right), \frac{1}{2}\left(\frac{\partial u}{\partial z} - \frac{\partial w}{\partial x}\right), \frac{1}{2}\left(\frac{\partial v}{\partial x} - \frac{\partial u}{\partial y}\right)\right]. \tag{2.76}$$

The *information* $\vec{\Theta}$ represents the angle of rotation of the fluid particle and $\vec{\Omega}$ represents its angular velocity. The factor that appears by multiplying the components of this vector is annoying. To avoid this, the vorticity vector, $\vec{\omega}$, is defined as twice the angular velocity $\vec{\Omega}$:

$$\vec{\omega} = 2\vec{\Omega}. \tag{2.77}$$

The vorticity $\vec{\omega}$, of great interest in Fluid Mechanics, serves to quantify the rate and direction of rotation of a fluid particle within a flow. Investigating Eqs. (2.76) and (2.77), it is concluded that the vorticity vector can be obtained by applying the rotational operator on the velocity field $\vec{V}$, Haller [8]:

$$\vec{\omega} = \vec{\nabla} \times \vec{V}, \tag{2.78}$$

where

$$\vec{\omega} = \vec{\nabla} \times \vec{V} = \begin{bmatrix} \vec{\imath} & \vec{\jmath} & \vec{k} \\ \frac{\partial}{\partial x} & \frac{\partial}{\partial y} & \frac{\partial}{\partial z} \\ u & \vartheta & w \end{bmatrix}. \tag{2.79}$$

Applying the divergence operator to the vorticity vector results in a field with zero divergence, and therefore, vorticity is a solenoidal field:

$$\vec{\nabla} \cdot \vec{\omega} = \vec{\nabla} \cdot \left(\vec{\nabla} \times \vec{V}\right) = 0. \tag{2.80}$$

The flows for which $\vec{\omega} = \vec{0}$ are said to be irrotational. Another important concept is the strain rate tensor of a fluid particle, given by:

$$\overline{\overline{S}} = \begin{bmatrix} S_{xx} & S_{xy} & S_{xz} \\ S_{yx} & S_{yy} & S_{yz} \\ S_{zx} & S_{zy} & S_{zz} \end{bmatrix}, \tag{2.81}$$

where each component of this tensor is given, in index notation, by:

$$S_{ij} = \frac{1}{2}\left(\frac{\partial u_i}{\partial x_j} + \frac{\partial u_j}{\partial x_i}\right), \tag{2.82}$$

or in tensor notation:

$$\overline{\overline{S}} = \frac{1}{2}\left(\vec{\nabla}\vec{V} + \vec{\nabla}\vec{V}^T\right), \tag{2.83}$$

where the tensor $\vec{\nabla}\vec{V}^T$ is the transpose of the tensor $\vec{\nabla}\vec{V}$. It is noted that,

$$\left(\vec{V} \cdot \vec{\nabla}\right)\vec{V} = \vec{\nabla}\left(\frac{V^2}{2}\right) - \vec{V} \times \vec{\omega}, \tag{2.84}$$

$$\nabla^2\vec{V} = \vec{\nabla}\left(\vec{\nabla} \cdot \vec{V}\right) - \vec{\nabla} \times \vec{\omega}, \tag{2.85}$$

$$\rho\vec{g} = -\rho g\vec{\nabla}z. \tag{2.86}$$

Rewriting Eq. (2.49) for μ constant and for incompressible flows, we have:

$$\rho\left[\frac{\partial\vec{V}}{\partial t} + \vec{\nabla}\cdot\left(\vec{V}\vec{V}\right)\right] = -\vec{\nabla}p + \rho\vec{g} + \mu\nabla^2\vec{V}. \tag{2.87}$$

Replacing Eqs. (2.84)–(2.86) in Eq. (2.87) and remembering that for incompressible flows $\vec{\nabla}\cdot\vec{V} = 0$, the following equation is obtained:

$$\rho\frac{\partial\vec{V}}{\partial t} + \vec{\nabla}\left(p + \frac{1}{2}\rho V^2 + \rho gz\right) = \rho\vec{V}\times\vec{\omega} - \mu\vec{\nabla}\times\vec{\omega}. \tag{2.88}$$

In this equation, the right side depends on vorticity. For irrotational flows, that is, with zero vorticity, the right-hand side of this equation cancels out, regardless of the viscosity of the fluid. This type of flow is called potential and, in this case, $\vec{V} = \vec{\nabla}\phi$, where ϕ is the velocity potential. Therefore, Eq. (2.88) can be rewritten in the form of the transient Bernoulli equation for incompressible flows:

$$\rho\frac{\partial\phi}{\partial t} + p + \frac{1}{2}\rho V^2 + \rho gz = cte. \tag{2.89}$$

Note that this constant is not restricted to a streamline. With this demonstration of the Bernoulli Equation, it can be applied between any two points of a flow, passing from one streamline to another, if the flow is irrotational. With this equation, it is shown that any solution for potential flows is also a solution for the Navier–Stokes Equations. However, no solution of this kind can address problems involving contact between a fluid and a solid, since the non-slip condition cannot be completely imposed. In fact, the solution to Eq. (2.88), for irrotational flows, $\left(\mu\vec{\nabla}\times\vec{\omega} = \vec{0}\right)$, only requires an initial condition and a boundary condition, as it only has first-order spatial derivatives. Thus, on a wall, the condition of zero velocity component normal to the wall is imposed. The tangential component must be free, since there are no viscous effects to be modeled close to the wall. Boundary layer flows can be modeled as potential flows on the outside of the thin viscous layer that forms adjacent to the wall. Viscous effects must be computed by solving the Navier–Stokes equations in the parietal region. The coupling between the two regions leads to plausible solutions and can be used for certain engineering applications. From the point of view of studying turbulence, the establishment of a balance equation for vorticity is of great interest. The deduction will be presented in detail in the following section.

2.8.2 Balance Equation and Vorticity Dynamics

We start with the Navier–Stokes equations, for incompressible flows of Newtonian fluids, with constant physical properties, transcribed below, in index notation:

$$\frac{\partial u_i}{\partial t} + u_j \frac{\partial u_i}{\partial x_j} = -\frac{1}{\rho_0}\frac{\partial p}{\partial x_i} + \nu \frac{\partial^2 u_i}{\partial x_j \partial x_j} + g_i. \tag{2.90}$$

The vorticity vector results, mathematically, from the application of the rotational operator to the velocity field. The component "p" is presented as follows:

$$\omega_p = \varepsilon_{pqi}\frac{\partial u_i}{\partial x_q}, \tag{2.91}$$

where ε_{pqi} is the tensorial operator ε of Levi–Civita, given by:

$$\varepsilon_{ijk} = \begin{cases} 0 & \text{if at least two of the indices } i, j, k \text{ are equal to each other} \\ 1 & \text{if } i, j, k \text{ is a pair permutation of } 1, 2, 3 \\ -1 & \text{if } i, j, k \text{ is an odd permutation of } 1, 2, 3 \end{cases}. \tag{2.92}$$

Thus, using Eqs. (2.91) and (2.92), the following equation is obtained:

$$\begin{aligned} \omega_1 &= \varepsilon_{123}\frac{\partial u_3}{\partial x_2} + \varepsilon_{132}\frac{\partial u_2}{\partial x_3} = \left(\frac{\partial u_3}{\partial x_2} - \frac{\partial u_2}{\partial x_3}\right), \\ \omega_2 &= \varepsilon_{213}\frac{\partial u_3}{\partial x_1} + \varepsilon_{231}\frac{\partial u_1}{\partial x_3} = \left(\frac{\partial u_1}{\partial x_3} - \frac{\partial u_3}{\partial x_1}\right), \\ \omega_3 &= \varepsilon_{312}\frac{\partial u_2}{\partial x_1} + \varepsilon_{321}\frac{\partial u_1}{\partial x_2} = \left(\frac{\partial u_2}{\partial x_1} - \frac{\partial u_1}{\partial x_2}\right). \end{aligned} \tag{2.93}$$

Isolating the advective term from Eq. (2.90), realizing that it can be written as a function of the strain rate and rotation rate tensors, we have:

$$u_j\frac{\partial u_i}{\partial x_j} = u_j\left(S_{ij} + \Omega_{ij}\right) = \frac{1}{2}u_j\left\{\left(\frac{\partial u_i}{\partial x_j} + \frac{\partial u_j}{\partial x_i}\right) - \varepsilon_{ijk}\omega_k\right\}, \tag{2.94}$$

since the strain rate S_{ij} is calculated with Eq. (2.82) and the rotation rate is calculated with:

$$\Omega_{ij} = -\frac{1}{2}\varepsilon_{ijk}\omega_k. \tag{2.95}$$

Equation (2.94) can be rewritten as follows:

$$2u_j\frac{\partial u_i}{\partial x_j} = u_j\left\{\left(\frac{\partial u_i}{\partial x_j} + \frac{\partial u_j}{\partial x_i}\right) - \varepsilon_{ijk}\omega_k\right\} = u_j\frac{\partial u_i}{\partial x_j} + u_j\frac{\partial u_j}{\partial x_i} - \varepsilon_{ijk}u_j\omega_k$$
$$\Rightarrow \tag{2.96}$$
$$2u_j\frac{\partial u_i}{\partial x_j} - u_j\frac{\partial u_i}{\partial x_j} = \frac{1}{2}\frac{\partial(u_ju_j)}{\partial x_i} - \varepsilon_{ijk}u_j\omega_k \Rightarrow u_j\frac{\partial u_i}{\partial x_j} = \frac{\partial e}{\partial x_i} - \varepsilon_{ijk}u_j\omega_k,$$

where $e = u_ju_j/2$ is the kinetic energy of a fluid particle. Replacing Eq. (2.96) in Eq. (2.90), the following equation is obtained:

$$\frac{\partial u_i}{\partial t} + \frac{\partial e}{\partial x_i} - \varepsilon_{ijk}u_j\omega_k = -\frac{1}{\rho_0}\frac{\partial p}{\partial x_i} + \nu\frac{\partial^2 u_i}{\partial x_j\partial x_j} + g_i. \tag{2.97}$$

By Eq. (2.91) the rotational operator is given by $\varepsilon_{pqi}\partial()/\partial x_q$. Applying this operator to Eq. (2.97), the following equation is obtained:

$$\begin{aligned}\varepsilon_{pqi}\frac{\partial^2 u_i}{\partial t\partial x_q} + \varepsilon_{pqi}\frac{\partial^2 e}{\partial x_i\partial x_q} - \varepsilon_{ijk}\varepsilon_{pqi}\frac{\partial}{\partial x_q}(u_j\omega_k) \\ = -\frac{1}{\rho_0}\varepsilon_{pqi}\frac{\partial^2 p}{\partial x_i\partial x_q} + \nu\varepsilon_{pqi}\frac{\partial^3 u_i}{\partial x_j\partial x_j\partial x_q} + \varepsilon_{pqi}\frac{\partial g_i}{\partial x_q}.\end{aligned} \tag{2.98}$$

Each term in this equation will be analyzed separately. The properties of the tensors involved will be explored. A tensor a_{ij} is said to be symmetrical if $a_{ij} = a_{ji}$ and it will be antisymmetric if $a_{ij} = -a_{ji}$. Therefore, any antisymmetric tensor must have its diagonal zero, given that $a_{pp} = -a_{pp}$ implies that $a_{pp} = 0$. The product of a symmetric tensor and an antisymmetric tensor is zero. Being the tensor ε_{ijk} antisymmetric and being a tensor $\partial^2()/\partial x_i\partial x_q$ symmetrical, it follows that $\varepsilon_{pqi}\partial^2 e/\partial x_i\partial x_q = 0$ and $\varepsilon_{pqi}\partial^2 p/\partial x_i\partial x_q = 0$. On the other hand, since the vector g_i is a constant, so, $\varepsilon_{pqi}\partial g_i/\partial x_q = 0$ also. With the so-called identity $\varepsilon - \delta$, it is established that:

$$\varepsilon_{pqi}\varepsilon_{ijk} = \delta_{pj}\delta_{qk} - \delta_{pk}\delta_{qj}. \tag{2.99}$$

Thus,

$$\begin{aligned}\varepsilon_{ijk}\varepsilon_{pqi}\frac{\partial}{\partial x_q}(u_j\omega_k) &= (\delta_{pj}\delta_{qk} - \delta_{pk}\delta_{qj})\frac{\partial}{\partial x_q}(u_j\omega_k) = \frac{\partial}{\partial x_k}(u_p\omega_k) - \frac{\partial}{\partial x_j}(u_j\omega_p) \\ &= u_p\frac{\partial \omega_k}{\partial x_k} + \omega_k\frac{\partial u_p}{\partial x_k} - u_j\frac{\partial \omega_p}{\partial x_j} - \omega_p\frac{\partial u_j}{\partial x_j}.\end{aligned} \tag{2.100}$$

It is observed that the divergence of the vorticity vector is zero, since:

$$\partial\omega_i/\partial x_i = \partial\left(\varepsilon_{ijk}\partial u_k/\partial x_j\right)/\partial x_i = \varepsilon_{ijk}\partial^2 u_k/\partial x_i\partial x_j = 0,$$

given that $\varepsilon_{ijk}\partial^2()/\partial x_i\partial x_j = 0$, remembering, again, that the product of an antisymmetric tensor and a symmetric tensor is zero. Thus, Eq. (2.100) results in:

$$\varepsilon_{ijk}\varepsilon_{pqi}\frac{\partial}{\partial x_q}\left(u_j\omega_k\right)=\omega_k\frac{\partial u_p}{\partial x_k}-u_j\frac{\partial \omega_p}{\partial x_j}. \tag{2.101}$$

Using these results in Eq. (2.98), it is reduced to:

$$\frac{\partial}{\partial t}\left(\varepsilon_{pqi}\frac{\partial u_i}{\partial x_q}\right)-\omega_j\frac{\partial u_p}{\partial x_j}+u_k\frac{\partial \omega_p}{\partial x_k}=\nu\frac{\partial^2}{\partial x_j\partial x_j}\left(\varepsilon_{pqi}\frac{\partial u_i}{\partial x_q}\right), \tag{2.102}$$

or

$$\frac{\partial \omega_p}{\partial t}+u_k\frac{\partial \omega_p}{\partial x_k}=\omega_k\frac{\partial u_p}{\partial x_k}+\nu\frac{\partial^2\omega_p}{\partial x_j\partial x_j}. \tag{2.103}$$

Or even in vector notation:

$$\frac{\partial \vec{\omega}}{\partial t}+\left(\vec{V}\cdot\vec{\nabla}\right)\vec{\omega}=\left(\vec{\omega}\cdot\vec{\nabla}\right)\vec{V}+\nu\nabla^2\vec{\omega}. \tag{2.104}$$

The first term on the left side of Eq. (2.104) represents the rate of variation of the vorticity contained in a fluid particle. This rate of variation, if different from zero, is justified by some physical effects. The first of these is the net advective flux of vorticity through the surface that separates a fluid particle from its surroundings, represented by the second term on the left side of this equation. It is the non-linear interactions that represent the advective transport process. With the first term on the right side of the same equation, an important mechanism for promoting vorticity is modeled, that is, the stretching of turbulent filaments. When a filament is stretched, it will rotate more quickly, resulting in an increase in the vorticity contained in the fluid particle. In fact, this term has its origin in the nonlinear term of the Navier–Stokes Equations and represents strong nonlinear interactions between shear regions of the flows and the turbulent filaments inserted inside the flow. This term can be canceled under two conditions: the first being when the non-linear effects are small compared to the diffusive effects and, secondly, when the flows are two-dimensional, for which this stretching term is identically null. Based on this analysis, it can be said that a flow can only occur in a turbulence regime for three-dimensional flows and for high values of the Reynolds Number. The second term on the right side of Eq. (2.104) represents the net diffusive flux of vorticity between a fluid particle and its surroundings.

Equation (2.104) is also known as the Helmholtz equation for fluid dynamics, which is usually written without the viscous term. Some theorems were established based on the Helmholtz Equation. With the first of these, known as Helmholtz theorem, it is guaranteed that the intensity of a vorticity filament remains constant over time, since stretching and diffusive effects are null. With the second of them, Lagrange's Theorem, it is established that the vorticity will be null for any time, if it is null for $t=0$. With the third theorem, it is established that the circulation over a closed contour, moving with the fluid, will be constant over time.

Important transition mechanisms to turbulence, as well as the mechanisms for its maintenance, are modeled with the interaction between the vorticity vector and the velocity gradient, Truesdell [11]. These are the stretching and tilting mechanisms of turbulent structures. As already mentioned, these physical mechanisms are modeled with the first term on the right side of the vorticity equation: $\omega_k \partial u_p/\partial x_k$. To deepen its physical interpretation, it will be written in an extended form, as follows:

$$\omega_k \frac{\partial u_p}{\partial x_k} = \begin{cases} \omega_1 \frac{\partial u_1}{\partial x_1} + \omega_2 \frac{\partial u_1}{\partial x_2} + \omega_3 \frac{\partial u_1}{\partial x_3}, \ p = 1, \\ \omega_1 \frac{\partial u_2}{\partial x_1} + \omega_2 \frac{\partial u_2}{\partial x_2} + \omega_3 \frac{\partial u_2}{\partial x_3}, \ p = 2, \\ \omega_1 \frac{\partial u_3}{\partial x_1} + \omega_2 \frac{\partial u_3}{\partial x_2} + \omega_3 \frac{\partial u_3}{\partial x_3}, \ p = 3. \end{cases} \tag{2.105}$$

Some of these terms will be interpreted below. The term $\omega_1 (\partial u_1/\partial x_1)$ is the product of the component ω_1 of vorticity by the component $\partial u_1/\partial x_1$ of the velocity gradient. In Fig. 2.4, the physical representation of this product is schematically shown. This component of the velocity gradient stretches the turbulent filament in the indicated direction and illustrates that it will stretch, under the action of this component of the velocity gradient, passing from the configuration presented in Fig. 2.4a for the configuration shown in Fig. 2.4b.

Considering the absence of compressibility effects, if the filament lengthens due to deformation and its volume is constant, then its diameter decreases. Neglecting viscous effects, it can be assumed that the moment of linear momentum of the filament remains constant, that is,

$$\underbrace{r \cdot \underbrace{(r\omega_1)}_{\text{linear momentum}}}_{\text{moment of the linear momentum}} = r^2 \omega_1 = cte,$$

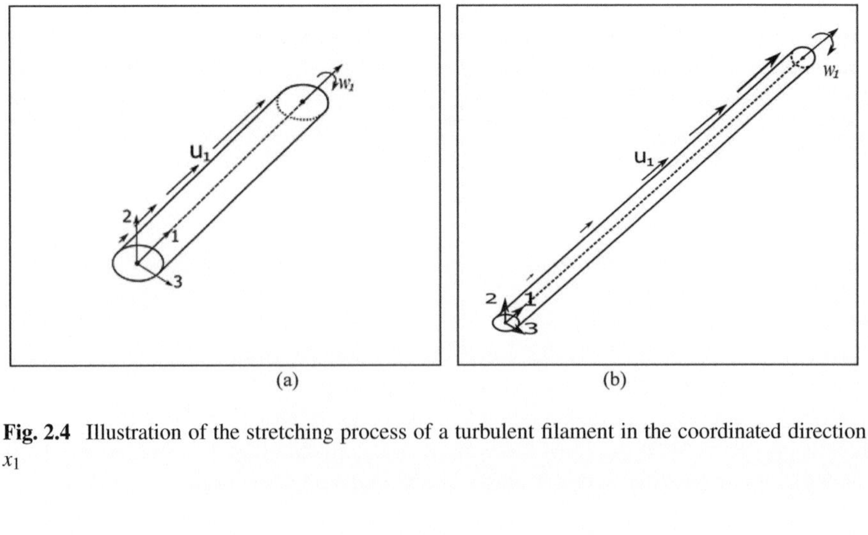

Fig. 2.4 Illustration of the stretching process of a turbulent filament in the coordinated direction x_1

at the same time as its radius decreases. This implies that the angular velocity increases. Thus, it is an example of the fact that the stretching of a filament, promoted by velocity gradients, increases vorticity.

With the diagonal terms of the matrix $\omega_k(\partial u_i/\partial x_k)$ this effect of stretching filaments of turbulent structures is modeled, in three directions, simultaneously, in accordance with the gradients of the velocity components. With the off-diagonal terms of this matrix, the tilting effects of the turbulent filaments are modeled. Taking the term $\omega_2(\partial u_1/\partial x_2)$, for example, it is clear that the cross-interaction between the vorticity component ω_2 and the velocity gradient component $\partial u_1/\partial x_2$ promotes bending of the filament, in the plane (x_1, x_2), inducing vorticity ω_3 in the third direction. In Fig. 2.5, the physical situation modeled here is illustrated, with the term $\omega_2(\partial u_1/\partial x_2)$ of Eq. (2.105).

Assuming simultaneous interaction between vorticity components and velocity gradients in the three coordinated directions, stretching and tilting of turbulent filaments will occur in the three directions. This general situation is illustrated in Fig. 2.6. Both stretching and tilting will promote vorticity in all three directions simultaneously. Stretches are modeled with direct products $\omega_1(\partial u_1/\partial x_1)$, $\omega_2(\partial u_2/\partial x_2)$ and $\omega_3(\partial u_3/\partial x_3)$ with which the vorticity induction $(\omega_1, \omega_2, \omega_3)$ is modeled, respectively. With crossed products $\omega_2(\partial u_1/\partial x_2)$ and $\omega_1(\partial u_2/\partial x_1)$, the deformation and inclination of turbulent filaments are modeled, in relation to the direction x_3; with the products $\omega_2(\partial u_3/\partial x_2)$ and $\omega_3(\partial u_2/\partial x_3)$, the deformation and inclination of turbulent filaments are modeled, in relation to the direction x_1 and, finally, with the products $\omega_1(\partial u_3/\partial x_1)$ and $\omega_3(\partial u_1/\partial x_3)$, the deformation and inclination of turbulent filaments are modeled, in relation to the direction x_3.

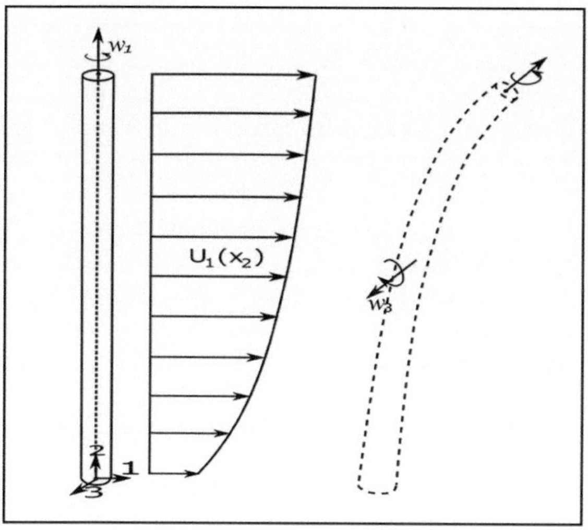

Fig. 2.5 Illustration of the bending and tilting process of a turbulent filament; promoting vorticity ω_3 according to the direction x_3

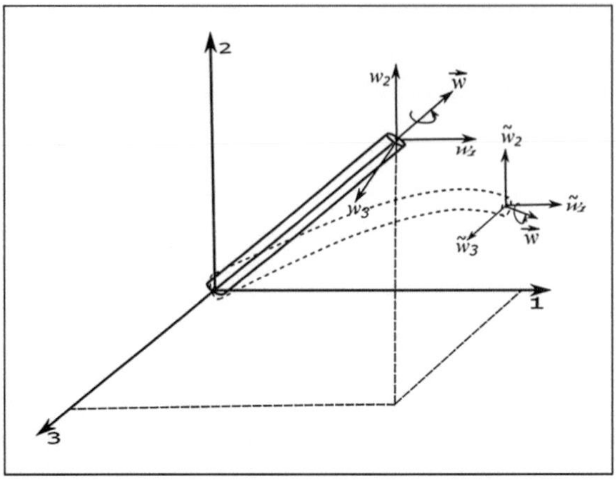

Fig. 2.6 Illustration of the stretching and tilting processes of a turbulent filament in all directions; vorticity promotion $(\omega_1, \omega_2, \omega_3)$

The process of promotion and distribution of vorticity over the three coordinated directions is illustrated at Fig. 2.6, through the simultaneous processes of stretching and tilting of turbulent structures. It can be seen, from the analysis presented, that the stretching process promotes vorticity in the direction of stretching. The tilting process promotes vorticity in the direction normal to the plane of rotation involved. For example, with the term $\omega_2(\partial u_1/\partial x_2)$, the tilting process of turbulent structures in the plane (x_1, x_2) is modeled, inducing and distributing vorticity to the component ω_3.

To improve understanding of the vorticity distribution process between the three directions, one of the components of the vorticity balance equation is analyzed, for example, the component in x_1, whose balance is expressed through Eq. (2.104).

The important conclusion that can be obtained from Eq. (2.104), representative of the vorticity balance, is that, for two-dimensional flows in the plane (x_2, x_3), for example, the vorticity vector has components $(\omega_1, 0, 0)$ and, consequently, the stretching and inclination term of turbulent structures is null, since $\omega_2 = \omega_3 = 0$. Thus, as will be better explored in Chap. 4. Turbulence can only be developed and maintained in three-dimensional flows. This is one of the main characteristics of turbulence: three-dimensionality with vorticity. The physical interpretations of the terms of this equation are presented in Eq. (2.106):

$$\underbrace{\frac{\partial \omega_1}{\partial t}}_{\substack{\text{Rate of change}\\ \text{of vorticity } \omega_1\\ \text{contained in a}\\ \text{article of fluid}}} + \underbrace{\left(\vec{V}\vec{V}\right)\omega_1}_{\substack{\text{Net advective}\\ \text{vorticity flux of}\\ \omega_1 \text{ between the fluid}\\ \text{particle and its}\\ \text{surroundings}}} = \underbrace{\omega_1 \frac{\partial u_1}{\partial x_1}}_{\substack{\text{Vorticity promotion}\\ \omega_1 \text{ by stretching of}\\ \text{turbulent structures}}} + \underbrace{\omega_2 \frac{\partial u_1}{\partial x_2}}_{\substack{\text{Vorticity distribution from}\\ \omega_1 \text{ to } \omega_3 \text{ due to the inclination of the}\\ \text{turbulent structure in the plane } (x_1, x_2)}}$$

$$+ \underbrace{\omega_3 \frac{\partial u_1}{\partial x_3}}_{\substack{\text{Vorticity distribution from} \\ \omega_1 \text{ to } \omega_2 \text{ due to the inclination of the} \\ \text{turbulent structure in the plane } (x_1, x_3)}} + \underbrace{\nabla^2 \omega_1}_{\substack{\text{Net diffusive} \\ \text{vorticity flux of} \\ \omega_1 \text{ between the fluid} \\ \text{particle and its} \\ \text{surroundings}}} \quad (2.106)$$

2.9 The Invariants

2.9.1 *Galilean Invariance*

The principle of Galilean relativity was established by Galilei [7]. The Florentine scientist observed changes in the movement of objects and animals, as well as in the act of jumping from a train, at first when the train was stopped and, in a second moment, when in non-accelerated movement. This principle highlights the importance of the fact that two observers can perceive different velocities and different trajectories when carrying out the same experiment. A consequence of this principle is that, once a property capable of being studied using the laws of classical mechanics has been defined, it can be observed without distinction by two inertial observers. One of them, in uniform movement relative to the other, in any direction, conducting the same experiment. In other words, mechanical entities that characterize the non-accelerated motion of objects should not undergo changes under Galilean transformations.

To exemplify this principle, consider a material particle located in an inertial reference frame, A, in rest. Two inertial observers are placed, one of which is in A, and the other is in an inertial reference frame, A', this one being in non-accelerated movement, with velocity U. For observer A, the material particle is at rest. For observer A', the particle is in non-accelerated motion, at a velocity U. The position of the particle relative to observer A, is $r(t)$, is constant. The position of the particle relative to observer A' is $r'(t)$, which is determined through the following equation:

$$r'(t) = r(t) - U \cdot t \quad (2.107)$$

The velocity of the particle is given by:

$$u'(t) = \frac{\mathrm{d}r'}{\mathrm{d}t} = \frac{\mathrm{d}r}{\mathrm{d}t} - U, \quad (2.108)$$

and its acceleration is given by:

$$a'(t) = \frac{\mathrm{d}u'}{\mathrm{d}t} = \frac{\mathrm{d}u}{\mathrm{d}t} - 0 = a(t). \quad (2.109)$$

It is noted that the position and velocity of the material particle are dependent on the observers, while its acceleration is independent of them. Thus, the particle's position and velocity are not Galilean invariant, while its acceleration is invariant. If the mass of the particle is constant over time, then Newton's Second Law is valid for any inertial reference frame, whether at rest or in non-accelerated motion. This fact shows the importance of defining a system, which is characterized by having an invariant mass over time. Thus, Newton's Second Law applies to any inertial system, whether at rest or in non-accelerated motion.

In a more general way, two systems of inertial axes are considered, (x_1, x_2, x_3) and $\left(x_1', x_2', x_3'\right)$. It is also considered that the second system of axes moves in relation to the first, with a velocity vector $\vec{\vartheta} = (U, V, W)$, constant over time. Then, the transformation between the two systems can be done through a transformation tensor $\overline{\overline{Q}}$, such that:

$$\vec{x}' = \overline{\overline{Q}}\vec{x} + \vec{\vartheta}t \tag{2.110}$$

Knowing that $\vec{x}$ and $\vec{x}'$ represents the position of a material particle in relation to two systems of effects; therefore, the transformation of a velocity in another can be obtained, deriving from Eq. (2.110) in relation to time:

$$\vec{\upsilon}' = \overline{\overline{Q}} \cdot \vec{\upsilon} + \vec{\vartheta} \tag{2.111}$$

Applying the gradient operator on both sides of Eq. (2.111), it can be demonstrated, as presented in Bacchi [2], that:

$$\vec{\nabla}\vec{\upsilon}' = QQ^T\vec{\nabla}\vec{\upsilon} = \vec{\nabla}\vec{\upsilon}. \tag{2.112}$$

Thus, it is demonstrated that the gradient operator, when applied to the velocity vector, is invariant to a Galilean transformation. The strain rate tensor $\overline{\overline{S}}$, given by Eq. (2.82), contains both the shear components, tangent to the area element, and the extensional components, normal to the same area element. The components of S_{ij} may undergo changes in magnitude, due to changes in the axis system (x, y, z), but the tensor itself satisfies the Galilean transformation invariance of symmetric tensors. Thus, three invariants, in the sense that they are independent of the direction or choice of axis system, are presented, without further mathematical demonstrations, in the present text. They are the Galilean invariants of the strain rate tensor:

$$I_1 = S_{xx} + S_{yy} + S_{zz}, \tag{2.113}$$

$$I_2 = S_{xx}S_{yy} + S_{yy}S_{zz} + S_{zz}S_{xx} - S_{xy}^2 - S_{yz}^2 - S_{zx}^2, \tag{2.114}$$

$$I_3 = \begin{bmatrix} S_{xx} & S_{xy} & S_{xz} \\ S_{yx} & S_{yy} & S_{yz} \\ S_{zx} & S_{zy} & S_{zz} \end{bmatrix}. \tag{2.115}$$

Other operators or mathematical entities, such as vorticity, stream function, among others, can be analyzed regarding Galilean invariance. This is of great importance when it becomes a criterion for visualizing turbulent flows. However, to achieve this, another invariance criterion must be established, still in this chapter. This is the criterion of objectivity or Euclidean Invariance.

2.9.2 *Euclidean Invariance or Objectivity*

Objectivity is a necessary property for Continuum Mechanics. Euclidean invariance requires that a property or operator be invariant to transformations between axis systems, where one of them must be inertial and the other may not be. Naturally, an objective property automatically satisfies Galilean invariance. Thus, objectivity requires invariance regarding the following transformation between axis systems:

$$\vec{x}' = \overline{\overline{Q}}(t)\vec{x} + \vec{c}(t) \tag{2.116}$$

where the tensor-transformation and the displacement velocity of one of the axis systems are no longer constant, as they depend on time. This is a Euclidean transformation. It is observed that a Euclidean transformation incorporates the effects of acceleration and rotation of axis systems. This is a general form of rigid transformation, that is, the coordinates of the axis systems are always orthogonal. It is observed that any scalar property, such as temperature, density or concentration of a chemical component, is always invariant with respect to a Euclidean transformation.

If the transformation of the base x to the base x' is done using Eq. (2.116). Then, objective vectorial quantities satisfy the following equation:

$$\vec{v}' = \overline{\overline{Q}}\vec{v}, \tag{2.117}$$

where $\overline{\overline{Q}}(t)$ is the second-order transformation tensor. On the other hand, a tensor *information* is objective if it obeys the following relationship between the tensor in the inertial axis system, $\overline{\overline{T}}$, and the same tensor in the non-inertial axis system $\overline{\overline{T}}'$:

$$\overline{\overline{T}}' = \overline{\overline{QTQ}}^T \tag{2.118}$$

To obtain the velocity, Eq. (2.116) is derived with respect to time, obtaining:

$$\vec{v}' = \overline{\overline{Q}}v + \dot{\overline{\overline{Q}}}\vec{x} + \dot{\vec{c}} \tag{2.119}$$

Comparing Eq. (2.119) with Eq. (2.117), velocity is not objective *information.* On the other hand, the distance between two points is objective *information*, since:

$$\vec{x}_a' - \vec{x}_b' = \overline{\overline{Q}}(\vec{x}_a - \vec{x}_b)\vec{c} - \vec{c} = \overline{\overline{Q}}(\vec{x}_a - \vec{x}_b). \tag{2.120}$$

As demonstrated by Bacchi [2],

$$\vec{\nabla}\vec{v}' = \overline{\overline{\dot{Q}Q}}^{T} + \overline{\overline{Q}}\vec{\nabla}\vec{v}\overline{\overline{Q}}^{T} \neq \overline{\overline{Q}}\vec{\nabla}\vec{v}\overline{\overline{Q}}^{T} \tag{2.121}$$

showing that the velocity gradient does not present the property of objectivity. Knowing that this velocity gradient tensor is Galilean invariant, the Euclidean transformation is more restrictive than the Galilean transformation.

Regarding the strain rate and rotation rate tensors, $\overline{\overline{S}}$ and $\overline{\overline{W}}$, Bacchi [2] demonstrates that:

$$\overline{\overline{S}}' = \frac{1}{2}\left(\vec{\nabla}\vec{v}' + \vec{\nabla}\vec{v}'^{T}\right) = \overline{\overline{QS}}'\overline{\overline{Q}}^{T} \tag{2.122}$$

and

$$\overline{\overline{W}}' = \frac{1}{2}\left(\vec{\nabla}\vec{\vartheta}' - \vec{\nabla}\vec{\vartheta}'^{T}\right) = \overline{\overline{\dot{Q}Q}}^{T} + \overline{\overline{QWQ}}^{T} \neq \overline{\overline{QWQ}}^{T} \tag{2.123}$$

It is thus demonstrated that the strain rate tensor $\overline{\overline{S}}$ is Euclidean invariant and therefore objective, while the rotation rate tensor $\overline{\overline{W}}$ is not Euclidean invariant. It is therefore concluded that, from a physical point of view, two observers, one inertial and the other non-inertial, observe the deformation process of a turbulent filament in the same way. On the other hand, these same observers observe different rotation processes of the same filament. The Euclidean invariance of the strain rate is consistent with the fact that the distance between two points is an objective entity, as demonstrated with Eq. (2.120).

Finally, Bacchi [2] demonstrates that the material derivative operator is not an objective operator. For this purpose, a generic objective tensor $\overline{\overline{A}}$ is considered. Therefore,

$$\overline{\overline{A}}' = \overline{\overline{QAQ}}^{T}. \tag{2.124}$$

But,

$$\dot{\overline{\overline{A}}}' = \overline{\overline{\dot{Q}AQ}}^{T} + \overline{\overline{Q\dot{A}Q}}^{T} + \overline{\overline{QA\dot{Q}}}^{T} \neq \overline{\overline{Q\dot{A}Q}}^{T}, \tag{2.125}$$

which demonstrates its non-objectivity.

All these properties are fundamental for choosing flow visualization criteria, in particular for the topological identification of turbulent structures that characterize the physical behavior of a turbulent flow. Naturally, an ideal visualization criterion should meet the two invariance principles: Galilean and Euclidean. If this is not the case, then two observers can visualize different vortex structures of the same flow.

2.10 Flows Visualization

Post-processing results is one way for researchers to understand the problem under analysis. Any type of physical problem can be understood, even partially, through visualizations. Flows can be understood when their fluid dynamic structures are visualized. From a material point of view, so-called tracers are used to make flows visible. When you look at the clouds, you can partially visualize how atmospheric movement is occurring. The particles that make up the clouds are the tracers, through which we visualize atmospheric flows.

In material experiments, various tracers are used, such as paints, smoke, hydrogen bubbles, tiny glass spheres and laser lighting, shadow visualization, among others. To visualize computational results, it is common to use all variables, such as the components of the velocity vector, pressure, temperature, vorticity, streamlines, emission lines and particle trajectories. However, what is observed physically, through the visualization of these quantities, may depend on the observer. For example, an inertial observer may perceive different things than a non-inertial observer. More recently, the scientific community has been concerned with the development of visualization techniques and has incorporated the notions of Galilean and Euclidean invariance for this purpose. Galilean invariance requires independence of observation made by more than one inertial observer. Euclidean invariance requires invariance on observation when viewed by non-inertial observers. Hunt et al. [9] proposed the criterion Q for flow visualization, which presents Euclidean invariance. This criterion is defined as the difference between the Euclidean norms of the vorticity tensor and the strain rate tensor:

$$Q = \frac{1}{2}\left[\left|\overline{\overline{\Omega}}\right|^2 - \left|\overline{\overline{S}}\right|^2\right], \tag{2.126}$$

where

$$S_{ij} = \frac{1}{2}\left(\frac{\partial u_i}{\partial x_j} + \frac{\partial u_j}{\partial x_i}\right),\ \Omega_{ij} = \frac{1}{2}\left(\frac{\partial u_i}{\partial x_j} - \frac{\partial u_j}{\partial x_i}\right),\ \left|\overline{\overline{\Omega}}\right|^2 = \Omega_{ij}\Omega_{ij} \text{ and } \left|\overline{\overline{S}}\right|^2 = S_{ij}S_{ij}.$$

With the criteria Q, regions of the flows can be identified where the rotation rate predominates over the deformation rate. Some researchers have worked on more elaborate visualization criteria, which meet the Euclidean invariance criterion. Among them, we can highlight the work of Haller [8]. Truesdell [11] presents in his book the vorticity number, given by the relation

$$N_k = \frac{\|\Omega\|}{\|D\|}. \tag{2.127}$$

This criterion represents the measure of the relative importance of vorticity and deformation in different regions of the flows. At the extremes of this criterion, $N_k \to \infty$ and $N_k \to 0$, there are respectively rigid body rotation and irrotational flows. An extensive review on the topic is presented by Bacchi [2].

2.11 Exercises

1. Develop the deduction of the continuity equation, the Navier–Stokes equation and the thermal energy equation, considering a rectangular fluid element, without using the Reynolds–Leibniz Transport Theorem. Make the deductions intuitively, using the notion of balancing the *information* of interest.
2. Consider a Circular Poiseuille Flow, laminar, steady state. Show that the pumping power required to maintain the fluid movement must be equal to the power to transform kinetic energy into thermal energy through viscous effects.
3. Consider a cylinder of infinite length rotating within an infinite fluid medium. Demonstrate, analytically, that in the steady state, the power delivered to the fluid by the cylinder must be equal to the power of transformation of kinetic energy into thermal energy by viscous effects. What happens in the transient regime?
4. Reflect and discuss the applicability of the differential model composed by the continuity, Navier–Stokes and thermal energy equations. What are the limitations regarding the type of fluid? Would they be enough to resolve any flow in the laminar and turbulent regimes and in the transition between these two regimes? Justify your answers.
5. Given a problem of an aircraft traveling at cruising velocity, choose the geometric and physical operating conditions; starting from the complete differential mathematical model for fluid dynamics, including non-inertial effects; perform the order of magnitude analysis and demonstrate which simplifications are appropriate. What is the differential, simplified but representative model for the problem?
6. Differentiate the models: physical, differential mathematical, numerical and computational. Talk about this topic.

References

1. Arpaci, V.S., Larsen, P.S.: Convection Heat Transfer Slattery. Prentice Hall, New York (1984)
2. Bacchi, R.D.A.: O que se deseja de uma definição de vórtices? Dissertação de Mestrado, UFF, Programa de Pós-Graduação em Engenharia Mecânica (2009)

3. Çengel, Y.A., Cimbala, J.M.: Mecânica dos Fluidos, Fundamentos e Aplicações. McGraw Hill, Londres (1997)
4. Chandrasekharaiah, D.S., Debnath, L.: Continuum Mechanics. Academic Press (1994)
5. Donald, F.M.: Introdução Mecânica do Fluidos. LTC, Rio de Janeiro-RJ (1998)
6. Faber, T.E.: Fluid Dynamics for Physicists. University Press, Cambridge (1995)
7. Galilei, G.: Dialogue Concerning the Two Chief World Systems. University (1953)
8. Haller, G.: An objective definition of a vortex. J. Fluid Mech. **525**, 1–26 (2005)
9. Hunt, J.C.R., Wray, A., Moin, P.: Eddies, stream, and convergence zones in turbulent flows. Center for Turbulence Research Report CTR-S88 (1988)
10. Kantha, L.H., Claysson, C.A.: Numerical Models of Oceans and Oceanic Processes. Academic Press (2000)
11. Truesdell, C.: The Kinematics of Vorticity. Indiana University (1953)
12. White, F.M.: Viscous Fluid Flow. McGraw-Hill Inc., London (1991)

Chapter 3
Transition to Turbulence

In this chapter, an interesting approach to the complex topic of transition to turbulence is presented. Flows are classified into canonical flows and complex flows. In Sect. 3.2, free shear flows are presented, and in Sect. 3.3, the Boundary Layer Transition is also presented. Following, still about canonical flows, the transition of Rayleigh–Bénard and Marangoni flows is presented, in Sect. 3.4. It is also presented, in Sect. 3.5, the transition of Taylor Couette flows and, in Sect. 3.6, we talk about Görtler instabilities. To conclude, the transition in complex flows is presented in Sect. 3.7, and, finally, a list of exercises is presented in Sect. 3.8.

3.1 Introduction

To study the transition to turbulence, consider a flow in a laminar regime, into which disturbances, with small amplitudes and a wide frequency band, are introduced. Thus, the transition to the turbulent regime can occur, as identified by several researchers, including Reynolds [27]. These disturbances, subjected to a nonlinear amplification process, generate fluid dynamic instabilities, which in turn develop turbulent structures. The use of the terms' disturbances, instabilities and turbulent structures will be done carefully in this text. Each of these words has a distinct physical meaning, so one should not be confused with the others. Disturbances can have different sources and physical natures: mechanical, dynamic, thermal or geometric. For example, the vibrations of the walls of a pipe in an industrial installation, which propagate into the flow, have a mechanical origin; oscillations in the flow propagate in the form of dynamic disturbances, while fluctuations of thermal energy, that can happen in a wall that delimits a flow, create thermal disturbances and, also, ripples or roughness in the walls characterize geometric disturbances.

In any flow, there are disturbances from different sources. These disturbances are characterized by a wide band of frequencies. A material experiment without

A. da Silveira Neto, *Turbulent Flows*, https://doi.org/10.1007/978-3-031-89222-6_3

disturbance is practically unrealizable. Vibrations arising from the movement of a car traveling outside a laboratory propagate through the ground and enter the environment where an experiment is being carried out. Pressure waves, coming from a researcher's breathing, propagate through the air in the room and reach the walls of a wind tunnel. The sources of disturbances that must be controlled are highly varied and diverse.

The process of transition to turbulence is still poorly understood and involves highly complex physical mechanisms. Until the publication of this book, no sufficiently comprehensive theory has yet been developed on the topic "transition to turbulence", with which it would be possible to represent, theoretically, its main physical aspects. Although the interaction mechanisms between the instabilities present in each flow are nonlinear, the first phase of the transition process to turbulence is weakly nonlinear and can be treated with the linear stability theory. Under these circumstances, the theory of linear stability offers great contributions to the understanding of the physical phenomena associated with the beginning of the transition process to turbulence. Nonlinear stability theory is also a powerful tool for understanding transition. With the use of these theories, linear stability and nonlinear stability, many important contributions to discovering what can happen in the transition process have been presented by the researchers, but they have contributed little to discovering the causes of the events. For example, before any discovery through material experimentation, Tollmien and Schlichting, Schlichting [29], affirmed the existence of instabilities, known today as "*T-S*" waves, in boundary layers transition, using the theory of linear stability. These instabilities were later identified through material experimentation. However, the physical causes for the development of the famous "*T-S*" waves remain unidentified. Another powerful form of analysis is associated with differential mathematical modeling for fluid dynamics. Significant discoveries have been made in understanding the topic of transition to turbulence and in understanding turbulence itself.

Turbulent flows can be treated as dynamic systems. The higher the Reynolds Number for a given turbulent flow, the greater the number of degrees of freedom that characterize it. The full solution of the differential mathematical model for this type of dynamic system requires that all degrees of freedom be solved simultaneously. This methodology is known as Direct Computational Simulation, whose acronym DNS (*Direct Numerical Simulation*) will be used throughout this text. The Brazilian turbulence community directly translates this designation, which results in Direct Numerical Simulation. It turns out that the simulation itself cannot be done numerically but computationally. A numerical model is not enough to carry out a simulation.

In fact, with DNS Methodology, we practice what, realistically, can be called virtual experimentation of transitional and turbulent flows. With this methodology, it has been possible to understand many physical aspects of flows, making it one of the two main methodologies for analyzing turbulence in fluids, complementing what the global scientific community calls physical experimentation and that's the reason in this book, it will be called material experimentation. This differentiation in nomenclature is due, in the author's understanding, to the fact that the theoretical analysis of a physical problem cannot be devoid of physical meaning. Therefore, both material experimentation and virtual experimentation are physical experiments. Several examples will be presented throughout this text to better clarify this point of view.

Despite all the advances made by the global community in understanding the topic of transition to turbulence, using the theories of linear and nonlinear stability, and material and virtual experiments, this topic continues to be one of the biggest challenges of contemporary physics, still lacking a theory to describe this matter. In this text, however, part of what is understood about the transition to turbulence so far is expressed. This topic, transition to turbulence, will be presented in such a way, to classify, by groups, the flows that transition from the laminar regime to the turbulent regime. This classification must follow standards, according to the physical nature of the formation and transport of fluid dynamic instabilities and turbulent structures that characterize the flow transition scenario. Although the Theorems of Rayleigh [26] and Fjørtoft [12] have been demonstrated for inflectional inviscid flows, in Drazin and Reid [8], their applications also extend to viscous flows. Due to these theorems, it was thought, for a long time, that only free shear flows were susceptible to transition. Until Prandtl [24] and his group of researchers discovered that viscous effects could destabilize boundary layer flows, in which there is no inflection in the average velocity field. In addition to boundary layer flows, there are other viscous flows susceptible to transition to turbulence, which also do not present inflectionality. Most of these flows will be dealt with in this text.

Flows can be grouped according to the physical nature of the instabilities that characterize the transition, that is: free shear flows; viscous flows; centrifugal flows; natural convection flows; Marangoni and Görtler flows. These are named canonical flows and are studied separately below, each one followed by illustrative examples. Finally, there are complex flows that result from the combination of two or more types of canonical flows, mentioned above, which will also be studied in the present chapter.

3.2 Free Shear Flows

Free shear flows are understood as flows that occur without the presence of walls. Furthermore, in this type of flow, there are regions with spatial variations in the velocity of the fluid particles, that is, regions with shear. Of particular interest are those that exhibit inflectional shear and, in particular, those whose mean velocity fields satisfy the Theorems of Rayleigh [26] and Fjørtoft [12]. This typology of flows can be subdivided into three distinct groups, although the transition process is similar in all of them, namely: mixing layers, jets and wakes. Next, each of them will be discussed separately.

3.2.1 *Mixing Layers*

A mixed layer originates from a shear layer that is located between fluid flows with different velocities. As will be presented in this chapter, this group of flows can be subdivided into two subgroups: temporally developing mixing layers and spatially developing mixing layers. The evolution towards the transition to turbulence will occur in two distinct ways: temporal evolution and spatial evolution. Both cases will be discussed here.

3.2.1.1 Temporal Mixing Layers

In Fig. 3.1a, an inflectional distribution of velocity at a given time is shown, which is repeated, similarly, throughout the entire domain. In Fig. 3.1b, there is an illustration of the thickness of the shear layer. When the average velocity fields are established over the entire layer, Fig. 3.1a, it satisfies Rayleigh's [26] and Fjørtoft's [12] Theorems. The necessary conditions are guaranteed for this layer to become unstable in the presence of external and internal disturbances, but this is not sufficient. Depending on the Reynolds Number, Re_δ, and on the band of frequencies that characterize the disturbances, a transition to turbulence may occur. The inflectional velocity field amplifies the injected disturbances. Thus, initially stable shear layers transform into unstable shear layers, developing Kelvin–Helmholtz Instabilities, which evolve into dynamic structures composed of multiple wave numbers (inverse of the wavelength) and corresponding frequencies. In Fig. 3.1, the inflectional velocity distribution is illustrated. The profile presented repeats throughout the calculation domain, in a self-similar way. The inflectional variation of velocity occurs in the vertical direction, y. The velocity varies from a minimum, U_2, up to a maximum, U_1. The inflection point is in the center of the domain, for the case illustrated in this figure. The shear layer is delimited between the two dotted lines. Thickness δ is represented on the right in the figure, where a way to define it is diagrammed. A straight line was drawn passing through the inflection point and is, at the same time, tangent to the curve at that point. Two vertical lines were drawn, tangent to the inflectional curve at infinity. The thickness δ is defined as the distance between the two crossing points, between the three straight lines mentioned.

In Fig. 3.2, a sequence of phases in the formation process of Kelvin–Helmholtz Instabilities is shown, in a mixing layer in temporal development. At first, disturbances with a broad band of frequencies are injected onto the inflectional velocity field, illustrated with streamlines, in Fig. 3.2a. In this figure, the process of selecting the most amplified wavelength can be seen, represented by λ_{max}. Subsequently, instabilities are observed on the shear layer, illustrated in Fig. 3.2b, which results from the amplification process of the disturbances injected into the flow. These instabilities appear with a wavelength λ_{max}, illustrated in Fig. 3.5, associated with which the most amplified wavenumber is defined, $k_{max} = 2\pi/\lambda_{max}$.

The process of amplification of disturbances can be better analyzed and understood using the pressure field. It is known that the local velocity is inversely proportional to

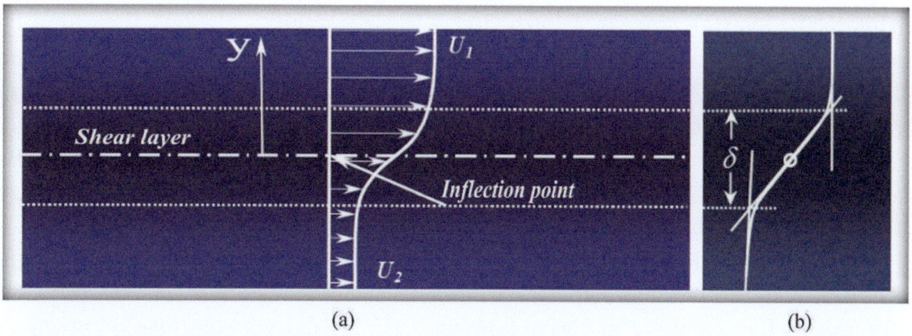

Fig. 3.1 Domain in which the initial inflectional velocity field is qualitatively represented, for mixing layers in temporal development

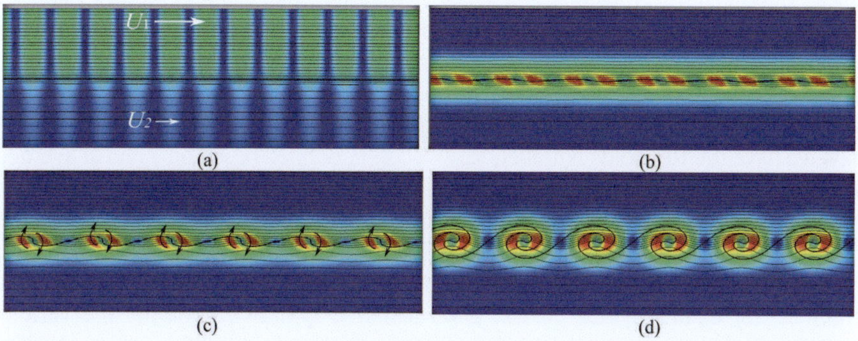

Fig. 3.2 Process of amplification of disturbances and formation of Kelvin–Helmholtz instabilities. Streamlines (black lines); iso-value of a passive scalar (central black line) and module of the pressure gradient field (colors varying between blue and red). Personal collection, with thanks to Dr. Rafael Romão

the distance between the streamlines. It is also known that static pressure is greater in regions of lower velocities and lower in regions of higher velocities. The velocity and pressure fields are coupled over time and space. Gradients appear in the pressure field, which in turn promote upward and downward vertical forces. As the system is being disturbed, the nonlinear process of disturbance amplification is responsible for promoting the first fluid dynamic instabilities, which have wavelength $\lambda_{\max}$, as illustrated in Fig. 3.2b.

From the moment these first instabilities manifest themselves, an imbalance in the pressure field appears, as illustrated in Fig. 3.2b. In fact, with the appearance of instabilities in the form of waves, crests and valleys appear, with regions of compression and expansion of the streamlines, indicating regions of acceleration and deceleration, respectively. Thus, the pressure gradients change sign alternately, causing an imbalance in the originally uniform force field. The modulus of the pressure gradient is illustrated in Fig. 3.2a–d. This process of amplification of disturbances and formation of instabilities is nonlinear and irreversible, if the system's operating conditions are maintained. Once the process of formation of these instabilities begins, they amplify, promoting a strong interaction of crests and valleys with velocity currents, U_1 and U_2, marked on Fig. 3.2a. The crests, which enter the upper part of the flow, are faster than the valleys, in relation to the central horizontal line, where it will be transported, in the direction of the flow, it starts a process of coiling of the shear layer, as illustrated in Fig. 3.2c, d. This process results in instabilities of a periodic nature, known as Kelvin–Helmholtz Instabilities.

The most amplified wavelength value, $\lambda_{\max}$, is selected by the flow, among all other wavelengths that make up the spectrum of injected disturbances. The physical mechanisms for selecting this wave number, $k_{\max}$ are not yet completely known. A very important study was carried out by Betchov and Szewcyk [4], with a view to quantifying the most amplified wave number, as a function of the Reynolds Number, for mixed-layer flows in temporal development. These authors used the theory of linear stability [9, 10], to obtain a stability diagram, like the one seen in Fig. 3.3. This diagram is a qualitative adaptation of the original publication, very useful for better understanding the process of transition to turbulence. The following dimensionless parameters are presented: on the vertical axis, there is the dimensionless wave number, $k\delta$, where δ is the thickness of the mixing layer; on the horizontal axis we have the Reynolds Number, $\mathrm{Re}_\delta = U\delta/\nu$, defined based on δ and at average velocities $U = (U_1 + U_2)/2$. The lines that appear in the diagram are iso-values of the dimensionless rate of amplification, $kc_i\delta/U$, of disturbances, where c_i is the transport velocity of disturbance waves.

To illustrate the usefulness of this diagram, a value is chosen for the Reynolds Number, for example, $\mathrm{Re}_\delta = 12$; a vertical line is drawn, crossing the various isolines of amplification rate. This vertical line touches the amplification rate isoline

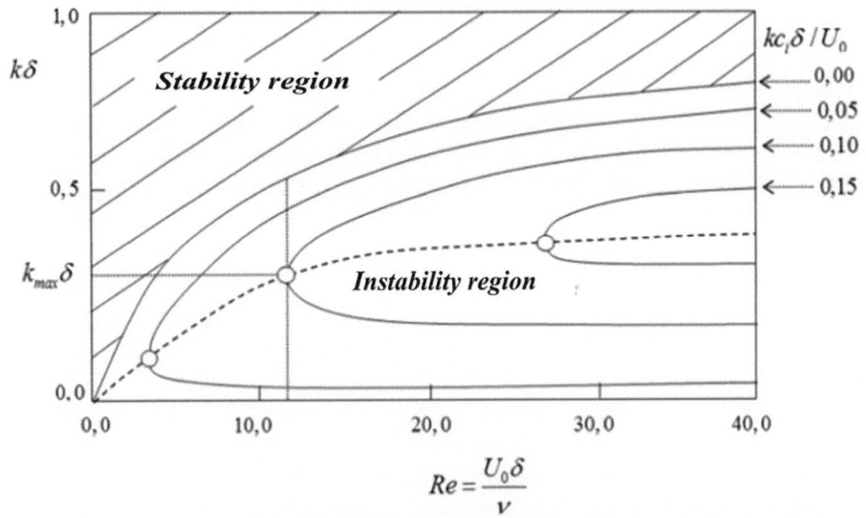

Fig. 3.3 Stability diagram. Adapted from Betchov and Szewcyk [4], schematic presentation

$kc_i\delta/U = 0.1$. The other amplification rate lines that these vertical crossing lines are related to amplification rates with values lower than this value ($kc_i\delta/U = 0.1$). Thus, for this value of the Reynolds Number, the maximum amplification rate is $kc_i\delta/U = 0.1$, which corresponds to $(k\delta)_{\max} = 0.25$. Therefore, the first instability to appear in this flow must have this wavelength. Other wavelengths would be amplified simultaneously, however, with lower amplification rates, which means that secondary instabilities will visibly manifest themselves later on.

It is important to emphasize that all wavelengths that characterize the band, below the neutral line ($kc_i\delta/U = 0.0$) of the stability diagram, amplify simultaneously, but with different amplification rates. Thus, in the evolution of the transition process, all wave numbers present in this amplified band create a wide variety of instabilities, which will make up the spectrum of turbulent flow, resulting from this transition process. Eventually, instabilities can pair up, resulting in longer wavelengths. On the other hand, disturbances of shorter wavelengths can be amplified, resulting in higher wavenumbers. It is worth noting that the mechanism for the formation of instabilities is the process of amplification of disturbances, with wavelengths different from the wavelength said to be the most amplified wavelength.

A more detailed analysis of the diagram can be useful for physically understanding the dynamics of a temporal mixing layer, as well as for validating results obtained from computational solutions of the fluid dynamic differential model, that is, the continuity equation and Navier–Stokes Equations. The base average velocity field, used to obtain Fig. 3.3, is given by Eq. (3.1):

$$U(y) = U\tanh(y/\delta), \tag{3.1}$$

where $U = (U_1 + U_2)/2$. In the diagram under study, two distinct regions are presented: one of stability, with negative amplification rates, and one of instability, with positive amplification rates, separated by a neutral line, given by a zero-amplification rate. The existence of lines of constant amplification rates is also observed. The diagram can also be used to determine, for each considered value of the Reynolds Number, a value $k_{\max}\delta$, corresponding to the maximum amplification rate. The dashed line joins the points of maximum amplification rates, determined as a function of the Reynolds Number. This curve can be given by a polynomial, as suggested by Betchov and Szewcyk [4]. Thus, it becomes possible to predict which wavelength will be most amplified in a temporal mixing layer, for a given value of the Reynolds Number.

The evolution of the process of amplification of disturbances and promotion of Kelvin–Helmholtz Instabilities is also illustrated in Fig. 3.4, obtained through the computational solution of the fluid dynamics equations, using the Pseudo-Spectral Fourier Methodology, Moreira and Silveira-Neto [19]. It is observed, in Fig. 3.4a, wavelength perturbation amplification $\lambda_{\max}$; in Fig. 3.4b the shear layer begins to curl, giving rise to the formation of Kelvin–Helmholtz Instabilities. In Fig. 3.4c, d, these instabilities are already developed and promote the transport of fluid rich in linear momentum (downward direction) and fluid poor in linear momentum (upward direction) into the turbulent structures. This transport process physically explains the

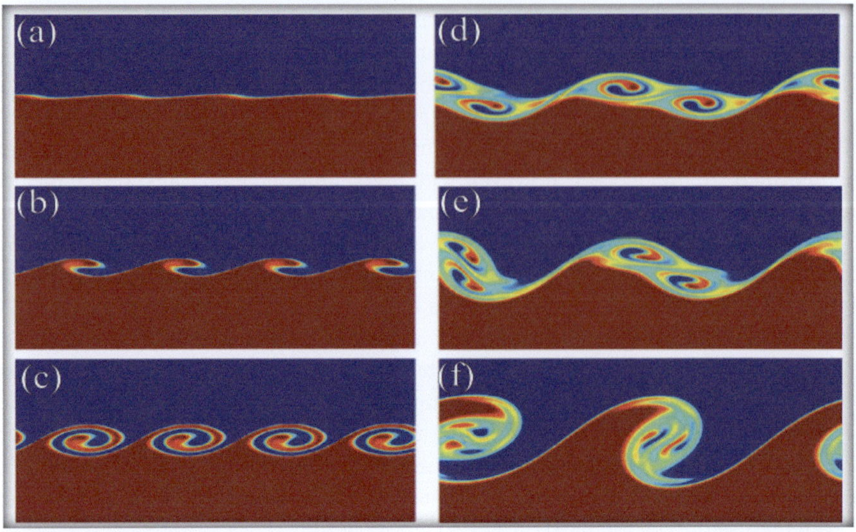

Fig. 3.4 Two-dimensional mixing layer in temporal development. Figure taken from Moreira and Silveira-Neto [19], personal collection. A passive scalar or numerical colorant is displayed

name "mixing layer" attributed to this type of flow. In this case, a two-dimensional mixed layer is shown in temporal development, since the turbulent structures are periodic and topologically similar throughout space. This implies that these turbulent structures evolve over time, preserving, at each moment, a spatial topological similarity.

The process of temporal evolution of a mixed layer occurs through the development of new instabilities, with wavelengths different from λ_{max}. These instabilities can be understood as harmonic and sub-harmonic, that is, with wave numbers that are multiples and sub-multiples of $k_{max} = 2\pi/\lambda_{max}$. They develop with an amplification rate lower than the maximum rate corresponding to the wavelength λ_{max}, with which primary instabilities appear. It means that the other instabilities that will make up the spectrum indicated in Fig. 3.3 will amplify concomitantly with the primary instability, considered the most amplified. However, they only manifest themselves, visually and quantitatively, later, because they are amplified with amplification rates lower than the maximum rate. These harmonic instabilities can be in the amplification rate spectrum illustrated in Fig. 3.3. Physically, these instabilities manifest as pairings of primary instabilities. These pairings occur because the spatial sequence of primary instabilities oscillates with a wavelength $\lambda > \lambda_{max}$, causing Kelvin–Helmholtz Instabilities to alternately enter faster and slower regions of the flow, creating a new winding and doubling the size of the fluid dynamic structures. This oscillation process of the Kelvin–Helmholtz Instability sequence is illustrated in Fig. 3.4; the same should occur when passing through Fig. 3.4c, d.

The structures twice the size is shown in Fig. 3.4d–f. This mechanism is similar to Primary Kelvin–Helmholtz Instability, through which primary structures form.

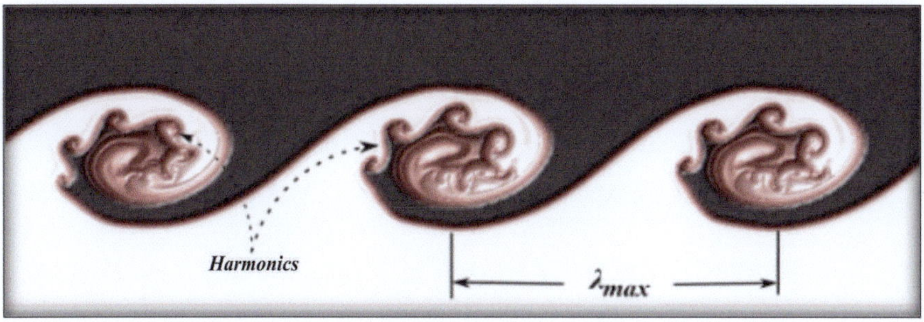

Fig. 3.5 Two-dimensional mixing layer in temporal development. Formation of harmonic Kelvin–Helmholtz instabilities within primary Kelvin–Helmholtz instabilities. Figure from personal collection

Thus, larger structures are formed, characterized by wavenumbers smaller than the most amplified wavenumber, $k < k_{\max}$. Likewise, harmonics with $k > k_{\max}$ form within the primary Kelvin–Helmholtz Instabilities, as illustrated in Fig. 3.5. These instabilities also make up the spectrum of turbulent structures that characterize the flow in transition to the full turbulence regime. The process of transition to turbulence necessarily involves the formation of three-dimensional instabilities.

Flows of this type occur frequently in nature and can often be visualized with the help of some natural tracer, for example, clouds in the Earth's atmosphere, Fig. 3.6a, or the existence of impurities where the waters of Rio Negro meet Rio Solimões, Fig. 3.6b. To illustrate, in Fig. 3.6a, a mixed layer in temporal development is shown, visualized in the earth's atmosphere, thanks to the presence of clouds, positioned precisely over the free shear layer, a region that is located between the two adjacent air currents, which have different velocities. In this case, the air currents move from right to left, with the faster current being above the Kelvin–Helmholtz Instabilities, while the slower current is below. This causes the Kelvin–Helmholtz Windings to occur in a counterclockwise direction. Observing this figure in the transverse direction, the three-dimensional nature of this flow can also be seen qualitatively. Figure 3.5 shows what is called, in this book, dynamic structure, for laminar flows or turbulent structure, for turbulent flows.

Three-dimensional instabilities can be studied in detail using computer simulations. In this sense, simulations were carried out in the Fluid Mechanics Laboratory at MFLab-FEMEC-UFU, using the Multi Physics Simulator—MFSim, an in-house software developed by the group of researchers at MFLab. It is a mixing layer in three-dimensional temporal development. The domain is Cartesian, defined by $D = [0; 8] \times [0; 4] \times [0; 8]$. This domain was discretized using a mesh composed of (64, 32, 64) cells. With this computational code, it is possible to use adaptive mesh refinement. In the present case, three refinement levels were used, with the most refined level being positioned in the region of greatest shear. The base inflectional velocity profile used is modeled with the equation below:

$$(u, \vartheta, w) = \left(20 \times \tanh\left(\frac{y-2}{\delta}\right), 0, 0\right). \tag{3.2}$$

(a) (b)

Fig. 3.6 Mixing layer in temporal development: **a** three-dimensional atmospheric flow; **b** meeting of the waters of the Negro and Solimões Rivers (personal collection)

As can be deduced from Eq. (3.2) and the domain specification, the velocities of the free currents, in the longitudinal direction of the domain, are − 10 and 10 [m/s]. White noise with the following intensities was superimposed on the three initial velocity components: 1% the velocities of free currents in the direction of flow, 0.5% in the vertical direction and 0.25% in the transverse direction. These intensities of different values in different directions allow the modeling of physical differences in the three directions. In this example, these differences have a purely didactic sense. They are not related to any specific material experiment and this noise is random in space. The initial thickness was set at $\delta_0 = 0.1\,[\mathrm{m}]$. The kinematic viscosity of the fluid is $\nu = 10^{-2}\left[\mathrm{m}^2/\mathrm{s}\right]$. Thus, the Reynolds Number based on the initial thickness of the mixed layer is $\mathrm{Re}_{\delta_0} = 200$. Periodic boundary conditions are used in the flow direction and transverse direction. In this way, the instabilities that leave are reinjected into the domain entry. In the vertical direction, the free flow condition was used.

Figure 3.7 shows four stages of flow evolution. For Fig. 3.7a, the time is $t = 0\,[\mathrm{s}]$, in which only the noise imposed on the components of the initial velocity appears, Eq. (3.2). A white noise modeler was used. Disturbances that are always present in material experiments were modeled with this noise, such as the one presented in Fig. 3.6. In Fig. 3.7b, the flow selected a wavelength λ_{max}, which, in accordance with the linear stability theory, was amplified with the maximum amplification rate. This process creates the so-called Kelvin–Helmholtz Instabilities. It is observed that the length of the domain in the flow direction is $L_x = 8\,[\mathrm{m}]$ and that, in Fig. 3.7b, five Kelvin–Helmholtz Structures appeared, which results in a computational simulation wavelength of $\lambda\,[\mathrm{m}]_{\mathrm{max.sc}}$. Using the Betchov and Szewczyk Stability Diagram [4], qualitatively represented by Fig. 3.3 and considering that the Initial Reynolds Number assumes the value $\mathrm{Re}_{\delta_0} = 200$, the prediction of the most amplified wavelength is obtained, by linear stability theory, $\lambda_{\mathrm{max.lts}} = 1.6\,[\mathrm{m}]$. With this comparison, the computational simulation is partially validated against the theory of linear stability.

The process of inducing longitudinal instabilities is also noticeable in Fig. 3.7b, which can be seen more clearly in Fig. 3.7c.

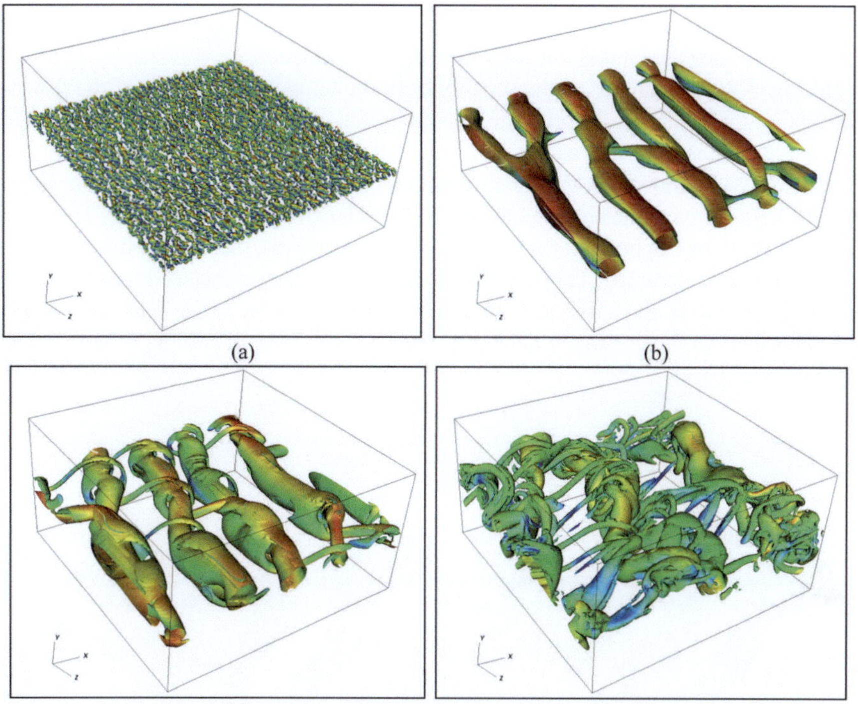

(a) (b) (c) (d)

Fig. 3.7 Mixing layer in temporal development; noise superimposed on the initial condition; temporal sequence of the flow, visualized with iso-values of the Q criterion: **a**–**d**. Simulation carried out in *MFLab-FEMEC-UFU*, by Rafael Romão, using the MFSim computational platform. Personal collection

These longitudinal filaments, whose origin is not yet understood, coil over the Kelvin–Helmholtz Instabilities, resulting in strong mechanical effects and strong deformations. It contributes to the process of three-dimensionalization of the flow. It's important to notice the physical process of stretching (pulling), rotating filaments, as commented on in Chap. 2, when the vorticity balance equation was deduced. These longitudinal instabilities manifest themselves after the Kelvin–Helmholtz instabilities. However, they are all being amplified at the same time, but at lower amplification rates.

In Fig. 3.7d, a greater wealth of longitudinal instabilities can be seen, with a variety of wavelengths. The transition begins with the selection of four wavelengths (five turbulent structures), illustrated in Fig. 3.7b. In Fig. 3.7c, there are four wavelengths, formed by five Kelvin–Helmholtz Structures. It is also shown that two Kelvin–Helmholtz Structures are paired. In Fig. 3.7d, there are only two wavelengths, showing that two more pairings have occurred. At this moment, the flow is already in an advanced state of three-dimensionalization, which can be characterized as an evolving turbulence regime. The aim of these figures was to illustrate the process of transition to turbulence and not fully developed turbulence.

In Fig. 3.8, the analysis of the sensitivity of the temporal evolution of the flow to the type of injected noise is presented, in the initial condition of the simulation. The simulated case is the same as Fig. 3.7, but with changes in noise. In Fig. 3.8a, it is observed that the noises in the three directions have different seeds and intensities. In Fig. 3.8b, we have the same seeds, but lower intensities than in case (a). The same transition pattern can be seen for the same time; in case (a), there is a more advanced evolution than in case (b). This pattern was expected, since if the seeds are the same, the frequencies are the same in both cases. However, as the noise intensity in case (b) is lower than in case (a), it is natural that the transition occurs more slowly and later in case (b).

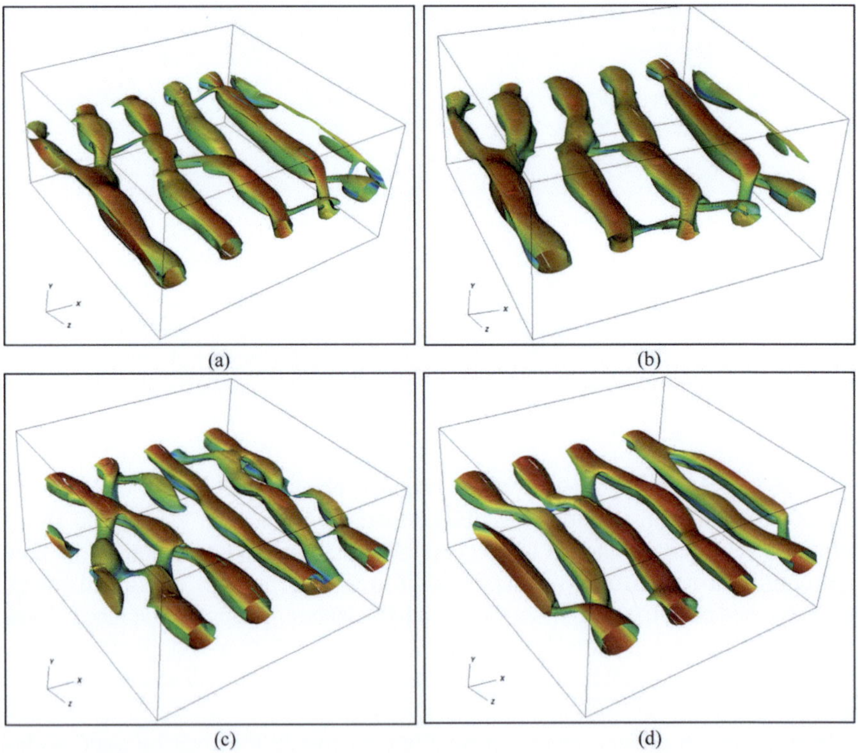

Fig. 3.8 Mixing layer in temporal development; visualization using the Q criterion iso-values; **a** noises with the seeds and different intensities in the three directions; **b** noises with the same seeds as in case (a) and with intensities equal to each other and lower than in case (a); **c** noises with seeds different from case (b), but with the same intensities as case (b); **d** noise only in the longitudinal direction, new seed, same intensity as in cases (b) and (c). Simulation carried out in *MFLab-FEMEC-UFU*, by Rafael Romão, using the computational platform *MFSim*

From Fig. 3.8b to c, the seeds were changed, but the intensities were maintained. Thus, the injected frequencies will have different energy levels; however, the frequency band widths were kept the same. The flow patterns are different; however, the most amplified wavelength was maintained, since the number of dynamic structures remained equal to five.

In Fig. 3.8d, the same flow as the previous sections is shown. However, the noise was imposed only in the longitudinal direction and with a different seed and the same intensity. Again, the transition occurs, but with a different pattern. It is interesting to note that, although there was no noise in the direction transverse to the flow, instabilities in this direction also have developed. The fact that no perturbations were explicitly injected in the transverse directions does not imply the absence of perturbations in these directions. A simulation was carried out without any explicit disturbance imposed on the initial condition, and yet the flow transitioned, a consequence of the existence of numerical noise associated with errors inherent to temporal and spatial discretization methods.

Numerical experiments showed that the transition to turbulence is highly sensitive to small changes in the initial condition, even to changes in the noise seed initially injected. However, the "choice", by the flow, of the most amplified wavelength, λ_{max}, on part of the flow, it only depends on the Reynolds Number that characterizes the flow.

3.2.1.2 Spatial Mixing Layers

A spatially developing mixed layer is characterized by the formation of Kelvin–Helmholtz Instabilities, through mechanisms like those already described in the previous section, for a temporally developing mixed layer. In Fig. 3.9, the illustrative scheme of this class of flow is presented, showing an inflectional velocity profile in Fig. 3.9a and the process of spatial expansion in Fig. 3.9b. It also shows the thickness $\delta(x)$ of the mixed layer, which varies in the main flow direction. In the upper and lower parts of the domain, there are uniform velocities U_1 and U_2 which are imposed at the entrance of the domain and remain outside the shear region, internal to the mixing layer.

While in a temporal mixing layer, the instabilities develop simultaneously and are distributed similarly throughout space, in a spatially developing mixing layer, the Kelvin–Helmholtz Structures are formed successively, with a more amplified wavenumber, k_{max}, in the vicinity of the domain entrance, which are spatially transported. As they are transported, they pair up, forming even larger dynamic structures with wavenumbers $k < k_{max}$. This pairing process characterizes the widening of the spatial mixing layer, as illustrated in Fig. 3.10. The initial conditions of these two types of flow are different. For a temporal mixing layer, the initial velocity field is inflectional throughout the domain. The flow develops from this initial condition. For a spatial mixing layer, the initial condition may be, for example, zero velocity throughout the domain, except at the entrance, where an inflectional profile

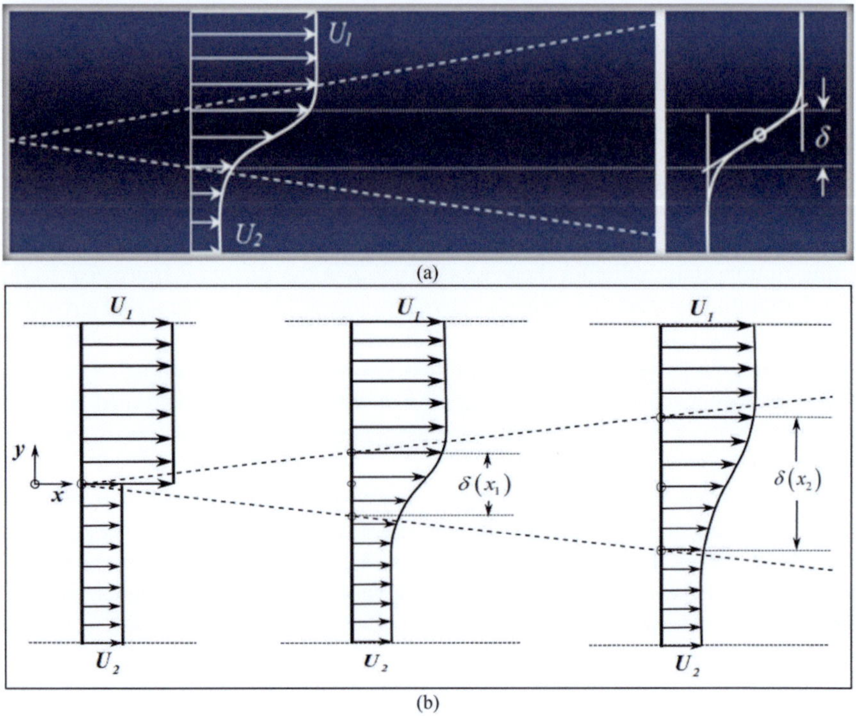

Fig. 3.9 Domain in which the initial inflectional velocity field is qualitatively represented, for mixing layers in spatial development

is imposed. This profile imposed at entry, as illustrated in Fig. 3.9b, remains the same throughout the simulation time.

The development of Kelvin–Helmholtz Instabilities can be observed in Fig. 3.10. These instabilities evolve into turbulent structures, whose size increases as they are transported spatially, through the structures' pairing mechanism, discussed in the previous paragraph. It is observed that the layer, composed of these structures, oscillates, with increasing wavelength along the main flow direction, as illustrated with the dashed line, drawn over the Kelvin–Helmholtz Structures, in Fig. 3.10. These oscillations transport the Kelvin–Helmholtz Structures up and down, causing them

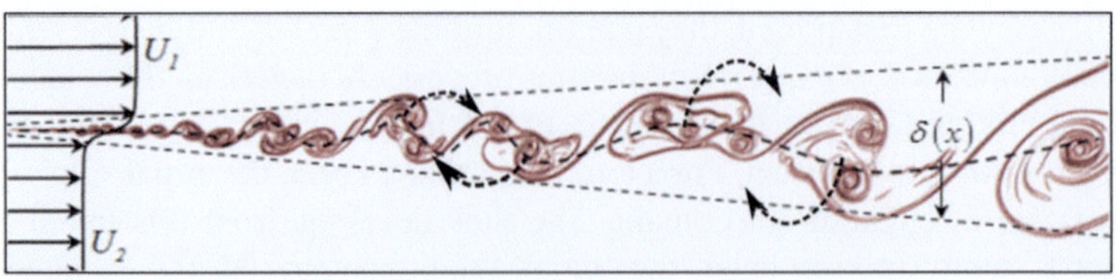

Fig. 3.10 Mixing layer in spatial development, personal collection

to enter faster and slower regions, respectively. As a result, one structure reaches the other, promoting successive pairing. This involves structures of increasingly longer wavelengths as they are transported toward the exit of the domain. This advective mechanism of pairing of dynamic structures is nonlinear, constituting the main mechanism of spatial widening of the mixed layer. In addition to this, there is also the linear momentum diffusion mechanism, with predominance in the direction transverse to the mixing layer. The thickness $\delta(x)$ has a linear behavior throughout space, as illustrated in Fig. 3.10, through the dashed boundary lines.

These turbulent structures, well organized in the largest sizes, are also known as coherent structures. Coherence is because they have a lifetime greater than their characteristic time, that is, the time needed to perform a complete rotation. The discovery of these structures led to the belief that, at least at the level of large structures, turbulence presents a given level of organization and coherence, from a statistical point of view. In other words, they are structures that maintain a given topological shape for a time longer than their characteristic rotation time. The transition to turbulence necessarily goes through physical processes of a three-dimensional nature. The most basic notion of this process is understood, so far, by the presence of fluid dynamic instabilities in the direction transverse to the flow, which manifest themselves on the Primary Kelvin–Helmholtz Structures. Concomitantly, counter-rotating longitudinal filaments appear, which intertwine with Kelvin–Helmholtz Instabilities, promoting strong mechanical deformation effects. This scenario can be viewed in Fig. 3.11, where the Primary Kelvin–Helmholtz Structures, transverse structures and longitudinal counter-rotating filaments are presented. This is the qualitative scenario that characterizes the beginning of the transition process to turbulence, in a spatially developing mixing layer.

From this scenario, instabilities and structures with other wavelengths will appear, whose formation occurs through nonlinear processes, which are still little known. These new wavelengths make up a broad, full and continuous spectrum, which characterizes the completely developed turbulent flow regime. This is one of the main characteristics of turbulent flows, that is, the multiplicity of turbulent structures, forming a dense and wide band. This stage can be seen in the second half of the domain presented in Fig. 3.11.

The presence of turbulent structures is observed, composed of many other filamentary structures. This is a good example that shows the importance of the nomenclature "turbulent structures", rather than "vortices" or "eddies". In fact, in this figure, it is shown that turbulence is composed and characterized by the presence of turbulent structures (compositions) and not vortices or eddies. Despite the complexity of these structures, they retain, when observed as a whole, forms of Kelvin–Helmholtz structures, which were evidenced using a blue line manually superimposed on the computational image, as presented in Fig. 3.12. The largest structures present in this flow have a lifetime, t, longer than their characteristic rotation time; that is, they can experience several rotations around their centers of rotation, maintaining their topological form, until they are transformed into other structures. Therefore, they are also characterized as coherent structures.

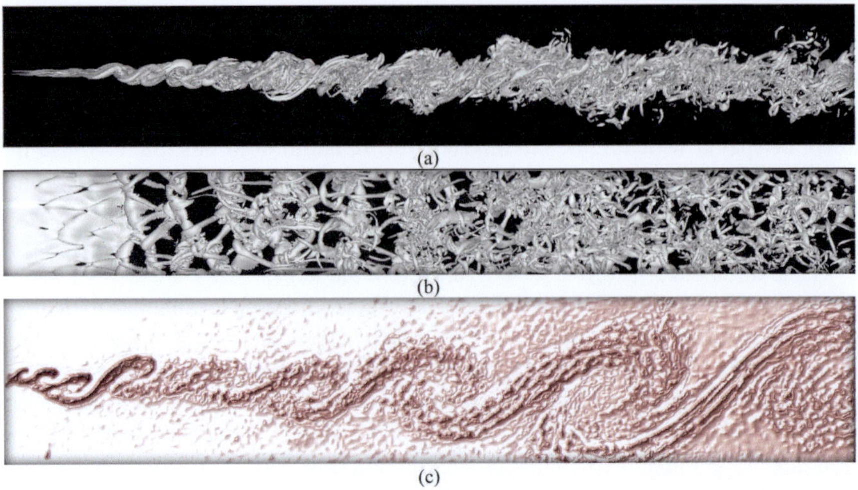

Fig. 3.11 Mixed layer in spatial development: Kelvin–Helmholtz turbulent structures; transverse instabilities and longitudinal filaments; three-dimensionalization and development of turbulence. Figure **c** was adapted from Brow and Roshko [5]

Fig. 3.12 Illustration of a coherent turbulent structure and the movement of a material particle, traveling a complete trajectory within this structure. Personal collection, thanks to Prof. Vedovoto from *MFLab*

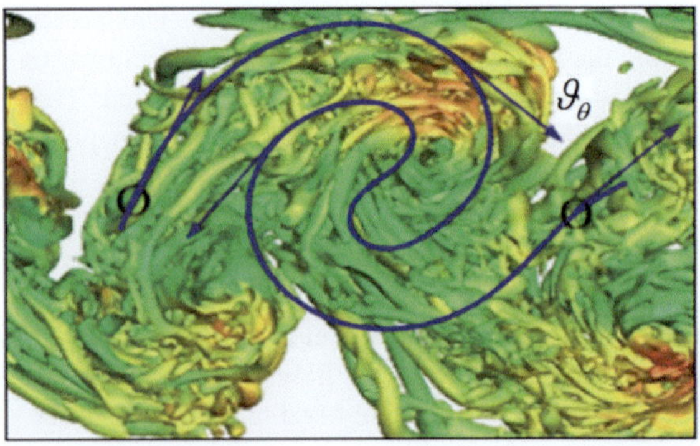

In Fig. 3.12, the trajectory of a material particle, transported by the global movement of the coherent structure, is also illustrated. The material particle moves with a velocity ϑ_θ and, to complete a turn, it takes a turn time τ. If $t > \tau$, then the turbulent structure is said to be coherent. There is also an internal movement to it, composed of other smaller structures. However, this internal movement is not clearly represented in Fig. 3.12.

The concept of coherent structure was launched by Brow and Roshko [5], who observed these turbulent structures using material experiments. Comparing Fig. 3.11a, b with Fig. 3.11c, there is a great qualitative similarity between the results obtained with computational experimentation and material experimentation.

3.2.2 *Jets*

Jet-type flows can be classified as temporally developing jets and spatially developing jets. Temporal jets are less frequent in nature and in industrial applications. Even so, their study is of great interest, given the importance of unraveling the intricate physical aspects of the process of transition to turbulence and of the turbulence that develops. Other equally interesting aspects refer to statistically transient turbulence, that is, statistical quantities are transient. In particular, he is interested in the energization processes of a spectrum, as well as the turbulence decay process. Energization occurs through the injection of kinetic energy into the spectrum of turbulent structures. The decay is a consequence of the viscous transformation of turbulent kinetic energy into thermal energy. Following these two groups of flows, temporal jets and space jets, will be studied in the introductory aspects of the topic.

3.2.2.1 Temporal Jets

Temporal jets are established whenever inflectional velocity fields are available, with a surplus, in relation to the environments that surround them, as illustrated in Fig. 3.13, through a profile of the horizontal component, distributed vertically. It is repeated throughout the entire domain, which characterizes its temporal behavior.

For the transition from the laminar regime to the turbulent regime to occur, the presence of disturbances superimposed on the inflectional base velocity field is necessary. In material experiments, these disturbances are due to effects intrinsic to the flow itself or come from the environment, such as pressure waves and structural vibrations, which are amplified and promote fluid dynamic instabilities.

These perturbations must be modeled mathematically. The numerical errors inherent to the process of discretization and solution of systems of algebraic equations are disturbances inherent to simulations, in general. However, these disturbances are undesirable, since there is no control over them. Mathematical models of disturbances, over which you have full control, are desirable. Mathematical noise can be

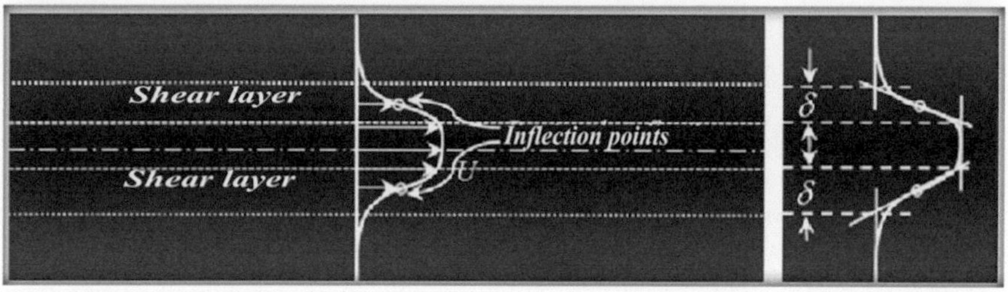

Fig. 3.13 Illustration of an inflectional base profile of velocities. The flow illustrated occurs from left to right

injected with controlled amplitudes and frequencies, using computational routines developed with the purpose of modeling different types of noise.

With linear stability theory, it also becomes possible to show, in a similar way to mixed layer flows, that disturbances will be amplified and that Kelvin–Helmholtz instabilities will form in the most amplified mode, first, and, in a broadband of modes, concomitantly, but with amplification rates lower than that of the most amplified mode.

In Fig. 3.14, the temporal evolution of three-dimensional images of a temporal-type jet is shown. The vorticity module was used for visualization. This flow is characterized, from a dynamic point of view, by the Reynolds Number, which, in the present case, $\mathrm{Re}_\delta = 5000$. The thickness $\delta(t)$ of the inflectional shear velocity field is illustrated in Fig. 3.13. The temporal sequences of images illustrate the formation and evolution of Kelvin and Helmholtz Instabilities, as well as the appearance of secondary filaments, induced by Primary Kelvin–Helmholtz Instabilities. These filaments are distributed approximately longitudinally. They will be called longitudinal filaments.

The following images presented in Fig. 3.14a, three-dimensional views of the flow and a possible turbulent regime. On the other hand, in the sequence of images from Fig. 3.14b, vertical cuts are shown that pass through the jet symmetry axis, highlighting the Kelvin–Helmholtz Instabilities in the first two views and the three-dimensionalization of the flow in the other views. Following images from Fig. 3.14c, horizontal cuts are shown in the last two views, showing the presence of longitudinal filaments. The presence of counter-rotating filaments, with the formation of small mushrooms, can even be seen in this figure.

The simulations were carried out using the Pseudo-Spectral Fourier Methodology, which will be presented in Chap. 5 of Isotropic Turbulence. This was possible considering that these flows are periodic in all directions. This method is characterized by spectral convergence rate, and, as a result, spatial derivatives are calculated with machine error. With this methodology, numerical diffusion and dispersion are very small. Therefore, numerical errors are not sufficient to model the disturbances existing in physical problems, necessary for the transition from the laminar regime to the turbulent regime to occur.

Therefore, controlled injection of noise in the simulations was necessary. For simulations performed without explicit noise injection, there is no transition. In the present work, noise was modeled using random number simulators, as described by Souza and Silveira-Neto [36]. The noise was superimposed on the initial velocity field. Fluid dynamic instabilities developed, reaching a state of turbulence. Subsequently, turbulence enters the decay regime. This means that the energy injected through the initial condition is used to promote the development of turbulence, which is responsible for transforming, through viscous effects, the turbulent kinetic energy into thermal energy, which is stored by the flowing fluid. The author of this book prefers the term energy transformation to the term energy dissipation, as will be seen throughout the text.

Each turbulent structure present in this flow transports a given amount of turbulent kinetic energy. These structures evolve over time and are characterized by a given

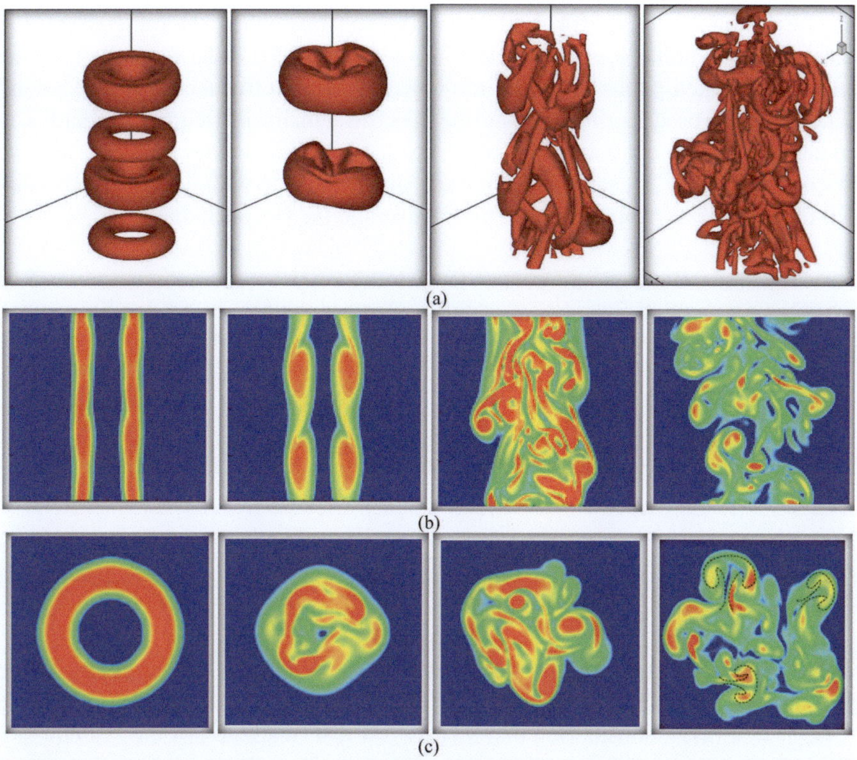

Fig. 3.14 Evolution in time of a jet in temporal development. Formation of Kelvin–Helmholtz Instabilities [36], longitudinal filaments and transition to turbulence; vorticity module field; **a** temporal evolution of three-dimensional views; **b** temporal evolution of a vertical axial slice; and **c** temporal evolution of a horizontal slice in half of the domain

wave number. The amount of energy per unit wave number can be associated with the energy transported by each turbulent structure. This amount of energy per unit wavenumber is called the spectral density of turbulent specific kinetic energy, denoted by $E(k,t)$, given in $\left[(\mathrm{J/kg})/(1/\mathrm{m})\right]$. The name specific refers to the energy per unit of mass, while the name density refers to the specific energy per unit wave number. From now on, this *information* will be called the spectral density of turbulent specific kinetic energy, which can be obtained using the Fourier Transform of the fluctuations of the components of the velocity vector. The above refers to the concept of wave number. To this end, spatial sampling is necessary. Time is fixed and spatial samples are taken. These samples are transformed into Fourier space, which leads to obtaining complex wavenumber functions, $\vec{k}$.

A related development can be made when sampling is done over time. This means that sampling is done at specific points in space, and information is collected as a function of time. In this case, the spectral density of turbulent specific kinetic energy will be obtained, denoted by $E(x,f)$, given in $\left[(\mathrm{J/kg})/(1/\mathrm{s})\right]$. Frequency f is the transformation parameter for the Fourier Space. Thus, when sampling is done in space, for a fixed time, it is transformed into a wave number. When sampling is done over time, at fixed points in space, it is transformed into frequency. It is necessary to develop the process of obtaining the spectral density in frequency, relative to simulations carried out using URANS, LES or Hybrid modeling. To do so, we take the component $\overline{u}_i(\vec{x}, t)$, sampled at a fixed point $\vec{x}$, varying with time. Its fluctuation $\overline{u}_i'$ is determined in relation to its temporal average $\langle\overline{u}_i\rangle(\vec{x})$, such that:

$$\overline{u}_i(\vec{x}, t) = \langle\overline{u}_i\rangle(\vec{x}) + \overline{u}_i'(\vec{x}, t). \tag{3.3}$$

The variance of the velocity's fluctuation is given by the temporal average correlation:

$$\begin{aligned}\left\langle\overline{u}_i'\overline{u}_i'\right\rangle(\vec{x}) &= \frac{1}{T}\int_0^T \overline{u}_i'(\vec{x}, t)\overline{u}_i'(\vec{x}, t)\mathrm{d}t = \frac{1}{T}\int_0^T \overline{u}_i'^{*}(\vec{x}, t)\overline{u}_1'(\vec{x}, t)\mathrm{d}t \\ &= \int_0^F \widehat{\overline{u}_i'}^{*}(\vec{x}, f)\widehat{u_i'}(\vec{x}, f)\mathrm{d}f\end{aligned} \tag{3.4}$$

The super index (*), which appears in Eq. (3.4), indicates the conjugate of the corresponding function. The second equality that appears in this equation is because $\overline{u}'^{*}(\overline{x}, t) = \vec{u}'(\overline{x}, t)$, considering that $\vec{u}_i'(\vec{x}, t)$ is a real function. For the last equality of this equation, we relied on Parseval's Theorem. It is observed that all terms of Eq. (3.4) have a specific power unit $\left[\mathrm{J/kg}\right]$. The specific energy spectral density is defined as:

$$\begin{aligned}E_{\overline{u}_i\overline{u}_i'}(\vec{x}, f) &\equiv \left(\frac{1}{2}\right)\widehat{\overline{u}'}^{*}(\vec{x}, f)\widehat{\overline{u}'}(\vec{x}, f) = \left(\frac{1}{2}\right)\left(\widehat{\overline{u}_{\mathrm{iR}}'} - \iota\widehat{u_{\mathrm{iI}}'}\right)\left(\widehat{u_{\mathrm{iR}}'} + \iota\widehat{u_{\mathrm{iI}}'}\right) \\ &= \left(\frac{1}{2}\right)\left(\widehat{u_{\mathrm{iR}}'}^2 + \widehat{u_{\mathrm{iI}}'}^2\right).\end{aligned} \tag{3.5}$$

Thus, Eq. (3.4) can also be written as follows:

$$\left\langle\overline{u}_i'\overline{u}_i'\right\rangle(\vec{x}) = \int_0^{F_\mathrm{c}} \hat{\overline{u}}_i'^{*}(\vec{x}, f)\widehat{u_i'}(\vec{x}, f)\mathrm{d}f \equiv \int_0^{F_\mathrm{c}} E_{\overline{u}_i'\overline{u}_i'}(\vec{x}, f)\mathrm{d}f, \tag{3.6}$$

where F_c represents the cutoff frequency, determined with the time step used in the simulation performed with LES, URANS or a hybrid modeling. The concepts of turbulence modeling will be covered in future chapters.

It is useful to remember here the definition of turbulent kinetic energy and how to express it, involving a dualism between physical space and Fourier Space, using Parseval's theorem:

$$
\begin{aligned}
k(\vec{x}) &\equiv \frac{1}{2}\left[\left\langle(\overline{u}')^2\right\rangle + \left\langle(\overline{\vartheta}')^2\right\rangle + \left\langle(\overline{w}')^2\right\rangle\right] \\
&= \int_0^F \left[E_{\overline{u}'\overline{u}'}(\vec{x}, f) + E_{\overline{\vartheta}'\overline{\vartheta}'}(\vec{x}, f) + E_{\overline{w}'\overline{w}'}(\vec{x}, f)\right] df \\
&\equiv \int_0^F E(\vec{x}, f) df.
\end{aligned} \tag{3.7}
$$

From Eq. (3.7) the definition of the spectral density of turbulent specific kinetic energy is also removed, in the spectral space of frequencies, $E(\vec{x}, f)$, given via Eq. (3.8):

$$
E(\vec{x}, f) \equiv \left[E_{\overline{u}'\overline{u}'}(\vec{x}, f) + E_{\overline{\vartheta}'\overline{\vartheta}'}(\vec{x}, f) + E_{\overline{w}'\overline{w}'}(\vec{x}, f)\right]. \tag{3.8}
$$

In the case of spatial sampling, statistical treatment must be carried out on spatial variations. Thus, the decomposition into a spatial average plot $\langle\overline{u}_i\rangle(t)$ and a floating portion $\overline{u}'(\vec{x}, t)$ is given by:

$$
\overline{u}_i(\vec{x}, t) = \langle\overline{u}_i\rangle\left(\vec{t}\right) + \overline{u}_i'(\vec{x}, t). \tag{3.9}
$$

The variance of the velocity fluctuation is given by the spatial average correlation:

$$
\begin{aligned}
\langle\overline{u}_i'\overline{u}_i'\rangle(t) &= \frac{1}{V}\int_V \overline{u}_i'(\vec{x}, t)\overline{u}_i'(\vec{x}, t)dV = \frac{1}{V}\int_V \overline{u}_i'^*(\vec{x}, t)\overline{u}_i'(\vec{x}, t)\mathrm{d}t \\
&= \int_0^{k_c} \widehat{\overline{u}_i'}^*\left(\vec{k}, t\right)\widehat{u_i'}\left(\vec{k}, t\right)d\vec{k}
\end{aligned} \tag{3.10}
$$

where k_c is the cutoff wavenumber, determined with the size of the discretization cell. The spectral density of the turbulent specific kinetic energy is defined:

$$
\begin{aligned}
E_{\overline{u}_i'\overline{u}_i'}(k, t) &\equiv \left(\frac{1}{2}\right)\widehat{\overline{u}'}^*(k, t)\widehat{\overline{u}'}(k, t) = \left(\frac{1}{2}\right)\left(\widehat{\overline{u}'_{iR}} - \iota\widehat{u'_{iI}}\right)\left(\widehat{u'_{iR}} + \iota\widehat{u'_{iI}}\right) \\
&= \left(\frac{1}{2}\right)\left(\widehat{u'_{iR}}^2 + \widehat{u'_{iI}}^2\right).
\end{aligned} \tag{3.11}
$$

Equation (3.10) can be rewritten as follows:

$$
\langle\overline{u}_i'\overline{u}_i'\rangle(t) = \int_V \hat{\overline{u}}_\iota'^*\left(\vec{k}, t\right)\widehat{\overline{u}_\iota'}\left(\vec{k}, t\right)\mathrm{d}f \equiv \int_V E_{\overline{u}_i'\overline{u}_i'}\left(\vec{k}, t\right)\mathrm{d}\vec{k} \tag{3.12}
$$

Turbulent kinetic energy is expressed by involving a dualism between physical space and Fourier Space:

$$k(t) = \frac{1}{2}\left[\left\langle (u')^2 \right\rangle + \left\langle (\vartheta')^2 \right\rangle + \left\langle (w')^2 \right\rangle\right]$$
$$= \int_0^{k_c} \left[E_{\overline{u}'\overline{u}'}(k,t) + E_{\overline{\vartheta}'\overline{\vartheta}'}(k,t) + E_{\overline{w}'\overline{w}'}(k,t)\right] d\vec{k}$$
$$\equiv \int_0^{k_c} E(k)dk. \tag{3.13}$$

From Eq. (3.13) one also obtains the definition of the spectral density of turbulent specific kinetic energy in the spectral space of wave numbers, $E(k,t)$, given via Eq. (3.14):

$$E(k,t) \equiv \left[E_{\overline{u}'\overline{u}'}(k,t) + E_{\overline{\vartheta}'\overline{\vartheta}'}(k,t) + E_{\overline{w}'\overline{w}'}(k,t)\right] \tag{3.14}$$

The temporal evolution of the spectral density of the turbulent specific kinetic energy, $E(k,t)$, is presented in Fig. 3.15. These spectra are relative to the temporal evolution of jets. The spectra are presented for three values of the Reynolds Number, whose calculation is based on the initial thickness, $\delta(0)$. For each value of the Reynolds Number, there is a temporal sequence of spectra, showing the evolution of turbulence. The flows were initialized in the laminar regime, with inflectional velocity fields. As mentioned in the previous paragraph, perturbations were superimposed on the initial base fields. Thus, the different wavelengths were amplified over time, establishing the spectra. It can be seen that, as time passes, the spectra tend to a slope of – 5/3, following the Law of Kolmogorov [15] and Kolmogorov [16].

It is also observed that the higher the Reynolds Number, the closer the slope of the spectra becomes to the slope expected by the Kolmogorov Theory, for fully turbulent flows. This is because the higher the Reynolds Number, the more intense the turbulence becomes and the closer it comes to the condition of developed three-dimensional turbulence. As expected for flows energized only by the initial condition, the state of turbulence is established, but soon after it begins to decay. This becomes clear by observing, for example, the spectra for $Re_\delta = 5,000$. In $t = 20\,[s]$, the spectrum contains the highest levels for all wave numbers. For $t = 30\,[s]$ and $t = 40\,[s]$, and for $k > 10\,[1/m]$, turbulence has lower energy levels than for $t = 20\,[s]$, thus illustrating the beginning of the decay process.

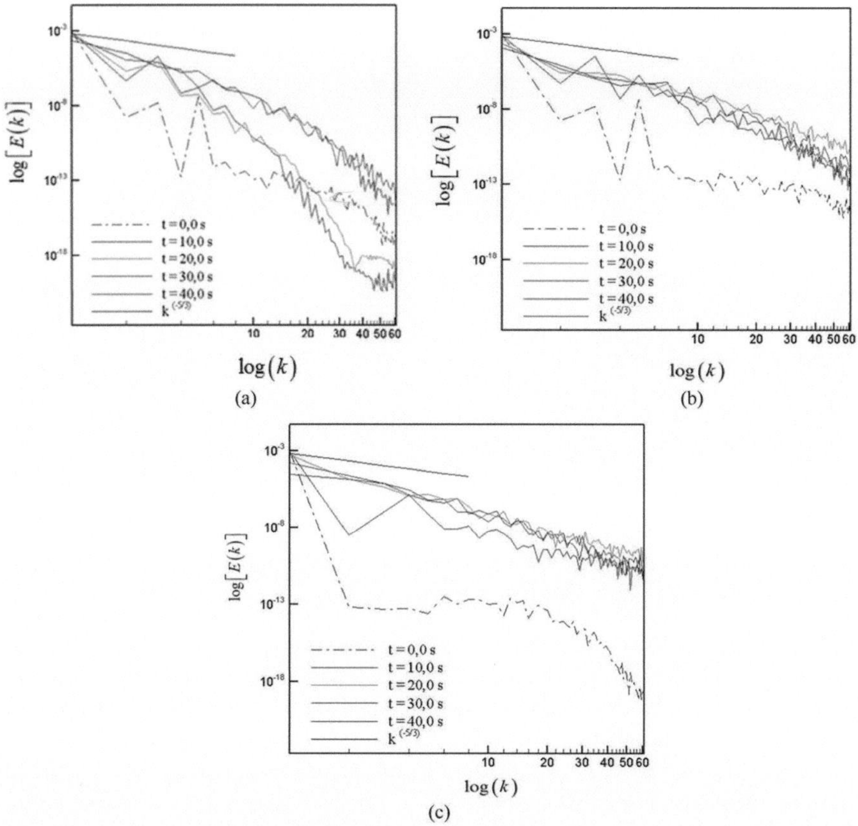

Fig. 3.15 Turbulent specific kinetic energy spectral density for temporally developing jets: **a** $\mathrm{Re}_\delta = 1,600$; **b** $\mathrm{Re}_\delta = 5,000$ and **c** $\mathrm{Re}_\delta = 10{,}000$

3.2.2.2 Spatially Developing Jets

Jets in spatial development are a kind of flow resulting from the injection of a fluid with a given velocity into a medium that may be at rest or even with a velocity different from the velocity of the jet itself. In this case, an average velocity profile is formed. This happens because the velocity of the jet is different from the velocity of the environment, as illustrated in Fig. 3.16a, b, for cases where the jet velocity is greater than the velocity of the environment.

Jets can be classified according to the geometry that forms them. We speak of a round jet if it originated from a circular orifice and a flat or rectangular jet if it originated from a flat or rectangular nozzle. In any of these types of jets, the transition to turbulence is characterized, like mixing layers, by the formation of primary instabilities of the Kelvin–Helmholtz type, which will induce the formation of secondary filaments. The interactions of the counter-rotating longitudinal filaments

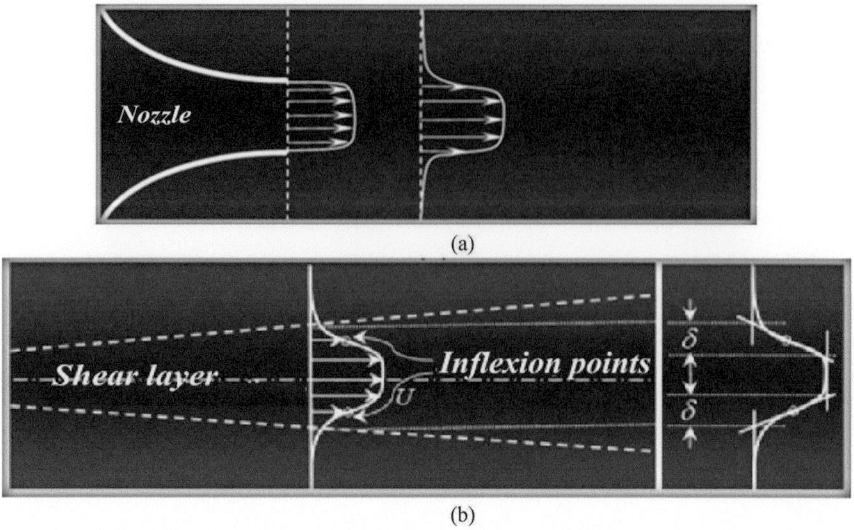

Fig. 3.16 Illustration of a nozzle forming a jet in a medium at rest (**a**) and characterization of the initial conditions to form a jet in spatial development (**b**)

with the Kelvin–Helmholtz primary turbulent structures will induce the formation of transverse oscillations, which are amplified and set the stage for the transition to three-dimensional turbulence.

In Fig. 3.17, the process of transition to turbulence is schematically illustrated, with the different phases that characterize it: (1) jet originating from a converging nozzle, in its development phase; (2) formation of Kelvin–Helmholtz Instability and promotion of axisymmetric two-dimensional toroidal instabilities, for the cases of circular jets; (3) growth and deformation of toroid's, through pairing mechanisms and the development of counter-rotating longitudinal filaments. The pairings are not visible in this figure, but the presence of longitudinal filaments can be inferred by the presence of the mushrooms shown in Fig. 3.17b; (4) the development of three-dimensional turbulence, which is organized into a vast band of turbulent structures, as illustrated in Fig. 3.17a.

The counter-rotating longitudinal filaments, present in the space jet, can be visualized in Fig. 3.17b, in a horizontal plane, transverse to it, as indicated by the cutting plane drawn in Fig. 3.17a. Mushroom-shaped turbulent structures are seen distributed around the jet. These mushrooms are formed by the presence of counter-rotating longitudinal filaments. The filaments interact with the primary toroidal structures, generating strong deformations and contributing to the transition process. This scenario has been observed both in material experimentation work and in computational experimentation work.

As in previous cases, the first instabilities appear in a sinusoidal shape, indicating a selective amplification process of disturbances. The jets transition at low values of Reynolds Number, as happens with all free shear flows. For circular jets, the

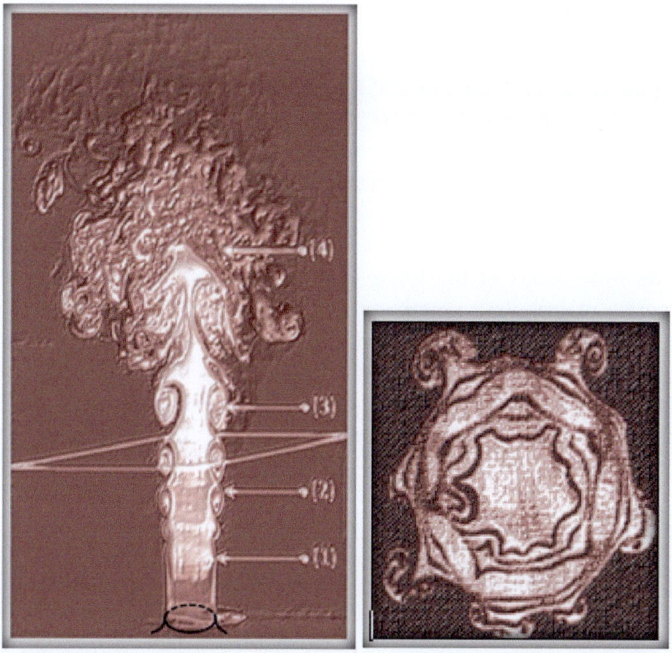

(a) (b)

Fig. 3.17 Circular jet in transition: **a** visualization in a vertical plane; **b** visualization in a horizontal plane, transverse to the jet, as indicated by the cutting plane drawn in figure (**a**). Figure adapted from Moreira and Silveira-Neto [19]

transition begins at $\mathrm{Re_d} = 10\,(\mathrm{Re_d} = Ud/\nu)$, according to Drazin and Reid [8]. It is worth noting that the transition at low Reynolds Number values should not be confused with trivialities in flow modeling and simulation. Transition to jet turbulence is a highly complex problem, since it happens very abruptly in space and is highly sensitive to the boundary conditions at the entrance to the domain.

The transition from a laminar jet to a turbulent jet occurs downstream of the nozzle. It depends on the geometry of the orifice or nozzle, and on the upstream flow conditions. Therefore, the experiments are difficult to compare with each other from the point of view of flow dynamics and topology. It is also difficult to compare data from material experiments with data from computational experiments. However, even from a statistical point of view, detailed knowledge of the fluid dynamics upstream of the nozzle is necessary.

In Fig. 3.18, the result of a computer simulation of large structures (LES), carried out by Moreira and Silveira-Neto, is presented [19]. In Fig. 3.18a, the circular jet is visualized in spatial development, which is in transition from the laminar regime to the turbulent regime. Kelvin–Helmholtz Instabilities develop, followed by consecutive toroidal turbulent structures. At Fig. 3.18b, what is happening is visualized in the horizontal plane indicated in Fig. 3.18a. The presence of structure pairs like mushrooms can be seen, illustrating the existence of counter-rotating turbulent structures, which develop in the axial direction of the jet.

The state of full turbulence can be verified quantitatively, using the spectral density of turbulent kinetic energy, relative to the region of turbulence developed, as can be seen in Fig. 3.18. A computational probe was placed approximately 20 times the diameter of the jet exit.

The three velocity components were sampled as a function of time. Fourier transforms of the time series of the three velocity components were performed. The spectral density of turbulent kinetic energy was obtained using the real parts and imaginary parts of the transforms, as presented in Eq. (3.5). The function $E(f)$ is presented in Fig. 3.19. It is observed that it tends to reach a slope of $-5/3$, as expected.

The distribution of the average (temporal) velocity on the center line of the jet, as a function of the axial coordinate, is presented in Fig. 3.20, for three values of the

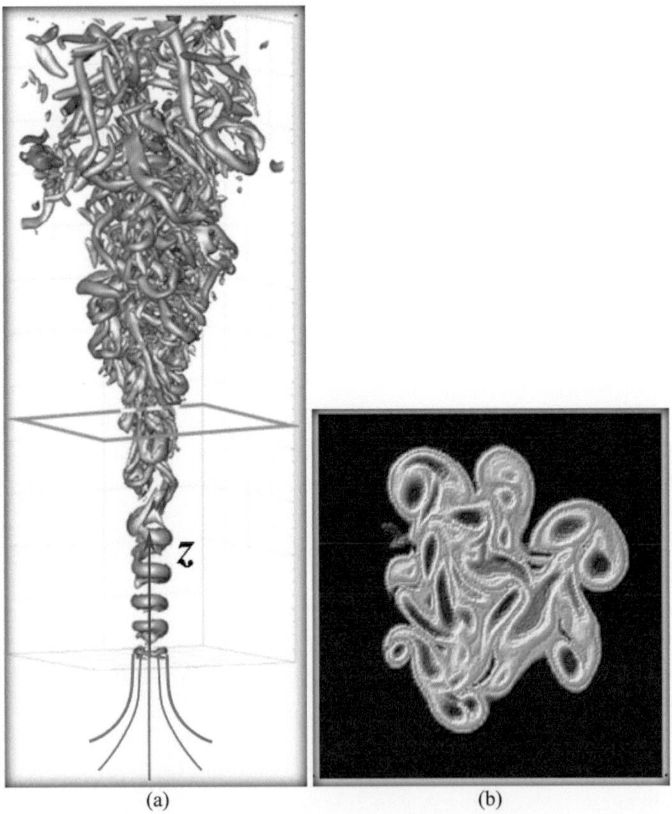

Fig. 3.18 Computational simulation of large turbulent structures, related to the transition process of a round jet [19]: **a** jet in spatial development in the vertical direction and **b** horizontal viewing plane in the position indicated in (**a**)

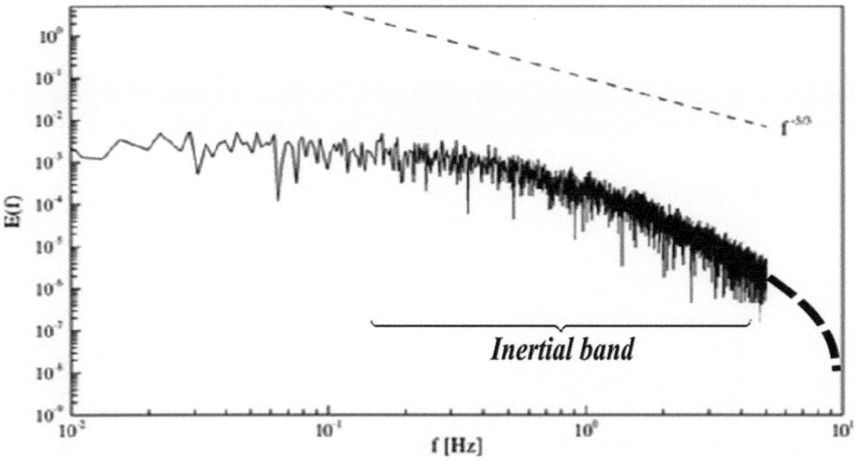

Fig. 3.19 Computational simulation of large structures of the transition process of a round jet. Spectral density of the turbulent kinetic energy, extracted from the turbulence zone developed in the jet in Fig. 3.18 [19]

Reynolds Number and the results are compared with data from material experiments. The velocity falls with the axial coordinate at the center of the jet, due to the process of its expansion that develops spatially, as illustrated in Fig. 3.16b. It is observed that the higher the value of the Reynolds Number, the shorter the jet transition length. This is consistent as nonlinear interactions become stronger.

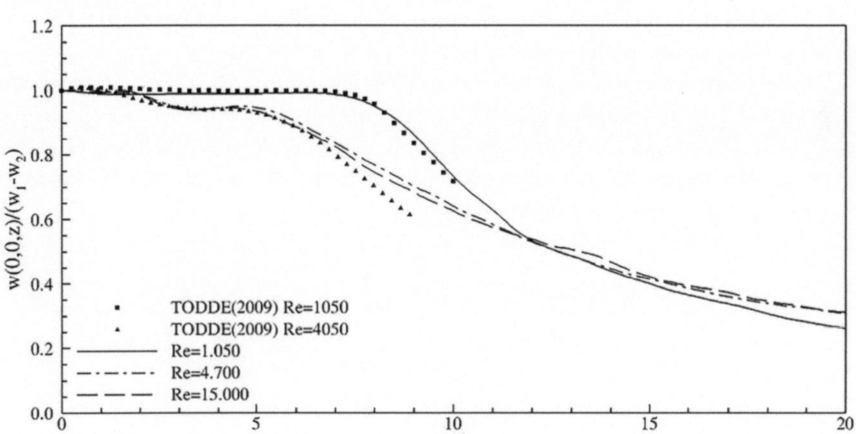

Fig. 3.20 Spatial evolution of the jet velocity on its center line, function of the axial coordinate, *z*, for three values of the Reynolds Number [19]

3.2.3 Wakes

Wake-type flows originate from inflectional velocity fields that are deficient in relation to the environment that surrounds them. The sources of this type of velocity field are linked to free current flows, faster than intermediate current, as illustrated in Fig. 3.21. Shear provides the formation of deficient inflectional velocity fields. This is a form that can be found in nature or can be induced, in a controlled manner, in laboratories. The other source, more frequent in nature and in industrial applications, occurs when a fluid flow passes around an obstacle, for example, an automobile, an aircraft, or any kind of body immersed in a fluid. This is, without a doubt, the transition flow most familiar to most people, even to laypeople who have already had the opportunity to observe the classic von Kármán wake, downstream of the pillars of a bridge. Like the flows presented in the previous sections, wake-type flows can also be classified into temporal wakes and spatial wakes. Its characteristics will be presented below.

3.2.3.1 Temporal Wakes

Temporal wakes are formed from inflectional velocity fields in deficit, as illustrated in Fig. 3.21. This figure shows a velocity distribution, where inflectional shear is promoted by two fast fluid currents, adjacent to a relatively slow core. Two inflection points and the thickness, δ, of the shear layers. The velocity field presented there is the initial field, which will give rise to the Kelvin–Helmholtz Instabilities. This field is self-similar throughout the domain, causing the repetition of instabilities in space and evolution in time. This justifies the name temporal wake, analogous to what was discussed for mixing layers and temporal jets.

We start from an initial velocity field, as illustrated in Fig. 3.21. Perturbations are superimposed on this initial field, which will be amplified by nonlinear mechanisms, inherent to inflectional shear flows. As a result, the formation of fluid dynamic instabilities of the Kelvin–Helmholtz type will be induced. It was already described in the canonical flows presented previously.

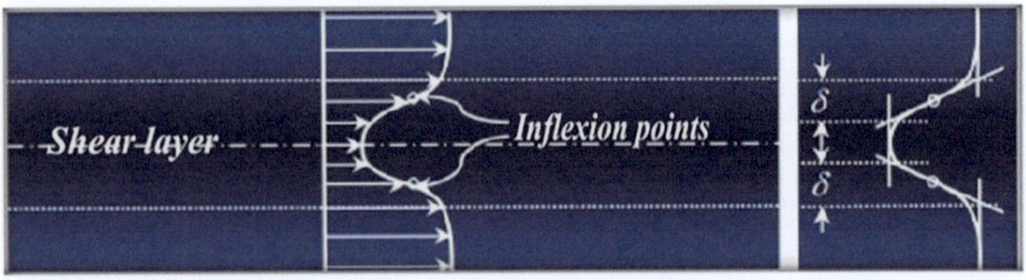

Fig. 3.21 Illustration of a base inflectional velocities field, in deficit. The illustrated flow occurs from left to right and evolves over time

To illustrate this process, simulations were carried out in MFLab of wake-type flows, in temporal development, using computational models completely developed in-house. A Fourier Pseudo Spectral Computational Code was used, with which the algebraic-differential model was solved, composed of the continuity equation, the Navier–Stokes equations and the boundary and initial conditions, for incompressible flows of Newtonian fluids. The average initial condition was established using the following equation:

$$\overline{u}(y, t = 0) = U\left(1 - \mathrm{e}^{-c_1 y^2}\right), \tag{3.15}$$

where U is the velocity of the medium surrounding the wake, c_1 is a constant that is used to model shear. With this profile, a thickness of $\delta = 1\,[\mathrm{m}]$, of the shear layers, was imposed, as proposed by Sondergaard [35]. The average components of velocity, $\overline{\vartheta}$ and $\overline{w}$, are null in $t = 0\,[\mathrm{s}]$. To mathematically model the physical perturbations, present in any material laboratory, random perturbations in space are superimposed on the initial average components at all points of the domain:

$$u(x, y, z, t = 0) = \overline{u}(y, t = 0) + a_u u'(x, y, z), \tag{3.16}$$

$$\vartheta(x, y, z, t = 0) = a_\vartheta \vartheta'(x, y, z), \tag{3.17}$$

$$w(x, y, Z, t = 0) = a_w w'(x, y, z). \tag{3.18}$$

The amplitudes of random disturbances, a_u, a_v and a_w are modeled, whose spatial variations are modeled by u', ϑ' and w'. The superposition of this noise on the initial velocity fields is necessary when using a highly accurate computational model, with numerical errors on the order of machine error. Thus, with very small numerical errors and without mathematical modeling of the disturbance to be superposed on the initial base field, the transition from the laminar regime to the turbulent regime would not occur, as occurs in material experiments and even in computational experiments, using conventional numerical methods. In fact, in this case, the numerical errors, inherent to low-convergence-rate methods, serve as a model for the noise, guaranteeing the flow transition, even without the superposition of explicit mathematical noise. However, this cannot be understood as some kind of advantage over less accurate numerical methods, because there is little control over numerical errors. On the other hand, when the noise is modeled with a dedicated model to this purpose, you have total control over the noise injected into the flow, being able to control amplitudes, frequency bands or wave number bands.

There are several libraries available in the literature, open to the public, aimed at mathematical modeling of noise, many of which are based on pseudorandom number modelers. In the present work, the mathematical digital noise generator proposed by Smirnov was chosen for the simulation of the flat temporal wake. Shi and Celik [34], called this model RFG (*Random Flow Generator*). It was considered by Moreira

Table 3.1 Values of parameters used in the simulation of a flat conveyor belt

Number of processes	128
Fourier modes, nx, ny, nz	$128 \times 128 \times 256$
Closing templates	Smagorinsky
Domain dimensions (Lx, Ly, Lz)	$(40\delta,\ 20\delta,\ 20\delta)$
Total simulation time	300 [s]
CFL	0.8
Characteristic velocity	$U = 1$ [m/s]
Thickness of shear layers, δ	1 [m]
Reynolds Number	$\text{Re}_\delta = 346$
Time integration method	*RK46*
(a_u, a_ϑ, a_w)	$\left(10^{-3}, 10^{-3}, 10^{-3}\right)$

and Silveira-Neto [20] to simulate the turbulent flat jet. Among the advantages of using RFG, it is easy to implement, and it has a low computational cost. Details of the parameters of the RFG method, implemented in the IMERSPEC3DP code, can be found in Moreira and Silveira-Neto [20]. For the Reynolds Number, $\text{Re}_\delta = 346$, was chosen, estimated to be sufficient to simulate the transition to turbulence. The parameters adopted for the simulation are defined in Table 3.1. Temporal integration was performed with the fourth-order, six-step Runge–Kutta Method, RK46 [1].

In Fig. 3.22, the flow is visualized at different times, starting from the initial condition, Fig. 3.22a. The vorticity fields and isolines are visualized. In Fig. 3.22b, after a few seconds, it is observed that the shear layer became wider, showing that there was diffusion of linear momentum, kinetic energy and vorticity, in the direction transverse to the shear layer. In fact, the vorticity is maximum at the inflection lines, tending to zero when it moves away from these lines, which establishes a gradient of this *information* in the direction normal to the flow. It can also be seen in this figure that the flow became unstable, showing that the disturbances superimposed on the initial condition were amplified, promoting a spatial sequence of four wavelengths, characterized by the presence of sequential turbulent structures. This wavelength is called λ_{max}, in the literature on linear stability analysis. This wavelength was selected by the physical mechanisms inherent to the flow to be amplified with maximum amplification rate. Thus, it is the first to appear in the form of fluid dynamic Kelvin–Helmholtz instabilities.

These instabilities that appear in the wake presented in Fig. 3.22 are of the same nature as the instabilities that were presented and discussed for mixing layers and jets. The formation mechanisms are the same as those previously described. So, these are also called Kelvin–Helmholtz instabilities structures. It is interesting to note that, as time passes, two sets of turbulent structures are formed, one upper (anticlockwise rotation) and one lower (clockwise rotation). These structures are out of phase with each other, characterizing the well-known sinuous way of formation and transport of dynamic structures. They rotate clockwise, on the upper set, and counterclockwise, on the lower set. As they rotate, they transport fluid from the layers outside them,

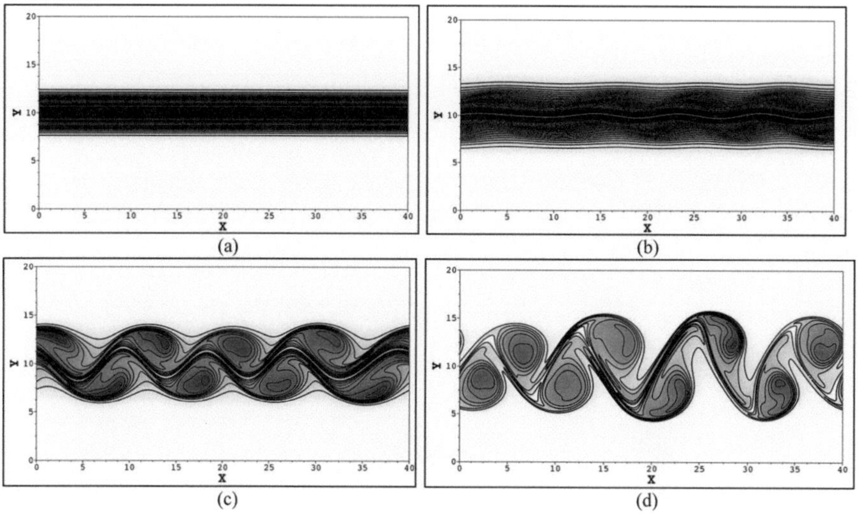

Fig. 3.22 Wake in temporal development: vorticity isolines; temporal evolution, starting from the initial condition; $\mathrm{Re}_\delta = (U\delta)/\nu = 346$

regions rich in linear momentum, which energizes them and causes them to grow over time. These structures increase in size over time. In the previously studied mixing layers, the main mechanism for increasing the size of these structures was pairing. Here, in the case of the simulated wake, the pairings are not present, and the main growth mechanism is the advective transport of linear momentum, promoted by the rotation of the dynamic structures.

Other wavelengths amplify concomitantly with λ_{max}; however, with lower amplification rates. They can be observed in a three-dimensional view, as illustrated in Fig. 3.23. This figure shows the flow, at four different times and from two viewing angles: Fig. 3.23 (left column) is a perspective view and Fig. 3.23 (right column) is a top view.

Shortly after the initial condition, the Kelvin–Helmholtz Instabilities were formed with wavelength λ_{max} and then, other instabilities of different wavelengths can be found, superimposed over these primary instabilities. This illustrates that an entire band of wavelengths is being amplified simultaneously. Instabilities of different wavelength λ values amplify concomitantly in a nonlinear way, however, with different amplification rates. In the second line, strong three-dimensional effects are observed in the two figures, as longitudinal filament-type instabilities, with strong interaction with the Kelvin–Helmholtz Instabilities. On the second and third lines, strong mechanical effects can be identified because of the nonlinear interactions between the turbulent structures, resulting in the three-dimensionalization of the flow. In the fourth line, there is great *information* on turbulent structures, and a completely three-dimensional and turbulent flow can be identified.

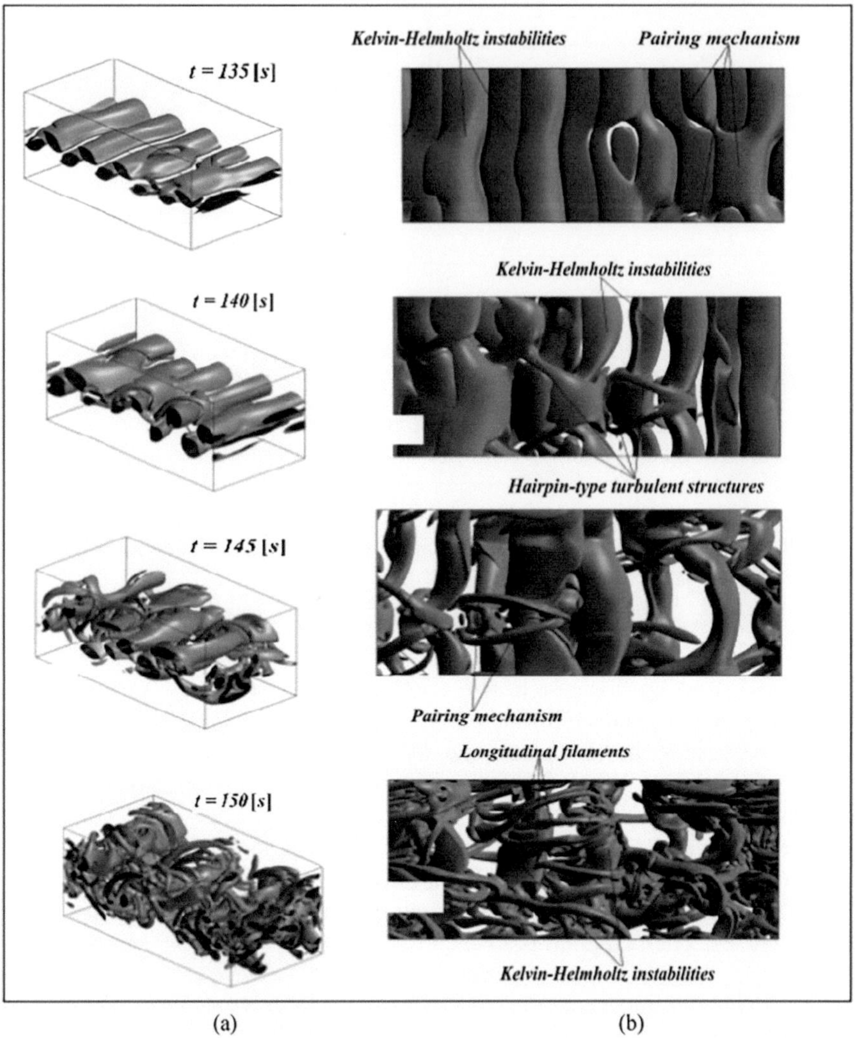

Fig. 3.23 Temporal wake in temporal development. Q criterion iso-surfaces; temporal evolution; $\text{Re}_\delta = (U\delta)/\nu = 346$ (own collection)

3.2.3.2 Spatial Wakes

Space wake-type flows appear downstream of an obstacle, where recirculating mean flows are established, which give rise to deficit inflectional velocity fields, as illustrated in Fig. 3.24. There is the formation of a shear layer in spatial development. A disturbance with a wide range of frequencies, or wave numbers, must be superimposed on this base inflectional field to develop Kelvin–Helmholtz type instabilities, which will be transported and will evolve spatially.

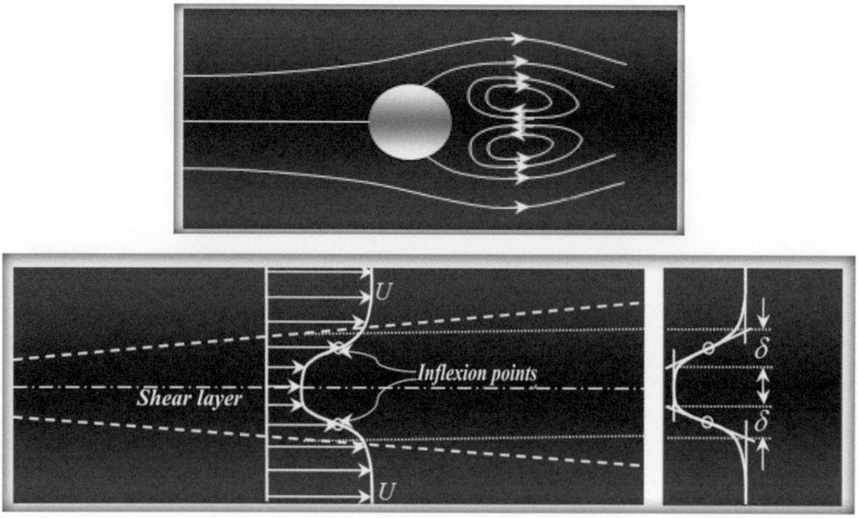

Fig. 3.24 Illustration of an inflectional field of base velocities, in deficit. The illustrated flow occurs from left to right and evolves spatially

In Fig. 3.25a, the flow over a sphere is visualized. The flow occurs from left to right, being characterized by $Re_d = 10{,}000$, where d is the diameter of the sphere. The turbulent structures formed downstream of the sphere are presented, using the Q Criterion as a way of flow visualization. The formation of hairpin-like structures is observed with strong nonlinear interaction. The formation of these structures occurs from the recirculation that occurs after the sphere, which becomes asymmetric and triggers the formation of these Kelvin–Helmholtz structures. Thus, these structures are formed from an inflectional velocity field, promoted by recirculation. Before hairpin-like structures form, Kelvin–Helmholtz structures form, as shown in Fig. 3.25b.

In Fig. 3.25a, right after the sphere, a Kelvin–Helmholtz structure deformed, forming an elongated flap, which gives rise to a "hairpin" structure. Thus, the transition process to turbulence involves the formation of instabilities and Kelvin–Helmholtz structures, as in previous shear flows.

The statistical treatment of the flow over the sphere allows the determination, via computational experimentation, of all statistical parameters of interest. In particular, it is presented in Fig. 3.26, the drag coefficient on a sphere, obtained by computational simulations of flows, carried out by Melo and Silveira-Neto [18]. It is also observed that for $Re_d = 2 \times 10^6$, because of the material experiment, a sudden drop in the drag coefficient appears. This effect is known as the "drag crisis", which is a consequence of the transition of the boundary layer that forms in the hemisphere upstream of the boundary layer detachment line. The transition to turbulence, before the detachment line, energizes the boundary layer, delaying the detachment line and, consequently, reducing the wake force behind the sphere. This is known as the "Eiffel Effect".

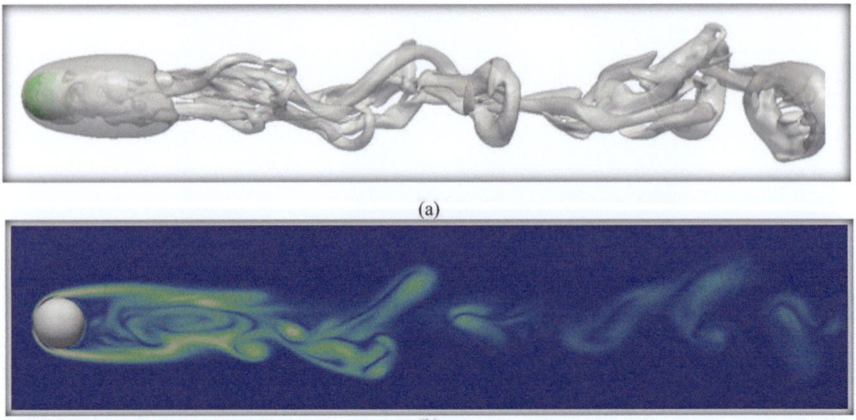

(a)

(b)

Fig. 3.25 Flow over a sphere: $Re_D = 10{,}000.0$. The Q criteria are displayed. Personal collection [11]

It was not possible to obtain this effect with the computational simulation of the present work, since the boundary layer transition cannot be captured with the level of mesh refinement used. However, it is worth noting here that good statistics can only result from good instantaneous results. Thus, it is concluded that what was presented in the form of instantaneous results represents well the physical reality, experienced in material wind tunnels, as presented in White [41], used as a reference for comparison in the present text. Worth reinforcing that the effect of the "drag crisis", a consequence of the boundary layer transition, was not captured in the present computational experiment.

A wake can form in two modes of instability, the most common being the sinuous mode, presented in Fig. 3.25. The varicose mode, presented below, is a mode rarely obtained, both in computational experiments and in material experiments.

In Fig. 3.27, the flow promoted by the movement of a rising bubble in a liquid medium, originally at rest, is presented. The dimensionless parameters that characterize this two-phase flow are the following: dynamic viscosity ratio, $\lambda = \mu_c/\mu_d = 54.3$ (ratio between the dynamic viscosities of the continuous and dispersed phases); $\eta = \rho_c/\rho_d = 100$ (ratio between the specific masses of the continuous and dispersed phases); $Eo = \left(g\Delta\rho d_b^2\right)/\sigma = 3.6$ (Number of Eötvos); $Mo = \left(g\Delta\rho\mu_c^4\right)/\left(\rho_c^2\sigma^3\right) = 2.581 \times 10^{-11}$ (Morton number) and finally $Re_d = (\rho_c U_b d_b)/\mu_c = 1,329.0$ (Reynolds Number). The indices "c", "d" and "b", which appear in these dimensionless parameters, refer respectively to the continuous phase and dispersed phase of the two-phase flow and to the bubble. The wake left by the rising bubble can be observed in a pulsating movement. In the first three situations presented in Fig. 3.27, there is the varicose mode, with the presence of approximately axisymmetric toroidal instabilities.

The formation and persistence of fluid dynamic instabilities in the so-called varicose mode, in the problem under study, are due to the pulsating movement of the

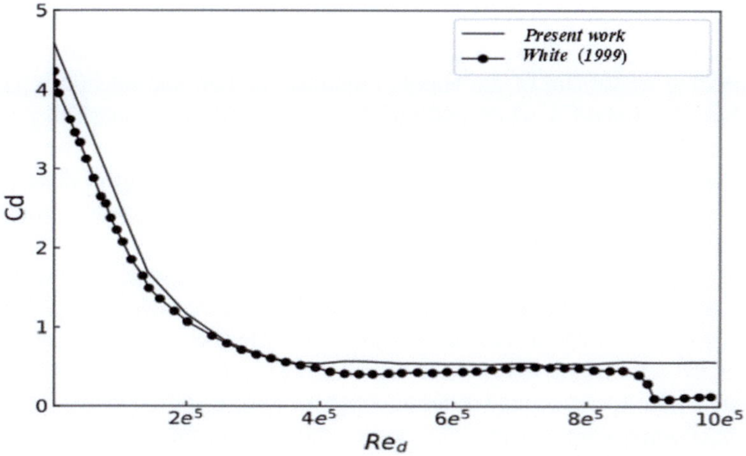

Fig. 3.26 Flow over a sphere: $\mathrm{Re_d} = 1,329.0$. Comparison of the drag coefficient on a sphere. Personal collection [18]

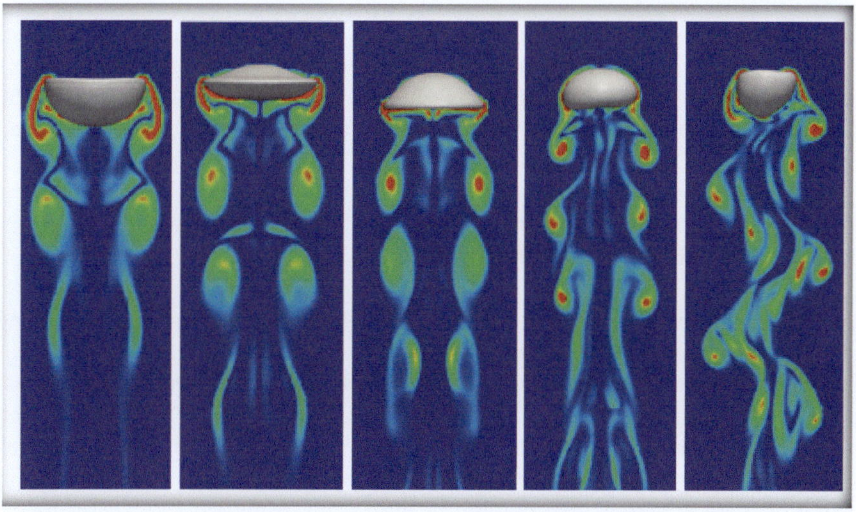

(a) (b) (c) (d) (e)

Fig. 3.27 Flow downstream of an upward-moving bubble. Temporal evolution: $\lambda = \mu_c/\mu_d = 54.3$; $\eta = \eta_c/\eta_d = 100.0$; $\mathrm{Mo} = 2.581 \times 10^{-11}$; $\mathrm{Re_d} = 1,329.0$. Transition from varicose to sinuous mode. Personal collection, Pivello et al. [23]

bubble. The varicose mode is extremely unstable and is difficult to establish for moments when the body downstream from which the wake is formed is at rest. The pulsating movement of the bubble promotes the periodic release of axisymmetric toroidal structures, which make up the wake in varicose mode. This pulsating movement can be observed in Fig. 3.27a–c. Over time, these toroids lose their axis symmetry, which gives rise to the bifurcation into the sinuous mode of emission and transport of dynamic Kelvin–Helmholtz turbulent structures. The sinuous mode can be clearly viewed in Fig. 3.27e, when looking at the wake.

In Fig. 3.28, a temporal sequence is shown, where the spatial evolution of the wake of turbulent structures is visualized. The assembly of the sequential elements in the figure was done to detach or separate the bubble itself from the wake formed behind it, in such a way as to be able to visualize the geometry of the bubble, which is deformable. The sinuous mode of instability formation is present in this figure. The turbulent structures are coherent, a fact that can be observed by their permanence, in the experimental domain, for a long distance, equivalent to thirty diameters of the bubble, $30d_\text{b}$. However, due to the three-dimensionality of the flow, the vorticity is attenuated by viscous effects, and the turbulent structures disappear as they move away from the bubble.

It is worth highlighting again the importance of the nomenclature "turbulent structure". The wake is formed by composite and complex entities and not by simple isolated "vortices". This composition is named a turbulent structure. In Fig. 3.27e, structures composed of diverse instabilities are also observed. They are called "coherent", since they have a longer life than their characteristic time.

3.2.3.3 Three-Dimensional Effects

In Fig. 3.29, a wake composed of Kelvin–Helmholtz Structures and counter-rotating longitudinal structures is evident. They characterize the initial phase of transition to turbulence. Two iso-values of the vorticity axial component were plotted, $\omega_z = +5\,[1/\text{s}]$ and $\omega_z = -5\,[1/\text{s}]$. This can be viewed in Fig. 3.29a. The blue and red colors correspond to these vorticity iso-values, with which the turbulent structures that rotate clockwise and counterclockwise can be visualized. The counter-rotating nature of these structures can be seen in Fig. 3.29b, using velocity-vector fields. Horizontal cuts on the wake are also shown. In Fig. 3.29a, a cutting plane illustrates that sections were made in four vertical positions. The cutting plane (1) is to visualize what is happening in the horizontal plane that passes through the center of the bubble; the planes (2), (3) and (4) were located on the wake, below the bubble. Several counter-rotating pairs are identified on the four horizontal planes. They are all correlated, showing similar counter-rotating structures.

The interactions of these longitudinal structures with the Kelvin–Helmholtz Structures, which are not shown in Fig. 3.29, but can be viewed in Fig. 3.27, represent the initial phase of the entire complex process of transition to turbulence. Understanding this transition process has not yet been developed by the scientific community. Only

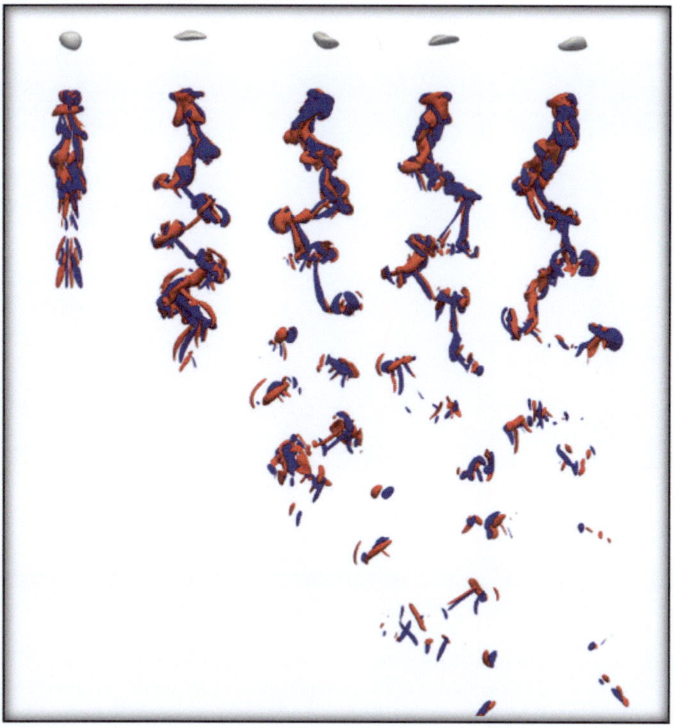

Fig. 3.28 Flow downstream of an upward-moving bubble. Temporal evolution: $\lambda = \mu_c/\mu_d = 54.3$; $\eta = \eta_c/\eta_d = 100.0$; $\mathrm{Mo} = 2.581 \times 10^{-11}$; $\mathrm{Re_d} = 1,329.0$. Transition to turbulence, personal collection, Pivello et al. [23]

isolated aspects, such as those presented here, have been noticed so far. They are parts of a big puzzle, which will be put together over time.

In Fig. 3.30, flow over a very long horizontal pipe is shown. The aspect ratio is defined as L/D, where L and D are, respectively, the length and diameter of the duct. For the present results, the ratio $L/D = 56$. The Reynolds Number, based on the diameter of the duct, is given by $\mathrm{Re_D} = 5 \times 10^5$ and the vorticity field is visualized, in Fig. 3.30a in a perspective view, and Fig. 3.30b shows a top view of the flow, with the cylinder at the bottom of the figure.

The physical times in Fig. 3.30a, b are, respectively, $t = 40\,[\mathrm{s}]$ and $t = 60\,[\mathrm{s}]$. This result is part of the solution of a complex fluid–structure interaction problem, studied by the author, in cooperation with Petrobras. Pipes laid out at the bottom of the Atlantic Ocean can be suspended in sections, between dunes, where the pipe is anchored. The pipe vibrates under the action of the loading induced by the turbulent wake, which, in turn, is affected by the vibration of the pipe. It should be noted that the pressure field, downstream of the cylinder, induces a force field on the cylinder. This is a highly nonlinear problem, which requires coupled solutions of the flow and the

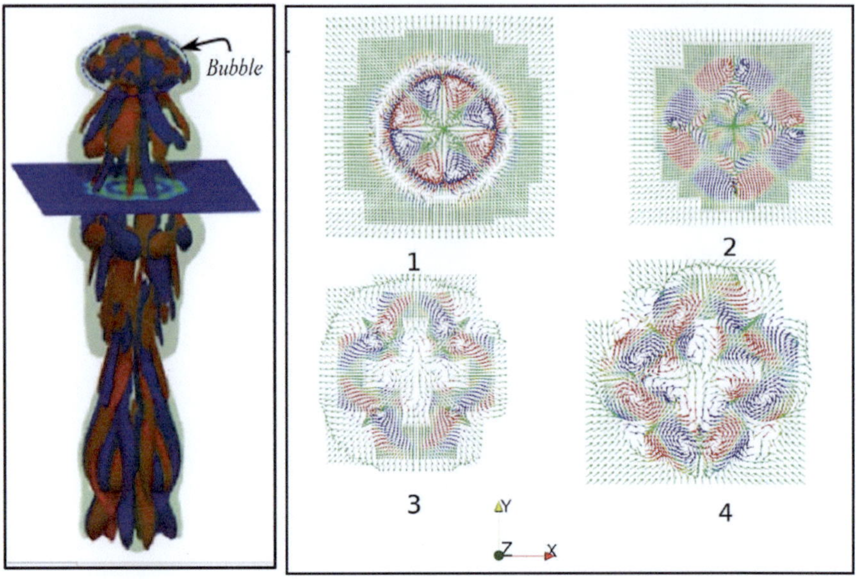

(a) (b)

Fig. 3.29 Flow downstream of a rising bubble: $\lambda = \mu_c/\mu_d = 54.3$; $\eta = \rho_c/\rho_d = 100$; $Mo = 2.581 \times 10^{-11}$; $Re = 1,329.0$. Presence of longitudinal rotating filaments. Personal collection [23]

solid structure (pipe). This simulation was carried out by the MFLab-FEMEC-UFU research group, using the MFSim computational fluid dynamics tunnel.

In Fig. 3.30a, the cylinder is visualized in gray color, as well as Kelvin–Helmholtz Structures downstream of the cylinder. These structures are transported and three-dimensionalized, giving rise to longitudinal filaments, as well as hairpin-like structures. The turbulent structures with the largest characteristic size, Kelvin–Helmholtz, are the carriers of the greatest amount of turbulent kinetic energy. They receive this energy from the mean flow and excite the cylinder to vibrate in a band of frequencies. A fluid–structure interaction analysis highlights the nonlinear behavior of this interaction. This is an industrial example that testifies to the enormous importance of understanding the physical process of the transition to turbulence.

3.3 Boundary Layer

When a fluid medium moves over a body, the fluid molecules in the vicinity of the body is affected by its presence, in a process of mutual interaction. Fluid dynamic forces between the body and the fluid are promoted by flow. This influence propagates inside the fluid through diffusive and advective transport processes simultaneously. This transfer of information between the fluid and the body results in dynamic and

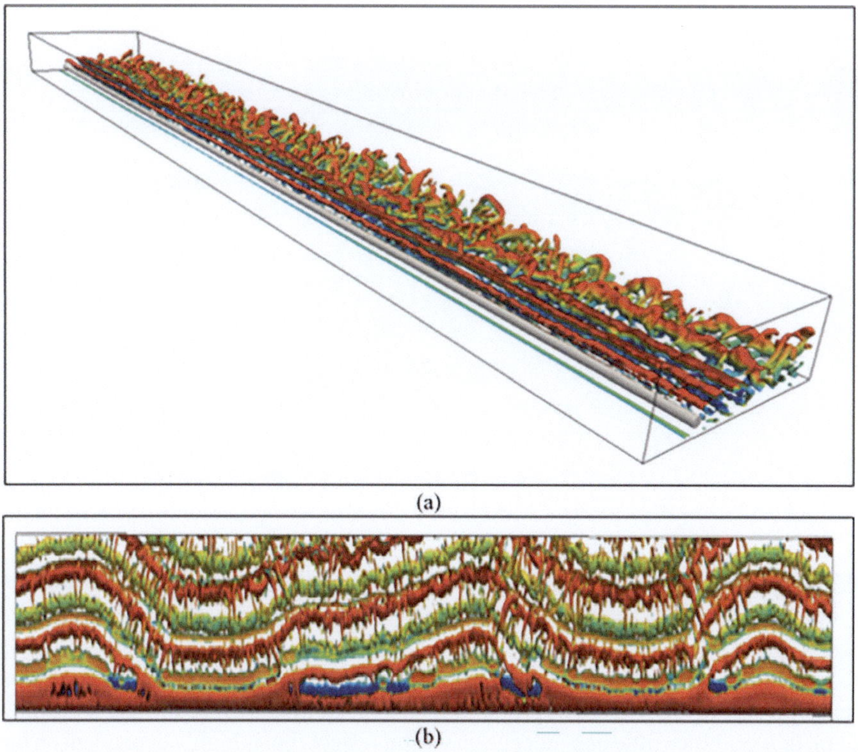

Fig. 3.30 Flow over a horizontal cylinder; the cylinder is at the bottom of the figure; aspect ratio $L/D = 56$; $\mathrm{Re_D} = 5 \times 10^5$; the vorticity field is visualized: **a** perspective view of the flow and **b** upper view. Personal collection, thanks to the researcher Hélio Ribeiro-Neto

thermal effects between the two material media. Fluid dynamic forces and thermal flux will arise because of this interaction. It is known that the flow regime, laminar or turbulent, dramatically affects the information transfer process between the two media. Turbulence increases the fluid dynamic force and thermal flux between the fluid and the solid. The boundary layer theory was initiated by the great German Professor and Researcher, Ludwig Prandtl, shortly after the year 1900. Details about this theory can be found in the book of one of his advisors, Schlichting [29], as well as in Balint et al. [3].

Viscous friction and the exchange of mass (in case of evaporation or condensation) and thermal energy between a body and a fluid that surrounds it increase significantly with the transition of the flow regime, from laminar to turbulent. Understanding and controlling the physical phenomena involved in the transition of a boundary layer arouses many practical interests, due to the desire to reduce drag effects on aircraft, ships, and submarines, among others. Many interests also arise in reducing pumping costs and maximizing the efficiency of thermal energy exchangers, as well

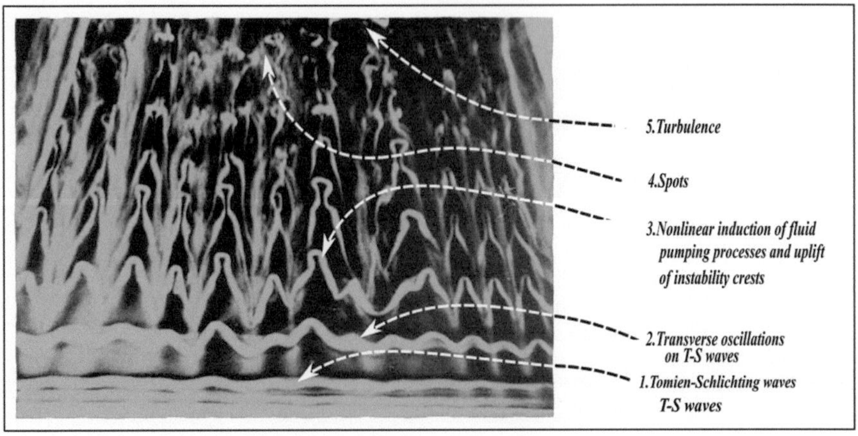

Fig. 3.31 Diagrams illustrating the phases of transition to turbulence of a boundary layer on a flat plate. Figure adapted from Perry and Li [22]

as obtaining benefits in the mixing of chemical components and in reactive thermodynamic processes. The topic of boundary layer regime transition will be discussed here.

In the previous sections, we commented on free shear flows, in which the transition occurs thanks to the process of promoting Kelvin–Helmholtz instabilities. This transition process holds little dependence on viscous effects but is highly dependent on the behavior of the average velocity fields, which must be inflectional. It is emphasized again that, in free shear flows, the fluid dynamic instabilities are of a shear nature. In the case of boundary layers, as there is no inflection in the mean velocity field, the origin of turbulence becomes strongly dependent on viscous effects. The stages of transition, recognized until the publication of this book, are illustrated schematically in Fig. 3.31, for the case of a flat plate. In this figure, the flow is occurring in an upward direction.

The flow preceding the transition is assumed to be laminar and there is no turbulence in the flow upstream of the plate. The phase (1), indicated in Fig. 3.31, corresponds to the formation of the first small-amplitude instabilities, close to the wall, which are called Tollmien–Schlichting waves or T-S waves. These instabilities are the first to appear in the flow and are therefore amplified at a maximum rate. The wavelength measured in the flow direction is called λ_{max}. Then, transverse oscillations appear over the primary instabilities, parallel to the plate, phase (2), with a harmonic wavelength, also selected from all wavelengths injected in the form of disturbances. This gives rise to the turbulent structures known as hairpin, horseshoe and lambda. These structures are counter-rotating filaments that rise into the boundary layer. According to Schlichting [29], White [41] and Lesieur [17], this process of lifting the counter-rotating filaments can be explained by the process of pumping fluid from the parietal region towards the central region of the flow, as illustrated in Fig. 3.32.

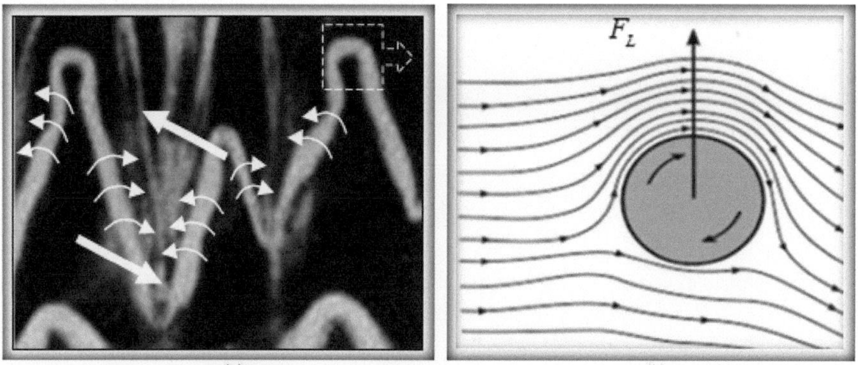

(a) (b)

Fig. 3.32 Transition of a boundary layer over a flat plate. Figure adapted from Perry and Li [22], with diagrams illustrating the transition phases

In the present text, this argument is complemented by considering the combined translational movements of the flow, relative to the filament, and the rotational movement of the filament itself. This effect of combined movements, relative translation and rotation, promotes a support force known as the Magnus effect. This can be viewed in Fig. 3.32a, b. On the crests, the upward pumping and Magnus effects combine to lift the filament into the flow. On the other hand, in the valleys, the downward pumping effect, promoted by counter-rotating filaments, keeps these filaments glued to the wall.

The turbulent structures that characterize this phase of the transition have at least three distinct topologies, which lead to the following nomenclature: "lambda"-type structures (Λ), "hairpin" structures and "horseshoe" structures. These structures are illustrated in the two situations presented in Fig. 3.33. Observing strictly, the three topologies can be identified in the transition phase of a boundary layer on a flat plate, as can be seen in Fig. 3.31.

In the indication (4) in Fig. 3.31, because of filament-type instabilities, turbulent spots appear, which represent strong concentrations of vorticity, which promote intense transport of matter from the wall to the interior of the boundary layer, classically visualized through smoke injection. In the last phase (5) of the transition, it is also illustrated in detail in Fig. 3.34, which represents a kind of reorganization of the flow in turbulent spots with strong concentrations of turbulent kinetic energy, which gives the transition process a strongly intermittent character. At this stage (5), the state of complete turbulence is represented. As illustrated in the case of wake-type flows, a boundary layer also develops over different types of geometries.

In Fig. 3.35a, the boundary layer transition over the surface of a sphere is shown. The boundary layer has its origin at the stagnation point, upstream of the sphere. As the observer moves over the surface of the sphere, Tollmien–Schlichting Waves

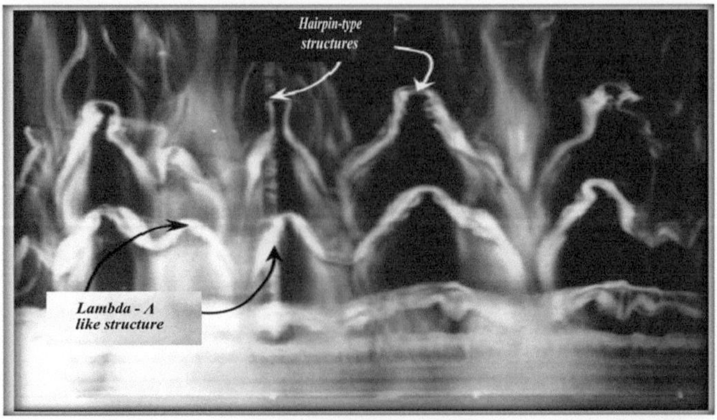

(a)

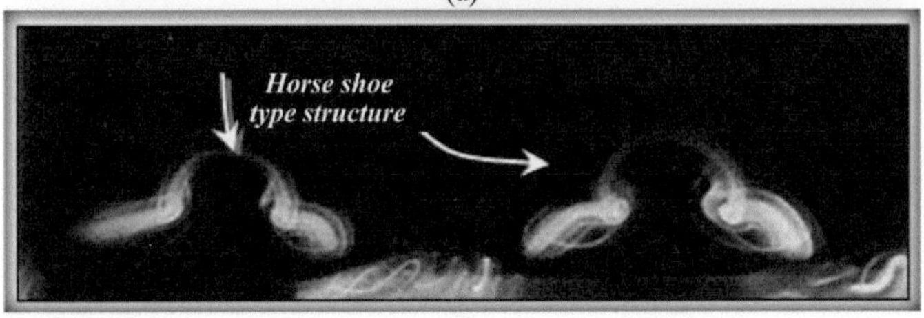

(b)

Fig. 3.33 Cross-section of horseshoe instabilities. Adapted from Taneda [37]

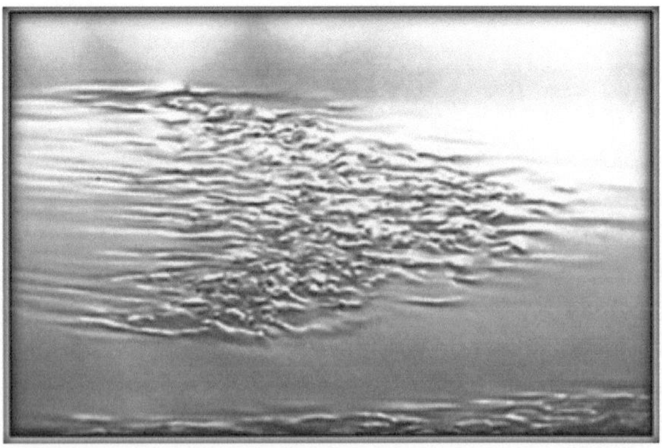

Fig. 3.34 Turbulent spot inside a transitional flat plate boundary layer [6]

and structures of other types appear: "hairpin", "lambda" and even the "horseshoe", typical of the process of transition in boundary layers. These details can be viewed in Fig. 3.35b.

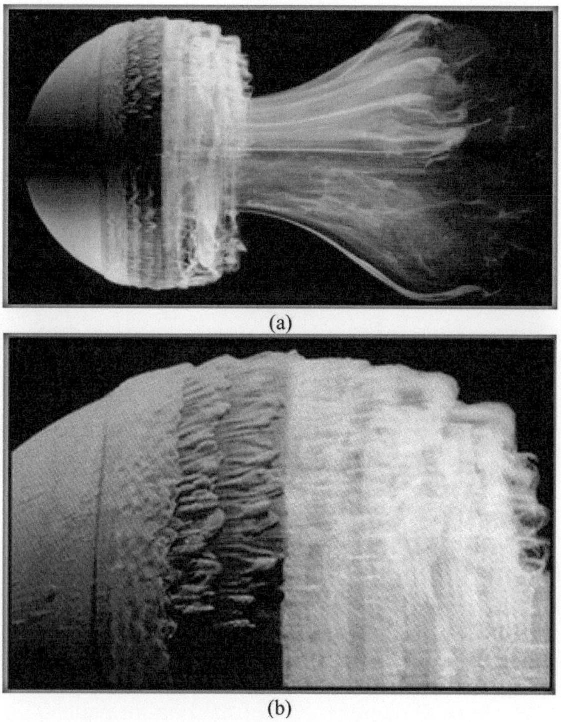

Fig. 3.35 Boundary layer over the first left half of a sphere (**a**); illustration of Tollmien–Schlichting waves and "hairpin" instabilities, lambda and horseshoe (**b**)

3.4 Rayleigh–Bénard Convection and Marangoni Convection

Still within the category of canonical flows, in addition to the flows already presented, other equally important types can be found that deserve comments and analysis. Below are examples of flows and their characteristic instabilities, which may eventually lead them to the transition process to turbulence. These flows have sources of movement linked to gravitational effects and interfacial tension effects.

Suppose a layer of fluid between two horizontal plates, separated by a distance, is subjected to a temperature difference ΔT. If the top plate is hotter than the bottom plate, as illustrated in Fig. 3.36a, a stable thermally stratified fluid medium is established. By stable stratification, we mean that the lighter layers of fluid will be on top of the heavier layers. The fluid particles will remain at rest once the weight, ρg, is balanced by buoyancy, $\rho g - \mathrm{d}p/\mathrm{d}z$. In this case, the gravity effects are given by the weight of fluid displaced by the particle, plus the upward vertical force due to the high-pressure gradient caused by the confinement of the layer between the two plates. The weight of the particle is given by ρg. In the case of a confined fluid layer, the vertical pressure distribution becomes nonlinear. This requires additional treatment. To calculate this pressure gradient, the temperature distribution is first established. As the thermal energy transport process will occur only by diffusion, the temperature distribution will be linear, given by Eq. (3.19):

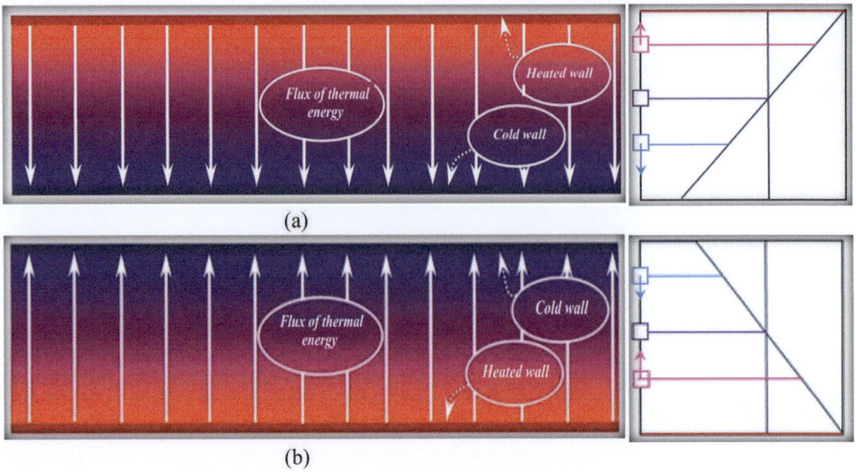

Fig. 3.36 Schematic representation of the Rayleigh–Bénard experiment. Two parallel plates with different temperatures; **a** estratificação estável: placa superior mais aquecida; and **b** estratificação instável: placa superior menos aquecida

$$T(z) = T_i + (T_s - T_i)(z/h), \tag{3.19}$$

where T_i is the temperature of the bottom plate, T_s is the temperature of the upper plate and z is the vertical coordinate, originating at the bottom plate. The pressure distribution is determined under the condition of equilibrium of the fluid particles:

$$\begin{aligned} &\frac{dp}{dz} = \rho_o \beta (T - T_o) g \Rightarrow \frac{dp}{dz} = \rho_o \beta [(T_i - T_o) + (T_s - T_i)(z/h)] g \\ &\Rightarrow p(z) = \rho_o \beta \left[(T_i - T_o) z + (T_s - T_i)\left(z^2/2h\right) \right] g \end{aligned} \tag{3.20}$$

where $T_o = (T_i + T_s)/2$. The integration constant for Eq. (3.20) was determined with the arbitrary condition $p(0) = 0$. It is observed that, as a result, $p(h) = 0$ and $p(h/2) < 0$. Therefore, $\frac{dp}{dz}(0 \le z < h/2) < 0$; $\frac{dp}{dz}(h/2 < z \le h) > 0$ and $\frac{dp}{dz}(h/2) = 0$. Thus, $p(h/2)$ is minimal. This ensures the balance of fluid particles throughout the range $0 \le z \le h$. It is observed that the pressure field is purely static in nature since there is no movement of the particles. The pressure distribution depends only on the temperature distribution.

If the bottom plate is the most heated, Fig. 3.36b, an unstable thermal stratification is established, that is, taking into account the definition of specific reference mass, the fluid contained in the lower half of the channel, $0 \le z < h/2$, tends to move in an upward vertical direction, while the second upper half, $h/2 < z \le h$, the fluid tends to descend. If the system were free from any disturbance, it would remain at rest, even though it is an unstable system. It turns out that, in any experiment, there will always be the presence of disturbances, whether mechanical, dynamic, thermal or geometric. Once disturbed, the system will select a wavelength to be most amplified, and the formation of fluid dynamic instabilities will begin. The fluid from the lower half of the domain will rise in specific regions, mediated by regions where the fluid from the second upper half of the domain will descend. Rayleigh–Bénard Cells will be formed. It is observed that the only way for the fluid to move is this, since complete layers of fluid cannot move without another layer replacing it. Advective currents are established, allowing the heated fluid present in the lower region to move to the cold upper region, while cold fluid from the upper wall will descend towards the lower wall, establishing the advective transport component of thermal energy. It is observed that, unlike the first configuration in which the pressure depends only on the temperature field, in this second configuration, the pressure starts to vary according to the coupling with the velocity field, once the particles start moving. Changing the velocity field changes the pressure field, which in turn accelerates the particles. This is a nonlinear process that, once initiated by disturbances, does not return to equilibrium.

Through linear stability theory, it can be demonstrated that the most amplified wavelength, in this type of problem, is equal to the height of the fluid layer, which is equal to the distance between the plates. Thus, when this system becomes unstable, advective cells will form, so that updrafts and downdrafts will occur in the fluid domain. In this way, the mass balance will be met with this fluid dynamics problem. The most likely configuration to occur will be in the form of hexagonal cells, with upward and downward movements. These are the well-known Rayleigh–Bénard cells, which appear in natural thermal convection flows. We appropriately speak here of "convection", since two processes of energy transport are actuating, that is, advection (transport of information through the movement of matter) and diffusion (transport of *information* through the existence of gradients of a potential, in a material medium).

These instabilities will only appear under critical conditions, involving: the operating mode (characterized by the temperature difference between the plates, ΔT, and by the gravitational field, of the place where the experiment is being carried out), the type of fluid (characterized by its physical properties) and the geometry of the system (characterized by the distance h between the plates). These operational, physical and geometric parameters, when grouped via dimensional analysis, lead to the definition of the Rayleigh Number, given by the following equation:

$$\mathrm{Ra} = \frac{\beta g \Delta T h^3}{\alpha \nu}, \tag{3.21}$$

where β, α and ν are, respectively: the volumetric expansion coefficient, the molecular diffusion coefficient of thermal energy and the molecular diffusion coefficient of linear momentum. This last property is better known as the fluid's kinematic viscosity coefficient. The characteristic length, h, is the distance between the two plates. Thus, if this dimensionless parameter is greater than a critical value, $\mathrm{Ra} > \mathrm{Ra_c}$, the process of amplification of disturbances and the formation of the so-called Rayleigh–Bénard Instabilities begins. Two transition modes can be identified: longitudinal rolls and hexagonal cells, as illustrated in Fig. 3.37a, b.

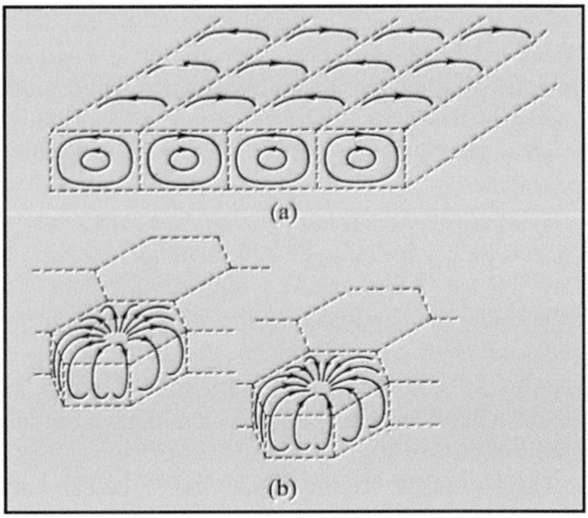

Fig. 3.37 Schematic representation of Rayleigh–Bénard Instabilities: **a** roll mode and **b** hexagonal cell mode

The critical value of the Rayleigh Number, to which the transition occurs, according to the theory of linear stability [10], is $Ra = 1,707.0$. Once these primary instabilities manifest, other wavenumbers will be amplified, and the process of transition to turbulence takes hold.

In Fig. 3.38, three flow configurations in a horizontal fluid layer, relating to three different flow regimes, are presented. It is observed that, as the Rayleigh Number increases, the flow becomes more unstable. In Fig. 3.38a, the Rayleigh–Bénard Cells are still well organized. In Fig. 3.38b, c, these cells have disappeared, giving way to a more disorganized stage, where transition regimes to turbulence, and possibly turbulence, can be found.

The movement present in this type of flow is due to the process of transforming potential energy into kinetic energy and vice versa, which feeds the advective currents. There are also viscous effects, which require the replacement of the kinetic energy of fluid particles with potential energy.

Problems like this can be encountered, in which the fluid movement is powered by another driving force. Imagine, for example, a thin layer of liquid placed on a flat horizontal plate, with an interface between this layer of liquid and the fluid that surrounds it, above it. This interface is also known as a free surface, which replaces the second plate in the case of the previous experiment. Interfacial tension effects will appear at this interface. By heating the lower surface of the fluid, as already discussed, vertical instabilities may arise.

Hotter regions then appear at the fluid–fluid interface, where ascending currents arrive, and colder regions, where descending currents begin. As a result, horizontal temperature gradients manifest at the interface. The interfacial tension coefficient varies with temperature, decreasing towards the hottest points. In this case, a force field appears, in line with the temperature gradients, which will promote horizontal movement of the fluid particles located at the interface. This is the so-called

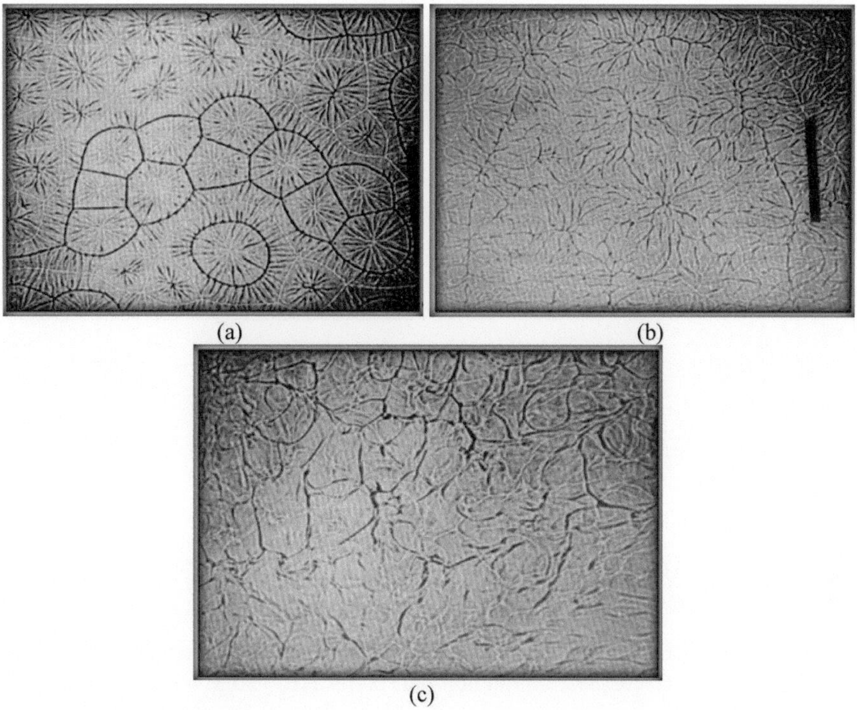

Fig. 3.38 Rayleigh–Bénard convection in a horizontal fluid layer, for Pr = 0.7: **a** Ra = 4.8×10^4; **b** Ra = 1.3×10^5; and **c** Ra = 1.7×10^5

Maragoni Effect, which, combined with the Rayleigh–Bénard Effect, fuels the movement that characterizes the so-called Maragoni Cells. The Marangoni Effect can be characterized by the Dimensionless Marangoni Number:

$$\mathrm{Ma} = -[(\mathrm{d}\sigma/\mathrm{d}T)(\rho h \Delta T)]/(\alpha \nu), \quad (3.22)$$

where h is a characteristic length, i.e., the thickness of the fluid layer; α is the thermal diffusion coefficient; ν is the diffusion coefficient of linear momentum and ΔT is the difference of temperatures of the bottom plate and the fluid–fluid interface.

In Fig. 3.39, the traces of Marangoni Cells formed in the soil of a lake (Salt Lake), after the drying process promoted by water evaporation, are illustrated. These cells, whose marks were left on the lake floor, are very regular and cannot characterize a turbulent regime, but it is certain that, being instabilities, they characterize the beginning of the transition process.

It is observed that, when there is a fluid–fluid interface, in natural convection problems, the two sources of maintenance of advection (interfacial energy and potential energy) coexist. The relationship between the Marangoni Effect and the

Fig. 3.39 Traces of hexagonal convective Marangoni cells generated at the bottom of a lake

Rayleigh–Bénard Effect can be quantified by the relationship between the dimensionless parameters, Ma (Marangoni number) and Ra (Rayleigh number), according to Eq. (3.23):

$$\frac{\mathrm{Ma}}{\mathrm{Ra}} = -\frac{(\mathrm{d}\sigma/\mathrm{d}T)}{\left(\rho g \beta h^2\right)}. \tag{3.23}$$

It is observed that the Marangoni effects will be more predominant over the Rayleigh–Bénard effects when: (a) the greater the variation in the interfacial tension coefficient in relation to the temperature variation; (b) the smaller the specific weight of the fluid particles; (c) the smaller the coefficient of thermal expansion, and (d) the smaller the thickness of the fluid layer. As an example, if silicone oil is placed on a horizontal surface and the thickness h of the liquid layer does not exceed 3 [mm], then, Marangoni effects will be predominant.

The typology of canonical flows studied in this section is marked by the presence of the turbulence phenomenon. This typology is very present in different industrial and technological applications, within the scope of engineering that involves fluid flow. For this reason, such flows have been the object of researchers' attention, as they are a field very suitable for the analysis of instabilities, with evolution towards turbulent flows.

3.5 Taylor–Couette Flows

Taylor–Couette Flows are another type of canonical flow. The instabilities that characterize the beginning of the transition process to turbulence, typical of this group of flows, are known as Taylor–Couette Instabilities. They are like Rayleigh–Bénard Instabilities and occur in flows inside cavities formed by two rotating concentric cylinders. This type of flow is known as Taylor–Couette Circular Flow since the only source of movement of the fluid is the movement of its boundaries. In this case, the centrifugal forces promoted by rotation are linked to the formation and persistence of Taylor–Couette Instabilities, to are described and discussed in this section. The physical model for this class of flows is presented in Fig. 3.40.

If the rotation of the external cylinder is greater than the rotation of the internal cylinder, the flow induced by the movements of the cylinders, inside the cylindrical cavity, will be unconditionally stable. This is because the distribution of centrifugal force will increase with the positive direction of the coordinate axis r. On the other hand, if the rotation of the internal cylinder is greater than the rotation of the external cylinder, the flow will be unstable and, if disturbed, fluid dynamic instabilities may develop. In a similar way to what was described for Rayleigh–Bénard flows, the disturbances injected into the flow can be geometric, dynamic and/or mechanical in nature.

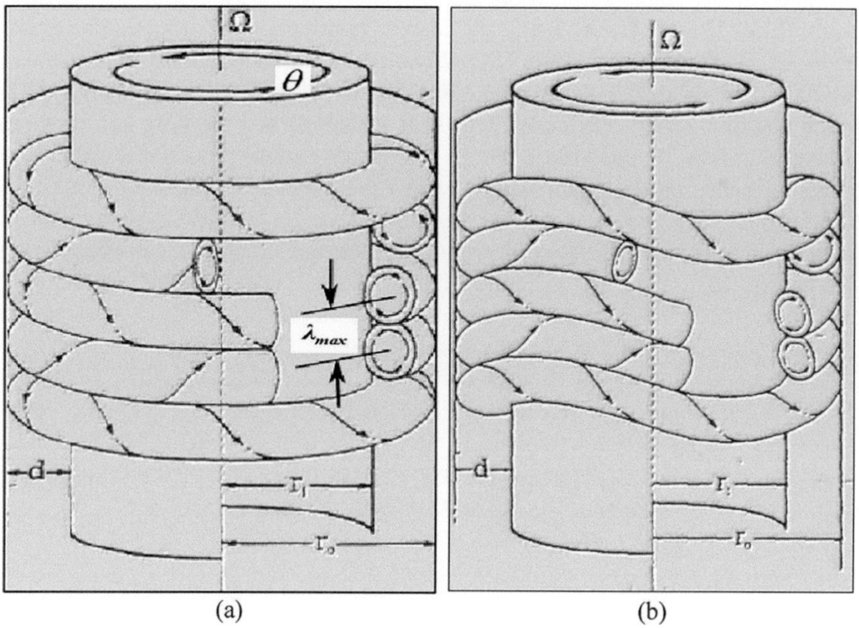

Fig. 3.40 Illustrative diagram of the Couette Circular Flow and the Taylor–Couette turbulent structures: **a** inner cylinder rotating angular velocity Ω_1; and **b** inner cylinder rotating at an angular velocity $\Omega_2 > \Omega_1$

It can be demonstrated, using linear stability theory, that the most amplified wavelength for this type of flow is equal to the thickness of the annular channel, defined as $d = r_0 - r_\mathrm{i}$. Fluid dynamic instabilities transform into primary dynamic structures with toroidal geometries, which can be seen in an axial section of the system, as illustrated in Fig. 3.40a, b. In these figures, these toroids are shown for two different operating regimes, that is, for two different values of rotation, with the rotation of the external cylinder being zero and the rotation of the internal cylinder being different from zero. Two values of the rotation of the internal cylinder were used. The rotation related to Fig. 3.40a is smaller than the rotation related to Fig. 3.40b. In Fig. 3.40a, the Taylor–Couette Contra-rotating Toroidal Structures are presented. These structures remain axisymmetric. The most amplified wavelength is given by $\lambda_{\max} = d$, observed and measured in the axial direction. When the rotation is increased, a new instability appears, of wavelength $\lambda > \lambda_{\max}$, measured in the azimuthal direction θ. In experiments, whether material or computational, as the rotation of the internal cylinder increases, new instabilities are formed by the nonlinear process of amplification of disturbances. As these new fluid dynamic instabilities evolve, new and more complex turbulent structures emerge, composed of different wave numbers and different frequencies. This process of multiplication of wave numbers characterizes the transition from the laminar regime to the turbulent regime.

In Fig. 3.41, visualization is presented, via material experimentation, of the process of formation of instabilities, as well as the process of transition to turbulence, for three different inner cylinder angular velocities. This angular velocity increases from Fig. 3.41a to c. At Fig. 3.41a, the counter-rotating toroids are visualized, with the flow still laminar, but they are three-dimensional and transient. As the rotation of the inner cylinder increases, harmonic instabilities arise on the toroids, Fig. 3.41b, which amplify, giving rise to new turbulent structures and the flow transitions to turbulence, as illustrated in Fig. 3.41c. It is observed that the transition development process depends on the relationship between the cylinder rotations, $\beta = \Omega_\mathrm{i}/\Omega_\mathrm{o}$, being Ω_i and Ω_o the angular velocities of the inner cylinder and the outer cylinder, respectively. Consider β_cr the critical ratio of rotations, for which the transition to turbulence occurs. If $0 \le \beta \le \beta_\mathrm{cr}$, the system will always be stable. If $\beta_\mathrm{cr} \le \beta < \infty$, then the system will be unstable. It should also be noted that it also depends on the type of disturbances injected into the system.

In a generic way, it is possible to obtain continuous solutions, using applicable simplifying assumptions for the differential model composed of the continuity equation and the Navier–Stokes equations, in laminar and turbulent regimes for canonical flows. In particular, the continuous solution for laminar flows between two concentric rotating cylinders, that is, Taylor–Couette laminar flows, is presented:

$$\vartheta_\theta(r) = Ar + B/r, \tag{3.24}$$

where $A = \left(\Omega_\mathrm{o} R_\mathrm{o}^2 - \Omega_\mathrm{i} R_\mathrm{i}^2\right)/\left(R_\mathrm{o}^2 - R_\mathrm{i}^2\right)$ and $B = (\Omega_\mathrm{o} - \Omega_\mathrm{i}) R_\mathrm{i}^2 R_\mathrm{o}^2/\left(R_\mathrm{o}^2 - R_\mathrm{i}^2\right)$. Equation (3.24) is a function of the radius only. This solution is composed of a first portion relating to solid body rotation and a second portion relating to potential flow.

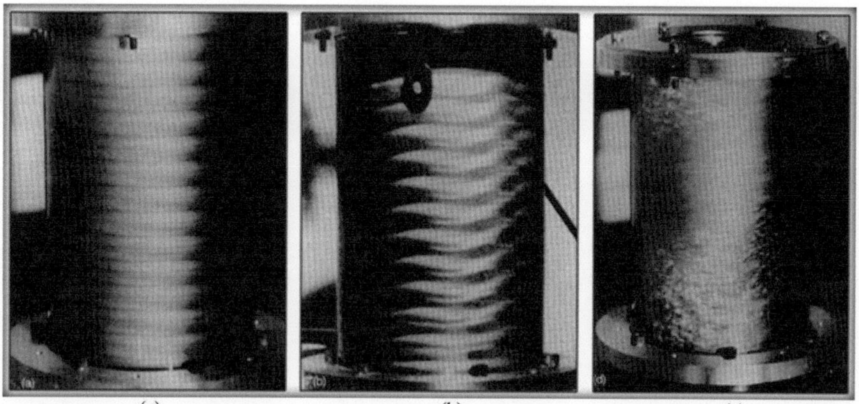

(a) (b) (c)

Fig. 3.41 Taylor–Couette Instabilities (**a**); harmonic oscillations (**b**) and transition to turbulence (**c**). Figure taken from Coles [7]

Fluid dynamic instabilities, promoted by small disturbances in this type of flow, were studied by Taylor [38], through theoretical theory and through material experiments. The notable British researcher determined, for small spacings between cylinders, $d = r_o - r_i$, compared to r_i, that the problem is simplified, and the flow depends on the Taylor Number, given by the following equation:

$$\mathrm{Ta} = r_i L^3 \left(\Omega_i^2 - \Omega_o^2\right)/\nu^2. \tag{3.25}$$

When this parameter increases above the critical value, $\mathrm{Ta}_c = 1,708.0$, Taylor–Couette structures arise, which are axisymmetric toroidal counter-rotating cells with a wavelength equal to d. These cells are called "Taylor Vortices". In the present text, they are called Taylor–Couette structures. The greatest interest in this section centers on the case in which the outer cylinder remains without rotation and the inner cylinder rotates at an angular velocity $\Omega_i \neq 0$, aiming to highlight the Turbulent structures of Taylor–Couette.

Simulations were conducted in MFLab, considering two geometric configurations: in the first, the geometric parameters of the study were considered via Material Experimentation. Andereck et al. [2] used radius ratio $\eta = R_o/R_i = 1.133$ and aspect ratio $\Gamma = L/d = 30$, where L is the length of the annular canal and $d = r_o - r_i$.

In Fig. 3.42, the comparison of the tangential velocity distribution between the computational results and the continuous solution is presented, Eq. (3.24), for two values of the Reynolds Number: $\mathrm{Re}_d = 50$ and $\mathrm{Re}_d = 100$. The tangential velocity profile, determined by computer simulation, coincides exactly with the profile via the continuous solution. These are cases with Reynolds Number values below the critical value, $\mathrm{Re}_{dc} = 114$, value equivalent to $\mathrm{Ta}_c = 1708.0$, for which the flow regime will be laminar. The other velocity components are practically zero for stable laminar flows.

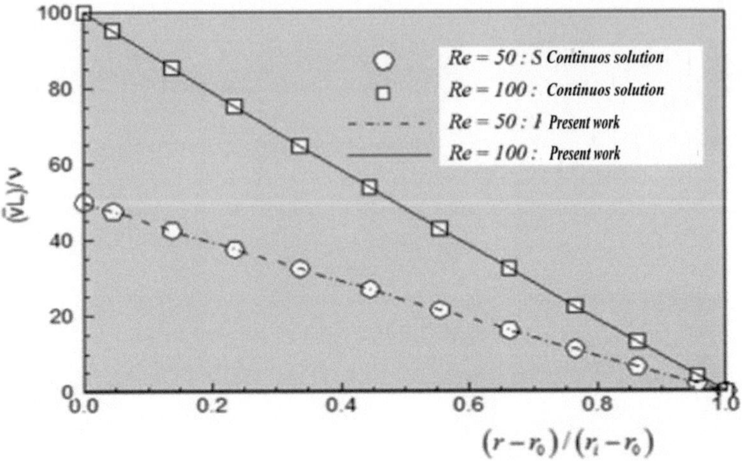

Fig. 3.42 Tangential velocity distribution for Taylor–Couette annular flow; $Re_d = 50$ and $Re_d = 100$. Personal collection [21]

The value of the Reynolds Number was increased to $Re_d = 118$, slightly above $Re_{dc} = 114$, and the formation of counter-rotating toroidal structures, axially arranged, was obtained, in a total of five cells. It can be seen in the three situations presented in Fig. 3.43. This is the formation of Taylor–Couette Structures, designed with the fields of non-dimensionalized velocity components, $\vartheta_r d/\nu$, $\vartheta_\theta d/\nu$ and $\vartheta_z d/\nu$, plotted on a plane (r, z) arbitrary. With the radial component of velocity, Fig. 3.43a, alternating dynamic structures (positive, negative) are designed and axially ordered in pairs. With the tangential component of the velocity, Fig. 3.43b, waves are perceived with peaks coinciding with the positive parts of the radial velocity, and with valleys coinciding with the negative parts. At axial velocity, Fig. 3.43c, alternate cells are shown, ordered in an axial direction. The velocity vector field is shown in Fig. 3.44, showing the presence of square cells counter-rotating, which shows excellent agreement with the experiments of Schultz-Grunow [30], Hein [14], Coles [7] and Wereley and Lueptow [40]. The most amplified axial wavelength is $\lambda_{max} = d$, as predicted by the theory of linear stability.

A seguir, apresentam-se as *estruturas do escoamento*, nas três situações analisadas na Fig. 3.45. The components of radial velocity, $\vartheta_r d/\nu$, Fig. 3.45a; axial, $\vartheta_z d/\nu$, Fig. 3.45b and the axial component of the vorticity, $\omega_z d^2/\nu$, Fig. 3.45c, were used for visualization. The corresponding values of the radial component of vorticity: $\omega_z d^2/\nu = 2$ (gray color) and $\omega_z d^2/\nu = -2$ (green color) were plotted. Thus, considering that the values displayed are equal in modulus but of opposite signs, it can be observed, through the cut, that the Taylor–Couette Structures are axisymmetric and counter-rotating.

In Fig. 3.46, the simulation results are presented for $Re_d = 2, 190.0$. It is observed that the transition to turbulence occurs. Three-dimensional toroidal turbulent structures can be seen, characterizing a state of high level of instabilities in fluid dynamic

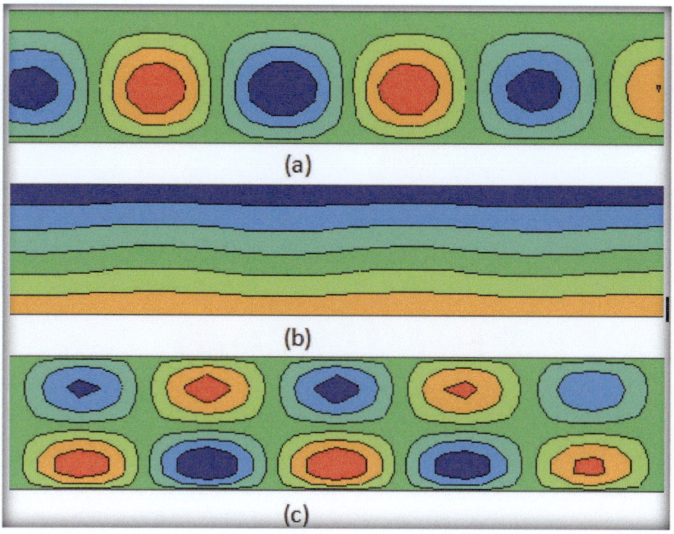

Fig. 3.43 Velocity fields for $\mathrm{Re_d} = 118$, $\eta = 1.133$ and $\Gamma = 10$: **a** radial, $\vartheta_r d/\nu$; **b** tangential, $\vartheta_\theta d/\nu$; and **c** axial, $\vartheta_z d/\nu$. Personal collection [21]

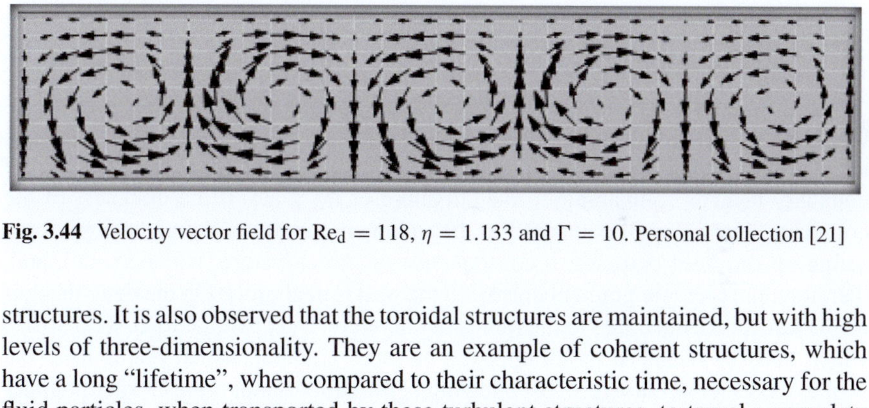

Fig. 3.44 Velocity vector field for $\mathrm{Re_d} = 118$, $\eta = 1.133$ and $\Gamma = 10$. Personal collection [21]

structures. It is also observed that the toroidal structures are maintained, but with high levels of three-dimensionality. They are an example of coherent structures, which have a long "lifetime", when compared to their characteristic time, necessary for the fluid particles, when transported by these turbulent structures, to travel a complete turn.

The temporal distribution of the radial component of velocity, $(\vartheta_r d/\nu)(\vec{r}, t)$, sampled at a point $\vec{r}$, in the central plane of the annular channel, is presented in Fig. 3.47a. A wide band of frequencies is perceived in the signal, which can be visualized quantitatively in Fig. 3.47b. The process of transforming the temporal signal into power spectral density was exposed in Sect. 3.2.2.1. This is a transition regime to turbulence, where a wide band of frequencies is already established, but it is not yet a band of frequencies wide and continuous enough to characterize the turbulence regime. This is because, as can be seen, the spectral density $E_{\vartheta'\vartheta'_r}(\vec{r}, f)$ is composed of discrete frequencies, while in a turbulence regime, there would be a broad and continuous spectrum.

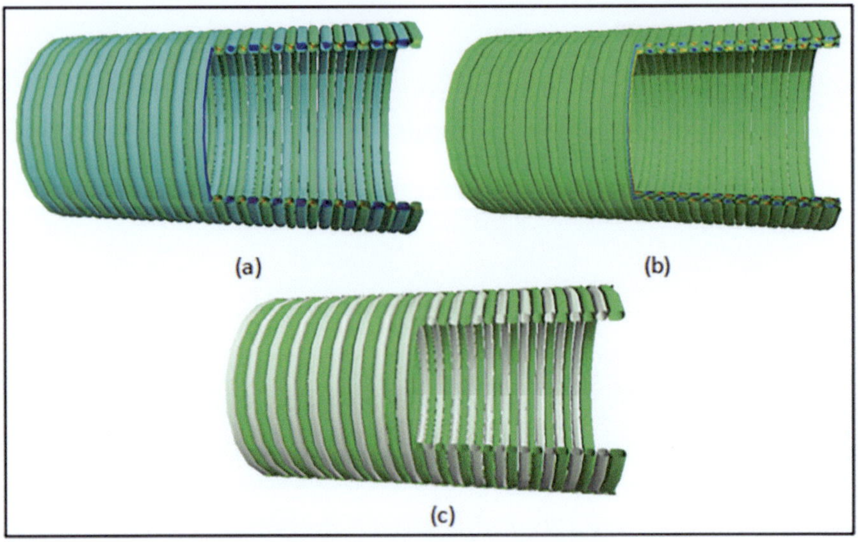

Fig. 3.45 Perspective view of the isosurfaces of: **a** radial velocity, $\vartheta_r d/\nu$; **b** axial velocity, $\vartheta_z d/\nu$; and **c** vorticity $\left(\omega_z d^2/\nu\right) = 2$ (gray color) and $\left(\omega_z d^2/\nu\right) = -2$ (green color); for $\mathrm{Re_d} = 118$, $\eta = 1.133$ and $\Gamma = 30$. Personal collection [21]

3.6 Görtler Structures

Görtler Structures appear in secondary flows that occur within boundary layers, under suitable conditions, that is, when the wall is concave and the thickness of the boundary layer is comparable to the curvature of the plate. If the thickness of the boundary layer is small compared to the radius of curvature, the centrifugal force acting on the fluid particles, a consequence of the curvature, will be small, and, therefore, the pressure gradient normal to the wall is negligible. On the other hand, if its thickness is comparable to the radius of curvature, a centrifugal force field arises, which must be balanced by an opposite pressure gradient, which points from the wall to its center of curvature. By injecting perturbations and, if the Görtler Number (G), defined in Eq. (3.26), exceeds 0.3, then centrifugal instabilities may appear, as illustrated in Fig. 3.48. Therefore, Görtler Structures appear within the boundary layer. This dimensionless parameter is defined by:

$$G = \frac{U_\infty \theta}{\nu}\left(\frac{\theta}{R}\right)^{1/2}, \tag{3.26}$$

where U_∞ is the velocity outside the boundary layer, θ is the linear momentum thickness is the kinematic viscosity and R is the radius of curvature of the wall. This dimensionless number represents the relationship between centrifugal effects and viscous effects. Part of these variables and Görtler's structures are illustrated in

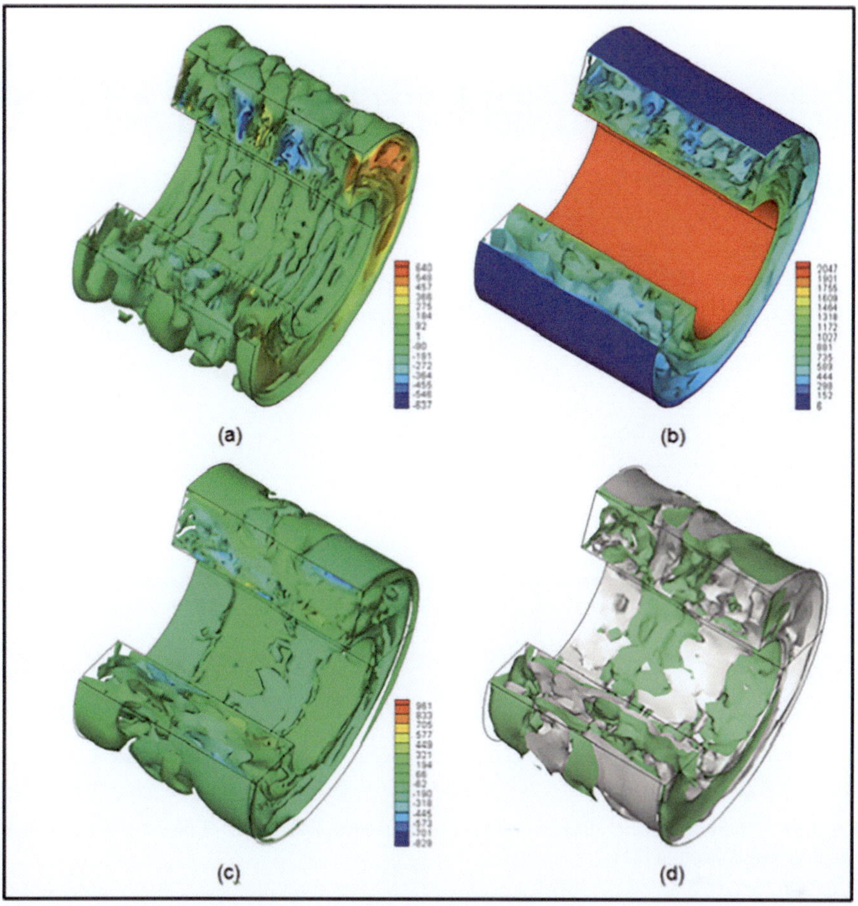

Fig. 3.46 Instantaneous isosurfaces: **a** radial component of velocity, $\vartheta_r d/\nu$; **b** axial component of velocity, $\vartheta_z d/\nu$; **c** azimuth component of velocity, $\vartheta_\theta d/\nu$; and **d** vorticity $\omega_z d^2/\nu = 2$ (grey color) and $\omega_z d^2/\nu = -2$ (green color). Personal collection [21]

Fig. 3.48. It is observed that, like the Taylor–Couette structures, the Görtler structures are also counter-rotating. For more details on this topic, the references Görtler [13] and Saric [28] can be consulted.

3.7 Complex Flows

Complex flows are not canonical and are the most frequent in nature and in industrial and technological applications. They can be understood as being a composition of the canonical flows presented in the previous sections. From the point of view of

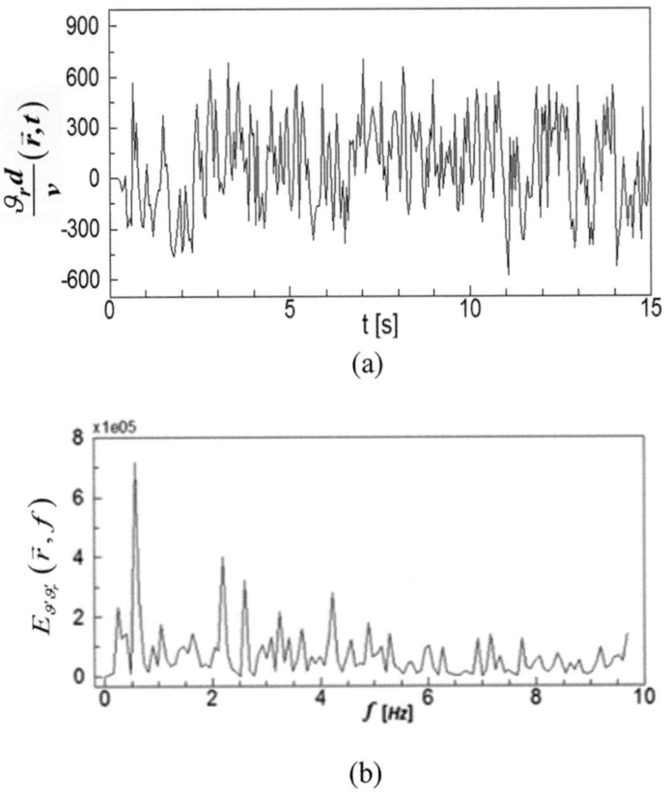

Fig. 3.47 Temporal velocity distribution (**a**) and frequency distribution of power spectral density (**b**). Personal collection [21]

the transition to turbulence, much of what was presented about canonical flows, regarding the physical nature of turbulent structures, can be found within the family of complex flows.

The case of a sudden expansion in a rectangular channel is presented here as an example of this class of flows. The geometry is very simple and is Cartesian, which is illustrated in Fig. 3.49. Despite the simplicity of the geometry, the related flows are characterized by high complexity. Another characteristic that makes this geometry a widely used test case, both by material experimentalists and computational experimentalists, is the fact that the detachment of the lower boundary layer occurs at a fixed point, independent of the flow conditions upstream of the expansion.

As illustrated in Fig. 3.49, in region (1), there are two boundary layers, one upper and one lower. Among them, the flow can be turbulent, laminar and even potential, depending on the characteristics of the experimental device. On the step, at point (2), the lower boundary layer detaches, promoting a shear zone (3), with the formation of Kelvin–Helmholtz Structures, which are subjected to the confinement effect of the channel and are transported towards the region of boundary layer adhesion, in

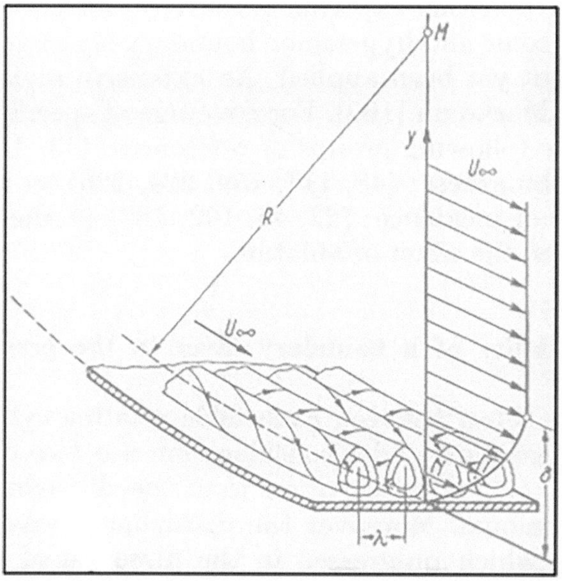

Fig. 3.48 Görtler instabilities formed within a boundary layer on a concave curved plate

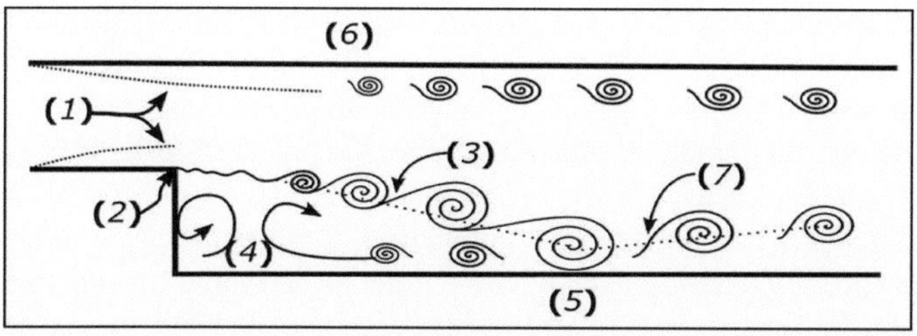

Fig. 3.49 Qualitative geometric and physical characteristics of flow in transition over a sudden expansion (personal collection)

the region (5). This region is of great physical complexity due to the collision of the turbulent structures with the lower wall of the channel. There is also a region of recirculating flow (4), interacting with the shear layer (3). After the reattachment region, there is the redevelopment region of the boundary layer (7), with which the turbulent structures that are transported to the channel exit interact. In short, this is simple geometry, as it is Cartesian. However, flows are very complex and difficult to predict by computer simulation. Among the statistical quantities that we seek to obtain in the analysis of a flow, we can highlight the average profiles of the velocity components, the pressure coefficient, and the components of the Boussinesq-Reynolds Tensor, as well as the turbulent kinetic energy. Another very important *information* in this problem is the boundary layer reattachment length, given by the distance between the step and point (5).

Illustrative results of computer simulations, carried out by Silveira-Neto et al. [33], are presented in this section. This work is considered the first computational simulation applied to a complex flow in transition, using the LES Methodology (Large Eddy Simulation). The transition to turbulence of a complex flow was studied computationally in this work.

In Fig. 3.50, the result of a two-dimensional simulation of the flow inside a channel with a sudden expansion is shown. The flow is oriented from left to right. The Reynolds Number based on step height is $Re_H = 38{,}000.0$ and the aspect ratio is $R = W/H = 2.5$, where W is the height of the channel downstream of the step and H is the height of the step. The transverse component of vorticity, ω_z, was used to visualize the flow. The red color indicates negative vorticity (clockwise) and the blue color indicates positive vorticity (counterclockwise). The formation of the lower shear layer can be seen after the detachment point of the boundary layer, illustrated by point (2) in Fig. 3.49. Kelvin–Helmholtz structures are formed and transported toward the bottom wall, as illustrated by the number (3) of Fig. 3.49. After colliding in the region of boundary layer reattachment, point (5) of Fig. 3.49, the structures are transported in the direction of the channel exit. The formation of Kelvin–Helmholtz structures occurs due to the recirculation that forms downstream of the step, creating an inflectional velocity field.

At the top of the channel, Fig. 3.50, the boundary layer has detached, giving rise to inflection in the velocity field, creating, consequently, Kelvin–Helmholtz structures. They are transported sequentially in the direction of the channel exit. It is curious to notice the detachment of a boundary layer on a flat plate! This occurs because the flow is confined between two plates, in a region where the pressure field is variable, with adverse pressure gradients below the upper flat plate. More details can be found at Silveira-Neto et al. [32].

Despite being a two-dimensional simulation, important details were captured, such as the vorticity bursting phenomenon in the lower wall of the channel, as indicated by the white arrows in Fig. 3.50. This type of physical effect was also captured in material experiments by Pronchick and Kline [25] using a wind tunnel, in which visualizations were made with smoke injection.

In Fig. 3.51, the results of a three-dimensional simulation of the flow in the channel, with a sudden expansion, are presented. In Fig. 3.51a, through an inclined view of the domain, after the sudden expansion, iso-surfaces of vorticity and in Fig. 3.51b, the iso-surfaces of the pressure field are visualized, in this case, a top view of the domain. The transition process can be seen in both figures, with the presence of Primary Kelvin–Helmholtz structures, which experience strong three-dimensional effects, as they are transported to the right, toward the exit of the channel. The presence of longitudinal contra-rotating structures, induced by the primary structures already described, can be clearly seen. The intense mechanical action of these longitudinal filamentary structures on the primary structures can be seen in Fig. 3.51b. Their importance is recognized in the process of transition to turbulence and in the establishment of turbulence itself, in later stages not shown in these figures. This configuration refers to a relatively small step, and therefore, the bottom wall exerts a

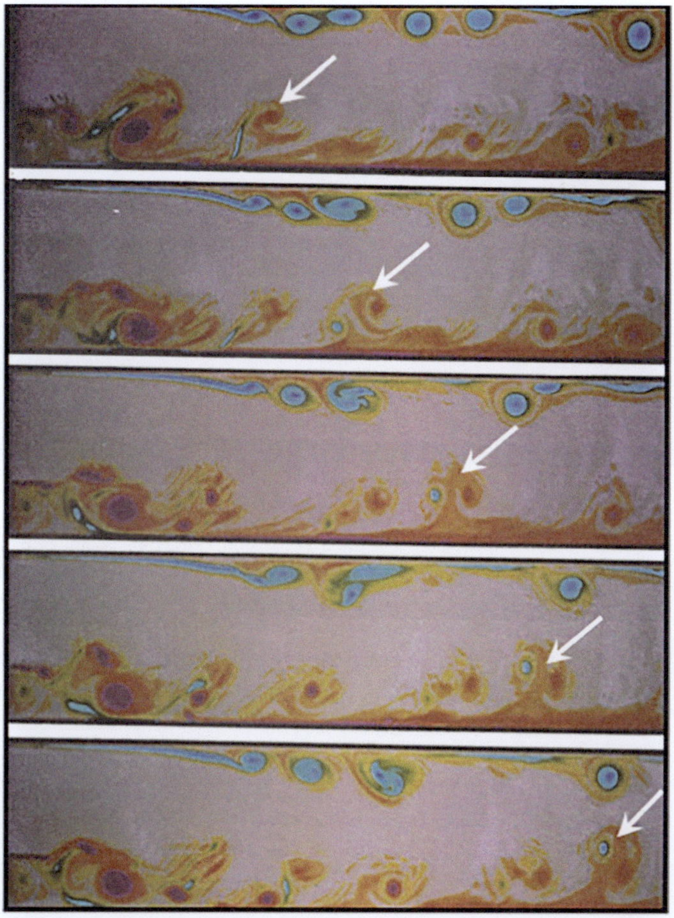

Fig. 3.50 Two-dimensional flow downstream of a sudden expansion. Time sequence of transverse vorticity fields (ω_z). Reynolds Number based on step height, $\mathrm{Re_H} = 38{,}000.0$; aspect ratio $R = W/H = 2.5$. Personal collection [32]

strong influence on the transition process. More details about this study can be found in Silveira-Neto et al. [31].

Seeking to better highlight the transition to turbulence downstream of the step, another simulation was carried out in a geometry with a width greater than the previous case, presented in Fig. 3.51. In this way, there is the opportunity to highlight a greater number of contra-rotating longitudinal structures. The Kelvin–Helmholtz structures were drawn with transverse vorticity, $\omega_z = 3U_0/H$, in lilac color, and the longitudinal filaments were drawn with longitudinal vorticity, in green for $\omega_x = +2.5U_o/H$ and in yellow for $\omega_x = -2.5U_o/H$. These last two colors were not chosen by chance! Clearly, the longitudinal filaments wrap around the Primary Kelvin–Helmholtz structures, as can be seen in Fig. 3.52.

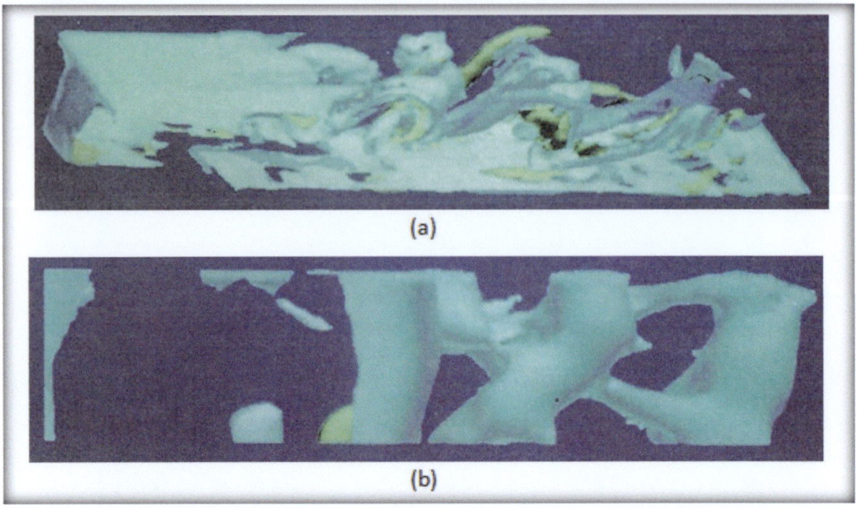

Fig. 3.51 Three-dimensional flow over a sudden expansion; Reynolds Number based on step height, $\mathrm{Re_H} = 38{,}000.0$; aspect ratio $R = W/H = 2.5$: **a** longitudinal vorticity isovalues, $\omega_x = \pm 2.5U_o/H$ and **b** isobaric $p = -0.3\rho U_o^2$. Personal collection [31]

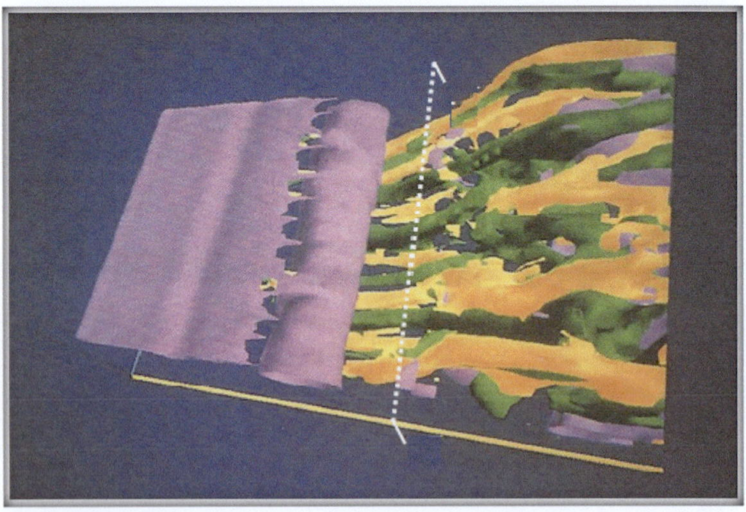

Fig. 3.52 Three-dimensional flow over a sudden expansion; Reynolds Number based on step height, $\mathrm{Re_H} = 6,000.0$; aspect ratio $R = W/H = 1.25$; transverse vorticity isovalue $\omega_z = 3U_0/H$; longitudinal vorticity isovalues, $\omega_x = \pm 2.5U_o/H$. Personal collection [33]

Another fact that became evident was the reconnection observed between filaments of the same sign, on the second Kelvin–Helmholtz structure. Two strands of opposite signs cannot reconnect. Therefore, they jump over each other to find a partner with the same sign so that reconnection can occur between them. This physical effect can be seen in Fig. 3.52, on the Second Kelvin–Helmholtz structure, to the right of this figure.

To better visualize the counter-rotating longitudinal filaments, a cross-section was made across the flow, as indicated in the plan drawn with a white line in the center of Fig. 3.52. The velocity vector field was plotted, which is presented in Fig. 3.53. The mushroom-shaped structures can be clearly observed, drawn with the velocity vectors and reinforced with the dashed line. The colors, ranging from red to blue, represent the velocity sign, with red being positive (ascending) and blue being negative (descending). The arrows also point to the directions of movement of the fluid particles.

With the results of this computational experiment, it was possible to establish a scenario of the transition process from the laminar regime to the turbulent regime in complex flows. This scenario begins with the formation of Kelvin–Helmholtz structures, which are transported sequentially from the entrance to the exit of the domain. The next event identified was the induction of counter-rotating longitudinal filaments by the action of the fluid that intermediates the Kelvin–Helmholtz structures. The possibility of pairing between longitudinal filaments of the same sign was also identified. Between this scenario described and the full turbulence, there is a list of events not yet known and which are the subject of scientific investigation.

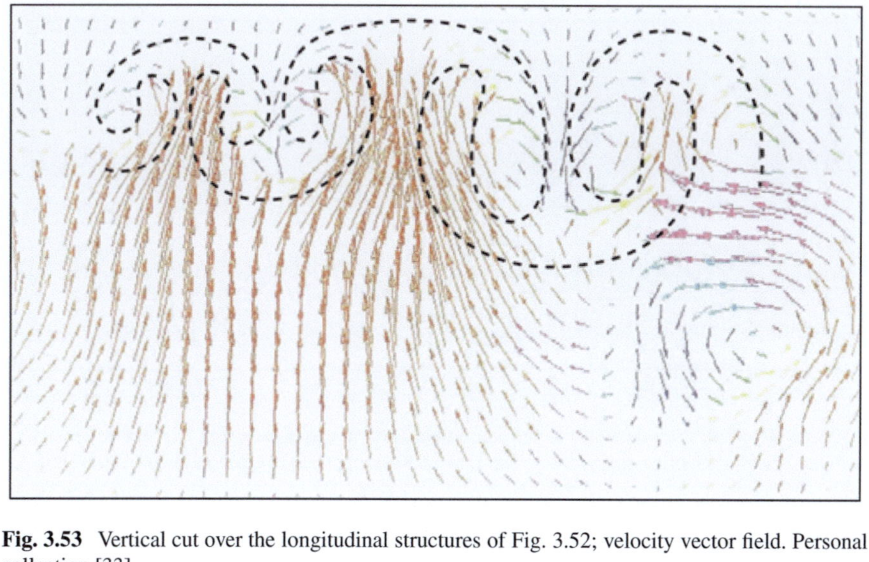

Fig. 3.53 Vertical cut over the longitudinal structures of Fig. 3.52; velocity vector field. Personal collection [33]

3.8 Preliminaries on Homogeneous and Isotropic Turbulence

There are many theoretical investigations developed about homogeneous and isotropic turbulence. In this type of turbulent flow, the statistical properties are invariant due to observer translation (homogeneity) and invariant due to rotation (isotropy). An experimental approach to this type of turbulence is the case of the flow of a fluid that passes through a thin screen, when observed downstream of it. Thus, theoretical findings can be proven experimentally. Theories arising from this type of flow lead to several predictions, not only regarding power spectral densities, but also to diffusion constants, with which the transport rates of linear momentum, thermal energy and passive and active scalars, within non-isotropic turbulent flows. It is within this context that the Theory of Kolmogorov [15], which will be addressed in Chap. 4.

In Fig. 3.54, the structures that result from a direct computer simulation (DNS) of a homogeneous and isotropic turbulent flow, inside a periodic cubic box, are presented. This flow is the academic exercise that comes closest to the isotropy hypothesis, due to it being periodic in all directions. It can be seen visually, Fig. 3.54a, that the turbulent structures, represented with the vorticity field, are similar, but not equal, throughout the calculation domain. The spectral density of turbulent specific kinetic energy in wavenumber, corresponding to this state of turbulence, is presented in Fig. 3.54b. There is a slope of – 5/3, which meets the Law of Kolmogorov [15, 16]. The theory for this class of flow will be presented in Chap. 4.

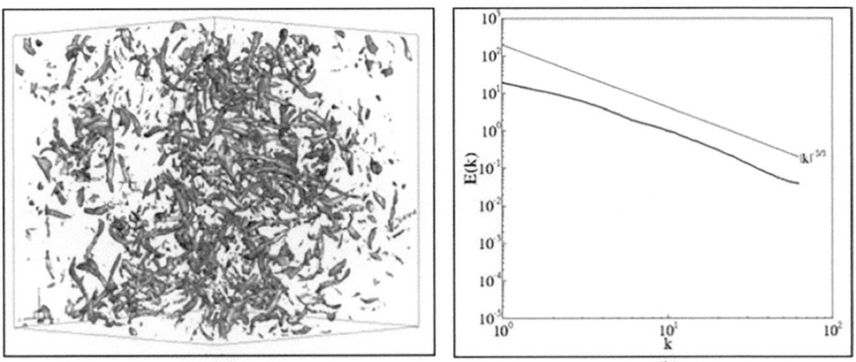

(a) (b)

Fig. 3.54 Isotropic turbulence in a periodic cubic box: **a** vorticidade and **b** spectral density of turbulent specific kinetic energy, $E(k)$. The Law of Kolmogorov [15]; personal collection [20]

3.9 Turbulence and Chaos

To every effect there follows a cause! Just one wrong line in the computer programs used in stock exchanges and we will see a wave of global financial instability. This is what those interested in one of the most important disciplines today say: chaos theory. For too long, science has functioned under the shelter of the general, comfortable satisfaction of predictable phenomena and linear equations. Knowing the flow rate of the tap and the volume of the bathtub, the time needed to fill it can quickly be determined. Knowing the weight of the satellite and the height of the desired orbit, the propulsion power required by the launcher to perform the task of placing it into orbit is calculated. Knowing the number of hungry people spread across the vast Brazil, the amount of food needed to satisfy them for a determined period is calculated. But what can we say about the dynamic behavior of world stock markets in the days following the so-called liberation of Kuwait in 1991? What can we say about what will happen, in the coming decades, to the population of specific Brazilian regions, due to the rural exodus that has occurred in recent decades? The answer is: almost nothing!

Scientists can say little about complex natural phenomena, which is why delving into them is necessary to understand them better. The discipline that deals with chaos is nothing new. Still in 1889, Henri Poincaré had already discovered the notion of deterministic chaos in the context of a three-body celestial mechanics problem. The French scientist moved away from his equations, considering the scarce calculation resources at that time. Today, however, the progress of mathematical sciences and extraordinary calculation resources points to a new horizon.

Turbulent flows are dynamic systems characterized by fantastic numbers of degrees of freedom. They have been seen as one of the most important and least understood domains of fluid dynamics. It has also been verified that many dynamic systems with fewer degrees of freedom present characteristics like turbulent movements. Some of these systems are much simpler than flows and understanding their behavior can help us understand increasingly complex systems.

The word turbulence has been associated only with flows, but this does not correspond to reality. Several other dynamic systems, which are characterized by a sufficiently high number of degrees of freedom, can reach the turbulence regime. It is observed that all dynamic systems are mathematically modeled with deterministic equations. It is also observed that a dynamic system with a low degree of freedom may not reach the turbulent regime, but may, however, achieve highly unpredictable or chaotic behaviors. It is in this sense that we speak of deterministic chaos. To better understand this important topic, a digression from fluid systems will be made, and some examples of non-fluid dynamic systems will be discussed.

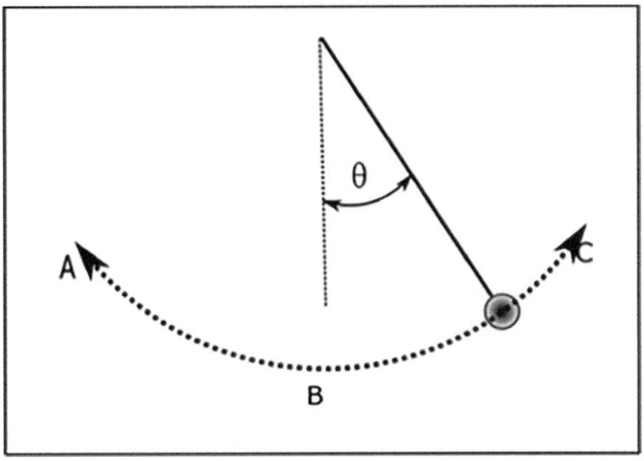

Fig. 3.55 Simple pendulum dynamic system

3.9.1 Simple Pendulum Dynamic System

Consider a simple pendulum with two degrees of freedom: circumferential rotation, moving along two of the coordinate axes. In Fig. 3.55, the movement of the sphere in a plane is illustrated, but it moves along two coordinate axes, which characterizes two degrees of freedom. Initially, a situation is considered in which the dampening effects associated with the viscous transformation of energy are present. Position the pendulum at point C and releasing, it will accelerate and when passing through point B, its acceleration is zero and its velocity is maximum. When heading to point A, the behavior is reversed, and the process repeats until reaching equilibrium at point B. At this point, the sphere has a lower level of potential energy than at point C, due to the damping effects caused by viscous energy transformation. Throughout the cycles, the pendulum's initial energy will be transformed until it reaches rest at point B. The so-called phase space is defined as being composed of two variables, acceleration and velocity. The behavior of this system is plotted in phase space in Fig. 3.56a. The point to which the system converges is an equilibrium point, which in the case under discussion corresponds to rest. For the case in which the illustrated system is not subject to damping effects due to the viscous energy transformation effects, or if the system is moved by an external force, the equilibrium will be represented by a limit circle, to which the system must converge, as illustrated in Fig. 3.56b.

3.9.2 Chaotic Movement of a Pendulum

The concept of chaotic movement is something quite abstract. Therefore, a simple dynamic system will be used here, which, even though it is not a fluid system, is an arrangement that exhibits this type of behavior. Suppose the pendulum of the previous case, however, with three degrees of freedom: circular movement and vertical movement. In this way, the pendulum sphere can travel curved paths with a

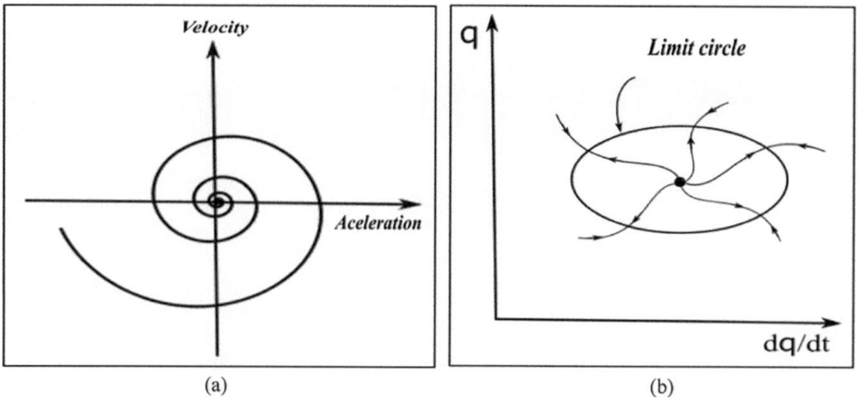

Fig. 3.56 Dynamic representation in phase space

variable radius. This is an arrangement known as a spherical or conical pendulum. The pendulum's fulcrum can oscillate according to a sinusoid, with a small amplitude compared to the length of the anchor wire. If the forcing frequency f is close to the natural frequency of the system, the oscillations will amplify and give rise to an unstable regime. As a result, there is a range of excitation frequencies for which the pendulum sphere orbits within a spherical shell with a given maximum radius, rather than traveling in an arc over a spherical shell.

This orbital motion exhibits the phenomenon with which the present study is related. It is shown, with experiments, that the system exhibits surprising behavior in the form of oscillations as the excitation frequency is changed. It is interesting to note that, when the excitation frequency is slightly higher than the natural frequency, the system acquires chaotic behavior, whereas if it is slightly lower, this does not happen. In Fig. 3.57, a set of 16 orbits assumed by the system, at different times, is shown. It should be noted that it is completely unpredictable which one the system will assume at a given time, despite this system being modeled with deterministic equations. Thus, its extreme sensitivity to injected disturbances is observed. In this sense, it is said that the system evolved into a regime of deterministic chaos. More details about this study can be found in Tritton [39].

3.9.3 Chaos in Fluid Dynamics

It is evident, based on the arguments presented in the previous sections, that the same dynamic system can be in both a chaotic and ordered regime, for different values of the forcing parameter. The system then follows a path towards chaos as this parameter varies. The processes of transition to chaos and transition to turbulence will be qualitatively compared.

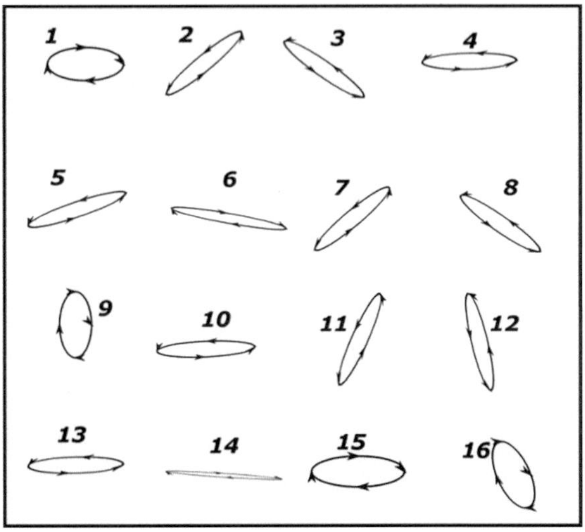

Fig. 3.57 Chaotic behavior of the orbits of a spherical pendulum. Figure of the author of this book

The first reason chaos theory might be important for fluid dynamics is turbulence. It should be emphasized that there are some ways in which a flow can exhibit chaotic behavior. The question is, when a flow becomes turbulent, can it be different from when it becomes chaotic? A Taylor–Couette Flow, for example, can acquire several combinations of regimes: laminar, laminar-chaotic and turbulent-chaotic. Following modern ideas about chaos, turbulent movements can be considered as examples of deterministic chaos, if it is assumed that they are modeled with the Deterministic Continuity and Navier–Stokes Equations.

The two most studied configurations have been: the case of Rayleigh–Bénard convection (Sect. 3.4), between two horizontal flat plates, and the Taylor–Couette flow (Sect. 3.5), between two concentric rotating cylinders. In the case of Rayleigh–Bénard convection, two transition routes are identified: the frequency multiplication route and the intermittency route. In Fig. 3.58, the power spectra for different values of the Rayleigh number are illustrated. It is observed in Fig. 3.58a, b that there is a multiplication of frequencies, passing by Fig. 3.58c, where the presence of intermittency is observed, across two energy levels of various frequencies. Finally, it is observed in Fig. 3.58d a continuous energy spectrum, showing the presence of a wide band of frequencies, and characterizing a turbulent scouring regime. It is observed that small changes in the experimental bench can promote the appearance of a different break for chaos. The reasons are not well known. There is, still, a question of a difficult answer: can a broken pattern for chaos be identified in the transition to turbulence?

A final issue to be addressed here is the possibility of a return to order or a relaminarization process of an effluent. With some stabilizing effects, the sludges can be relaminarized and returned to order. For example, it is the case of restabilization of turbulent sediments, when subjected to strong stable stratification effects at a specific mass. Rotation effects, as well as compressibility effects, exert similar patterns.

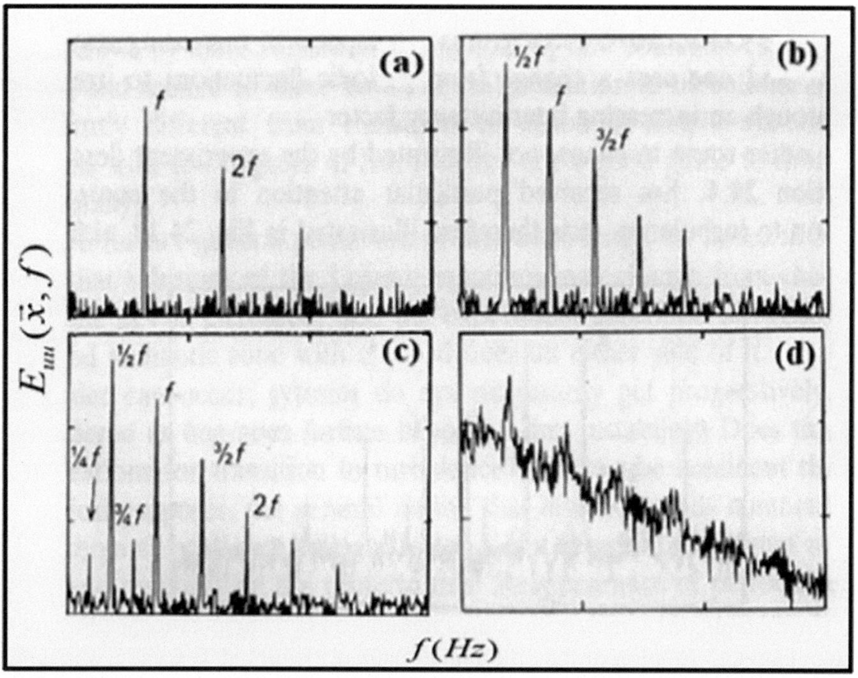

Fig. 3.58 Power spectral density of the horizontal component of velocity, $E_{uu}(\vec{x}, f)$, in an experience of Rayleigh–Bénard: **a** $R_a/R_{ac} = 21$; **b** $R_a/R_{ac} = 26$; **c** $R_a/R_{ac} = 27$ and **d** $R_a/R_{ac} = 36.9$. Adapted from Tritton [39]

3.10 A Deterministic and a Statistical View of Turbulence

Turbulence is a regime of operation of some dynamical systems of great complexity, which present physical characteristics necessary to operate in such a regime. Much is already understood about this regime, but no doubt, much more remains to be understood. According to the view already acquired, everything indicates that turbulence is a deterministic regime and, as such, is, in its essence, predictable. The issue of unpredictability, already discussed earlier, is associated with scientific impossibilities linked to the lack of ability to accurately provide initial and boundary conditions, as well as errors associated with discrete models of solution of two different mathematical models.

Another limiting factor is the available computational resources. In fact, very small errors in initial conditions, boundary conditions, or errors promoted with the use of discrete models will be amplified by nonlinear interactions, promoting instabilities that are dependent on these noises. Any variation in these conditions will determine completely different states in the forecasts. One of the first two scientists to perceive this fact was Henri Poincaré, who discovered, in the early twentieth century, that a system such as the sun-earth-moon, interacting gravitationally, can have unpredictable or chaotic behavior. This is, however, a deterministic point of view, since it refers to a dynamical system modeled with deterministic equations. It is in this sense that Einstein says that "God does not decide by dice", because, of course, He knows in all detail the tiny perturbations that vibrate in the Universe.

A fantastic example of a chaotic system is the dynamic behavior of the earth's atmosphere: the colossal challenge is to meteorological forecasting of the earth's climate. Surprisingly, too, the advances made in this domain, thanks to increased accuracy in measurements from experimental weather stations. These measurements are provided to increasingly powerful computers, with computer codes being processed every day that are more representative of the physics of these flows. This set of factors has allowed forecasting of climate behavior, with a good level of reliability for a few days into the future.

Fortunately, for most engineering applications, and even for the search for phenomenological understanding, accurate prediction of the position, at a given time, of a turbulent structure is not as indispensable as one might imagine. However important, it is a good prediction of the statistical behavior of a dynamical system. This means that, even if a material experiment carried out in a laboratory cannot be exactly reproduced, it is known to be reproducible, not referring to its statistical behavior, both with material and computational experiments.

For example, the turbulent structures reproduced in a computational experiment (computational solution of the differential mathematical model) do not correspond exactly to the turbulent structures observed in a material experiment, not referring to their position in space and time, as close as they may be initial and boundary conditions. However, when statistical information is extracted from these theoretical results, excellent agreements with material experiment statistics are obtained. Another important fact is that physical phenomena are also correctly represented, which allows interpretations and understanding of the results of virtual experiments. It presents, then, the idea that the behaviors of dynamical systems can be unpredictable, even when they are modeled with deterministic equations. In fact, it is not possible to provide the same initial and boundary conditions present in the material experiments. On the other hand, the statistical behavior of two dynamical systems is independent of these initial perturbations, for most practical applications. The statistical tools of turbulence modeling are, however, indispensable and allow, too, to aid in the question of unpredictability. It allows us to further aid in the limiting question of available computational capacity to resolve consistently, turbulent flows. It will be seen, from Chaps. 6 to 10, that turbulence modeling is directly associated with these characteristics.

3.11 Exercises

1. Simulate a temporary mixing layer and compare the most amplified wave compression, obtained via computer simulation, with that predicted using Fig. 3.3. Apply FFT at wave number, over a velocity field, at a fixed time, in the transition phase.
2. Simulate a spatial mixing layer. Search for a material experiment for reference and comparison. Make a comparison between the results obtained with computational experiments and with material experiments.
3. Repeat Exercise 2 for a jet at spatial development.
4. Simulate a boundary layer on a flat plate and compare the computational results with the Blasius solution.
5. Note: In the simulations proposed above, no turbulence model is used. Would you be doing a DNS? Reflect and discuss that question!

References

1. Allampalli, V., Hixon, R.: High-accuracy large-step explicit Runge–Kutta (HALE-RK) schemes for computational aeroacoustics. J. Comput. **14**, 68–83 (2009)
2. Andereck, C.D., Liu, S.S., Swinney, H.L.: Flow regimes in a circular Couette system with independently rotating cylinders. J. Fluid Mech. **164**, 155–185 (1985)
3. Balint, J.L., Vukoslavcevic, P., Wallace, J.M.: The velocity and vorticity fields of a turbulent boundary layer. Part. 2 statistical property. J. Fluid Mech. **228**, 53–86 (1991)
4. Betchov, R., Szewczyk, G.: Stability of a shear layer between parallel steams. Phys. Fluids **6**, 1391–1396 (1963)
5. Brow, G.L., Roshko, A.: On density effects and large structures in turbulent mixing layers. J. Fluid Mech. **64**, 775–816 (1974)
6. Cantwell, Coles, D.B., Dimotakis, P.: Structure and entrainment in the plane of symmetry of a turbulent spot. J. Fluid Mech. **87**, 641–672 (1978)
7. Coles, D.: Transition in circular Couette flow. J. Fluid Mech. **21**, 385–425 (1965)
8. Drazin, P.G., Reid, W.H.: Hydrodynamic Stability. Cambridge University Press (1981)
9. Drazin, P.G.: Introduction to Hydrodynamic Stability. Cambridge University Press (2002)
10. Drazin, P.G., Reid, W.H.: Hydrodynamic Stability. Cambridge University Press (2004)
11. Duarte, B.A.F., Silveira-Neto, A.: Modelagem matemática e simulação computacional para o problema de interação fluido-estrutura em uma plataforma parcialmente imersa no oceano. Início: 2014. Tese (Doutorado em Engenharia Mecânica), Universidade Federal de Uberlândia (2016)
12. Fjørtoft, R.: Application of integral theorems in deriving criteria of stability for laminar flows and for the barocline circular vortex. Geofys. Pub. Oslo **17**(6), 1–52 (1950)
13. Görtler, H.: Dreidimensionales zur Stabilitätstheorie laminarer Grenzschichten. J. Appl. Math. Mech. **35**, 362–363 (1955)
14. Hein, H., Shultz-Grunow, F.: Z. Flugwiss. **4**, 28–30 (1956)
15. Kolmogorov, A.N.: The local structure of turbulence in incompressible viscous fluid for very large Reynolds number. Dokl. Akad. Nauk SSSR **30**, 301–305 (1941)
16. Kolmogorov, A.N.: On degeneration of isotropic turbulence in an incompressible viscous liquid. Dokl. Akad. Nauk SSSR **31**, 538–541 (1941)
17. Lesieur, M.: Turbulence in Fluids. Kluwer Academic Publishers (1995)

18. Melo, R.R.S., Silveira-Neto, A.: Desenvolvimento de uma ferramenta computacional educacional para simulação de problemas de dinâmica dos fluidos. Dissertação Mestrado em Engenharia Mecânica, Universidade Federal de Uberlândia (2012)
19. Moreira, L.Q., Silveira-Neto, A.: Simulação de grandes escalas de jatos periódicos temporais utilizando a Metodologia Pseudo-Espectral de Fourier. Dissertação de Mestrado-FEMEC-UFU (2007)
20. Moreira, L.Q., Silveira-Neto, A.: Modelagem Matemática de Jatos em Desenvolvimento espacial usando a Metodologia Pseudo-Espectral de Fourier. Tese (Doutorado em Engenharia Mecânica), Universidade Federal de Uberlândia (2011)
21. Padilla, E.L.M., Silveira-Neto, A.: Simulação de Grandes Escalas da transição à turbulência em sistemas rotativos com transferência de calor. Tese de Doutorado em Engenharia Mecânica, Universidade Federal de Uberlândia, Conselho Nacional de Desenvolvimento Científico e Tecnológico (2004)
22. Perry, A.E., Li, J.D.: Experimental support for the attached-eddy hypothesis in zero-pressure gradient turbulent boundary layers. J. Fluid Mech. **218**, 405–438 (1990)
23. Pivello, M.R., Villar, M.M., Serfaty, R., Roma, A.M., Silveira-Neto, A.: A fully adaptive front tracking method for the simulation of two phase flows. Int. J. Multiph. Flow **1**, 72–82 (2013)
24. Prandtl, Z.A.: Bericht über undersuchugen zur ausgebildeten turbulenz. Zs Angew. Math. Mech. **5**, 136–169 (1925)
25. Pronchik, S.W., Kline, S.J.: An experimental investigation of the structure of a turbulent reattaching flow behind a backward-facing step. Department of Mechanical Engineering, Stanford University, Report MD-42 (1983)
26. Rayleigh, L.: 1880, Scientifics Papers, vol. 1, pp. 474–487. Dover, New York (1964)
27. Reynolds, O.: An experimental investigation of the circumstances which determine whether the motion of water shall be directly and sinuous, and the law of resistance in parallel channels. Phil. Trans. R. Soc. 55–105 (1883)
28. Saric, W.S.: Görtler vortices. Annu. Rev. Fluid Mech. **26**, 379–409 (1994)
29. Schlichting, H.: Boundary Layer Theory. Pergamon Press, London (1968)
30. Shultz-Grunow, F.: Jahrd. Schffbautech. Ges. **39**, 176–198 (1938)
31. Silveira-Neto, A., Dominique, G., Olivier, M., Lesieur, M.: Large-eddy simulation of the turbulent flow in the downstream region of a backward-facing step. Phys. Rev. Lett. **66**(18), 2320–2323 (1991)
32. Silveira-Neto, A., Dominique, G., Lesieur, M.: Simulation numérique d'un écoulement tourbulent stratifié derrière une marche. Int. J. Heat Mass Transf. **8**, 1999–2001 (1991)
33. Silveira-Neto, A., Grand, D., Lesieur, M., Metais, O.: A numerical investigation of coherent vortices in turbulence behind a backward-facing step. J. Fluid Mech. **256**, 1–25 (1993)
34. Smirnov, A., Shi, S., Celik, I.: Random flow generation technique for large eddy simulations and particle-dynamics modeling. J. Fluids Eng. **123**, 359 (2001)
35. Sondergaard, R., Cantwell, B., Mansur, N.: Direct Numerical Simulation of a Temporally Evolving Incompressible Plane Wake: Effect on Initial Conditions on Evolution and Topology, pp. 187–258. Stanford, CA (1997)
36. Souza, A.M., Silveira-Neto, A.: Simulação Computacional de Grandes Escalas de Jatos Planos em Transição à Turbulência. Tese de Doutorado em Engenharia Mecânica, Universidade Federal de Uberlândia (2005)
37. Taneda, S.: J. Phys. Soc. Jpn. **50**(4) (1981)
38. Taylor, G.I.: The motion of a sphere in a rotating liquid. Proc. R. Soc. Lond. Ser. A **102**, 213–215 (1922)
39. Tritton, D.J.: Physical Fluid Dynamics. Clarendon Press, Oxford (1988)
40. Wereley, S.T., Lueptow, R.M.: Velocity field for Taylor Couette flow with an axial flow. Phys. Fluids **11**, 3637–3649 (1999)
41. White, F.M.: Viscous Fluid Flow. McGraw-Hill, Inc., New York (1991)

Chapter 4
Homogeneous and Isotropic Turbulence

In this chapter, the kinematics of homogeneous and isotropic turbulence is presented in Sects. 4.1 and 4.2. In Sect. 4.3, the transformation of the differential model into Fourier Space is presented in detail, which resulted in a powerful tool for the physical interpretation of the nonlinearities of turbulent flows. In Sect. 4.4, the concept of Spectral Tensor and Turbulent Kinetic Energy Spectrum is presented, whose balance equation is presented in Sect. 4.5. In Sect. 4.6, the Kolmogorov theory is presented, including the characteristic quantities of turbulence, in Sect. 4.7. Finally, a list of exercises is presented in Sect. 4.8.

4.1 Introduction

Flows that occur in nature, in industrial applications and in laboratories are anisotropic. Isotropic flows cannot be found naturally in the strong sense of this word, but flows that approximately meet this hypothesis can be found. Scientists are interested in studying, from a theoretical point of view, this class of flows, to enable developments whose applications can be extended to real situations.

Part of the turbulence closure models requires the assumption of isotropy (which will be introduced in the following paragraphs). For example, Boussinesq's proposal [2] is used to solve the famous closure problem of turbulence. It requires the use of the concept of turbulent viscosity, which is a scalar quantity. However, a conceptual limitation of this proposal arises, since the aim is to model an anisotropic phenomenon, that is, turbulent flows, using the concept of turbulent viscosity, which is a scalar quantity.

Most flows are anisotropic and inhomogeneous (shear flows), as can be seen schematically in Fig. 4.1a and b. Shear is a requirement for the formation of instabilities, as illustrated in the diagram of Fig. 4.2, which implies anisotropy and

A. da Silveira Neto, *Turbulent Flows*, https://doi.org/10.1007/978-3-031-89222-6_4

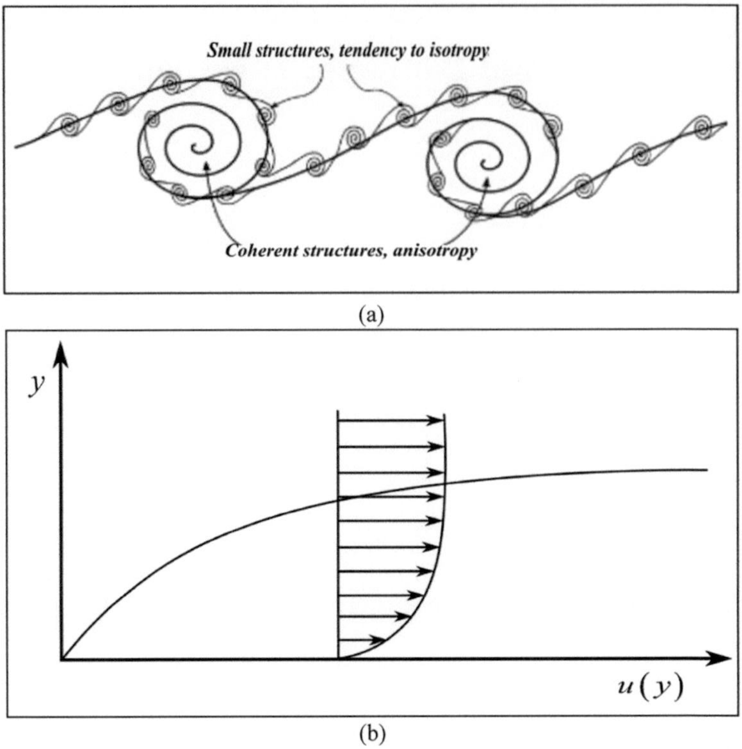

Fig. 4.1 Examples of shear flows. Free shear mixing layer (**a**); and boundary layer, with shear near the vicinity of walls (**b**)

non-homogeneity. Coherent structures are anisotropic, but the smaller the turbulent structures, the greater the tendency or degree of isotropy and homogeneity.

At the level of large structures, transitional and turbulent flows are anisotropic and inhomogeneous. However, even in these flows, at the level of small structures, the existence of isotropy and homogeneity can be approximately considered. In addition, one can also find isotropic flows that can be experimented with in laboratories. Flows downstream of a fine grid are examples. This type of flow is schematically represented in Fig. 4.3.

Despite the previous examples, greater rigor is needed in the definition of these two concepts, which will be presented using statistics. Homogeneity is the statistical invariance of flow properties when a translation of the axis system is promoted. This translation is illustrated in Fig. 4.4. If no variations are observed in any statistical property, represented here by the function $f(\vec{x}, t)$, then:

$$f(\vec{x} + \vec{r}, t) = f(\vec{x}, t). \tag{4.1}$$

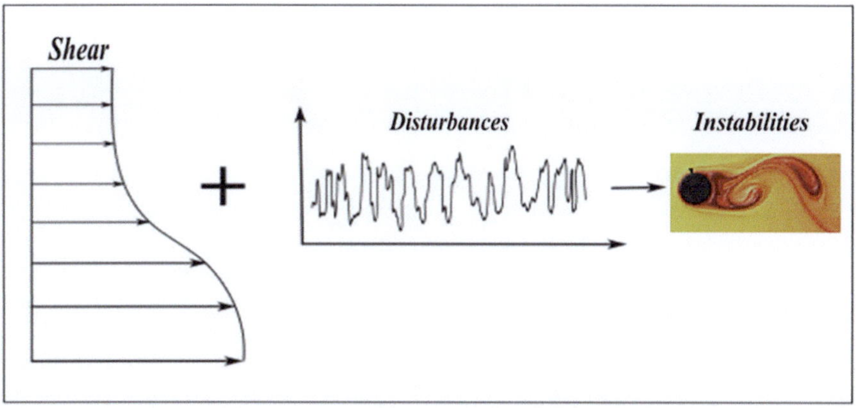

Fig. 4.2 Illustrative diagram of the formation process of instabilities in shear flows

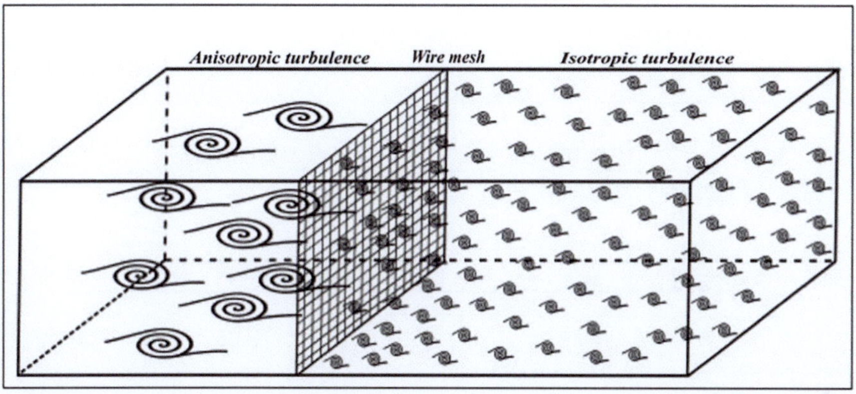

Fig. 4.3 Illustrative diagram of grid turbulence: presence of anisotropic coherent structures upstream and a tendency to isotropy downstream

In this case, the flow presents homogeneity in relation to the directions in which the translation took place. Therefore, homogeneity is a directional property. Figure 4.5 shows an example of flow developing at the channel composed of two distinct regions: the entrance region, where there is no homogeneity direction, and the fully developed flow region, where there is homogeneity in the axial direction of the pipe.

Isotropy is the statistical invariance of the properties of a flow in relation to rotation in the system of axes. It is understood, then, that isotropy implies homogeneity, but the contrary is not true. In Fig. 4.6, the rotation process of a system of axes is illustrated.

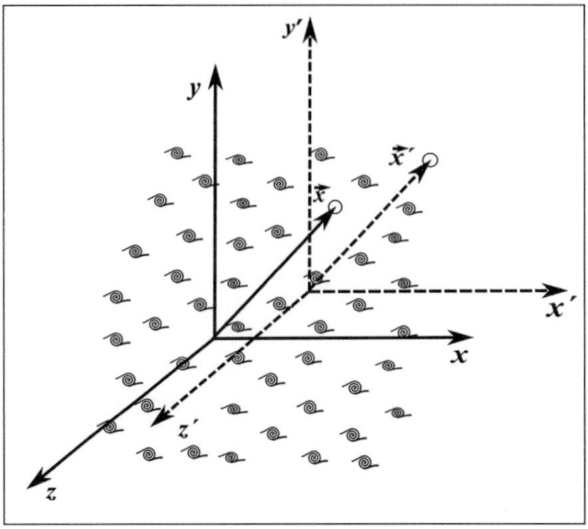

Fig. 4.4 Illustrative diagram of the axis translation process

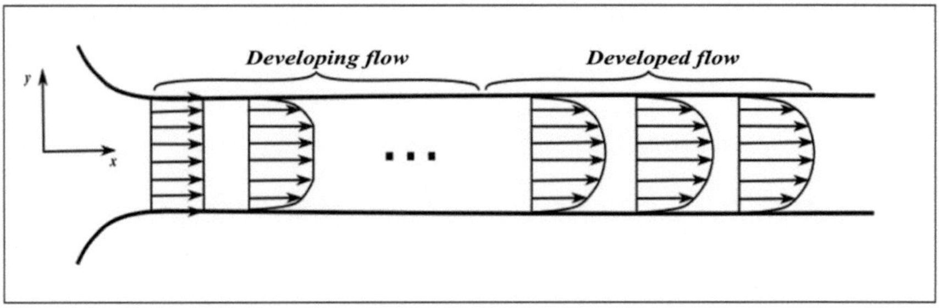

Fig. 4.5 Illustrative diagram of a flow in a pipe, with a region of inhomogeneity and a region of directional homogeneity

4.2 Kinematics of Homogeneous and Isotropic Turbulence

The study of isotropic turbulence requires statistical concepts, which will be briefly introduced in the following items.

4.2.1 Statistical Formalism

In any experiment, we are interested in a generic property $f(\vec{x}, t, w)$, where $w \in \Omega$, is a sample in the sample space Ω. The function $f(\vec{x}, t, w)$ is a property relating to a

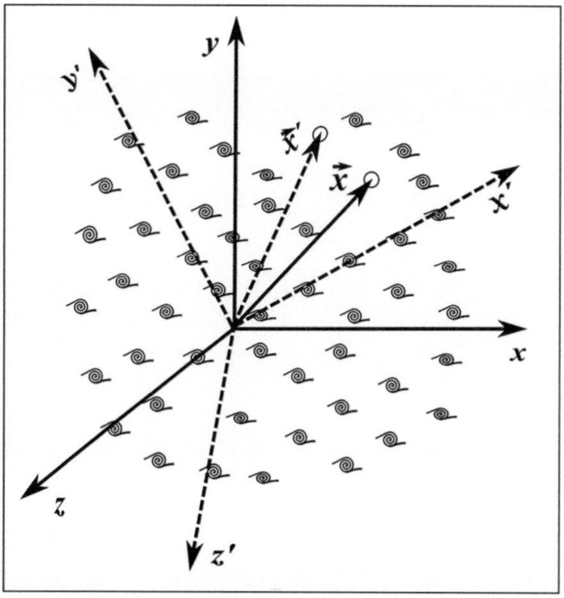

Fig. 4.6 Illustrative diagram of the axis system rotation process

given flow under analysis. In Fig. 4.7, a set of n temporal samples is illustrated, for a fixed point $\vec{x}$, belonging to the spatial domain of interest.

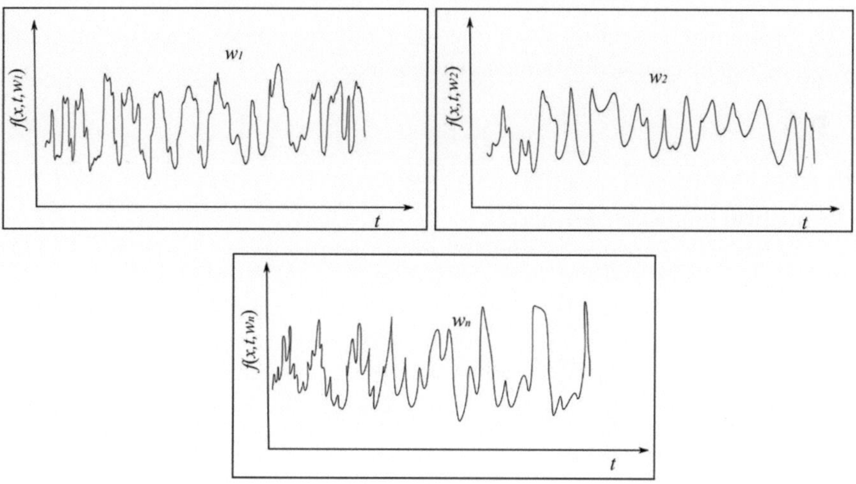

Fig. 4.7 Examples of n samples w

- **Temporal average**

The temporal average is calculated, using each sample individually, integrating the variable $f(\vec{x}, t, w)$ over time, according to Eq. (4.2):

$$\langle f\rangle(\vec{x}, w_i) = \frac{1}{T}\sum_{j=1}^{M} f\left(\vec{x}, t_j, w_i\right)\Delta t_j, \tag{4.2}$$

where M is the number of time steps, ΔT_j, used to divide the temporal sample w_i. Thus, the temporal average is a function of the point $\vec{x}$ and the sample w_i. Given the n samples, statistical definitions can be made as follows:

- **Ensemble average**

The ensemble average is calculated using all N samples w_i, illustrated in Fig. 4.7. This average will be performed according to Eq. (4.3):

$$[f](\vec{x}, t) = \frac{1}{N}\sum_{i=1}^{N} f(\vec{x}, t, w_i). \tag{4.3}$$

It is observed that the sum over the N samples is made for a fixed time, and if the average is made again for another time, the value obtained may be different from the previous one. Thus, the ensemble average, made over temporal samples, is a function of time and space.

- **Ergodicity hypothesis**

A set of samples is said to satisfy the ergodicity hypothesis when the ensemble averaging can be obtained by temporal averaging:

$$[f](\vec{x}) = \langle f\rangle(\vec{x}). \tag{4.4}$$

- **Statistical moments of order n**

Consider a set of variables listed below, representative of an arbitrary flow:

$$\begin{cases} f^{a_1}(\vec{x}, t, w) \Rightarrow \text{velocity} \\ f^{a_2}(\vec{x}, t, w) \Rightarrow \text{pressure} \\ f^{a_3}(\vec{x}, t, w) \Rightarrow \text{temperature} \\ \vdots \\ f^{a_n}(\vec{x}, t, w) \Rightarrow \text{concentrations, etc.} \end{cases} \tag{4.5}$$

A statistical moment of order n is defined as the ensemble average of the product of the n variables involved:

$$\left[f^{a_1}f^{a_2}\ldots f^{a_n}\right](\vec{x},t)=\lim_{N\to\infty}\frac{1}{N}\sum_{i=1}^{N}f^{a_1}(\vec{x},t,w_i)f^{a_2}(\vec{x},t,w_i)\ldots f^{a_n}(\vec{x},t,w_i). \quad (4.6)$$

For example, you can do $n=2$ and $f^{a_1}=f^{a_2}=u'$, where u' is the fluctuation of one of the components of the velocity vector. With this variable, the turbulent intensity relative to one of the flow directions is obtained:

$$\left[u'u'\right]=\left\langle u'^2\right\rangle. \quad (4.7)$$

This is a second-order statistical moment. In this case, the ergodicity hypothesis was used, establishing equality between the ensemble average $\left[u'u'\right]$ and the temporal average $<u'u'>$. A second-order moment is thus obtained, which represents the variance of the velocity fluctuation, as illustrated in Fig. 4.8. Extending this example to the three velocity components, one can obtain the Boussinesq–Reynolds Tensor, composed of nine second-order statistical moments, as defined below.

$$\overline{\tau}(\vec{x})=\begin{bmatrix}\left\langle u'^2\right\rangle & \left\langle u'v'\right\rangle & \left\langle u'w'\right\rangle\\ \left\langle v'u'\right\rangle & \left\langle v'^2\right\rangle & \left\langle v'w'\right\rangle\\ \left\langle w'u'\right\rangle & \left\langle w'v'\right\rangle & \left\langle w'^2\right\rangle\end{bmatrix}. \quad (4.8)$$

This tensor is symmetric because of the definition of statistical moment, according to Eq. (4.6). It can also be rewritten, in indexical notation, as follows:

$$\tau_{ij}=\left\langle u'_i u'_j\right\rangle. \quad (4.9)$$

With the trace of the tensor, given by Eq. (4.9), turbulent kinetic energy is obtained, in accordance with its definition:

$$k\equiv\frac{1}{2}\left\langle u'_i u'_i\right\rangle=\frac{1}{2}\left(\left\langle u'^2\right\rangle+\left\langle v'^2\right\rangle+\left\langle w'^2\right\rangle\right). \quad (4.10)$$

Einstein's Sum rule was used for repeated indices.

4.2.2 *Turbulence Classification*

Turbulence can be classified according to homogeneity and isotropy as follows: non-homogeneous and non-isotropic turbulence; homogeneous and non-isotropic turbulence and homogeneous and isotropic turbulence. In the latter case, it follows that:

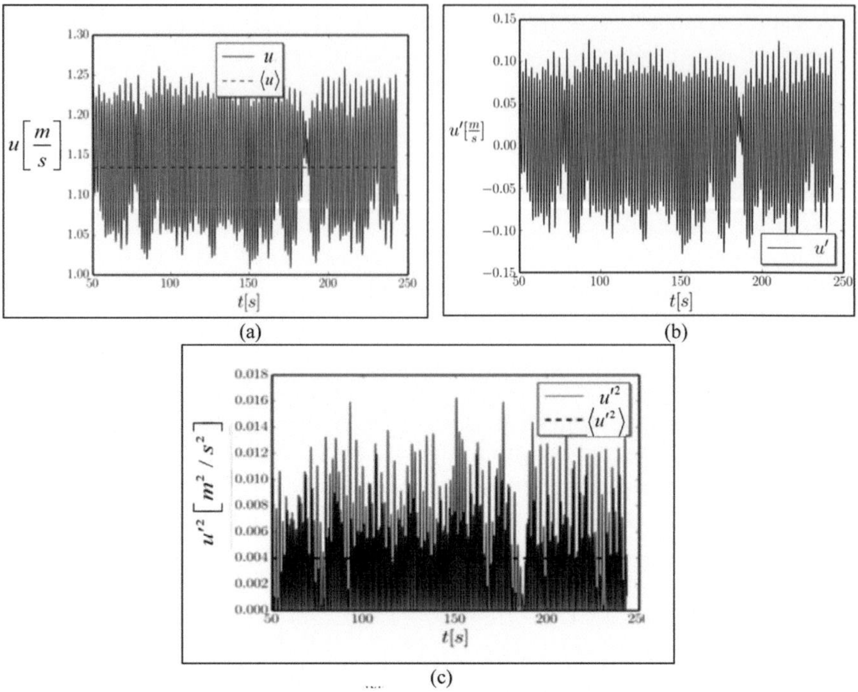

Fig. 4.8 Temporal distribution of a velocity component and its temporal average (**a**); its fluctuation (**b**); and its variance (second order momentum) (**c**)

$$\langle f^{a_1} f^{a_2} \ldots f^{a_n} \rangle (\vec{x}) = \langle f^{a_1} f^{a_2} \ldots f^{a_n} \rangle (\vec{x} + \vec{r}). \tag{4.11}$$

The vector $\vec{r}$ represents a rotation or translation of the axis system.

4.3 Transformation of Balance Equations from Physical Space to Fourier Space

Isotropic flows are also periodic. A temporal mixing layer is an example of a periodic flow, which is illustrated in Fig. 4.9. The periodicity hypothesis is not necessary for the formalism of transforming a function into Fourier Space. Only the Discrete Fourier Transform requires the property of periodicity. The use of the Discrete Fourier Transform (DFT) is frequent when solving partial differential equations, using the facilities inherent to this operator.

Fig. 4.9 Periodic flow in the horizontal direction

4.3.1 Fourier Transform

The Spatial Fourier Transform of a function $f(\vec{x}, t)$ is defined as follows:

$$\hat{f}\left(\vec{k}, t\right) \equiv \left(\frac{1}{2\pi}\right)^3 \int\limits_V \mathrm{e}^{-i\vec{k}.\vec{x}} f(\vec{x}, t)\mathrm{d}\vec{x}, \tag{4.12}$$

where $\vec{k} = 2\pi(1/\lambda_1, 1/\lambda_2, 1/\lambda_3)$ is the wave number and $\vec{\lambda} = (\lambda_1, \lambda_2, \lambda_3)$ is the vector wavelength, as illustrated in Fig. 4.10. The symbol $i = \sqrt{-1}$ is an imaginary number. The wavenumber vector has three components corresponding to the three coordinate directions $\vec{k} = (k_1, k_2, k_3)$. The inverse transform is defined as:

$$f(\vec{x}, t) \equiv \int\limits_{\hat{V}} e^{i\vec{k}\cdot\vec{x}} \hat{f}\left(\vec{k}, t\right) d\vec{k}. \tag{4.13}$$

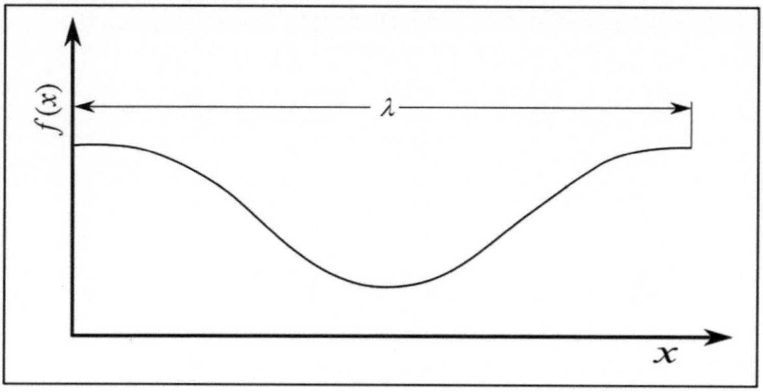

Fig. 4.10 Periodic one-dimensional signal, wavelength λ

- **Operators of Interest for Transforming Balance Equations**

The objective is to transform the Continuity, Navier–Stokes and Thermal Energy Equations into Fourier Space. It is necessary to know the transformation of the differential operators that make up these equations.

- **Transformation of the Derivative of a Function**

Let a function $g(\vec{x}, t)$ be equal to the derivative of the function $f(\vec{x}, t)$. It is proposed to obtain the transform of $g(\vec{x}, t)$:

$$g(\vec{x}, t) = \frac{\partial f(\vec{x}, t)}{\partial x_\ell} = \frac{\partial}{\partial x_\ell}\left(\int_{\hat{V}} e^{i\vec{k}\cdot\vec{x}}\hat{f}\left(\vec{k}, t\right)d\vec{k}\right) = \int_{\hat{V}} e^{i\vec{k}\cdot\vec{x}} ik_\ell \hat{f}\left(\vec{k}, t\right)d\vec{k}. \tag{4.14}$$

Therefore,

$$\begin{aligned}
\hat{g}\left(\vec{k}, t\right) &= \widehat{\frac{\partial f}{\partial x_\ell}} = \left(\frac{1}{2\pi}\right)^3 \int_V e^{-i\vec{k}.\vec{x}}\left[\frac{\partial f}{\partial x_\ell}\right]d\vec{x} \\
&= \left(\frac{1}{2\pi}\right)^3 \int_V e^{-i\vec{k}.\vec{x}}\left\{\int_{\hat{V}} e^{i\vec{k}.\vec{x}}\left[ik_\ell \hat{f}\left(\vec{k}, t\right)\right]d\vec{k}\right\}d\vec{x} = \text{TF}\left\{\text{TF}^{-1}\left[ik_\ell \hat{f}\left(\vec{k}, t\right)\right]\right\} \\
&= ik_\ell \hat{f}\left(\vec{k}, t\right).
\end{aligned} \tag{4.15}$$

The terms FT and FT^{-1} in Eq. (4.15) represent, respectively, direct and inverse Fourier transforms.

- **The Fourier Transform of the gradient of *f***

Using Eq. (4.15), the transform of the gradient of a function is given by

$$\widehat{\vec{\nabla} f} = \left(\widehat{\frac{\partial f}{\partial x}}, \widehat{\frac{\partial f}{\partial y}}, \widehat{\frac{\partial f}{\partial z}}\right) = i\left(k_x, k_y, k_z\right)\hat{f}\left(\vec{k}, t\right) = i\vec{k}\hat{f}\left(\vec{k}, t\right). \tag{4.16}$$

- **The Fourier Transform of a divergent of a vector**

$$\widehat{\vec{\nabla} \cdot \vec{V}} = i\vec{k} \cdot \hat{\vec{V}} or \widehat{\frac{\partial u_i}{\partial x_i}} = ik_i\hat{u}_i. \tag{4.17}$$

- **The Fourier Transform of the Laplacian of a function**

Using Eq. (4.15), the transform of the Laplacian operator is obtained:

$$\widehat{\frac{\partial^2 u_i}{\partial x_j \partial x_j}} = -k^2\widehat{u_i}. \tag{4.18}$$

- **The Fourier Transform of the product of two functions**

The transform of the product of two functions: $f(\vec{x}, t)$ and $g(\vec{x}, t)$, results in a convolution integral, involving their transforms individually:

$$\widehat{f(\vec{x}, t)g(\vec{x}, t)} = [\hat{f} * \hat{g}]\left(\vec{k}, t\right) = \int_{\vec{k}=\vec{p}+\vec{q}} \hat{f}(\vec{p}, t)\hat{g}(\vec{q}, t)\mathrm{d}\vec{p}\mathrm{d}\vec{q}$$

$$= \int_{\vec{k}=\vec{p}+\vec{q}} \hat{f}(\vec{p}, t)\hat{g}\left(\vec{k} - \vec{p}, t\right)\mathrm{d}\vec{p}, \tag{4.19}$$

where $[\hat{f} * \hat{g}]\left(\vec{k}, t\right)$ denotes the convolution product of the two transformed functions $\hat{f}(\vec{p}, t)$ and $\hat{g}(\vec{q}, t)$. The transformation parameters are respectively $\vec{p}$ and $\vec{q}$, where $\vec{k} = \vec{p} + \vec{q}$. This convolution integral represents the triadic interactions between the three wavenumbers, relative to the transforms of the two functions separately, and the transform of the product of the two functions.

4.3.2 *The Fourier Transform of the Balance Equations for Incompressible Flows*

4.3.2.1 Preliminary

The equations used to model incompressible flows of Newtonian fluids, with thermal effects, in physical space: continuity or mass balance, Navier–Stokes equations or linear momentum balance and balance of thermal energy, are presented below, Eqs. (4.20):

$$\begin{cases} \frac{\partial u_i}{\partial x_i} = 0 \\ \frac{\partial u_i}{\partial t} + \frac{\partial}{\partial x_j}\left(u_i u_j\right) = -\frac{1}{\rho}\frac{\partial p}{\partial x_i} + \nu\frac{\partial^2 u_i}{\partial x_j \partial x_j} + \frac{f_i}{\rho} \\ \frac{\partial T}{\partial t} + \frac{\partial}{\partial x_j}\left(u_j T\right) = \alpha\frac{\partial^2 T}{\partial x_j \partial x_j} + \frac{\phi}{\rho C_p} \end{cases}. \tag{4.20}$$

For the process of transforming these equations, physical properties such as viscosity, specific mass, as well as thermal capacity and conductivity were considered unitary and constant.

Fig. 4.11 Orthogonality π-plane in Fourier Space

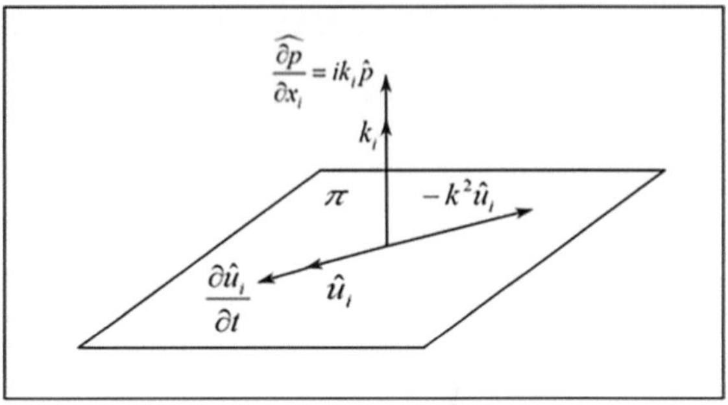

4.3.2.2 The Fourier Transform of the Balance of Mass (Continuity Equation)

Using Eq. (4.17), we have the following equation, in Fourier Space:

$$\widehat{\frac{\partial u_i}{\partial x_i}} = ik_i\hat{u}_i = 0. \tag{4.21}$$

In Fourier Space, a plane π is plane defined perpendicular to which lies the wavenumber vector k_i, also known as the plane of zero divergence, since, for incompressible flows, the transformed velocity field $\hat{u}_i\left(\vec{k}, t\right)$, is orthogonal to the wavenumber vector k_i. This property is illustrated in Fig. 4.11.

4.3.2.3 The Fourier Transform of the Linear Momentum Balance Equation (Navier–Stokes Equation)

- **Term of the rate of change of the linear momentum contained in a fluid particle**

$$\widehat{\frac{\partial u_i}{\partial t}} = \frac{\partial \widehat{u_i}}{\partial t}. \tag{4.22}$$

It is observed that, from Eq. (4.21):

$$k_i\hat{u}_i = 0. \tag{4.23}$$

Therefore:

$$\frac{\partial}{\partial t}\left(k_i\widehat{u_i}\right) = k_i\frac{\partial \widehat{u_i}}{\partial t} = 0. \tag{4.24}$$

Observing that Eq. (4.24) represents the scalar product between the wavenumber k_i and the vector $\partial\widehat{u_i}/\partial t$, and that this scalar product turned out to be null, results in the term $\partial\widehat{u_i}/\partial t$ also belongs to the plane π.

- **Diffusive net flux term of linear momentum**

Based on Eq. (4.18) and considering that k^2 is a scalar quantity, it is concluded that the transform of the diffusive term of the Navier–Stokes equations also belongs to the plane π, as illustrated in Fig. 4.11.

$$\widehat{\frac{\partial^2 u_i}{\partial x_j \partial x_j}} = -k^2 \hat{u}_i. \tag{4.25}$$

- **Pressure gradient**

Using Eq. (4.16), the transform of the pressure field gradient is given by:

$$\widehat{\frac{\partial p}{\partial x_i}} = ik_i\hat{p}. \tag{4.26}$$

Note that the transformed pressure field is collinear with the wavenumber vector, therefore being perpendicular to the plane π.

- **Nonlinear term that models the net advective flux of linear momentum**

With the results above, we have the following:

$$\underbrace{\left(\widehat{\frac{\partial u_i}{\partial t}} - \widehat{\nabla^2 u_i}\right)}_{\in\,\pi} \underbrace{+}_{\Rightarrow} \underbrace{\left(\widehat{\frac{\partial}{\partial x_j}\left(u_i u_j\right)} + \widehat{\frac{\partial p}{\partial x_i}} - \hat{f}_i\right)}_{\in\,\pi} = 0 \tag{4.27}$$

In Eq. (4.27), there is a sum of two parentheses. The terms found inside the first parentheses belong to plane π, based on the results already presented. The sum of the terms inside the second parenthesis must also belong to the plane π, since the sum of the two parentheses is a vector and is identically zero. To transform the advective term together with the pressure gradient and the term f_i, a projection tensor $\wp_{ij}$ will be defined as follows:

$$\wp_{ij}\left(\vec{k}\right) = \delta_{ij} - \frac{k_i k_j}{k^2}, \tag{4.28}$$

where

$$\delta_{ij} = \begin{cases} 1 & \text{se} \quad i = j \\ 0 & \text{se} \quad i \neq j \end{cases}, \tag{4.29}$$

is the Kronecker delta. What is the function of this tensor $\wp_{ij}$? To check it, take any vector a_j, and make the projection of a_j, using $\wp_{ij}$, to obtain the following result:

$$\wp_{ij}a_j = a_j\delta_{ij} - a_j\frac{k_ik_j}{k^2} = a_i - a_jk_j\frac{k_i}{k^2} = a_{pi}, \tag{4.30}$$

where a_{pi} represents the projection of the vector a_j by the action of the projection tensor $\wp_{ij}$. Taking the scalar product of the projection a_{pi} by the wavenumber $\vec{k}$, we have:

$$a_{pi}k_i = a_ik_i - a_jk_j\frac{k_ik_i}{k^2} = 0. \tag{4.31}$$

Thus, it is verified that the tensor $\wp_{ij}$, defined with Eq. (4.28), projects any vector a_j onto the plane π. Returning to the transform of the advective term, we have that:

$$\widehat{\frac{\partial}{\partial x_j}\left(u_iu_j\right)} = ik_j \int\limits_{\vec{k}=\vec{p}+\vec{q}} \hat{u}_i(\vec{q})\hat{u}_j\left(\vec{k}-\vec{q}\right)d\vec{q}. \tag{4.32}$$

As already mentioned, the second parentheses of Eq. (4.27) must belong to the plane π, that is,

$$\hat{R}_{i\pi} = ik_i\hat{p} + ik_j \int\limits_{\vec{k}=\vec{p}+\vec{q}} \hat{u}_i(\vec{q})\hat{u}_j\left(\vec{k}-\vec{q}\right)d\vec{q} - \hat{f}_i \in \pi. \tag{4.33}$$

This vectorial sum is shown in Fig. 4.12, from which it can be concluded that this sum is equal to the projection, on the plane π, of the sum of the transformed advective term and the term $\hat{f}_i$:

$$\begin{aligned}\hat{R}_{i\pi} &= \left\{\imath k_i\hat{p} + \imath k_j \int\limits_{\vec{k}=\vec{p}+\vec{q}} \hat{u}_i(\vec{q})\hat{u}_j\left(\vec{k}-\vec{q}\right)\mathrm{d}\vec{q} - \hat{f}_i\left(\vec{k}\right)\right\}_\pi \\ &= \wp_{ij}\left(\vec{k}\right)\left\{\left[\imath k_m \int\limits_{\vec{k}=\vec{p}+\vec{q}} \hat{u}_j(\vec{q})\hat{u}_m\left(\vec{k}-\vec{q}\right)\mathrm{d}\vec{q} - \hat{f}_j\right]\right\}.\end{aligned} \tag{4.34}$$

Thus, substituting these results in Eq. (4.27) and considering constant physical properties, the Navier–Stokes Equation transformed into Fourier Space is obtained:

$$\left(\frac{\partial}{\partial t} + \nu k^2\right)\widehat{u_i}\left(\vec{k},t\right) = -\wp_{ij}\left(\vec{k}\right)\left\{\imath k_m \int\limits_{\vec{k}=\vec{p}+\vec{q}} \hat{u}_m(\vec{q},t)\hat{u}_j\left(\vec{k}-\vec{q},t\right)\mathrm{d}\vec{q} - \frac{\hat{f}_j}{\rho}\right\} \tag{4.35}$$

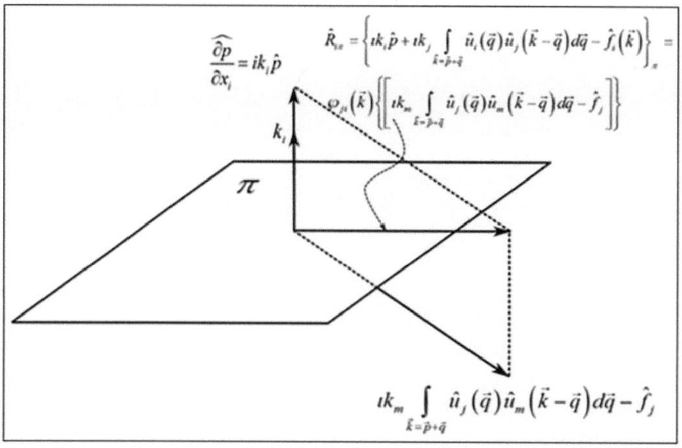

Fig. 4.12 Scheme of projections of transforms in Fourier Space, onto the π plane

It is observed that the Navier–Stokes Equations in Fourier Space do not depend on the concept of pressure, contrary to what happens in physical space. It is worth remembering that the hypothesis of incompressible flows used.

4.3.2.4 Transformation of the Thermal Energy Balance Equation

The transformation of the thermal energy balance equation, Eq. (4.20), is carried out using the previous process, which results in:

$$\left(\frac{\partial}{\partial t}+\alpha k^2\right)\hat{T}\left(\vec{k},t\right)=\imath k_j\int\limits_{\vec{k}=\vec{p}+\vec{q}}\hat{T}(\vec{q})\hat{u}_j\left(\vec{k}-\vec{q}\right)\mathrm{d}\vec{q}+\frac{\hat{\phi}\left(\vec{k},t\right)}{\rho C_p}. \tag{4.36}$$

To determine the transform $\hat{\phi}\left(\vec{k},t\right)$, this function is first transcribed into physical space:

$$\begin{aligned}\phi(\vec{x},t)=\mu\Bigg\{&2\left[\left(\frac{\partial u}{\partial x}\right)^2+\left(\frac{\partial \vartheta}{\partial y}\right)^2+\left(\frac{\partial w}{\partial z}\right)^2\right]\\&+\left(\frac{\partial u}{\partial y}+\frac{\partial \vartheta}{\partial x}\right)^2+\left(\frac{\partial u}{\partial z}+\frac{\partial w}{\partial x}\right)^2+\left(\frac{\partial \vartheta}{\partial z}+\frac{\partial w}{\partial y}\right)^2\Bigg\}.\end{aligned} \tag{4.37}$$

The transform of the first term is developed below:

$$\widehat{\left(\frac{\partial u}{\partial x}\right)^2}\left(\vec{k},t\right) = \widehat{\frac{\partial u}{\partial x}\frac{\partial u}{\partial x}}\left(\vec{k},t\right) = \int\limits_{\vec{k}=\vec{p}+\vec{q}} \widehat{\frac{\partial u}{\partial x}}(\vec{p},t)\widehat{\frac{\partial u}{\partial x}}(\vec{q},t)\mathrm{d}\vec{p}\mathrm{d}\vec{q}$$

$$= \int\limits_{\vec{k}=\vec{p}+\vec{q}} \iota\vec{p}\hat{u}(\vec{p},t)\iota\vec{q}\hat{u}(\vec{q},t)\mathrm{d}\vec{p}d\vec{q}$$

$$= -\int\limits_{\vec{k}=\vec{p}+\vec{q}} \vec{q}\left(\vec{k}-\vec{q}\right)\hat{u}(k-\vec{q},t)\hat{u}(\vec{q},t)\mathrm{d}\vec{q}. \quad (4.38)$$

The transforms of the other terms in Eq. (4.37) are obtained analogously to the procedure in Eq. (4.38).

4.3.3 Physical Interpretation of the Transformed Equations Terms

Each term of the transformed equations, in Fourier Space, can be interpreted physically; this interpretation is complementary and even richer in Fourier Space, when compared to the interpretations that can be obtained in physical space. These interpretations are presented below.

Velocity: In physical space, velocity is associated with a "fluid particle" in the sense of the continuum hypothesis. This particle is illustrated in Fig. 4.13a over which a velocity vector is placed. It depends on its position and on the time at which it is being observed. On the other hand, the transformed velocity $\hat{u}_\ell(\vec{k},t)$ represents an average velocity value, relative to a given wave number, $\vec{k} = 2\pi(1/\lambda_1, 1/\lambda_2, 1/\lambda_3)$, where $\vec{\lambda} = (\lambda_1, \lambda_2, \lambda_3)$ is the wavelength vector that represents the characteristic length of the turbulent structure in the three coordinate directions. This average is carried out by the transform that acts on the entire volume and "searches" each structure, present in the domain, to calculate the average value for the wavelength in question. In Fig. 4.12a and b, the coexistence of several turbulent structures of different characteristic sizes is observed. It is also clear that a dynamic structure of a given size can appear in different places in the considered volume. It is important to emphasize that this is a space where information is expressed through complex functions. Therefore, the interpretations presented here are intuitive in nature.

Temperature: the temperature in physical space, $T(\vec{x},t)$, represents the thermodynamic state of a fluid particle, Fig. 4.13a. In Fourier Space, $\hat{T}\left(\vec{k},t\right)$ represents the temperature of a turbulent structure with wave number $\vec{k}$, which is obtained by taking the average over all turbulent structures contained in the material volume under analysis, characterized by that wave number, as illustrated in Fig. 4.13b).

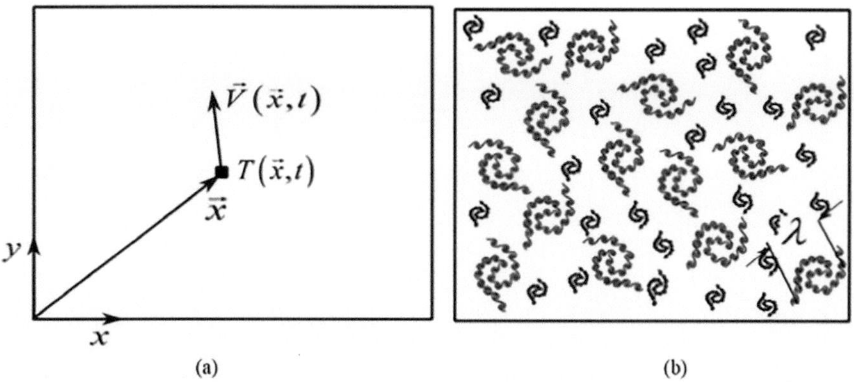

Fig. 4.13 Illustration of a fluid particle in physical space (**a**) and turbulent structures of different characteristic sizes or different wavelengths λ, in Fourier Space (**b**)

The term $\partial\hat{u}_\ell\left(\vec{k},t\right)/\partial t$: this term represents the rate of variation of the linear momentum contained in a turbulent structure characterized, statistically, by a wave number $\vec{k}$. In physical space, the corresponding term, that is, $\partial u_\ell(\vec{x},t)/\partial t$, represents the rate of variation of the linear momentum contained in a fluid particle, located in a spatial position $\vec{x}$ and at a time t. It should be noted that the concept of fluid particle is associated with the continuum hypothesis. In Fig. 4.13a and b, the two concepts are illustrated: fluid particle in physical space and turbulent structure in Fourier Space. In Fig. 4.14a, the fluid particle is illustrated exchanging linear momentum with its physical surroundings, which promotes, in part, the rate of variation of the linear momentum contained in that same particle. Figure 4.14b illustrates the nonlinear exchange of diverse physical quantities, for example, linear momentum, between a turbulent structure of wavelength λ and all other structures that compose the spectrum of turbulent structures.

The term $\partial\hat{T}\left(\vec{k},t\right)/\partial t$: represents the rate of change of thermal energy contained in a turbulent structure of wavenumber $\vec{k}$. In physical space, the corresponding term, that is, $\partial T(\vec{x},t)/\partial t$, represents the rate of variation of thermal energy contained in a fluid particle, located in a spatial position $\vec{x}$ and at a time t.

The terms $\nu k^2\hat{u}_l\left(\vec{k},t\right)$, $\alpha k^2\hat{T}\left(\vec{k},t\right)$ **and** $\hat{\phi}\left(\vec{k},t\right)$: the first two terms represent, respectively, the net diffusive flux of linear momentum and the net diffusive flux of thermal energy, between a turbulent structure of wavenumber $\vec{k}$ and all other wavelengths that make up the spectrum of structures. It is important not to confuse these terms, which represent diffusion processes, with the term that represents the transformation of energy through viscous effects, $\hat{\phi}\left(\vec{k},t\right)$, which appears only in the thermal energy equation. The Navier–Stokes equations do not account for this viscous transformation effect, but only the diffusion of physical quantities. In the

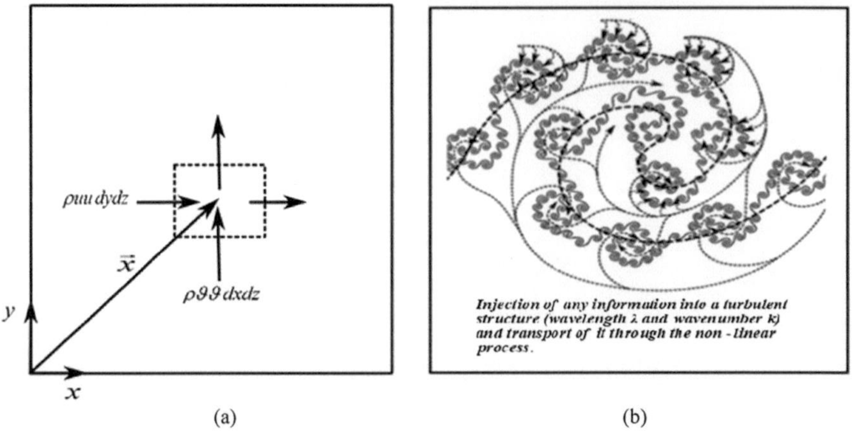

Fig. 4.14 Illustration of advective transport of linear momentum between a fluid particle and its surroundings, in physical space (**a**) and between a turbulent structure of wavelength λ and all other structures that interact with it non-linearly

thermal energy equation, the specific power of energy transformation due to viscous effects is represented by the term ϕ.

The transform of ϕ (the viscous transformation function) follows Eq. (4.37). The diffusion processes depend on the square of the wavenumber norm k^2: with this we conclude that the greater the wavenumber modulus of a turbulent structure, the greater the diffusive effects involving this turbulent structure. In other words, the smaller the turbulent structure, the higher the corresponding wave number and the more intense the diffusive effects of linear momentum and thermal energy. This same interpretation can be given to the specific power function of viscous transformation, $\hat{\phi}\left(\vec{k}, t\right)$.

The smaller the turbulent structures, the more intense the process of viscous transformation of kinetic energy into thermal energy. This is because the smallest structures, that is, shorter wavelengths, are characterized by higher wave numbers. These diffusive terms, in Fourier Space, depend on the square of the wavenumber norm, k^2. The terms $\nu\partial^2 u_\ell/\partial x_j \partial x_j$ and $\alpha\partial^2 T/\partial x_j \partial x_j$ in physical space, represent the net diffusive flux of linear momentum and thermal energy between a fluid particle and its surroundings. The specific viscous transformation power function, $\phi(\vec{x}, t)$, represents the transformation of kinetic energy into thermal energy in each fluid particle, due to the interaction with the neighboring particles.

The term $\wp_{ij}\left(\vec{k}\right)\left\{\iota k_m \int_{\vec{k}=\vec{p}+\vec{q}} \hat{u}_m(\vec{q}, t)\hat{u}_j\left(\vec{k}-\vec{q}, t\right)\mathrm{d}\vec{q} - \hat{f}_j\right\}$: this term is the projection, onto the plane π, of the transform into the Fourier Space of the sum $\partial\left(u_i u_j\right)/\partial x_j + \partial p/\partial x_i - f_i$. The term $\iota k_m\left(\vec{k}\right) \int_{\vec{k}=\vec{p}+\vec{q}} \hat{u}_m(\vec{q})\hat{u}_i\left(\vec{k}-\vec{q}\right)\mathrm{d}\vec{q}$ is the transform of the nonlinear term, $\partial\left(u_i u_j\right)/\partial x_j$, of the Navier–Stokes Equations. Both represent the net advective flux of linear momentum. In physical space, this is the liquid advective flux between a fluid particle and its surroundings, as illustrated in Fig. 4.14a. Despite being a differential quantity, it represents a net flux. The product $u_m u_j$ represents the linear momentum flux. The divergence of this flux, $\partial\left(u_i u_j\right)/\partial x_j$, represents the sum of what leaves and what enters the fluid particle, that is, it represents the net advective flux of the linear momentum through the control surface that delimits the fluid particle. In Fourier Space, there is a nonlinear exchange of linear momentum between a wavenumber $k_i = 2\pi/\lambda_{ki}$ and all other wavenumbers $p_i = 2\pi/\lambda_{pi}$ and $q_i = 2\pi/\lambda_{qi}$.These wavelengths are illustrated in Fig. 4.15, and the respective wavenumbers are illustrated in Fig. 4.16. The convolution integral, resulting from the transformation of this nonlinear term, represents the sum of all possible interactions, when p_i and q_i vary infinitesimally, with the restriction $k_i = p_i + q_i$. This vector sum is illustrated in Fig. 4.16.

The convolution integral is carried out in a fixed time. All non-linear interactions happen simultaneously and not progressively, as presented in the literature. The transfer of information is non-linear and, consequently, occurs simultaneously between all the turbulent structures that coexist in space, which is also illustrated in Fig. 4.16b. The idea of sequential information transfer leads to a misperception of linear transfer, instead of very strong nonlinear interactions.

These interactions between all the wavenumbers that make up the spectrum are known as triadic interactions. The convolution integral is the sum, at the same time, of all possible triadic interactions. The triadic interactions require the constraint that the wavenumber vector k_i remains fixed, while integration is being performed by sweeping the vectors p_i and q_i. This integral computes the net exchange of physical quantities between a selected fixed wavenumber, k_i, and all other possible

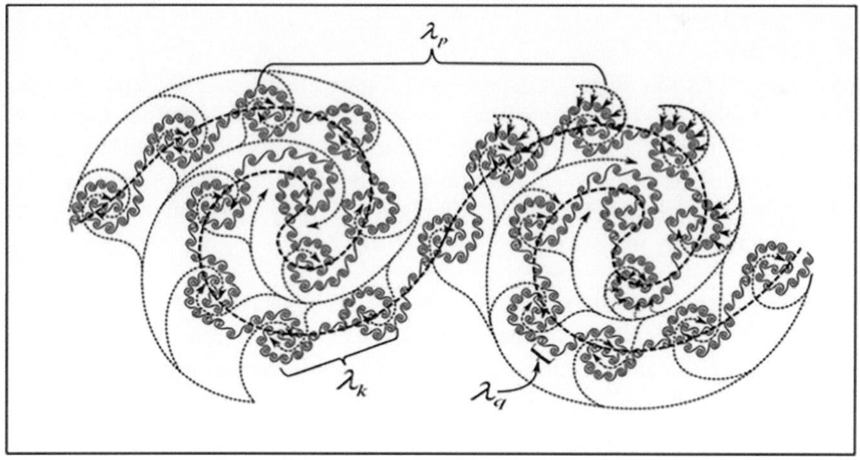

Fig. 4.15 Schematic representation of the existence of multiple turbulent structures, λ_p, λ_q and λ_k, and the nonlinear process of physical quantities exchange between them

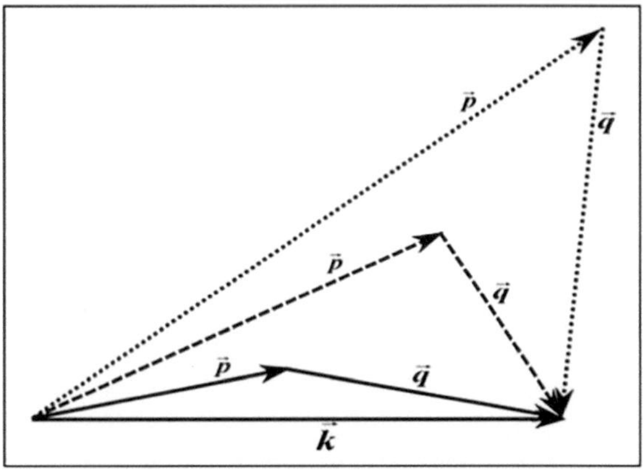

Fig. 4.16 Schematic representation of the existence of multiple turbulent structures and illustration of the triadic interactions between the wavenumber $\vec{k}$ and all other structures, $\vec{p}$ and $\vec{q}$, that interact with it

wavenumbers that are computed by the continuous variation of p_i and q_i, such that $k_i = p_i + q_i$.

Thus, in Fourier Space, the convolution integral relative to the nonlinear term of the Navier–Stokes Equations represents the net exchange of linear momentum between all existing turbulent structures, inside the analysis volume, and the chosen turbulent structure. As illustrated in Fig. 4.15, each structure feeds on linear momentum coming from its neighborhood, which is also the stage for other structures. If a given structure gains linear momentum through the process of advective transfer, other structures must lose it through the same process. This process is highly non-linear and represents the simultaneous exchange between all coexisting turbulent structures in the domain under analysis.

The idea of progressive transfer is expressed by "Richardson Parody" [10], which can be quoted as: "The largest turbulent structures feed the smaller ones, which in turn feed those even smaller and so on, until the Kolmogorov Structures". Many authors represent this interpretation through figures like Fig. 4.17. This interpretation contradicts the convolution integral, which clearly shows that the transfer of information between the multiple turbulent structures cannot be sequential but simultaneous between all the turbulent structures, involving them all at the same time. The reader must compare and identify the difference in the physical forms of information transfer qualitatively represented in Figs. 4.15 and 4.17.

The term $\imath k_j \int_{\vec{k}=\vec{p}+\vec{q}} \hat{T}(\vec{q})\hat{u}_j\left(\vec{k}-\vec{q}\right)\mathrm{d}\vec{q}$: Analogously to the preceding term, this term represents the net flux of thermal energy between a turbulent structure of wavenumber $\vec{k}$ and the other structures $\vec{p}$ and $\vec{q}$. Again, this is a non-linear process, represented by the drawing in Figs. 4.15 and 4.16.

The pressure gradient: the pressure field gradient, Eq. (4.20), represents an external force upon a fluid particle, promoted by neighboring particles acting on it. In Fourier Space, this term was eliminated by the process of projecting the nonlinear term onto the plane π. It should be noted that this is only valid for incompressible flows.

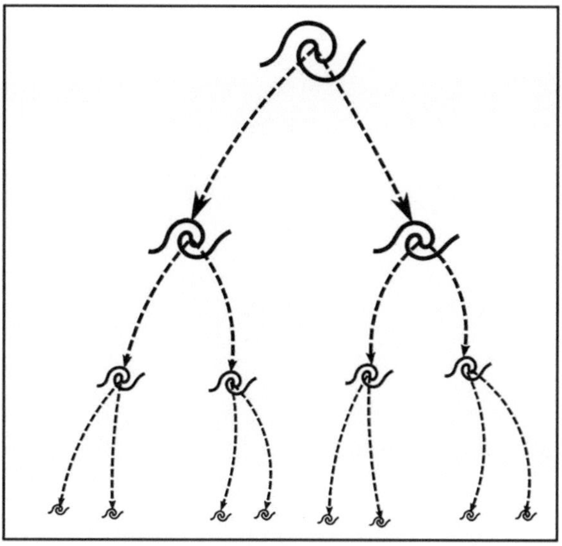

Fig. 4.17 Schematic representation of the existence of multiple turbulent structures and the linear process of exchanging information between them (physically inconsistent)

The solution to the System of Eqs. (4.35) and (4.36) requires the solution of the nonlinear convolution integrals. This process is computationally expensive. An alternative is the so-called pseudo-spectral methods: (a) the products $u_i u_j$ and $u_j T$ are resolved in physical space; (b) the products are transformed into Fourier Space; (c) the transformed velocity and temperature fields are obtained; (d) the inverse transform is carried out and (e) the velocity and temperature fields in the physical space are determined, and the process is restarted. The biggest limitation of this methodology is that it only applies to periodic flows, since the Fourier transforms are discrete. As an alternative, mixed finite difference schemes have been used for non-periodic directions and DFT for periodic flow directions.

4.4 Spectral Tensor and Turbulent Kinetic Energy Spectrum

The spectral tensor is defined as the Fourier transform of the second-order statistical moment, obtained from the correlation at two points, between the velocity fluctuations relative to the coordinated directions:

$$\hat{U}_{ij}\left(\vec{k}, t\right) = \left(\frac{1}{2\pi}\right)^3 \int\limits_{\Omega} e^{-i\vec{k}.\vec{r}} U_{ij}(\vec{r}, t) d\vec{r}, \tag{4.39}$$

where

$$U_{ij}(\vec{r}, t) = \left\langle u_i'(\vec{r}, t) u_j'(\vec{r} + \vec{x}, t) \right\rangle, \tag{4.40}$$

is a second-order statistical moment. The hypotheses of homogeneity and isotropy were used, since the left side of Eq. (4.40) was independent of position $\vec{x}$. Setting $i = j$, in Eq. (4.40), we obtain the trace of the tensor $\hat{U}_{ij}\left(\vec{k}, t\right)$:

$$\hat{U}_{ii}\left(\vec{k}, t\right) = \left(\frac{1}{2\pi}\right)^3 \int_{\Omega} \mathrm{e}^{-i\vec{k}.\vec{r}} U_{ii}(\vec{r}, t) \mathrm{d}\vec{r}. \tag{4.41}$$

Based on the spectral tensor trace, the spectral density of turbulent kinetic energy is defined, such that:

$$\frac{1}{2} U_{ii}(\vec{r} = 0, t) \equiv \int_0^{\infty} E(k, t) \mathrm{d}k, \tag{4.42}$$

where $E(k, t)$ is the spectral density of turbulent specific kinetic energy. This nomenclature is because $U_{ii}(t)(\mathrm{J/kg})$ must be obtained by integrating $E(k, t)\left[(\mathrm{J/kg})/(1/\mathrm{m})\right]$, over the band $0 \leq k \leq \infty$. In the denominator of this last term, the unit $[1/\mathrm{m}]$ appears referring to the fact that $E(k, t)$, in Fourier Space, represents specific energy $\left[\mathrm{J/kg}\right]$ per unit wavelength $[1/\mathrm{m}]$. To explain it another way, the word "density" refers to the fact that specific energy appears per unit wavenumber.

In Eq. (4.42), Einstein's sum rule is used for repeated indices. In Fig. 4.18, the spectral distribution of turbulent kinetic energy is visualized, which is a powerful way of understanding how the turbulent activity of a flow occurs as a function of the sizes of the different turbulent structures that characterize the flow. This figure shows a continuous spectrum of energy, which is characterized by a band of wavelengths over which the injection of turbulent kinetic energy takes place. This energy is also transferred to the largest structures or lowest wave numbers, as well as to the smallest structures, the highest wave numbers.

The transfer process is non-linear and should not be represented sequentially, which would give the impression of linear transfer. In Fig. 4.18, a composition of turbulent structures of different sizes is shown. Only three wavelengths were drawn so as not to overload the figure. However, when it comes to a turbulence regime, this figure must contain the entire spectrum, full and continuous, of turbulent structures. It is an enormous wealth of structures coexisting in space and exchanging information in a non-linear and simultaneous way. The transfer represented in this figure occurs in a fixed time, and this statistic, represented here, is carried out in space. This representation can also be made in the frequency domain, when statistics are performed using temporal samples, at a fixed point in space.

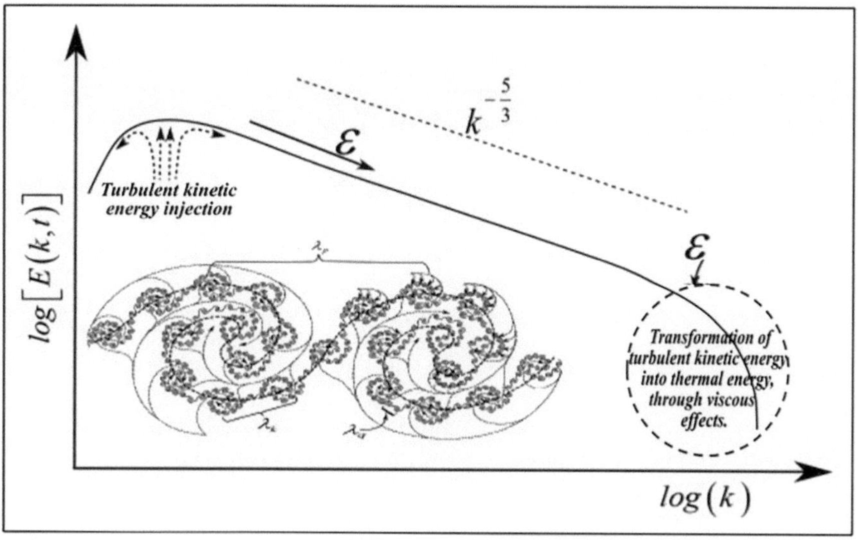

Fig. 4.18 Spectral density of turbulent specific kinetic energy; wavenumber band of energy injection; inertial band; viscous transformation band

In this spectrum, there is a distribution of energy over the turbulent structures of different wavelengths. Intuitively, the largest structures (smallest wave numbers) are the ones that carry the greatest amount of energy. The higher the wave number, the smaller the turbulent structure and the smaller the amount of energy transported by.

In Fig. 4.18, there are different characteristic bands of the spectrum. The first of these is the inertial band, where the process of non-linear energy transfer between different structures predominates. The second band is made up of the largest turbulent structures, which carry the greatest amount of energy. As the wave number increases, that is, the size of the turbulent structures decreases, the local Reynolds number also decreases, and the viscous transformation process begins to predominate over the nonlinear effects. This is the third band of interest that makes up the spectrum.

Each flow has a band of wavenumbers characteristic of the energy injection. This band of wavelengths can also be represented by a single length, also known as the integral characteristic length Due to the physical process of energy injection into the flow, it is divided following two courses, that is, the course of direct energy transfer, from the largest to the smallest structures, and the course of inverse energy transfer, that is, transfer from the smallest to the largest structures.

In Fig. 4.19, the spatial development of the flow inside a rectangular cavity is illustrated. This example allows us to understand how energy is injected into turbulent structures and how it is distributed across the energy spectrum.

The recirculating flow inside the cavity is fed by the average flow inside the main channel. The energy transfer process takes place through the shear layer that forms the interface between the two flows, the secondary one inside de cavity and the main flow, inside the channel. In fact, the Kelvin–Helmholtz Turbulent Structures capture

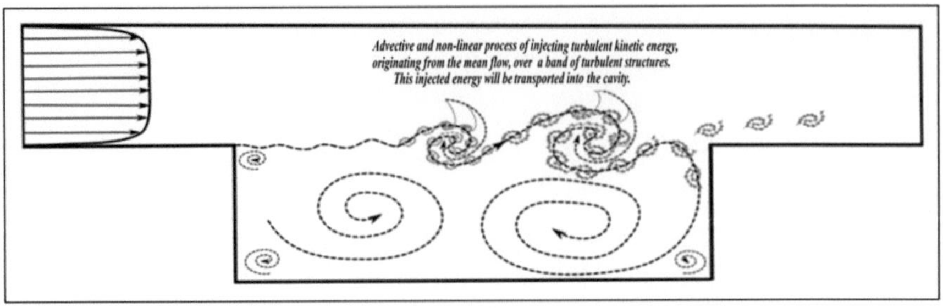

Fig. 4.19 Flow in a channel with a cavity, illustrating the process of energy transfer from the medium flow to the turbulent structures of the shear layer and into the cavity

energy from the main conduit and pump it into the cavity, inducing recirculations of large sizes and low wave numbers. This process is represented by the spectrum illustrated in Fig. 4.18.

Observing the Kelvin–Helmholtz turbulent structures on the shear layer, they are composed of other even smaller structures (harmonics), but of a similar nature to the larger ones. The energy for the formation and maintenance of these smaller structures must be supplied by the energy-receiving turbulent structures, as illustrated in Fig. 4.19. The physical process of the so-called direct cascade of energy is explained, that is, the nonlinear transport, which occurs from the largest to the smallest structures. A turbulent kinetic energy balance equation in Fourier Space can be useful to understand the energy transfer process along the spectrum.

4.5 Balance Equation for Spectral Density of Turbulent Specific Kinetic Energy

Starting from the Navier–Stokes equations, which are also valid for velocity fluctuations, multiplying them by $\widehat{u'_j}(k,t)$, averaging < >, manipulating algebraically, we obtain the balance equation for the spectral tensor and, consequently, for the trace of this tensor:

$$\frac{\partial \hat{U}_{ij}(k,t)}{\partial t} + 2\nu k^2 \hat{U}_{ij}(k,t) = \mathrm{P}_{ijm} \int_{\vec{k}=\vec{p}+\vec{q}+\vec{r}} I\left[\widehat{u'_i}(\vec{r},t)\widehat{u'_j}(\vec{p},t)\widehat{u'_m}(\vec{q},t)\right] \mathrm{d}\vec{p}\mathrm{d}\vec{q}\mathrm{d}\vec{r}. \tag{4.43}$$

The I function depends on third-order statistical moments. It is defined in Lesieur [6]. The statistical moments, arguments of this function, represent tetradic nonlinear interactions, that is, interactions between four simultaneous wavenumbers. By adding the three components of the spectral tensor trace $\hat{U}_{ii}$ and using the definition of the spectral density of turbulent kinetic energy, we obtain the balance equation:

$$\frac{\partial E(k,t)}{\partial t} + 2\nu k^2 E(k,t) = T(k,t). \tag{4.44}$$

The first term of this equation represents the rate of variation of the spectral density of turbulent kinetic energy, $E(k,t)$, contained in a turbulent structure with wave number, whose modulus is k; the second term represents the net flux of spectral density of turbulent kinetic energy, by the process of molecular diffusion, between the turbulent structure and the entire band of wavenumbers that make up the spectrum. With the term $T(k,t)$, we model the net flux of $E(k,t)$, between one given turbulent structure and the entire band of wavenumbers that make up the spectrum. The tensor $P_{ijm}\left(\vec{k},t\right)$ is given by:

$$P_{ijm}\left(\vec{k},t\right) = k_m \wp_{ij}\left(\vec{k},t\right) + k_j \wp_{im}\left(\vec{k},t\right). \tag{4.45}$$

In Fig. 4.18, the distribution of $E(k,t)$ is illustrated, as well as the nonlinear process of energy transfer between different turbulent structures that make up the spectrum. In Eq. (4.44), the temporal variation of the amount of energy contained in a turbulent structure with wave number k is assumed to be transient, $\partial E(k,t)/\partial t \neq 0$.

This requires that the spectrum $E(k,t)$ be non-stationary, that is, that the turbulence is statistically transient. For example, turbulence may be developing or may be in a temporal decay regime. In Fig. 4.20, the transient process of establishing a turbulent kinetic energy spectrum is illustrated. It starts with an initial energy distribution concentrated over a narrow spectral band over the energy injection wavenumbers. As time passes, the injected energy is distributed through the physical mechanisms modeled in the previous equation, establishing the spectrum corresponding to fully developed turbulence.

It will be seen that this steady state of fully developed turbulence follows Kolmogorov's law, which establishes that the inertial zone of the spectrum $E(k,t)$ must have a slope of -5/3, as illustrated in Fig. 4.20, for the distribution $\log[E(k,t)] \times \log[k]$.

4.6 Kolmogorov Theory

This is the best-known theory about isotropic turbulence, established by Kolmogorov [4, 5]. Its basis, simple but ingenious, is dimensional analysis. The existence of a specific power of viscous transformation $\varepsilon\left[\mathrm{W/kg}\right]$ is assumed. To have a steady state, it is necessary that all the injected energy, over the injection wavenumber band, is transformed by viscous effects. This is the equilibrium hypothesis. On the time-related curve t_4, Fig. 4.20, a turbulent flow is illustrated which is in a statistically

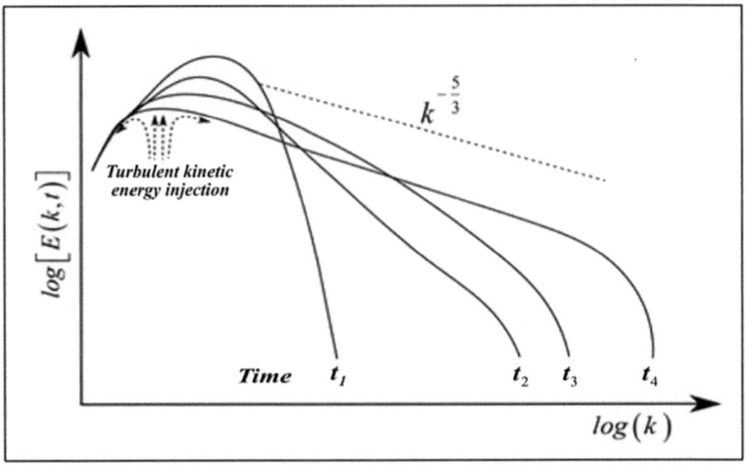

Fig. 4.20 Distribution of the spectral density of turbulent specific kinetic energy, illustrating the process of establishing a spectrum

stationary regime. The injection wave number, k_I, is defined such that it statistically represents the entire band of injection wavelengths. In Kolmogorov theory, it is assumed that the energy spectrum, for wavenumbers greater than k_I, depends only on the viscous transformation power ε and wave number k. Carrying out a dimensional analysis based on the Theorem of π's, of Vaschy-Buckingham, Kolmogorov obtained the following expression:

$$E(k) \approx \varepsilon^{\alpha} k^{\beta}. \tag{4.46}$$

Determining the values of α and β, by dimensional inspection and defining a proportionality constant, C_K, which results in the following expression:

$$E(k) = C_K \varepsilon^{2/3} k^{-5/3}, \tag{4.47}$$

where $C_K = 1,4$ is the Universal Kolmogorov constant, determined analytically. With this equation, when linearized through logarithm, the distribution illustrated in Fig. 4.20, the inertial zone has an inclination $-5/3$, Obukhov [7, 8]. For more details on the development of this theory, the reader can refer to the review article Frisch (1995)[1].

A variety of material and computational experiments have been carried out, aiming to prove this law, for different values of the Reynolds Number. All of them have resulted in the proof of Kolmogorov's law, expressed with Eq. (4.47).

To demonstrate this fact, the results of a large range of experiments, carried out for different physical situations, such as wakes, grid turbulence, channels, among others, are shown in Fig. 4.21. To allow all of them to be presented in a single figure, the spectral density of turbulent kinetic energy was scaled using the Kolmogorov theory.

So, $E_{11}(k_1)/(\varepsilon\nu^5)^{\frac{1}{4}}$, it is a dimensionless quantity with which all experiments are placed on the same scale. Similarly, the wavenumber is scaled with the Characteristic Length of Kolmogorov, $k_1\eta$, which will be defined, in Sect. 4.7.

The different symbols represent the different experiments. The number that appears in front of each work description represents the value of the **Reynolds Number**, Re_λ, based on the Micro Length of Taylor, λ (Eq. (4.75)). A range of results relative to various values of the **Reynolds Number**, $23 \leq Re_\lambda \leq 3,180$, is presented in this figure. The solid lines were obtained from a model proposed and presented by Pope [9], which expresses Kolmogorov's law, extended to the regions of large turbulent structures: low wavenumber values, which precede the energy injection region, as well as the viscous transformation region; high wave numbers, which follow the inertial zone of the spectrum. This equation is reproduced in this text, as follows:

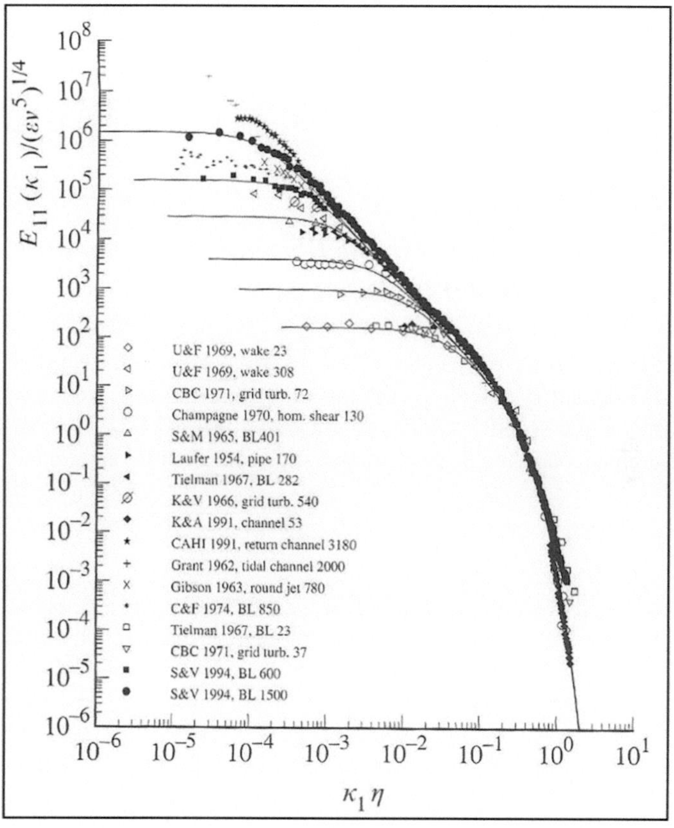

Fig. 4.21 Spectral density of turbulent specific kinetic energy, relative to the velocity component in the flow direction. Figure constructed with data from material experiments, Saddoughi and Veeravalli [11]

$$E(k) = C\varepsilon^{2/3}k^{-5/3}f_L(kL)f_\eta(k\eta), \tag{4.48}$$

where $f_L(kL)$ and $f_\eta(k\eta)$ are functions proposed to model, respectively, the region of so-called energy-carrying structures, large structures, and the region of viscous transformation of turbulent kinetic energy. Both functions must be in the inertial region of the energy spectrum and reproduce Kolmogorov's Law, presenting a slope in $-$ 5/3. For more details on these functions, the reader is invited to consult the Reference Pope [9]. For all these experiments, Kolmogorov's law is recovered. The dependence of the spectrum width on the **Reynolds Number** is presented. The greater the value of this dimensionless parameter, the greater the width of the spectrum.

4.7 Characteristic Quantities of Turbulence

Before delving into any type of study on turbulent flows, it is necessary to know the orders of magnitude of the variables involved in the phenomena, through the characteristic quantities of turbulence. These quantities are related to time, length, speeds, energy and vorticity.

4.7.1 Kolmogorov Characteristic Quantities—Smallest Turbulent Structures

To launch the concept of Kolmogorov Viscous Transformation Length, a turbulent structure of characteristic size r is introduced, originating in a fluid of viscosity ν, with a characteristic speed $\vartheta_r = \vec{V}$, as illustrated in Fig. 4.22. A local Reynolds Number is defined, based on these characteristic quantities of the turbulent structure:

$$\mathrm{Re}_r = \frac{\vartheta_r r}{\nu}. \tag{4.49}$$

The square of this parameter represents the relative importance between inertia forces and viscous forces. It is assumed that the size r is in a region of the spectrum where, according to Kolmogorov's Theory: $\vartheta_r = (\varepsilon r)^{1/3}$, Lesieur [6]. Replacing ϑ_r in the equation above, we have: $Re_r = (\varepsilon r^4)^{1/3}/\nu$. Considering that, for this size r, viscous effects are smaller than non-linear effects, it can then be stated that Re_r will be greater than 1. If r decreases, Re_r also decreases, and if $r = \eta$, where η is defined so that $Re_\eta = (\varepsilon\eta^4)^{1/3}/\nu = 1$, results in the well-known Kolmogorov Characteristic length:

$$\eta = \left(\frac{\nu^3}{\varepsilon}\right)^{1/4}. \tag{4.50}$$

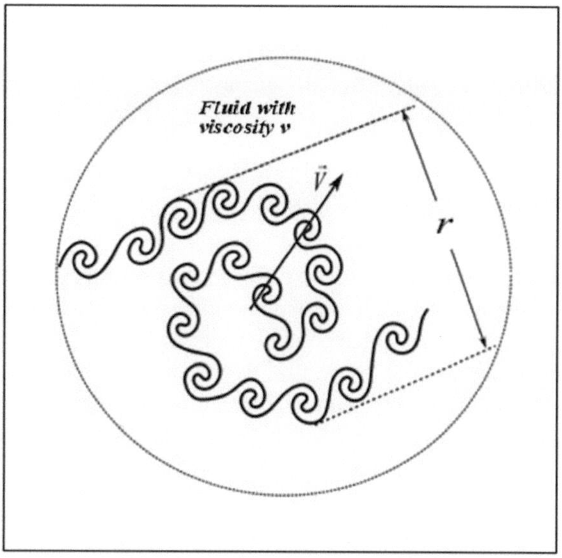

Fig. 4.22 Illustration of a turbulent structure of characteristic size r, characteristic speed $\vartheta_r = \vec{V}$, formed in a fluid of viscosity ν

So, for $r < \eta$, Re_r becomes smaller than 1, and viscous effects begin to predominate over nonlinear effects. This length η is, by definition, the Kolmogorov characteristic viscous transformation length. Thus, disturbances with wavelengths shorter than η cannot be amplified, due to the predominance of viscous effects. This analysis allows us to understand why the turbulent kinetic energy spectrum drops so quickly as it approaches the Kolmogorov wave number, $2\pi/\eta$. As an example, the Kolmogorov characteristic length within the atmospheric boundary layer is of the order of 1 mm, whereas in grid turbulence, it is of the order of 0, 1 mm.

Carrying out a dimensional analysis and expressing the characteristic time as a function of ν and ε, we arrive at the following expression for this parameter, which represents the Kolmogorov characteristics time,

$$\tau = \left(\frac{\nu}{\varepsilon}\right)^{1/2}. \tag{4.51}$$

In a similar way, the Kolmogorov characteristics speed, vorticity and turbulent kinetic energy may be derived:

$$\vartheta_r = (\nu\varepsilon)^4, \tag{4.52}$$

$$\omega = \left(\frac{\varepsilon}{\nu}\right)^{1/2}, \tag{4.53}$$

$$e = (\nu\varepsilon)^{1/2}. \tag{4.54}$$

These characteristic quantities refer to the smallest possible turbulent structures that make up the spectrum, relative to a turbulent flow. They are known as Kolmogorov characteristic quantities.

4.7.2 Large Structures

The largest structures of a flow are determined by the geometry of the problem. Consider L the characteristic length of large structures, which is of the order of the geometry over which the flow occurs. The turbulent structures known as the most energized, in the sense that they are those that receive energy from some source external to the problem of interest, have dimensions characterized by the so-called integral length, ℓ_I. In Fig. 4.23a, the flow downstream of a partially immersed plate is shown. The characteristic length of the geometry is L, which characterizes the part of the plate immersed in the flow. The integral length ℓ_I refers to the most energized turbulent structures. At Fig. 4.23b, the flow over a circular cylinder of diameter D, which represents the characteristic dimension of the problem. The integral structure is also marked in the drawing. In both cases, the integral quantities of the two problems are of the order of the magnitudes L and D.

The characteristic velocity and length of the integral structures are considered as U and ℓ_I. With these two characteristic quantities, ℓ_I and U, the other characteristic quantities are defined, in the following order: time, vorticity and energy, which are given by the following equations:

$$t = \frac{\ell_I}{U}, \tag{4.55}$$

$$W = \frac{U}{\ell_I}, \tag{4.56}$$

$$E = U^2. \tag{4.57}$$

These quantities are relative to the turbulent structures said to be the most energized. They are also those in which energy is injected into the spectrum. The largest turbulent structures are not necessarily present in the flow. This fact is illustrated in Fig. 4.19, from which it can be seen that the largest turbulent structures are inside the cavity, but the energy receiving structures are the Kelvin–Helmholtz Structures, which receive energy from the channel flow and transport it to the cavity, to feed the largest structures. In such a case, the integral length ℓ_I is not related to the largest turbulent structures but rather to the energy receiving structures, Kelvin–Helmholtz Structures.

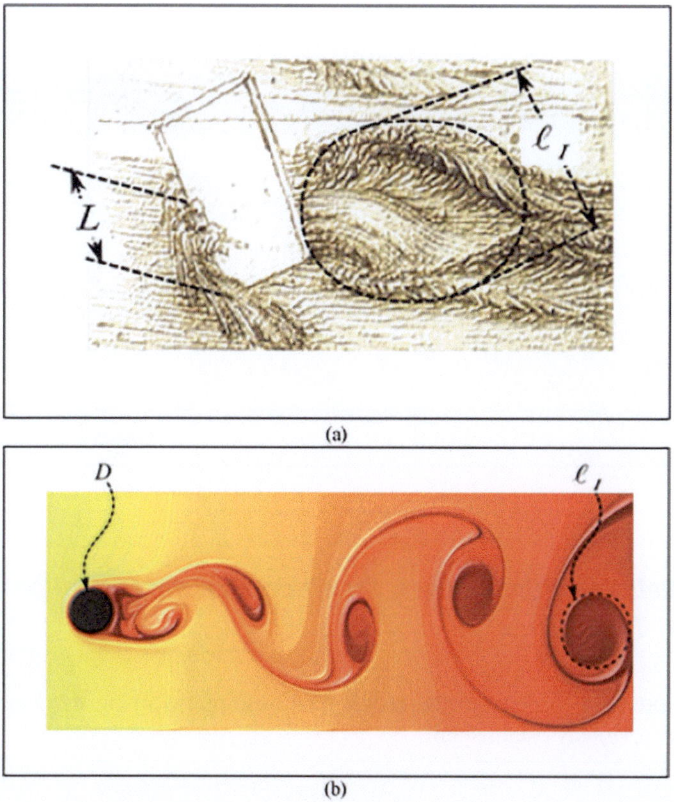

Fig. 4.23 Flows over immersed bodies: **a** plate partially immersed in a moving fluid and **b** flow over an immersed circular cylinder

4.7.3 Specific Transformation Power due to Viscous Effects

For fully developed turbulent flows, the equilibrium hypothesis can be made, whereby the specific power of viscous transformation, $\varepsilon[\mathrm{W/kg}]$, is equal to the specific power of kinetic energy injection in large structures: $(U^2/t)[\mathrm{W/kg}]$. Thus, the specific power of viscous transformation can be expressed as a function of viscosity-independent quantities, as schematically illustrated in Fig. 4.24. In this way, the energy transformation rate can be expressed as follows:

$$\varepsilon \approx \frac{U^2}{t} = \frac{U^3}{\ell_I}. \tag{4.58}$$

With this equation, the viscous transformation rate is modeled, based on parameters related to large structures and without the participation of viscosity.

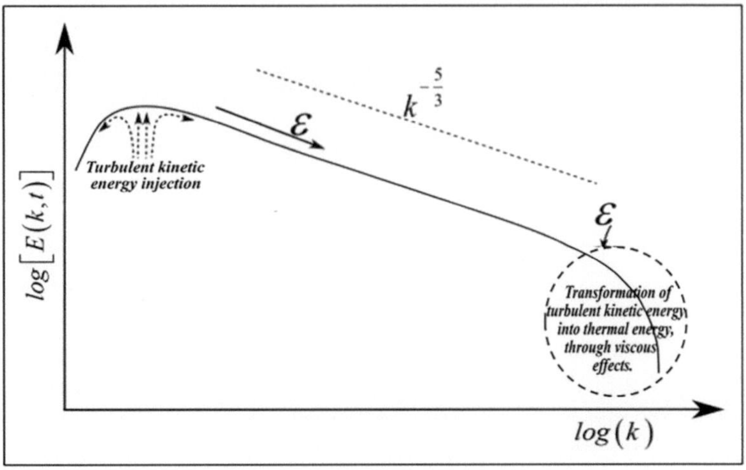

Fig. 4.24 Illustrative scheme of the equilibrium hypothesis

4.7.4 Relationships Between the Characteristic Quantities of Turbulence

It is now possible to deduce interesting relationships involving the characteristic quantities previously established:

$$\eta = \left(\nu^3 / \left(\frac{U^3}{\ell_I}\right)\right)^{1/4}. \tag{4.59}$$

so,

$$\frac{\ell_I}{\eta} = \mathrm{Re}_{\ell_I}^{3/4}. \tag{4.60}$$

Similarly,

$$\frac{t}{\tau} = \mathrm{Re}_{\ell_I}^{1/2}, \tag{4.61}$$

$$\frac{U}{\vartheta_r} = \mathrm{Re}_{\ell_I}^{1/4}, \tag{4.62}$$

$$\frac{\omega}{W} = \mathrm{Re}_{\ell_I}^{1/2}, \tag{4.63}$$

$$\frac{E}{e} = \mathrm{Re}_{\ell_I}^{1/2}. \tag{4.64}$$

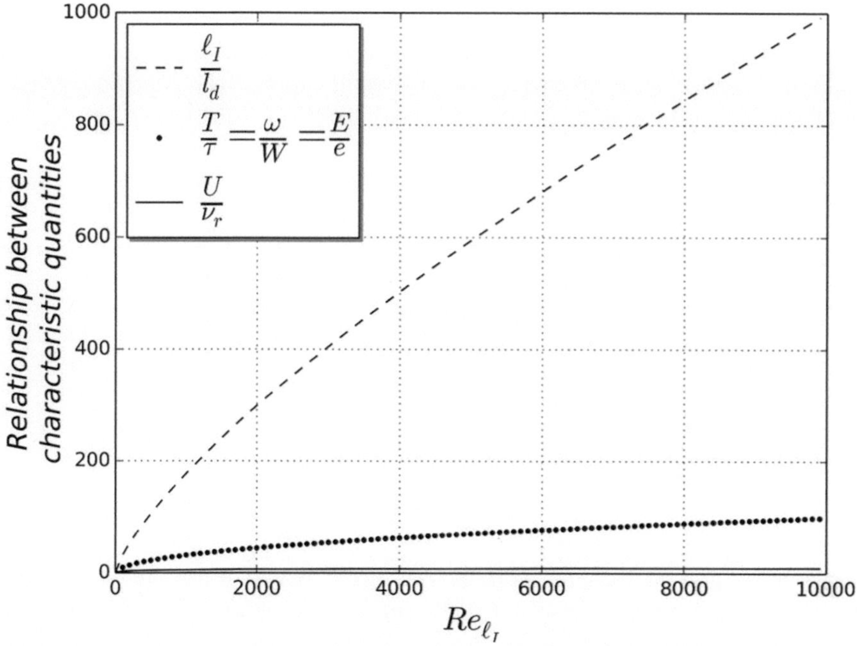

Fig. 4.25 Quantitative behaviors of characteristic magnitude relations

Some interesting facts are observed when analyzing these expressions that relate the characteristic quantities of turbulence with the Reynolds Number. For all of them, it is shown that the Kolmogorov characteristic quantities are much smaller than the integral quantities, except for the vorticity. Through Eq. (4.60), considering that the integral length is determined by the geometry of the problem, it is shown that the greater the value of the Reynolds Number, the smaller the Kolmogorov structures will become. The laws of variation with the Reynolds number are different, as illustrated in Fig. 4.25. The characteristic lengths (ℓ_I and η) move away from each other faster than the characteristic time, vorticity and energy. The characteristic velocities are those that move away from each other most slowly.

By the relation between the characteristic vorticities, Eq. (4.63), it is seen that the Kolmogorov turbulent structures have more vorticity than the integral structures and, conversely, for the characteristic energy, Eq. (4.64), integral structures carry a greater amount of energy when compared to Kolmogorov structures.

4.7.5 *Molecular Length Versus Kolmogorov Length*

Kolmogorov Structures are the smallest that can occur in a turbulent flow. It is important to check whether these structures can be influenced by molecular activity.

Consider ξ, the molecular mean free path. For gases, the characteristic molecular speed can be associated with the velocity of sound, c. From the kinetic theory of gases, it is shown that the kinematic viscosity can be expressed as a function of these two characteristic quantities, ξ and c, i.e., $\nu \approx c\xi$. So:

$$\xi \approx \frac{\nu}{c}. \tag{4.65}$$

It has already been shown that:

$$\eta = \left(\frac{U}{\nu}\right)^{-3/4} \ell_I^{1/4}. \tag{4.66}$$

Dividing Eq. (4.65) by Eq. (4.66), result:

$$\frac{\xi}{\eta} = \frac{M}{\left(Re_{\ell_I}\right)^{1/4}}, \tag{4.67}$$

where $M = U/c$ is the Mach number. In a first analysis of this equation, it can be said that the molecular characteristic length, ξ, is always much smaller than the characteristic length of the viscous transformation η, since, even for high values of the Mach Number, the **Reynolds Number** must still be much higher than that, so that this ratio would always be much smaller than unity. However, particular care must be taken because as the **Reynolds Number** increases, the Kolmogorov length, η, decreases, tending to molecular length. According to Lesieur [6], for Mach numbers above 15, these two lengths get closer to each other. This information is extremely important, as this would imply that the Navier–Stokes equations are no longer representative of the flows characterized by $M > 15$. In this case, alternative models that are not based on the continuum hypothesis would have to be used. For almost all of today's practical problems, the Mach Number does not exceed this mark, even for flows with strong compressibility and strong thermal effects.

The result of the relationships between molecular and Kolmogorov quantities allows us to ensure that turbulent flows are a problem that can be modeled considering that fluids are continuous media. Thus, for Newtonian fluids, the Navier–Stokes equations, the thermal energy balance equation and the continuity equation are adequate and sufficient for modeling turbulent flows. For non-Newtonian fluids, the Cauchy Equations can be closed with continuous models, specific to each family of fluids.

4.7.6 Full Length of Turbulence

So far, one-point correlations have been used, such as the intensity of turbulence, using temporal averages.: $\overline{u'(\vec{x})u'(\vec{x})}$ or the components of the Boussinesq-Reynolds tensor, such as $\overline{u_i'(\vec{x})u_j'(\vec{x})}$. These quantities are calculated by taking the product between the temporal distributions of two or more variables and, subsequently, taking the temporal average. To develop this unit in this book, it will be necessary to use two-point correlations. Turbulent flows are dominated by large and coherent turbulent structures, with a lifespan long enough to affect different regions of the flow, meaning that it is not enough to describe turbulence with only local information. To describe what happens at a point, it is necessary to establish correlations with what is happening in its neighborhood.

Townsend [12] established the following: differently from what occurs with molecular movement, the movement of a fluid, in a turbulent regime, affects the entire neighborhood of a point and not just the point itself. Therefore, an adequate description cannot be established with just average information associated only with one point. A turbulent flow is less random and is more organized than molecular movement. To describe a turbulent flow, correlations are necessary at two or more points within the flow.

Two-point correlations and the characteristic length that can be deduced from this correlation will be discussed here. Consider two points in space, $\vec{x}$ and $\vec{x} + \vec{r}$, where $\vec{r}$ is a separation in space between these two points. The two-point spatial correlation tensor for velocity fluctuations is given by the temporal average of the velocity fluctuation products, involving two points simultaneously:

$$R_{ij}(\vec{x}, \vec{r}) = \overline{u_i'(\vec{x}, t)u_j'(\vec{x} + \vec{r}, t)}, \tag{4.68}$$

where the operator $\overline{(\cdot)}$ represents the temporal average. The distance between the two correlated points is represented by the vector $\vec{r}$, and by way of example, the turbulent kinetic energy can be obtained by summing the elements of the main diagonal of the tensor $R_{ij}(\vec{x}, r)$, divided by 2, when $\vec{r}$ is equal to zero.

The function $f(\vec{x}, r)$ is defined with the purpose of proposing a mathematical model for the integral length:

$$f(\vec{x}, \vec{r}) = \frac{R_{ii}(\vec{x}, \vec{r})}{R_{ii}(\vec{x}, 0)}, \tag{4.69}$$

where "i" is the direction chosen for the statistical treatment. With this normalized correlation function, the integral length can be defined as being:

$$\ell_I(\vec{x}) = \int_0^\infty f(\vec{x}, r)\mathrm{d}r. \tag{4.70}$$

The integration variable is $r = \vec{r}$. The graphical interpretation of this definition of turbulence integral length can be viewed in Fig. 4.26. This length is defined such that the area under the distribution $f(\vec{x}, r)$ is equal to the area of the wide rectangle with width $\ell_I(\vec{x})$ and height "1".

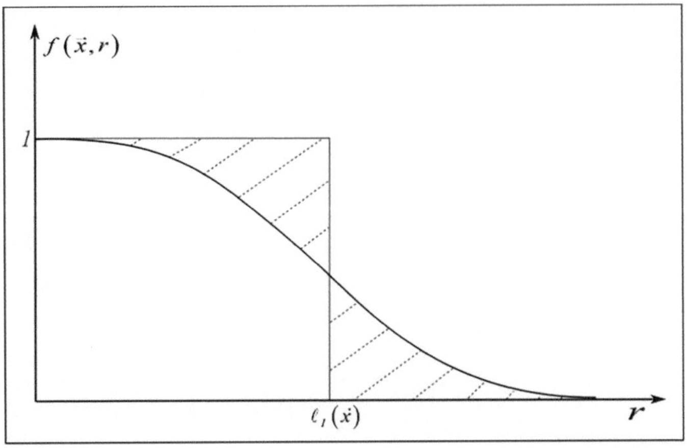

Fig. 4.26 Illustration of the correlation function $f(\vec{x}, r)$. Graphical interpretation of integral length $\ell_I(\vec{x})$

4.7.7 Taylor Micro Length

Based on experimental work for stationary turbulence, that is, turbulence for which all statistical moments are independent of time, the Taylor micro length is defined, based only on the component in the so-called predominant flow direction:

$$R_{11}(\vec{x}, r) = \overline{u'(\vec{x}, t)u'(\vec{x} + \vec{r}, t)}. \tag{4.71}$$

Normalizing this correlation with the intensity of turbulence in the flow direction, the following function is defined:

$$g(\vec{x}, r) \equiv \frac{R_{11}(\vec{x}, r)}{\overline{u'^2}(\vec{x}, 0)}. \tag{4.72}$$

The functions $f(\vec{x}, r)$ and $g(\vec{x}, r)$ are the same functions. The osculatory parabola $p(r)$ is determined relating to this function, Fig. 4.27, which has two properties: one is that the maximum points of the function $g(\vec{x}, r)$ and the parable $p(r)$ are coincident; the other is that this parabola crosses the axis $\vec{x}$ in a quota λ, which is defined as the Taylor micro length. With these two conditions, the determination of the parable $p(r)$ is presented below:

$$p(r) = ar^2 + br + c$$

$$p(0) = 1 \Rightarrow c = 1$$

$$p(\lambda) = 0 \text{ and } p(-\lambda) = 0 \Rightarrow \begin{cases} a = -1/\lambda^2 \\ b = 0 \end{cases}$$

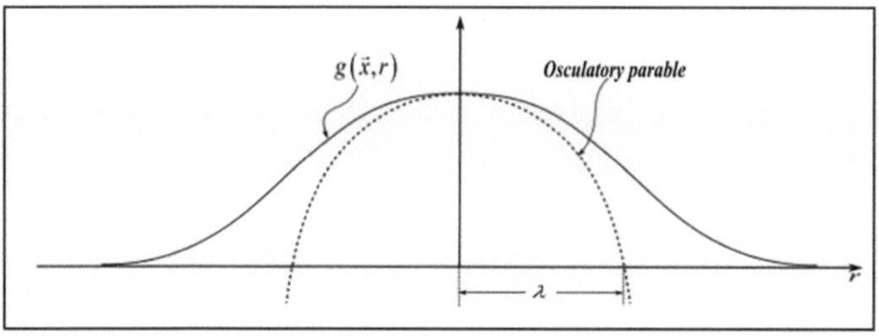

Fig. 4.27 Illustration of the correlation function $g(\vec{x}, r)$ and the osculatory parable

$$p(r) = 1 - \left(\frac{r}{\lambda}\right)^2. \tag{4.73}$$

Performing a Taylor series expansion of the function $g(\vec{x}, r)$, results:

$$\begin{aligned} g(\vec{x}, r) &= g(\vec{x}, 0) + \left(\frac{\partial g}{\partial r}\right)_{r=0} (r - 0) + \left(\frac{\partial^2 g}{\partial r^2}\right)_{r=0} \frac{(r-0)^2}{2} + \cdots \\ &= 1 + \left(\frac{r^2}{2}\frac{\partial^2 g}{\partial r^2}\right)_{r=0} + \cdots \end{aligned} \tag{4.74}$$

The second derivative of the function $g(\vec{x}, r)$, in $r = 0$, can be approximated by the second derivative of the osculatory parabola, also evaluated in $r = 0$. Therefore:

$$1 + \left(\frac{r^2}{2}\frac{\partial^2 g}{\partial r^2}\right)_{r=0} + \cdots \cong 1 + \frac{r^2}{2}\left(-\frac{2}{\lambda^2}\right) \Rightarrow \lambda \equiv \sqrt{\frac{-2}{\left(\frac{\partial^2 g}{\partial x^2}\right)_{x=0}}}. \tag{4.75}$$

The definition sign above is because the parabola was determined in such a way as to define this characteristic length. More details can be obtained from Hinze [3].

4.7.8 *Two-Point Correlation and Associated Characteristic Quantities*

With Taylor's hypothesis of displacements in time and space, it is established that:

$$\frac{\partial}{\partial t} = U\frac{\partial}{\partial x}. \tag{4.76}$$

From this hypothesis, it is established that the relationship between the Taylor micro time and micro length is given by:

$$\lambda = U\tau_T, \tag{4.77}$$

where λ is the Taylor micro length, U is the characteristic velocity of turbulent flow and τ_T is Taylor micro time. For homogeneous and isotropic turbulence, it is established, with Taylor Analysis, that the rate of decay of the turbulent kinetic energy is given by:

$$\frac{\mathrm{d}k}{\mathrm{d}t} = -\frac{10\nu k}{\lambda^2}. \tag{4.78}$$

To see how the Taylor and Kolmogorov microquantities interrelate, it is observed that:

$$\frac{\mathrm{d}k}{\mathrm{d}t} = -\varepsilon \Rightarrow \varepsilon = \frac{10\nu k}{\lambda^2}, \tag{4.79}$$

where ε is the specific power of viscous transformation of the turbulent specific kinetic energy. For high values of the Reynolds number:

$$\varepsilon \approx \frac{k^{3/2}}{\ell_I}. \tag{4.80}$$

Considering also empirical data:

$$\varepsilon \cong 0.09\frac{k^{3/2}}{\ell_I} \Rightarrow k \cong \left(\frac{\varepsilon \ell_I}{0.09}\right)^{2/3}. \tag{4.81}$$

The Kolmogorov characteristic length is given by $\eta = \left(\nu^3/\varepsilon\right)^{1/4}$. Combining these equations, we have:

$$\frac{\lambda}{\eta} \approx 7\left(\frac{\ell_I}{\eta}\right)^{1/3}. \tag{4.82}$$

But, according to Kolmogorov's theory, $(\ell_I/\eta) \approx (\mathrm{Re}_{\ell I})^{3/4}$, from where:

$$\left(\frac{\lambda}{\eta}\right) \approx 7(\mathrm{Re}_{\ell I})^{1/4}. \tag{4.83}$$

To have a well-defined inertial band, it is necessary that $\ell_I/\eta > 1000$. So, the Taylor Micro Length must be at least 70 times the Kolmogorov length. Thus, the Taylor micro length will always be within the inertial band, above the Kolmogorov length and below the integral length. These lengths are illustrated in Fig. 4.28.

As a quantitative example, for grid turbulence, the integral length, which is on the order of the grid mesh size, takes on the approximate value of $\ell_I \approx 40\,\mathrm{mm}$, the Kolmogorov length takes the value $\eta \approx 0.1[\mathrm{mm}]$ and the Taylor micro length takes the value $\lambda \approx 5$. It is observed that the Taylor Micro Length has a hybrid nature, depending on integral length and Kolmogorov Characteristic Length. In fact, as already presented:

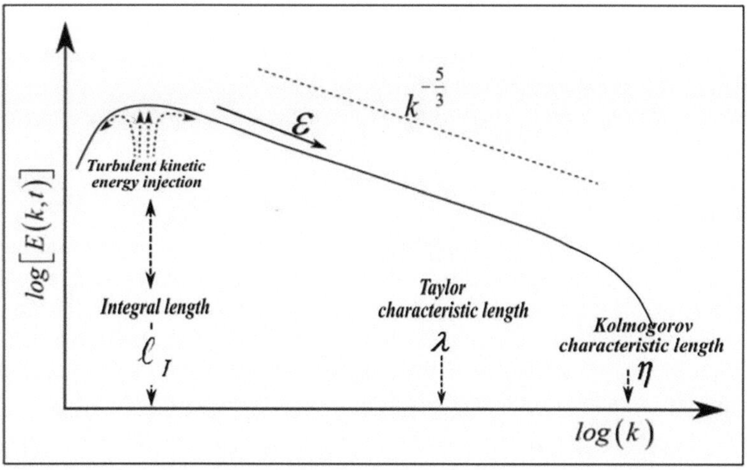

Fig. 4.28 Illustration of turbulence characteristic lengths: integral length ℓ_I; Taylor micro length λ and Kolmogorov length η

$$\varepsilon = \frac{10\nu k}{\lambda^2} \Rightarrow \lambda = \sqrt{\frac{10\nu k}{\varepsilon}}. \tag{4.84}$$

Thus, this length depends on the turbulent kinetic energy, k, which is a typical quantity of integral turbulent structures, as well as on Kolmogorov quantities, that is, the kinematic viscosity, ν, and the specific power of viscous transformation of turbulent kinetic energy, ε. The Taylor microquantities of length and time are too small to represent the integral structures and are too large to represent the Kolmogorov turbulent structures. Because of this, these Taylor microquantities have been ignored for the development of turbulence closure models.

4.8 Exercises

1. Consider flows of the following types: periodic box, Couette flow, temporal mixing layer, boundary layer over a flat plate, Taylor-Couette flow between two concentric cylinders and Rayleigh–Bénard flow between two horizontal and heated flat plates. Discuss each one separately, regarding the hypotheses of homogeneity and isotropy.
2. What is the ergodicity hypothesis? Search for a physical problem whose statistics are ergodic.
3. Search for examples of first, second, third and fourth order statistical moments. Interpret physically each of them.

4. Study the continuity, Navier–Stokes and energy equations in Fourier Space. Physically interpret each term of these equations in this space. Make a parallel with the corresponding interpretations in physical space.
5. Interpret the fact that, for incompressible flows, in Fourier Space, pressure is eliminated from the fluid dynamics formulation. Would it be possible to do the same thing in physical space? I would say yes! Look for a solution in the literature!
6. Transform the differential model for compressible fluid dynamics into Fourier Space. It will take a few days of work, but it's worth it! Pressure is eliminated?
7. Which of the two figures below do you choose to represent the nonlinear dynamics of information transfer between the turbulent structures that make up a turbulent flow? Justify, explaining at least one page. This requires a lot of thought!

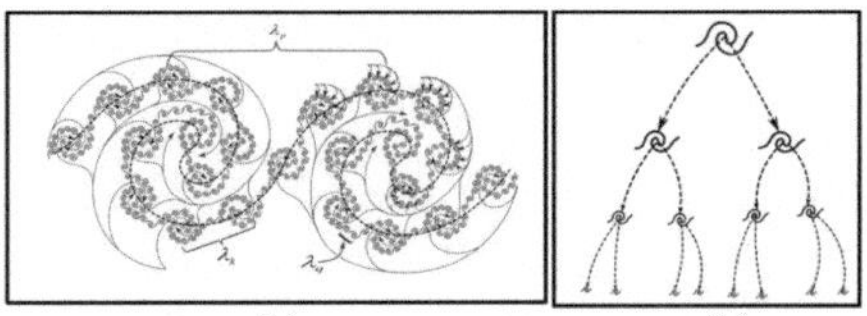

(a) (b)

8. Consider the flow in a cavity illustrated in the figure below. Identify and physically explain the integral length of energy injection in the turbulent flow inside the cavity. Specify the fluid, the velocity at the inlet of the channel upstream of the cavity, the height of the cavity, its length and the height of the inlet channel. Then calculate:

    (a) The integral length, time, kinetic energy and vorticity.
    (b) The characteristic Kolmogorov length, time, kinetic energy and vorticity.
    (c) Taylor's micro quantities of length and time.

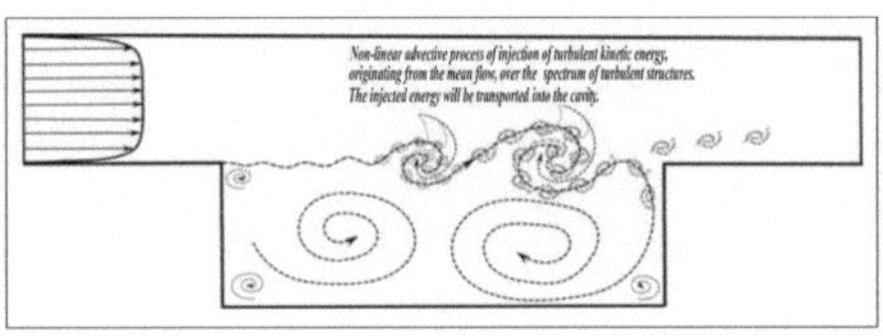

9. Repeat Exercise 8, considering a commercial aircraft that carries 500 passengers on a single trip. What is the *turbulent specific kinetic energy spectral density*? What is the physical meaning of each of these terms, highlighted in italics?
10. What is statistically transient and statistically permanent turbulence? Plot the turbulent specific kinetic energy spectral density to answer.
11. What is the basis for the development of Kolmogorov theory for homogeneous and isotropic turbulence? What is the most important result of your theory from your point of view?

12. Would the differential model based on the continuum hypothesis really be valid for modeling turbulent fluid dynamics? Are there limitations to the application of this model? Give examples of physical problems to which the continuum hypothesis could not be applied. Base your answers on the Kolmogorov Theory for homogeneous and isotropic turbulence.

References

1. Frisch, U., Turbulence, the Legacy of A. N. Kolmogorov. Cambridge University Press
2. Boussinesq, J.: Theorie de l´coulement tourbillonnant et tumultueux des liquides dans les lits rectilignes a grande section. Fautiers-Villars, Paris, 2nd edn (1877)
3. Hinze, J.O.: Turbulence. Mc Graw Hill (1975)
4. Kolmogorov.: The local structure of turbulence in incompressible viscous fluid for Large Reynolds numbers. Dokl. Akkad. Nauk SSSR. **30**, 301–305 (1941a)
5. Kolmogorov.: On degeneration of isotropic turbulence in an incompressible viscous liquid. Dokl. Akkad. Nauk SSSR **31**, 538–541 (1941b)
6. Lesieur, M.: Turbulence in Fluids. Kluwer Academic Publishers, Netherlands (1995)
7. Oboukhov, A.M.: On the distribution of energy in the spectrum of turbulent flow. Dokl. Akkad. Nauk SSSR **32a**, 22–24 (1941a)
8. Obukhov, A.M.: Spectral energy distribution of turbulent flow. Izv. Akkad. Nauk SSSR Ser. Geogr. Geophys. **5**, 453–466 (1941)
9. Pope, S.B.: Turbulent Flows. Cambridge University Press (2000)
10. Richardson, L.F.: Weather Prediction by Numerical Process. Cambridge University Press (1922)
11. Saddoughi, S.G., Veeravalli, S.V.: Local isotropy in turbulent boundary layers at High Reynolds number. J. Fluid Mech. **268**, 333–372 (1994)
12. Townsend, A.A.: The Structure of Turbulent Shear Flow. Cambridge University Press (1967)

Chapter 5
Equations for Turbulent Flows and the Closure Problem

In this chapter, the equations for turbulent flows are presented, as well as the famous closure problem. An introduction to the topic is presented in Sect. 5.1. The mathematical model is presented in Sect. 5.2, and the Boussinesq–Reynolds average equations are presented in Sect. 5.3. The filtered equations are presented in Sect. 5.4, and finally, a list of exercises is presented in Sect. 5.5.

5.1 Introduction

One of the main characteristics of turbulence is the multiplicity of degrees of freedom. This characteristic is illustrated again, in the present text, through Fig. 5.1. This is one of Leonardo da Vinci's famous paintings, in which the Renaissance genius clearly expressed the three-dimensional effects and the notion of the multiplicity of turbulent structures in a turbulent flow. To emphasize and enumerate some of the turbulent structures existing in this painting, the author of this book superimposed some lines on the drawing. Structures 1 to n, where n is the total number of structures that can be found in the flow itself and not in da Vinci's painting. Another very important notion that da Vinci expressed in his painting is the notion of the spatial coexistence of multiple turbulent structures and, somehow, even the notion of the non-linear interactions that occur in the observed turbulent flow.

It is known, through the Kolmogorov theory, [4, 5], that the number n of turbulent structures depends on the value assigned to the Reynolds Number. The relationship between the characteristic lengths of the most energized and smallest turbulent structures depends on the value of the Reynolds Number; this dependence is given by the following expression:

$$\left(\frac{\ell_I}{\eta}\right) \approx \mathrm{Re}_{\ell_I}^{3/4}, \tag{5.1}$$

A. da Silveira Neto, *Turbulent Flows*, https://doi.org/10.1007/978-3-031-89222-6_5

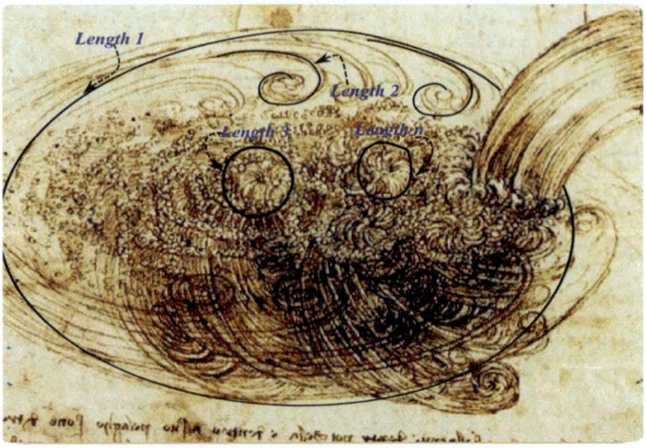

Fig. 5.1 Figure of da Vinci, used to illustrate the multiplicity of sizes and the nonlinear spatial–temporal coexistence of the multiple turbulent structures

where ℓ_I is the integral length of turbulence, which expresses the order of magnitude of the size of the most energized turbulent structure present in the flow. It is emphasized that it provides an order of magnitude! In a turbulent flow, one can find turbulent structures larger than ℓ_I. The length η represents an order of magnitude of the size of the smallest turbulent structures present in the flow. This length is known as the Kolmogorov Length. In one-dimensional problems, this length relationship gives the number of possible sizes present in the interval $[\eta - \ell_I]$. For example, looking at an instrument measuring length of 1(m), graduated in 1(mm), 10^3 possible sizes can be found, e.g., 1(mm), 2(mm), …, 1000(mm). This corresponds to $\left(10^3/1\right) = 10^3$ possible sizes. Taking a surface of $1 \times 1(\mathrm{m}^2)$, also graduated in 1 (mm) in both directions, there are $(10^3)^2 = 10^6$ possible geometric figures. Similarly, taking a volume of $1 \times 1 \times 1\,(\mathrm{m}^2)$, $(10^3)^3 = 10^9$ possible geometric figures will be found in this volume. Thus, in conclusion, in a volume ℓ_I^3 divided in η^3 smallest volumes (in terms of turbulence, they are the smallest possible turbulent structures!), $(\ell_I/\eta)^3$ volumes will be found. Associating these possible volume sizes with the geometric figures that represent the turbulent structures, and thinking about turbulent flows as dynamic systems, the number of degrees of freedom (dof) of these dynamic systems can be estimated with the relationship:

$$\left(\frac{\ell_I}{\eta}\right)^3 \approx \mathrm{Re}_{\ell_I}^{9/4}. \tag{5.2}$$

With this expression, we have an order of magnitude for the number of possible turbulent structures, per unit of volume, ℓ_I^3, present in a turbulent flow.

Two extreme examples of turbulent flows that can be found in nature will be analyzed. The first of them is extracted from an industrial problem of high dimensions, tens of meters, but whose physics refers to the analysis of the flow around and within a single droplet of a few mm in diameter. This is a gas washer, Fig. 5.2a, whose function is to extract contaminating gas, through a process called gas washing. The contaminating gas is hydrophilic and, when flowing relative to a water droplet, it will be absorbed. As the drops, injected through the sprayers, fall and the gas rises, a process of absorption of the toxic gas contained in the air is promoted by the falling drops, relative to the gas. The efficiency of the absorption process will depend on the transport of the gas absorbed by the surface of the droplets (by diffusion) to the interior of the droplets (by advection). Therefore, the more unstable the flow inside the droplet, the more efficient the gas washing process will be.

The example that will be presented is that of an air bubble moving in a water reservoir. The unstable and three-dimensional flow that can be found inside the air bubble, moving in a vertical and upward direction, can be visualized in Fig. 5.2b. The vorticity field was used to visualize the flow inside and outside the bubble. Focusing attention on its interior, it is possible to observe the existence of multiple nuclei with strong levels of vorticity, which highlights the dynamic structures present in the flow, inside the bubble. It can be stated that there are multiple dynamic structures inside the bubble, as well as in its wake. It is not possible to say, however, that it is a turbulent flow. It is safer to say that the flow is in transition from the laminar regime to the turbulent regime.

It is possible to estimate the maximum number of degrees of freedom that are present in this flow, if it reaches the turbulence regime. To do so, calculate the

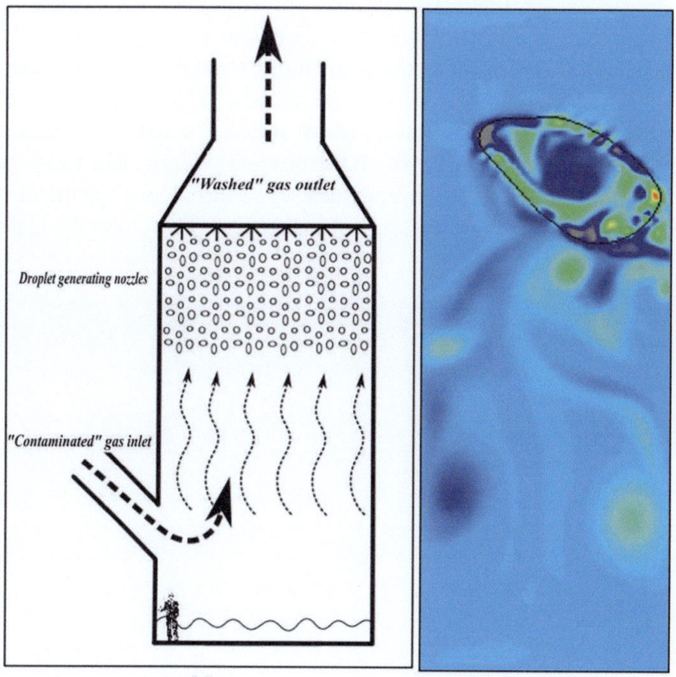

(a) (b)

Fig. 5.2 Illustration of: **a** gas washing tower using the movement of small drops and **b** detailed analysis of the movement of a small rising bubble

Reynolds Number, based on the diameter d the bubble, its ascent velocity U, in a fluid medium of kinematic viscosity v. Typical values of these variables can be obtained from material experiments: $d = 3 \times 10^{-3}$ m, $U = 0.5$ m/s and $\nu = 10^{-5}$ m^2/s. This data results in $\mathrm{Re}_d = 150$. Using Eq. (5.2), the degrees of freedom of this flow, per unit of volume, d^3, are obtained, dof = 78,000.

Another flow of interest, but which is at another extreme in terms of integral characteristic length, is a geophysical flow that occurs in the earth's atmosphere. In Fig. 5.3, an enormous turbulent geophysical structure is illustrated. It is not possible to identify the multiplicity of turbulent structures in this image, which does not imply the absence of other smallest structures. In fact, they are inside this enormous structure that is visualized. The order of magnitude of the Kolmogorov Characteristic Length can be calculated. A good estimate of the characteristic size of this large turbulent structure is the average radius of the Earth., $\ell_I = 6370$ (km). An estimate of the characteristic velocity of this problem provides $U = 80$(m/s). The kinematic viscosity of the fluid can be approximated by $\nu = 10^{-5}$(m^2/s). The Reynolds Number, under these conditions, assumes $Re_{RT} = 5.1 \times 10^{13}$ value. Thus, using Eq. (5.2), the dof $= 7 \times 10^{30}$, contained in the volume ℓ_I^3, is determined. This is the number of degrees of freedom per unit of volume. As a curiosity, the Kolmogorov length for this case is on the order of 0.3(mm). This means that between the integral length ($\ell_I = 6.370$ (km)) and the Kolmogorov length ($\eta = 0.3$ (mm)), there is a continuous spectrum of 7×10^{30} degrees of freedom.

Continuous solutions for the mathematical differential model for fluid flows are not yet known. Computational solutions are the sole way to solve such a complex mathematical model. To achieve this, many numerical methods require a mesh composed of finite volumes, with which the physical volume that encompasses the problem must be discretized. To correctly describe a turbulent structure of characteristic size ℓ, at least 8 finite volumes are required $(\ell/2)^3$. Therefore, the mathematical

Fig. 5.3 Illustration of a planetary-sized turbulent structure located in the earth's atmosphere

discretization, for the computational solution, of all the turbulent structures that make up a flow, requires a mesh with a characteristic size $(\eta/2)$. Thus, the number of cells required to resolve all the turbulent structures of a turbulent flow will be approximately $8 \times \left(\mathrm{dof}/m^3\right) \times V_d$, where V_d is the volume of the domain. As an approximation, the number of cells to be used is an order of magnitude times the dof of the flow, per unit volume, times the total volume.

When all the degrees of freedom of the flow are solved, it is said that the Direct Computational Simulation Methodology (DNS) is being practiced. For the first problem presented, the ascending bubble, approximately $10 \times 78{,}000 = 780{,}000$ cells per unit volume are required to solve this problem by DNS. This is a problem whose solution using DNS is now completely feasible with the computational resources available in high-performance computing laboratories. However, for the second problem presented, it would be necessary to solve linear systems with approximately 10^{32} simultaneous equations, each one corresponding to a finite volume that makes up the discretization mesh. This is a fantastically large number, which exceeds any computational capacity until the publication of this book, even exceeding any expectation, even the most optimistic of all!

Between these two extremes, there is a huge range of natural physical or engineering problems that require analysis via computer simulation. To solve this enormous problem, methodologies have been proposed with the aim of eliminating the need to solve all the *dof* that characterize the flows of interest.

It is a consensus that with the Navier–Stokes equations, coupled to the continuity and the energy equations, laminar and turbulent flows can be modeled without distinction. However, as demonstrated above, the solution of these equations for flows at high Reynolds Numbers is not yet possible with currently available machines, using currently existing numerical methods for solving PDEs. A palliative for this problem appeared with the work of Boussinesq [1] and Reynolds [8], who proposed, independently, the decomposition of variables (velocities, pressure, temperature, concentration, etc.) into average values and fluctuations.

In other words, these scientists proposed to determine the average behavior of turbulent flows but considering the interaction with fluctuating behaviors. Starting from this notion and submitting continuity, Navier–Stokes and energy equations to an average operator, the so-called "Boussinesq–Reynolds Averaged Equations" are obtained. This procedure gives rise to an additional tensor and an additional vector, composed of second-order statistical moments, involving velocity and temperature fluctuations. As will be detailed, this symmetric tensor and this vector are composed of nine new unknowns, which are incorporated into the differential model, which implies the famous turbulence closure problem.

Not satisfied with obtaining the average behavior of turbulent flows, the scientific community sought to determine more information, going beyond the concept of average. As a result, the concept of filtering was launched, more recently, by Smagorinsky [7], for the establishment of the large structures simulation methodology. In the present book, we propose to discuss, in a comparative way, these two ways of approaching the deduction of the turbulence equations. It is also proposed to standardize these two forms (averaged equations and filtered equations) in a general form and, also, to discuss the most important characteristics of the two main lines (Average Boussinesq–Reynolds equations and filtered equations), to then address the systematic solution of the closure problem of turbulence.

5.2 Differential Mathematical Model for Direct Computational Simulation of Turbulence

The continuity, Navier–Stokes and energy equations make up the turbulence model, which in itself is sufficient to solve turbulent flows. The hypotheses of the continuum, Newtonian fluids, incompressible flows, and constant physical properties will be considered here, except for the diffusion coefficients of linear momentum and thermal energy, which will be kept as variables. This model has already been presented, but for ease of reading, it is rewritten here:

$$\frac{\partial u_i}{\partial x_i} = 0, \tag{5.3}$$

$$\frac{\partial u_i}{\partial t} + \frac{\partial}{\partial x_j}\left(u_i u_j\right) = -\frac{1}{\rho_0}\frac{\partial p}{\partial x_i} + \frac{\partial}{\partial x_j}\left[\nu\left(\frac{\partial u_i}{\partial x_j} + \frac{\partial u_j}{\partial x_i}\right)\right] + \frac{f_i}{\rho_0}, \tag{5.4}$$

$$\frac{\partial T}{\partial t} + \frac{\partial}{\partial x_j}\left(u_j T\right) = \frac{\partial}{\partial x_j}\left(\alpha\frac{\partial T}{\partial x_j}\right) + \phi/\left(\rho_0 C_{p0}\right). \tag{5.5}$$

In these equations, $\nu = \mu/\rho_o$ is the kinematic viscosity or linear momentum diffusion coefficient, and $\alpha = k/\left(\rho_0 C_{p0}\right)$ is the thermal energy diffusion coefficient; both quantities are physical properties of the fluid. The linear momentum diffusion term, which appears in Eq. (5.4) represents the closure model for Cauchy Equations, Eq. (2.11). This closure was done using the Stokes model. The diffusion term that appears in Eq. (5.5), that is, the first term on the right side of this equation, represents the closure of the energy equation, for which Fourier's law for thermal diffusion was used.

These are the first two closure problems that arose due to the need to model molecular effects at the continuum level. This is the best model, at the continuum level, for non-isothermal turbulent flows. However, its solution requires numerical-computational methods and mesh refinement suitable for the Direct Computational Simulation (DNS) methodology of turbulent flows. As explained above, the higher the Reynolds Number that characterizes the flow, the more expensive the use of this methodology will become. These arguments justify the Boussinesq-Reynolds decomposition, which will be presented in the following item.

5.3 Averaged Boussinesq–Reynolds Equations

Boussinesq and Reynolds proposed applying the temporal average operator to Eqs. (5.3)–(5.5). This operator will be represented here by the symbol $\langle\ \rangle$. Knowing that the temporal average operator and the partial derivative operators are linear, they can be switched, resulting in the following equations:

$$\frac{\partial \langle u \rangle_i}{\partial x_i} = 0, \tag{5.6}$$

$$\frac{\partial}{\partial x_j}\left(\langle u_i u_j \rangle\right) = -\frac{1}{\rho_0}\frac{\partial \langle p \rangle}{\partial x_i} + \frac{\partial}{\partial x_j}\left[\nu\left(\frac{\partial \langle u_i \rangle}{\partial x_j} + \frac{\partial \langle u_j \rangle}{\partial x_i}\right)\right] + \frac{\langle f_i \rangle}{\rho_0}, \tag{5.7}$$

$$\frac{\partial}{\partial x_j}\left(\langle u_j T \rangle\right) = \frac{\partial}{\partial x_j}\left(\alpha \frac{\partial \langle T \rangle}{\partial x_j}\right) + \langle \phi \rangle / \left(\rho_0 C_{p0}\right). \tag{5.8}$$

The transient terms have disappeared since a time-averaged process is considered here. Physical properties, including kinematic viscosity (ν) and thermal diffusivity (α) were maintained within the divergent operator, in line with Eqs. (5.4) and (5.5). The averaged Eqs. (5.7) and (5.8), would be ready to be solved, if it weren't for the nonlinear terms $\langle u_i u_j \rangle$ and $\langle u_j T \rangle$, that appear in these averaged equations. The variables that are the subject of the balance procedure are $\langle u_i \rangle$ and $\langle T \rangle$, which are the unknowns of these equations, together with the mean pressure field $\langle p \rangle$. Thus, these equations cannot be solved in the forms in which they are written, since to determine their solutions, it would be necessary to know the full quantities u_i and T, which would only be possible via DNS (Direct Computational Simulation). These equations could be solved if instead of having $\langle u_i u_j \rangle$ and $\langle u_j T \rangle$ terms, we had the terms $\langle u_i u_j \rangle$ and $\langle u_j \rangle \langle T \rangle$. Thus, the decomposition proposed by Boussinesq–Reynolds is carried out. The total velocity and temperature fields are decomposed into their averages and their fluctuating parts. Mathematically, these decompositions are done as follows:

$$u_i = \langle u_i \rangle + u_i', \tag{5.9}$$

$$T = \langle T \rangle + T'. \tag{5.10}$$

It should be noted that the pressure field does not require decomposition, since its average $\langle p \rangle$ already appears in Eq. (5.7). Replacing Eqs. (5.9) and (5.10) in Eqs. (5.7) and (5.8) results in:

$$\frac{\partial}{\partial x_j}\left(\langle\langle u_i\rangle\langle u_j\rangle\rangle+\langle u_i'\langle u_j\rangle\rangle+\left\langle\langle u_i\rangle u_j'\right\rangle+\left\langle u_i'u_j'\right\rangle\right)$$
$$=-\frac{1}{\rho_0}\frac{\partial\langle p\rangle}{\partial x_i}+\frac{\partial}{\partial x_j}\left[\nu\left(\frac{\partial\langle u_i\rangle}{\partial x_j}+\frac{\partial\langle u_j\rangle}{\partial x_i}\right)\right]+\frac{\langle f_i\rangle}{\rho_0}, \quad (5.11)$$

$$\frac{\partial}{\partial x_j}\left(\langle\langle u_j\rangle\langle T\rangle\rangle+u_j'\langle T\rangle+\langle u_j\rangle T'+\left\langle u_j'T'\right\rangle\right)=\frac{\partial}{\partial x_j}\left(\alpha\frac{\partial\langle T\rangle}{\partial x_j}\right)+\langle\phi\rangle/\left(\rho_0C_{p0}\right). \quad (5.12)$$

The following properties, associated with the temporal average operator, were used:

$$\langle u_i\rangle \text{ is constant in time} \quad (5.13)$$

$$\langle T\rangle\text{is constant in time} \quad (5.14)$$

$$\left\langle u_i'\right\rangle=\langle u_i\rangle-\langle\langle u_i\rangle\rangle=\langle u_i\rangle-\langle u_i\rangle=0, \quad (5.15)$$

$$\left\langle T'\right\rangle=\langle T\rangle-\langle\langle T\rangle\rangle=\langle T\rangle-\langle T\rangle=0, \quad (5.16)$$

$$\langle\langle u_j\rangle\langle u_j\rangle\rangle=\langle u_j\rangle\langle u_j\rangle, \quad (5.17)$$

$$\langle\langle u_j\rangle\langle T\rangle\rangle=\langle u_j\rangle\langle T\rangle, \quad (5.18)$$

$$\left\langle u_i'\langle u_j\rangle+\langle u_i\rangle u_j'\right\rangle=\langle u_i'\langle u_j\rangle\rangle+\left\langle\langle u_i\rangle u_j'\right\rangle=\langle u_i'\rangle\langle u_j\rangle+\left\langle u_j'\right\rangle\langle u_i\rangle=0. \quad (5.19)$$

Thus, Eqs. (5.11) and (5.12), through the properties established with Eqs. (5.13)–(5.19), result in:

$$\frac{\partial}{\partial x_j}\left(\langle u_i\rangle\langle u_j\rangle+\left\langle u_i'u_j'\right\rangle\right)=-\frac{1}{\rho_0}\frac{\partial\langle p\rangle}{\partial x_i}+\frac{\partial}{\partial x_j}\left[\nu\left(\frac{\partial\langle u_i\rangle}{\partial x_j}+\frac{\partial\langle u_j\rangle}{\partial x_i}\right)\right]+\frac{\langle f_i\rangle}{\rho_0}, \quad (5.20)$$

$$\frac{\partial}{\partial x_j}\left(\langle u_j\rangle\langle T\rangle+\left\langle u_j'T'\right\rangle\right)=\frac{\partial}{\partial x_j}\left(\alpha\frac{\partial\langle T\rangle}{\partial x_j}\right)+\langle\phi\rangle/\left(\rho_0C_{p0}\right). \quad (5.21)$$

The tensor $\left\langle u_i'u_j'\right\rangle$, known as the Boussinesq–Reynolds Tensor, is symmetric, since the order of the components of the velocity fluctuations does not change the second-order statistical moments that make up this tensor. Thus, we have six additional unknowns, resulting from the correlations between the velocity fluctuations relative to the three coordinated directions. This tensor represents the entire exchange of linear momentum between the averaged part and the fluctuation part of the turbulent flow.

The vector $\left\langle u_j'T' \right\rangle$ represents the turbulent flux of thermal energy between the average part and the fluctuating part of the turbulent flow. There are three more unknowns, in addition to the average temperature. It is up to these two entities, a tensor, $\left\langle u_i'u_j' \right\rangle$, and a vector, $\left\langle u_j'T' \right\rangle$, the role of modeling the transfer of the information in question, i.e., linear momentum and thermal energy, between the average velocity field, $\langle u_i \rangle$ and the fluctuation velocity fields u' and between the average temperature field, $\langle T \rangle$, and the fluctuation temperature field T'. This information transfer activity is highly non-linear in nature, since these two terms, the Boussinesq–Reynolds tensor and the turbulent flux of thermal energy, have their origins linked to the non-linear terms of balance equations.

These two terms represent nonlinear transfer processes of advective-turbulent nature, but, for modeling reasons, they will be considered diffusive-turbulent nature. Thus, by analogy to the process of information transfer through molecular dynamics activity, they will be modeled as turbulent diffusive processes. To better understand this analogy, the two processes are illustrated in Figs. 6.4a and 6.4b. In Fig. 6.4a, the movement of the molecules that make up a fluid medium is illustrated. The red molecules carry a greater amount of linear momentum, as they are in a region of the flow in which the fluid moves with greater velocity. The blue molecules are in a region where the fluid moves at a slower velocity. Thus, the dashed plane is in a shear region of the flow. The viscous tension, modeled with the concept of molecular viscosity of the fluid, results from the molecular flux over this plane. A flux of red molecules occurs from left to right, which transports linear momentum to the slower region. This linear momentum flux will accelerate this region. Similarly, a flux of blue molecules occurs from right to left, carrying a smaller amount of linear momentum. The net flux of linear momentum on this plane will be translated, at the continuous level, by a viscous shear stress between the two regions of the flow.

Similarly, in Fig. 6.4b, there is a movement of turbulent structures, at the continuum level, due to the strong agitation promoted by turbulence. Turbulent structures highlighted in red, rich in linear momentum, move from left to right, while black structures, with a low level of linear momentum, move from right to left. Movements in both directions promote a net flux of linear momentum over this imaginary plane. At the mean flow level, this net flux of linear momentum will be modeled as a turbulent tension, expressed by the Boussinesq–Reynolds tensor, $\left\langle u_i'u_j' \right\rangle$.

It will be seen that this analogy also gave rise to the famous Boussinesq [1] Hypothesis, which led to the concept of turbulent viscosity. Stokes used the concept of molecular viscosity to model the viscous stress tensor that appears in the Cauchy equations. Boussinesq made an analogy to the Stokes model and proposed modeling the Boussinesq–Reynolds tensor with the concept of turbulent viscosity (Fig. 5.4). This model will be seen in more detail in Sect 7.2. Similarly, the turbulent flux vector of thermal energy, whose origin is advective, will also be considered as having a turbulent diffusive nature.

Due to this analogy, the Boussinesq–Reynolds tensor $\left\langle u_i'u_j' \right\rangle$ and the turbulent thermal energy flux $\left\langle u_j'T' \right\rangle$ will be transported, in Eqs. (5.20) and (5.21), from the left

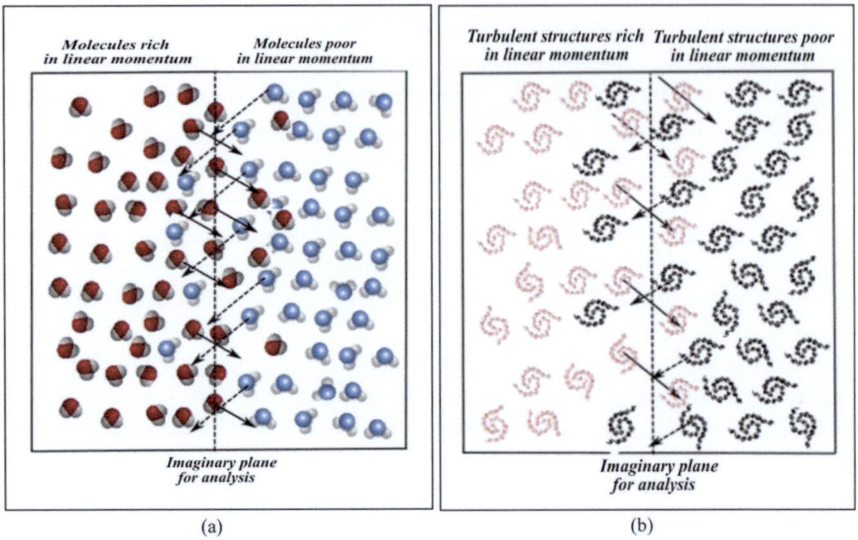

Fig. 5.4 Illustration of the net flux of linear momentum over an imaginary plane: **a** liquid flux due to molecular movement and **b** liquid flux due to the movement of turbulent structures

side to the right side and will be incorporated into the molecular diffusion terms of linear momentum and thermal energy. These two equations are rewritten with these modifications:

$$\frac{\partial}{\partial x_j}\left(\langle u_i\rangle\langle u_j\rangle\right) = -\frac{1}{\rho_0}\frac{\partial\langle p\rangle}{\partial x_i} + \frac{\partial}{\partial x_j}\left[\nu\left(\frac{\partial\langle u_i\rangle}{\partial x_j} + \frac{\partial\langle u_j\rangle}{\partial x_i}\right) - \left\langle u_i'u_j'\right\rangle\right] + \frac{\langle f_i\rangle}{\rho_0}, \tag{5.22}$$

$$\frac{\partial}{\partial x_j}\left(\langle u_j\rangle\langle T\rangle\right) = \frac{\partial}{\partial x_j}\left(\alpha\frac{\partial\langle T\rangle}{\partial x_j} - u_j'T'\right) + \langle\phi\rangle/\left(\rho_0 C_{p0}\right). \tag{5.23}$$

Equations (5.22) and (5.23) are called averaged Boussinesq–Reynolds equations, which have the acronym RANS, for Reynolds averaged Navier–Stokes equations. These equations, added to the continuity equation, make up the model for the average behavior of a turbulent flow. This mathematical model is an open model, having nine additional unknowns to the average velocities, temperature and pressure. There are six unknowns that make up the Boussinesq–Reynolds tensor and three unknowns that make up the turbulent flux of thermal energy. This is the turbulence closure problem that requires additional closure models. This discussion will be presented later, in Chap. 7.

5.4 Filtered Equations

With the solution of the average equations for turbulence, part of the engineering applications and other areas involving stationary fluid dynamics are covered. However, a large majority of problems remain unsolved, since the transient behavior of turbulent flows has important practical consequences. For example, climate prediction requires that at least part of the turbulent structures must be obtained from the solution of balance equations.

Fluid–structure interaction is another class of problems that requires the determination of the transient behavior of turbulent flows. In this sense, the proposal to decompose turbulent flows into an average part and a fluctuation part cannot meet the classes of problems described above. Smagorinsky [7] proposed the decomposition of turbulent flows, in such a way to decompose the complete band of frequencies into a band of lower frequencies or lower wave numbers and a band of higher frequencies or higher wave numbers. This requires the concepts of cutoff frequency and cutoff wavenumber.

In Fig. 6.5, the decomposition of a generic information $\phi(\vec{x}, t)$, is illustrated. The complete temporal distribution was obtained via computational experiment (DNS) of a turbulent flow or via a material experiment. The solid line represents the complete signal, $\phi(\vec{x}, t)$, containing all frequencies of physical phenomenon. The dashed line represents a filtered signal, $\overline{\phi}(\vec{x}, t)$, where a given cutoff frequency, f_c, was used to remove the filtered part of the complete signal. The difference between the complete signal and the filtered part provides the fluctuation of the information, in relation to the filtered part:

$$\phi(\vec{x}, t) = \overline{\phi}(\vec{x}, t) + \phi'(\vec{x}, t). \tag{5.24}$$

With the filtering process, the signal is decomposed into two bands of frequencies or wave numbers. The signals $\phi(\vec{x}, t)$ and $\overline{\phi}(\vec{x}, t)$, illustrated in Fig. 5.5, can be processed, using Fourier transform, to obtain the spectral densities of turbulent energy. At Fig. 5.6, the spectral densities of the function $\phi(\vec{x}, t)$ is shown as well for the filtered function, $\overline{\phi}(\vec{x}, t)$. The spectrum of function $\phi(\vec{x}, t)$ covers the entire band of frequencies associated with the physical problem under analysis. This spectrum has a width that is a function of the Reynolds Number.

The filtering process implies the decomposition of the energy spectrum into two bands of frequencies or wave numbers. The band $f < f_c$ or $k < k_c$ will, from now on, be called the band of large structures, while the band $f \geq f_c$ or $k \geq k_c$ will be called the band of small structures or the band of sub-grid structures, in allusion to the fact that the cutoff frequency, f_c, will be determined by the time step and k_c by the dimensions of the numerical cells that make up the discretization mesh. The grid size determines the cutoff wave number.

Mathematically, the decomposition is done in the form of Eq. (5.24), where the filtered variable will be given by a convolution integral between a filtering function and the function to be filtered. In the case of temporal filtering, this integral is given by:

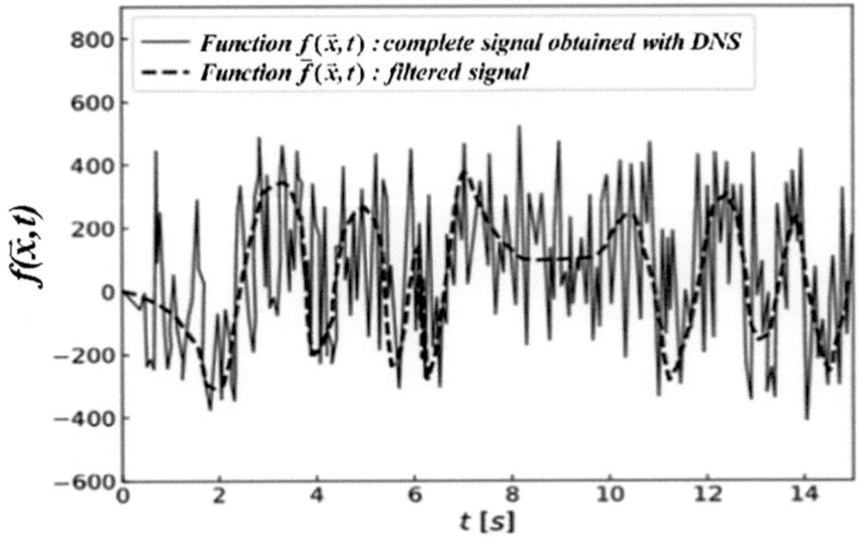

Fig. 5.5 Temporal distribution of an information $f(\vec{x}, t)$ obtained by measurements or by DNS; filtered signal distribution $\overline{f}(\vec{x}, t)$ and the fluctuation $f'(\vec{x}, t)$

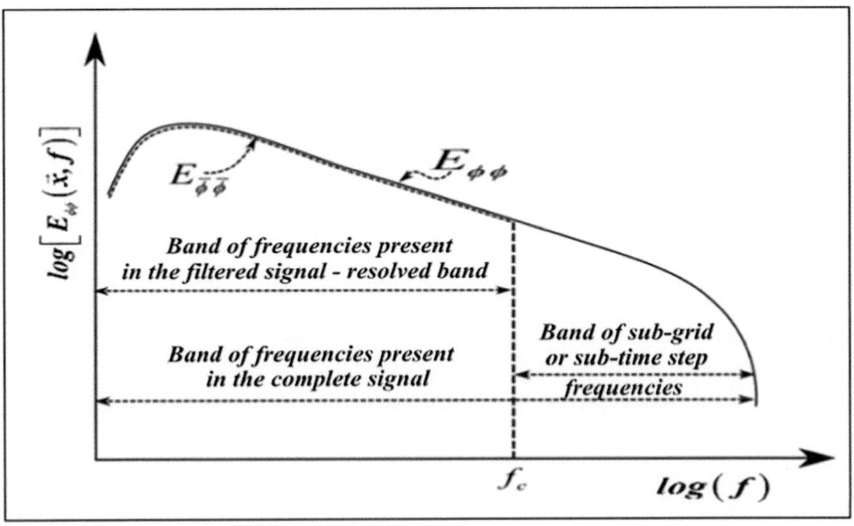

Fig. 5.6 Distribution of spectral density of turbulent specific energy associated with the temporal distribution of information $\phi(\vec{x}, t)$

$$\overline{\phi}(\vec{x},t) = \int\limits_{\Delta t} g(t')\phi(\vec{x},t-t')\mathrm{d}t', \tag{5.25}$$

where $g(t)$ is a normalized function, called the filter function and Δt is the integration time, which determines the cutoff frequency of the filtering process, $f_c = 2\pi/\Delta t$. For spatial filtering:

$$\overline{\phi}(\vec{x},t) = \int\limits_{\Delta V} h(\vec{x}')\phi(\vec{x}-\vec{x}',t)\mathrm{d}\vec{x}', \tag{5.26}$$

where $h(\vec{x})$ is the filter function and $\Delta V = \Delta x \Delta y \Delta z$ is the integration volume, with which the cutoff wave number of the filtering process is determined, $\vec{k}_c = 2\pi(1/\Delta x, 1/\Delta y, 1/\Delta z)$. The quantities Δx, Δy and Δz are the dimensions of the cells that make up the discretization mesh of the *PDE*.

In Fig. 5.7, the illustrative curves of two types of filter function are shown: one of them is like a Gaussian, but is what is called a Dirac Nucleus with compact support. This function cancels out outside the filtering range $\left[-f_c \le f \le f_c\right]$ or $[-k_c \le k \le k_c]$, defined by the period or cutoff wavelength of the filtering used. The second function is known as box filter, represented by a constant equal to unity, in the interval $\left[-f_c \le f \le f_c\right]$ or $[-k_c \le k \le k_c]$. The filter functions serve as a distribution of weights in the convolution integral. Therefore, the amplitude must be chosen in conjunction with the thickness of the compact support of the filter function. In other words,

$$\int\limits_{-\infty}^{+\infty} g(t)\mathrm{d}t = \int\limits_{-\infty}^{+\infty} h(x)\mathrm{d}x = 1.0. \tag{5.27}$$

The box-type filter functions can be given, for example, by:

$$h(x) = \frac{1}{\Delta V}, \tag{5.28}$$

$$g(t) = \frac{1}{\Delta t}. \tag{5.29}$$

Dirac Nucleus distributions are available in different forms, for example, in polynomial form:

$$h(r) = \begin{cases} 1 - \frac{1}{2}\|r\| - \|r\|^2 + \frac{1}{2}\|r\|^3 & \text{if } 0 \le \|r\| < 1 \\ 1 + \frac{11}{16}\|r\| + \|r\|^2 - \frac{1}{6}\|r\|^3 & \text{if } 1 \le \|r\| < 2 \\ 0 & \text{if } 2 \le \|r\| \end{cases}, \tag{5.30}$$

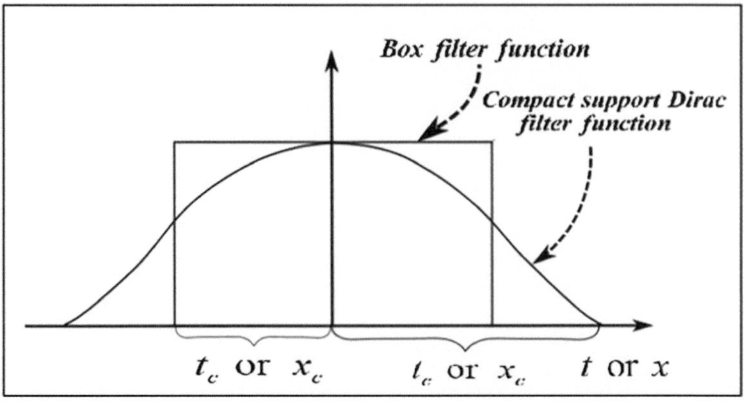

Fig. 5.7 Illustration of two types of filter functions: box filter and Dirac Nucleus filter with compact support

where $r_x = (x - X)/\Delta x$, $r_y = (y - Y)/\Delta y$ and $r_z = (z - Z)/\Delta z$. In a similar way, time filtering can be done.

The mathematical formalism of filtering the balance equations is done without explicitly specifying the filter function. It is a mere formalism, as will be seen. The hypothesis that the filter is a linear operator will be considered, and its properties will be used. With this formalism, applied to the balance equations, filtered equations for turbulence are obtained, resulting in the closure problem. Next, the equation filtering process is presented for modeling incompressible flows. Considering the linear filtering operator, the commutation and dissociation properties will be used between the filtering operator and the spatial and temporal derivative operators that appear in the balance equations. Thus, the filtering of the temporal derivative, for example, is equal to the temporal derivative of the filtered function. The same rule applies to spatial derivative operators. A filtering sum is the sum of the filtering. It was also considered that the physical properties: kinematic viscosity, ν, specific mass, ρ, and thermal diffusivity, α, are constant in relation to the filtering process. Under these hypotheses, the following filtering procedure for the balance equations is performed:

$$\frac{\partial \overline{u}_i}{\partial x_i} = 0, \tag{5.31}$$

$$\frac{\partial \overline{u}_i}{\partial t} + \frac{\partial}{\partial x_j}\left(\overline{u_i u_j}\right) = -\frac{1}{\rho_0}\frac{\partial \overline{p}}{\partial x_i} + \frac{\partial}{\partial x_j}\left[\nu\left(\frac{\partial \overline{u}_i}{\partial x_j} + \frac{\partial \overline{u}_j}{\partial x_i}\right)\right] + \frac{\overline{f}_i}{\rho_0}, \tag{5.32}$$

$$\frac{\partial \overline{T}}{\partial t} + \frac{\partial}{\partial x_j}\left(\overline{u_j T}\right) = \frac{\partial}{\partial x_j}\left(\alpha \frac{\partial \overline{T}}{\partial x_j}\right) + \frac{\overline{\phi}}{\rho_0 C_{p0}}. \tag{5.33}$$

As already discussed in Sect. 5.3, about the Average Boussinesq–Reynolds equations, these equations cannot be solved as they are an open equations system, due to the terms $\overline{u_i u_j}$ and $\overline{u_j T}$. Using the process of decomposition of turbulence variables, given by Eq. (5.24), Eqs. (5.32) and (5.33) are rewritten:

$$\frac{\partial \overline{u}_i}{\partial t} + \frac{\partial}{\partial x_j}\left(\overline{\overline{u}_i \overline{u}_j} + \overline{u_i' u_j'} + \overline{u_i' \overline{u}_j} + \overline{\overline{u}_i u_j'}\right) = -\frac{1}{\rho_0}\frac{\partial \overline{p}}{\partial x_i} + \frac{\partial}{\partial x_j}\left[\nu\left(\frac{\partial \overline{u}_i}{\partial x_j} + \frac{\partial \overline{u}_j}{\partial x_i}\right)\right] + \frac{\overline{f}_i}{\rho_0}, \tag{5.34}$$

$$\frac{\partial \overline{T}}{\partial t} + \frac{\partial}{\partial x_j}\left(\overline{\overline{u}_j \overline{T}} + \overline{u_j' T'} + \overline{u_j' \overline{T}} + \overline{\overline{u}_j T'}\right) = \frac{\partial}{\partial x_j}\left(\alpha \frac{\partial \overline{T}}{\partial x_j}\right) + \frac{\overline{\phi}}{\rho_0 C_{p0}}. \tag{5.35}$$

So far, the filtering process has led to the same terms obtained when using the averaging process, as can be seen by comparing Eqs. (5.34) and (5.35) to Eqs. (5.11) and (5.12), except for the transient terms, which do not appear in the average equations. However, the properties associated with second-order statistical moments involving the filtering process are different from the properties associated with the averaging process. All moments are non-zero, since the filtering of a fluctuation is not zero, the filtered function is not a constant and the filtering of the product of two filtered functions is different from the product of the two filtered variables. These statements are put in the following form: $\overline{u_i' u_j'} \neq 0$, $\overline{u_i' \overline{u}_j} \neq 0$, $\overline{\overline{u}_i u_j'} \neq 0$, $\overline{\overline{u}_i \overline{u}_j} \neq \overline{u}_i \overline{u}_j$, $\overline{u_j' T'} \neq 0$, $\overline{u_j' \overline{T}} \neq 0$, $\overline{\overline{u}_j T'} \neq 0$ and $\overline{\overline{u}_j \overline{T}} \neq \overline{u}_j \overline{T}$. Thus, the filtering process led to a higher number of unknowns than the averaging process. The problem of turbulence closure is also present here.

Statistical moments originating from correlations between fluctuations, $\overline{u_i' u_j'}$ and $\overline{u_j' T'}$ will be called, from now on: Boussinesq–Reynolds sub-grid tensor and the turbulent sub-grid flux of thermal energy, respectively. The terms $\overline{u_i' \overline{u}_j} + \overline{\overline{u}_i u_j'}$ and $\overline{u_j' \overline{T}} + \overline{\overline{u}_j T'}$ will be called crossed sub-grid tensor and the cross-turbulent sub-grid flux, respectively. These four second-order statistical moments must be modeled. However, even when these statistical moments are modeled, Eqs. (5.34) and (5.35) would not yet be ready for solution, due to the presence of the terms $\overline{\overline{u}_i \overline{u}_j}$ and $\overline{\overline{u}_j \overline{T}}$. These terms must be rewritten to give rise to the terms $\overline{u}_i \overline{u}_j$ and $\overline{u}_j \overline{T}$ in Eqs. (5.34) and (5.35). Therefore, Leonard [6] proposed the following mathematical definitions:

$$L_{ij} \equiv \overline{\overline{u}_i \overline{u}_j} - \overline{u}_i \overline{u}_j \tag{5.36}$$

$$Lq_j \equiv \overline{\overline{u}_j \overline{T}} - \overline{u}_j \overline{T}. \tag{5.37}$$

Equations (5.36) and (5.37) are the definitions of the Leonard tensor, L_{ij}, and the Leonard vector, L_{qj}. The cross-correlations were also added, to compose two unique terms, C_{ij} and C_{qij}:

$$C_{ij} = \overline{u'_i \overline{u}_j} + \overline{\overline{u}_i u'_j} \tag{5.38}$$

$$C_{qj} = \overline{u'_j \overline{T}} + \overline{\overline{u}_j T'} \tag{5.39}$$

Equations (5.38) and (5.39) were proposed by Clark et al. [2]. Replacing Eqs. (5.36) to (5.39) in Eqs. (5.34) and (5.35), and transporting all the additional terms, which appeared as a result of the decomposition of the non-linear terms, to the diffusive terms, we have:

$$\frac{\partial \overline{u_i}}{\partial t} + \frac{\partial}{\partial x_j}(\overline{u}_i \overline{u}_j) = -\frac{1}{\rho_0}\frac{\partial \overline{p}}{\partial x_i} + \frac{\partial}{\partial x_j}\left[\nu\left(\frac{\partial \overline{u}_i}{\partial x_j} + \frac{\partial \overline{u}_j}{\partial x_i}\right) - \overline{u'_i u'_j} - C_{ij} - L_{ij}\right] + \frac{\overline{f}_i}{\rho_0}, \tag{5.40}$$

$$\frac{\partial \overline{T}}{\partial t} + \frac{\partial}{\partial x_j}(\overline{u}_j \overline{T}) = \frac{\partial}{\partial x_j}\left(\alpha \frac{\partial \overline{T}}{\partial x_j} - \overline{u'_j T'} - C_{qj} - L_{qj}\right) + \frac{\overline{\phi}}{\rho_0 C_{p0}}. \tag{5.41}$$

These are the filtered equations, which are open in the sense that there are more unknowns than equations. The four tensors and four additional fluxes must be modeled, but this topic will only be addressed throughout Chap. 7. The separate modeling of each of these tensors and fluxes represented, for the global scientific community, an optimistic expectation of being able to better model turbulence. However, this expectation was betrayed when it was realized that the relative importance between the Boussinesq–Reynolds sub-grid tensor $\overline{u'_i u'_j}$ and the crossed tensors C_{ij} and Leonard L_{ij}, is very big. The same was observed comparing the relative importance of the turbulent sub-grid thermal flux $\overline{u'_j T'}$ and turbulent thermal cross flux C_{qj} and Leonard L_{qj}.

Modeling all these additional terms has added little in terms of the quality of results. On the other hand, this represented a significant increase in the cost of simulations. Therefore, the separate modeling of all these terms was abolished. However, these additional tensors and flux were not neglected, but were encompassed in a single term, as proposed by Germano [3]. He defined what can be called the sub-grid global tensor and flux. Germano proposed to define a tensor τ_{ij} and a vector q_j, rewriting the tensor $\overline{u_i u_j}$ and the vector $\overline{u_j T}$ in order to make appear the terms $\overline{u}_i \overline{u}_j$ and $\overline{u}_j \overline{T}$, in Eqs. (5.32) and (5.33). This proposal was transcribed in the present text, with the equations below:

$$\tau_{ij} \equiv \overline{u}_i \overline{u}_j - \overline{u_i u_j}, \tag{5.42}$$

$$q_j \equiv \overline{u}_j \overline{T} - \overline{u_j T}. \tag{5.43}$$

It is emphasized that Eqs. (5.42) and (5.43) are the definitions of the global tensor τ_{ij} and the global vector q_j. These equations measure the difference between all the nonlinear activity that would be calculated if the velocity and temperature fields were fully known $\overline{u_i u_j}$ and $\overline{u_j T}$ and the nonlinear activity that can be obtained with the filtered velocity and temperature fields $\overline{u}_i\overline{u}_j$ and $\overline{u}_j\overline{T}$. Replacing Eqs. (5.42) and (5.43) in Eqs. (5.32) and (5.33) and adding the continuity equation, we obtain the following equations:

$$\frac{\partial \overline{u}_i}{\partial x_i} = 0 \tag{5.44}$$

$$\frac{\partial \overline{u}_i}{\partial t} + \frac{\partial}{\partial x_j}(\overline{u}_i\overline{u}_j) = -\frac{1}{\rho_0}\frac{\partial \overline{p}}{\partial x_i} + \frac{\partial}{\partial x_j}\left[\nu\left(\frac{\partial \overline{u}_i}{\partial x_j} + \frac{\partial \overline{u}_j}{\partial x_i}\right) + \tau_{ij}\right] + \frac{\overline{f}_i}{\rho_0}, \tag{5.45}$$

$$\frac{\partial \overline{T}}{\partial t} + \frac{\partial}{\partial x_j}(\overline{u}_j\overline{T}) = \frac{\partial}{\partial x_j}\left(\alpha\frac{\partial \overline{T}}{\partial x_j} + q_j\right) + \frac{\overline{\phi}}{\rho_0 C_{p0}}. \tag{5.46}$$

It is worth noting that the average operator is the extreme of the filtering operator, with which all frequencies or all wave numbers are filtered, leaving only the average behavior of the filtered signal.

The averaging or filtering processes of the balance equations are a mere mathematical formalism that gives rise to the turbulence closure problem. These processes will not be explicitly applied before the spatial–temporal discretization of the PDEs. For discretization, it is necessary to use a grid of spatial volumes and a sequence of temporal steps. Within each discretization volume, variables are considered constant. Similarly, the temporal evolution in the numerical-computational solution cannot be obtained continuously. It will also be a discrete solution in time. Therefore, it is natural that all wavelengths related to the physical process, which are smaller than the size of the cells, are not captured using the discrete solution. Likewise, all physical effects with characteristic times smaller than the time step used in temporal discretization cannot equally be captured via a discrete solution. Thus, the discretization processes in time and space constitute space–time filters.

In Fig. 5.8, a function $\phi(\vec{x}, t)$ is illustrated, which was obtained using a discrete method and the solution, also discrete, obtained through DNS. The discrete solution is piecewise continuous and constant in each spatial interval. The same behavior occurs with respect to time. This way, whenever a differential equation is discretized, the discrete solution is filtered by the spatial grid and time step used. The discretization of a differential model is equivalent to applying spatial–temporal filtering processes, where the spatial and temporal filter functions are of the box type, as $h(x)$ and $g(x)$, presented through Eqs. (5.28) and (5.29).

Taking the limit of Eq. (5.25), when $\Delta t \to T$, where T is the total sampling time of the *information* of interest, it is observed that Eqs. (5.32) and (5.33) tend to Eqs. (5.7) and (5.8). It is concluded, then, that the filtered equations tend towards average equations when the filtering time limit tends to the total sampling time. On the other hand, taking the limits of Eqs. (5.25) and (5.26) as $\Delta t \to$ "τ" and/or when $\Delta V \to$ "η^3", where τ and η are the Kolmogorov characteristic time and length, respectively.

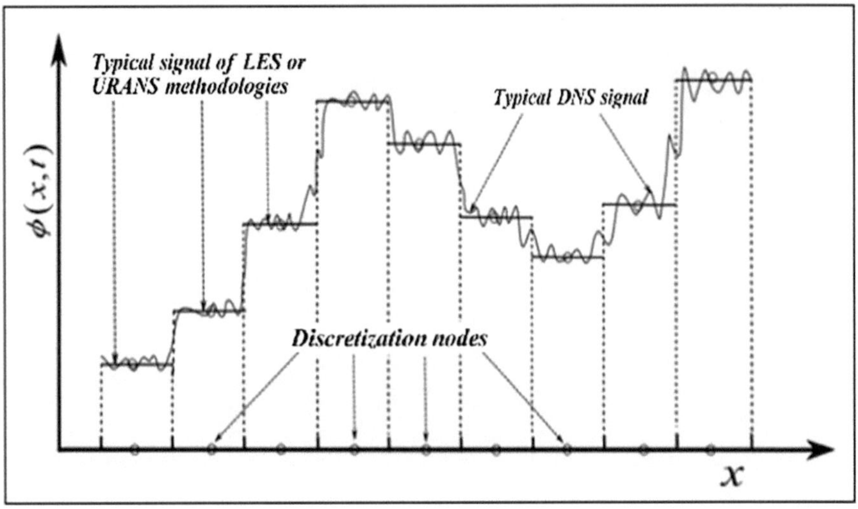

Fig. 5.8 Illustration of the filtering process originating from spatial discretization

Using Eqs. (5.42) and (5.43), then $\tau_{ij} \rightarrow 0$ and $q_j \rightarrow 0$. So, the Filtered Eqs. (5.44)–(5.46) tend, respectively, to the Continuity Eq. (5.3), to the Navier–Stokes Eqs. (5.4) and to the Energy Eq. (5.5). So, Eqs. (5.44)–(5.46) are the most general forms of equations for modeling turbulence. They are the filtered equations but tend towards the Boussinesq–Reynolds average equations: Continuity, Navier–Stokes and Energy equations, considering the limits already described above.

The Equations obtained via the Boussinesq–Reynolds average (Reynolds Averaged Navier–Stokes equations—RANS), Eqs. (5.22) and (5.23), are a model for a permanent regime, Reynolds [9]. However, with the application of the limit mentioned above, Eqs. (5.45) and (5.46) can be interpreted as the averaged Boussinesq–Reynolds equations, but in a transient regime (Unsteady Reynolds averaged Navier–Stokes equations—URANS).

Thus, it is clear that Eqs. (5.44) and (5.46) can be used along with the URANS Methodology for modeling turbulence and for the Methodology of Large Structures Simulation (LES) and even for the Direct Computational Simulation Methodology (DNS).

When presenting the turbulence closure models, these three methodologies will be characterized by the closure modeling and the numerical apparatus used for the computational solution of the differential model. Another question is: when should one choose and use one of these three methodologies? The choice depends on the objective. If the objective is the average results, then the most appropriate methodology is URANS. If the objective is the refined and detailed results about the flow, the DNS Methodology is the right choice. However, for high values of the Reynolds Number, the DNS Methodology is not viable, as already demonstrated. In this case, LES Methodology provides more details than URANS, and it is viable for high values of the Reynolds Number. Below, the characteristics of each of these turbulence modeling methodologies are described in more detail.

1. **URANS**: The basis of this methodology is Eqs. (5.44) to (5.46). The objective of this model is to simulate the low-frequency transient behavior of turbulent flows; that is, the objective is to obtain $\overline{u}_i$, $\overline{p}$ and $\overline{T}$. The average symbol $\langle \, \rangle$ will be replaced by the filter symbol $\overline{()}$. The low-frequency behavior of a turbulent flow can only be determined if the effects of turbulence are incorporated into the results. With the use of the terms τ_{ij} and q_j the transfer of linear momentum and thermal energy between turbulent fluctuations and the flow average quantities is modeled. This means that the closing models for τ_{ij} and q_j is a very complex task! Using closure models, almost all nonlinear interactions must be modeled, and the result will reflect in the low-frequency behavior of turbulence. Only low frequencies are calculated. These classes of models are difficult to develop, difficult to implement and difficult to obtain numerical-computational stability. With this methodology, little is calculated when solving Eqs. (5.44) to (5.46), and a lot of modeling is done, using the closing models, which will be presented later on. The practice of this methodology requires robust numerical methods, although it may have a low rate of convergence. The great advantage of using this methodology is that it can be practiced with spatial discretization with a low level of mesh refinement.
2. **DNS**: The basis of this methodology is also Eqs. (5.44) to (5.46), with $\tau_{ij} \rightarrow 0$ and $q_j \rightarrow 0$, i.e., the continuity, Navier–Stokes and energy equations are solved with cutoff frequency and cutoff wavenumber tending to Kolmogorov characteristic quantities. The objective of solving this model is to simulate the entire spectrum of turbulent structures, from integral structures to Kolmogorov structures. The aim is to determine the complete fields of $u_i(\vec{x}, t)$, $p(\vec{x}, t)$ and $T(\vec{x}, t)$. It will be seen that the numerical errors in the discretization process can be interpreted as a low-quality closure model, which does not consider the physical characteristics of the problem, but only the errors made in the discretization. Thus, the use of the DNS methodology requires numerical methods with a high rate of convergence, to minimize numerical errors. It also requires a very refined spatial mesh and very small time steps. What determines the size of the grid and the time step is the physical nature of the flow, i.e., the value attributed to the Reynolds Number. The use of this methodology does not require a turbulence closure model, since $\tau_{ij} \rightarrow 0$ and $q_j \rightarrow 0$, as the size Δx of the grid and the time step Δt tend respectively to the characteristic Kolmogorov length and time. With this methodology, 100% of the physics of the modeled flow can be directly calculated. Hence the name, Direct Computational Simulation. As the refinement of the grid and the time step depend on the Reynolds Number, the greater the value assigned to this parameter, the greater the refinement required for the spatial grid and the time step. Therefore, the application of DNS to simulate turbulent flows is limited to flows at low values of the Reynolds Number. This is the biggest limitation of using this methodology. The best feature of this methodology is that 100% of the information about a turbulent flow can be determined.

3. **LES**: The base of this methodology is, naturally, Eqs. (5.44) to (5.46), where τ_{ij} and q_j must be modeled. The closure models for this methodology are known as sub-grid models, since they model the interaction between the two parts of the spectrum, originating from the filtering process. In the low-frequency or low-wavenumber band, all large turbulent structures are calculated directly, and the interaction of this band of large structures with the band of small structures is modeled with the so-called sub-grid models. The finer the spatial mesh and the smaller the time step, the wider the resolved band becomes and the smaller the interactions to be modeled with the sub-grid models. Thus, for the LES methodology, the sub-grid closure modeling can be done using simple models, as their "task" is to model the interactions between the resolved band and the sub-grid band. The more refined the spatial grid and the smaller the time step, the more it is resolved and the less it is modeled. In the limit, when cell size tends to Kolmogorov characteristic length, η, and when the time step tends to Kolmogorov characteristic time, τ, then, LES tends to DNS. This methodology requires spatial and temporal discretization methods of at least second order for the convergence rate. The great advantage of this methodology is that it captures part of the spectrum, even for high values of the Reynolds Number.

In summary, the choice for one of the three methodologies (URANS, LES and DNS) depends on the objective sought with the simulation. If this objective is to determine the low-frequency behavior of a turbulent flow, the recommended methodology is URANS. If the objective is to obtain complete information, the recommended methodology is DNS. In this case, the application is limited to flows characterized by low values of the Reynolds Number. If you are looking for details of the solution, but without restricting to flows characterized by low values of the Reynolds Number, the methodology to be used is LES.

5.5 Exercises

1. What is the multiplicity of turbulent flow structures? Why is this characteristic important for establishing the turbulence closure problem? Comment on this topic.
2. Consider turbulent flow over a long cylinder 10 (cm) in diameter, over which water flows at an approaching velocity of 10 (m/s). Would this flow be in a turbulent regime? Calculate the number of degrees of freedom. Evaluate the calculation power necessary to completely solve (DNS) this flow. Considering that to practice LES it is necessary for the cutoff frequency or cutoff wave number to be in the inertial zone of the spectrum, estimate the reduction in computational cost, comparing LES with DNS.
3. Differentiate the averaging and filtering processes. Can the average be considered a particular case of filtering? Why?

4. Discuss, comparatively, the analogy between the Stokes closure model and the Boussinesq closure model.
5. What is turbulence modeling and what is turbulence closure modeling? Separate these two concepts.
6. Differentiate a Gaussian from a Dirac kernel. Why is the Dirac kernel preferred over a Gaussian kernel for filtering processes of differential equations?
7. Why separate modeling of tensors and crossflows and Leonard no longer done? Would they be neglected, with the modeling proposed in this book?
8. Why can filters in time and space be modeled with the spatio-temporal discretizations of PDEs? Disserts, illustrating with drawings.
9. How to differentiate the RANS and URANS methodologies? What about URANS, LES and DNS? How to choose one between them? What would be the criteria?
10. How to choose numerical methods for each of these methodologies? What criteria?

References

1. Boussinesq, J.: Essai sur la théorie des eaux courantes, Mém. Acad. Sci. Paris **22**, 46 (1877)
2. Clark, R., Ferziger, J., Reynolds, W.: Evaluation of sub grid-scale models using an accurately simulated turbulent flow. J. Fluid Mech. **91**, 1–16 (1979)
3. Germano, M.: A proposal for a redefinition of the turbulent stresses in Filtered Navier-Stokes equations. Phys. Fluids **29**(7), 2323–2324 (1996)
4. Kolmogorov: The local structure of turbulence in incompressible viscous fluid for Large Reynolds numbers. Dokl. Akad. Nauk SSSR **30**, 301–305 (1941a)
5. Kolmogorov: On degeneration of isotropic turbulence in an incompressible viscous liquid. Dokl. Akad. Nauk SSSR **31**, 538–541 (1941b)
6. Leonard, A.: Energy cascade in large-eddy simulations of turbulent fluid flows. Adv. Geophy. A **18**, 237–248 (1974)
7. Smagorinsky, J.: General circulation experiments with primitive equations. Mon. Weather Rev. **91**, 2501–2514 (1963)
8. Reynolds, O.: An experimental investigation of the circumstances which determine whether the motion of water shall be direct of sinuous, and of the law of resistance in parallel channels. Philos. Trans. R. Soc London **174**, 935–982 (1884)
9. Reynolds, O.: On the dynamical theory of incompressible viscous fluids and the determination of the criterion. Philos. Trans. R. Soc. London **186**, 123–161 (1884)

Chapter 6
URANS Methodology and Closure Models

In this chapter, the URANS methodology and closure modeling are presented, with an introduction to the topic presented in Sect. 6.1. The concepts of turbulent viscosity and diffusivity are presented in Sect. 6.2. The closing modeling is presented in Sect. 6.3. In Sect. 6.4, a classification of closure models is presented. To close the chapter, a list of exercises is proposed in Sect. 6.5.

6.1 Introduction

The mathematical model for fluid dynamics is composed of continuity, Navier–Stokes and energy equations. This model is sufficient to represent mathematically flows in any of the two regimes, laminar or turbulent, and for the transition between these two regimes, for any value of the Reynolds Number. The problem lies in the fact that the higher the value of this parameter, the wider the spectrum of turbulent structures associated with a given flow becomes. As the formation of the largest structures is regulated by the geometry of the problem, this means that large values of the Reynolds Number imply high frequencies and high wave numbers, or Kolmogorov structures that are smaller the higher the value of the Reynolds Number. To calculate all these structures, it is necessary to use extremely fine meshes, which implies high computational costs. Therefore, the practice of Direct Computational Simulation is limited to flows characterized by low values of the Reynolds Number.

The Reynolds Number values, for which DNS practice is possible, become higher every year: in 1991, the world record was $\mathrm{Re}_{\ell I} = 600$, which required the solution of 1.8×10^5 simultaneous equations, per unit volume; in 2018, 10^{12} simultaneous equations were resolved, for $\mathrm{Re}_{\ell I} = 10,000$. However, these values are still too modest to solve most engineering problems, which are characterized by high Reynolds Number values.

A. da Silveira Neto, *Turbulent Flows*, https://doi.org/10.1007/978-3-031-89222-6_6

As seen in the previous items, the alternative solution that has been used, historically, with relative success, involves the decomposition of the spectrum of turbulent structures into two bands, which leads to the concept of filtering of the balance equations. It was seen in Chap. 5 that this leads to the turbulence closure problem. This process led to Eqs. (5.44) to (5.46). These so-called turbulence equations, deduced with the filtering process of the continuity, the Navier–Stokes and the thermal energy equations, gave rise to the Boussinesq–Reynolds global tensor $\tau_{ij} = \left[\overline{u_i u_j} - \overline{u}_i \overline{u}_j\right]\left(\mathrm{m}^2/\mathrm{s}^2\right)$ as well as the global turbulent flux of thermal energy, $q_j = \left[\overline{u_j T} - \overline{u}_j \overline{T}\right](\mathrm{m}\,^\circ\mathrm{C/s})$. They are called global turbulent tensor and turbulent flux, because they incorporate all the tensors and vectors that appeared separately, that is, the Boussinesq–Reynolds sub-grid tensor, $\overline{u_i' u_j'}$, the crossed tensor, $C_{ij} = \overline{u_i' \overline{u}_j} + \overline{\overline{u}_i u_j'}$; Leonard's Tensor, $L_{ij} = \overline{\overline{u}_i \overline{u}_j} - \overline{u}_i \overline{u}_j$, the sub-grid turbulent flux, $\overline{u_j' T'}$, the cross flux, $C_{qj} = \overline{u_j' \overline{T}} + \overline{\overline{u}_j T'}$ and Leonard's flux, $L_{qj} = \overline{\overline{u}_j \overline{T}} - \overline{u}_j \overline{T}$. In other words, $\tau_{ij} = \overline{u_i u_j} - \overline{u}_i \overline{u}_j = \left(\overline{u_i' u_j'} + C_{ij} + L_{ij}\right)$ and $q_j = \overline{u_j T} - \overline{u}_j \overline{T} = \left(\overline{u_j' T'} + C_{qj} + L_{qj}\right)$. The tensor τ_{ij} is symmetric, therefore, it is composed of six unknowns. The vector q_j is made up of three unknowns. Nine more unknowns (six from $\overline{u_i' u_j'}$ and three from $\overline{u_j' T'}$) appear in the filtered equations. One of the ways to close this open problem is to write this tensor and this vector, as a function of the gradients of the filtered velocity and filtered temperature fields, which is known as the first gradient hypothesis proposed by Boussinesq [1]:

$$\tau_{ij} = f\left(\overline{u}_i, \overline{u}_j\right), \tag{6.1}$$

$$q_j = g\left(\overline{T}\right) \tag{6.2}$$

The distinguished French researcher proposed closing the system of filtered equations of fluid dynamics, modeling the Boussinesq–Reynolds Tensor, using the concept of turbulent viscosity, $\nu_t(\vec{x}, t)$. Similarly, to close the filtered energy equation, the concept of turbulent thermal diffusivity, $\alpha_t(\vec{x}, t)$, was proposed. Boussinesq's proposal is discussed below.

6.2 Concepts of Turbulent Viscosity and Diffusivity

Boussinesq [1] proposed an analogy with the Stokes Model for molecular viscous stresses, which is illustrated in Fig. 5.4. With this analogy, the French scientist proposed the following model for the Boussinesq–Reynolds global tensor to close the filtered equations:

$$-\tau_{ij} = \nu_t\left(\frac{\partial \overline{u}_i}{\partial x_j} + \frac{\partial \overline{u}_j}{\partial x_i}\right) - \frac{2}{3}k\delta_{ij}, \tag{6.3}$$

where

$$k \equiv \frac{1}{2}\tau_{ii} = \frac{1}{2}(\tau_{11} + \tau_{22} + \tau_{33}), \tag{6.4}$$

is the turbulent specific kinetic energy, k. The term involving the Kronecker Delta, δ_{ij}, arises due to the need to make the definition of k, compatible with the Boussinesq–Reynolds Tensor trace. Considering that the Sub-Grid Boussinesq–Reynolds Tensor, $\overline{u_i'u_j'}$, predominates over the cross tensor, C_{ij}, and the Leonard Tensor, L_{ij}, we have that the classical definition of turbulent specific kinetic energy is recovered:

$$k \equiv \frac{1}{2}\overline{u_i'u_i'} = \frac{1}{2}\left(\overline{u'^2} + \overline{\vartheta'^2} + \overline{w'^2}\right) \tag{6.5}$$

It should be noted that turbulent kinematic viscosity is a property of the flow and not of the fluid, as for the molecular viscosity, for Newtonian fluids. This characteristic implies greater difficulty in evaluating turbulent viscosity, since it depends on the nature of the flow itself, constituting a strong non-linearity in the solution of the equations. It is a property that depends on space and time, $\nu_t(\vec{x}, t)$.

Turbulent kinetic energy k, which appears in the closure model proposed by Boussinesq, Eq. 6.3, is also an unknown. However, as it depends only on the trace of the Boussinesq–Reynolds Tensor, it has a similar nature to pressure and can be incorporated into the pressure gradient term.

Analogously to what was proposed by Boussinesq to model turbulent tensions, turbulent thermal flux, $q_j = \overline{u_j T} - \overline{u}_j \overline{T}$, is modeled making the analogy with the Fourier Model for the molecular diffusive transport of thermal energy, $-q_{j\text{mol}} = \alpha\left(\partial T/\partial x_j\right)$, obtaining a way to model the turbulent thermal flux:

$$-q_j = \alpha_t \frac{\partial \overline{T}}{\partial x_j} \tag{6.6}$$

where α is the molecular thermal diffusivity, and α_t is the turbulent thermal diffusivity. Molecular diffusivity α is a property of the fluid and turbulent diffusivity α_t is a property of the flow.

6.3 Turbulence Closure Modeling

Replacing Eqs. (6.3) and (6.6) in Eqs. (5.45) and (5.46), results in filtered Navier–Stokes and energy equations, prepared for closure modeling:

$$\frac{\partial \overline{u}_i}{\partial t} + \frac{\partial}{\partial x_j}\left(\overline{u}_i\overline{u}_j\right) = -\frac{1}{\rho_o}\frac{\partial \overline{p}}{\partial x_i} + \frac{\partial}{\partial x_j}\left[(\nu + \nu_t)\left(\frac{\partial \overline{u}_i}{\partial x_j} + \frac{\partial \overline{u}_j}{\partial x_i}\right) - \frac{2}{3}k\delta_{ij}\right], \tag{6.7}$$

$$\frac{\partial \overline{T}}{\partial t} + \frac{\partial}{\partial x_j}(\overline{u}_j \overline{T}) = \frac{\partial}{\partial x_j}\left[(\alpha + \alpha_t)\frac{\partial \overline{T}}{\partial x_j}\right] + \frac{\overline{\phi}}{\rho_o C_{po}} \tag{6.8}$$

Remembering, $\nu_t = \mu_t/\rho_o$ and $\alpha_t = k_t/(\rho_o C_{po})$, where μ_t and k_t are, respectively, the turbulent dynamic viscosity and the turbulent thermal conductivity. The divergent term involving turbulent kinetic energy $(2/3)k\delta_{ij}$ results in the gradient of this *information*:

$$\frac{\partial}{\partial x_j}\left(\frac{2}{3}k\delta_{ij}\right) = \frac{2}{3}\frac{\partial k}{\partial x_i}, \tag{6.9}$$

which allows incorporate this term into the pressure gradient, resulting in a modified pressure:

$$\overline{p^*} = \overline{p} + \frac{2}{3}\rho_o k, \tag{6.10}$$

$$\frac{\partial \overline{u}_i}{\partial t} + \frac{\partial}{\partial x_j}(\overline{u}_i \overline{u}_j) = -\frac{1}{\rho_o}\frac{\partial \overline{p^*}}{\partial x_i} + \frac{\partial}{\partial x_j}\left[(\nu + \nu_t)\left(\frac{\partial \overline{u}_i}{\partial x_j} + \frac{\partial \overline{u}_j}{\partial x_i}\right)\right] \tag{6.11}$$

The thermal energy equation is also rewritten, resulting in:

$$\frac{\partial \overline{T}}{\partial t} + \frac{\partial}{\partial x_j}(\overline{u}_j \overline{T}) = \frac{\partial}{\partial x_j}\left[(\alpha + \alpha_t)\frac{\partial \overline{T}}{\partial x_j}\right] + \frac{\overline{\phi}}{\rho_o C_{po}} \tag{6.12}$$

The sums $\nu_{ef} = (\nu + \nu_t)$ and $\alpha_{ef} = (\alpha + \alpha_t)$ are the effective viscosity and diffusivity, composed of a molecular part and another part of a turbulent nature. These balance equations for the filtered components of velocity and temperature would be closed, that is, they would contain the same number of equations as unknowns, if it were not for the presence of ν_t and α_t, which are unknown quantities. The calculation of turbulent viscosity and thermal diffusivity constitutes the main problem to be solved for modeling the closure of the filtered equations.

To recover the true static pressure field, it is also necessary to model the turbulent kinetic energy, which must be done simultaneously with the calculation of turbulent viscosity. This calculation is the fundamental role of the so-called turbulence closure models. So, an analogy is used between what happens at the level of fluid particles and what happens at the level of turbulent structures, as schematized and illustrated in Fig. 6.4. Observing this figure and Table 6.1, we have the basis for calculating turbulent viscosity and diffusivity. However, some shortcomings in this analogy stand out:

- In the kinetic theory of gases, it is assumed that the size of a fluid particle is in the order of the molecular mean free path. However, for turbulence, it is possible to

Table 6.1 Analogy between the process of turbulent diffusion and molecular diffusion

Analogy between turbulent diffusion and molecular diffusion processes	
Molecular diffusion process	Turbulent diffusion process
Movement and interactions between molecules	Nonlinear movement and interactions between turbulent structures
Transfer of linear momentum and thermal energy between fluid particles (continuum hypothesis) occurs through the movement of molecules between particles	Transfer of linear momentum and thermal energy between different flow sites occurs through nonlinear movement and interactions between turbulent structures
These transfers are modeled, at the continuum level, via molecular viscosity and diffusivity: $\nu \approx C\zeta$ and $\alpha \approx \chi\nu$ C: Molecular characteristic velocity ζ: Molecular characteristic length χ: Constant to be determined	These transfers are modeled, by analogy, with the concepts of turbulent viscosity and thermal diffusivity: $\nu_t \approx UL$ and $\alpha_t \approx \chi_t\nu_t$ U: Transport speed of turbulent structures χ: Length of turbulent structures χ_t: Function to be determined

have turbulent structures much larger than the characteristic length L, normally evaluated for each flow.

- Molecular viscosity is a scalar *information*, considered isotropic. However, turbulence is characterized by high levels of anisotropy, especially for large flow structures. This means that using turbulent viscosity, as it is an isotropic *information*, can be a compromising approximation.

Despite the unfavorable points highlighted, with the concept of turbulent viscosity, a huge advance was achieved in the field of modeling and simulation of turbulent flows. With the development of new modeling philosophies, this analogy has become much more realistic: this is the case of the Large Structures Simulation Methodology (LES), since, philosophically, we seek to resolve large structures and model the nonlinear interactions between the band resolved and the sub-grid band of turbulent structures. With this methodology, in which the turbulence closure model serves to deal with interactions with the smallest structures, the use of the isotropy hypothesis becomes more realistic.

6.4 Classification of Models for the Turbulence Closure Problem

Models for the closure problem of turbulence can be classified according to their direct or indirect dependence on the concepts of turbulent viscosity and turbulent diffusivity. In both groups, it becomes necessary to obtain additional algebraic or balance equations for the calculation of turbulent viscosity and turbulent diffusivity, or equations for the alternative closure of the filtered equations for turbulence. In this second case, the concepts of turbulent viscosity and diffusivity are used indirectly, that is, no longer for the direct modeling of the Boussinesq–Reynolds Tensor and turbulent thermal flux. Thus, there are two large groups of models, which are described below.

6.4.1 Models that Directly Depend on Turbulent Viscosity and Diffusivity

The following models depend directly on the concept of turbulent viscosity:

- Models based on zero balance equation

This is a whole family of models based on the Prandtl Mixing Length Hypothesis [8] and most sub-grid models, used for the Large Structures Simulation Methodology (LES).

- Models based on one balance equation

The models in this family are based on a balance equation for a given statistical property of turbulence. The first model based on a balance equation was proposed by Prandtl [9]. In this model, a balance equation was derived for the turbulent specific kinetic energy, k, and an algebraic relation for a characteristic length of the turbulence, L. Other examples of models for a balance equation are those proposed by Spalart and Allmaras [11, 12] and by Baldwin and Barth [12]. This model is characterized by a single balance equation for the turbulent viscosity itself, which eliminates the difficult task of evaluating a characteristic length for turbulence.

- Models based on two balance equations

Kolmogorov [6] proposed the first two-equation model. His first proposal was based on the concepts of turbulent specific kinetic energy, k, and the Kolmogorov Frequency, ω. This *information* is also called the specific power of viscous transformation of turbulent kinetic energy. However, effectively, this *information* has a frequency unit. With these two characteristic quantities of turbulence, k and ω, Kolmogorov modeled turbulent viscosity, ν_t. This model was then called the $k-\omega$ closure model. Subsequently, Prandtl and Kolmogorov proposed the concept of viscous transformation power of turbulent kinetic energy, ε. Again, the turbulent viscosity was then modeled as a function of these two parameters, k and ε. This model was called $k-\varepsilon$ closure model. Based on Kolmogorov's Proposal, several models of the type $k-\varepsilon$ were developed with a view to various improvements, such as the effect of the Reynolds Number value and the proximity or not of a solid boundary.

The Closing model $k-\omega$ proposed by Kolmogorov was taken up by Wilcox [13]. The two main differences of this model, proposed by Wilcox, as compared with the $k-\omega$ proposed by Kolmogorov, is that the Wilcox model is suitable for flows at low Reynolds Number values and was also developed for flows in the vicinity of walls. The greatest difficulty in using this model arises when there is a flow composed of a region known as a turbulent core, characterized by a high value of the Reynolds Number, and a layer close to a wall, characterized by a low value of the Reynolds Number. In this case, the proposed modeling becomes inadequate.

6.4.2 *Models that Are Independent or Indirectly Dependent on Turbulent Viscosity and Diffusivity*

- **Algebraic models**

These are models without balance equations. With these models, the components of the Boussinesq–Reynolds tensor are algebraically equated, without using the Boussinesq turbulent viscosity hypothesis. Sub-grid models are examples of algebraic models. RSM (Reynolds Shear Model) can also be of algebraic nature.

- **Balance Model for Boussinesq–Reynolds Tensor Components—RSM**

Many difficulties were encountered with complete two-equation models, such as: (a) lack of generality; (b) difficulties in modeling (with the same model) flows with characteristics close to walls, where there are low values of the local Reynolds Number and, at the same time, in the turbulent core, with high values of the Reynolds Number; (c) flows with strong effects of adverse pressure gradients, and (d) flows with strong effects of curvature and rotation. Given this, the community sought to develop models based on balance equations for the components of the Boussinesq–Reynolds tensor, τ_{ij}.

Donaldson [4], Hanjalik and Launder [5], and Bradshaw [2], showed progress in this class of closure models, based on the solution of approximate equations for the components of τ_{ij}. The first model in this line was proposed by Rotta [10]. Preliminary predictions were made by Daly and Harlow [3]. These models still carried approximations that took away the generality necessary for solving complex turbulent flows. The most complete model in this line was proposed by Launder, Reece and Rodi [7]. These models have meant major advances in terms of generality and quality of results. The biggest difficulties with this family of models are implementation, numerical stability and computational cost. In fact, this type of closure model is composed of seventeen balance equations for isothermal flows. Another set of PDEs must be added for flows with thermal effects, as will be presented in Chap. 9. As will be seen, when this model is presented, the Boussinesq–Reynolds Tensor will be modeled with additional balance equations for each of its components. The necessary closures will be made at the level of third-order statistical moments. At this level of closure, the concept of turbulent viscosity will still be necessary. This justifies the fact that, in this item, models that indirectly depend on the concept of turbulent viscosity have been classified.

6.5 Exercises

1. What is the first gradient hypothesis for modeling turbulence closure?
2. What are the main weaknesses associated with modeling turbulence closure by analogy between turbulent diffusion and molecular diffusion processes?
3. Discuss closure models that depend directly on the concept of turbulent diffusion and models that do not depend directly on this concept.
4. How many balance equations make up the RSM closure model for isothermal flows? Closing will occur at the level of statistical moments of which order?
5. What is an algebraic closure model? Give examples.

References

1. Baldwin, B.S., Barth, T.J.: A one-equation turbulence transport model for High Reynolds number wall-bounded flows, NASATM-102847 or AIAA paper 91–0610 (1990)
2. Bradshaw, P.: The understanding and prediction of turbulent flow. Aero. J. **76**, 403 (1972)
3. Daly, B.J., Harlow, F.H.: Transport equations of turbulence. Phys. Fluids **13**, 2634 (1970)
4. Donaldson, C.: A progress report on an attempt to construct an invariant model of turbulent shear flows. In: Proceedings AGARD Conference on Turbulent Shear Flows, London, paper B-1 (1971)
5. Hanjalic, K., Lauder, B.E.: A Reynolds stress model of turbulence and its application to thin shear flows. J. Fluid Mech. **52**, 609 (1972)
6. Kolmogorov, A.N.: The equation of turbulent motion in an incompressible fluid. Izvestia Acad. Sci USSR Phys. **6**, 56–58 (1942)
7. Launder, B.E., Reece, G.J., Rodi, W.: Progress in the development of a Reynolds-stress turbulence closure. J. Fluid Mech. **68**, 537–566 (1975)
8. Prandtl, Z.A.: Bericht über Untersuchugen zur Ausgebildeten Turbulenz. As Angew Math. Mech. **5**, 136–169 (1925)
9. Prandtl, Z.A.: Über ein neues Formelsystem für die ausgebildete Turbulenz. Nachr. Akad. Wiss. Göttingen Math-Phys. **K1**, 6–19 (1945)
10. Rotta, J.C.: Statistische Theorie nichthomogener Turbulenz. 2. Phys. **129**, 547 (1951)
11. Spalart, P.R., Allmaras, S.R.: A One-Equation Turbulence Model for Aerodynamic Flows, AIAA Paper 92–0439 (1992)
12. Spalart, P.R., Allmaras, S.R.: A One-Equation Turbulence Model for Aerodynamic Flows, La Recherche Aerospatiale, N^{o}, pp. 5–21 (1994)
13. Wilcox, C.W.: Turbulence modeling for CFD, La Cañada, Ca: DCW Industries (1998)

Chapter 7
URANS Closure Models Based on Zero Balance Equation

In this chapter, the URANS closure modeling based on the zero balance equation is presented. It begins with an introduction to the topic in Sect. 7.1. In Sect. 7.2, the Prandtl Mixing Length is presented. In Sect. 7.3, continuous solutions for a mixing layer in temporal development in laminar and turbulent regimes are presented. In Sect. 7.4, a mixing layer in spatial development is also presented. The continuous solution for a turbulent distant wake is presented in Sect. 7.5. Continuous solutions for flat laminar and turbulent jets are presented in Sect. 7.6. In Sects. 7.7 and 7.8, a theoretical analysis is presented for turbulent plane Poiseuille flows with thermal effects. This chapter ends with a list of exercises in Sect. 7.9.

7.1 Introduction

Turbulence closure models based on zero balance equations are algebraic models. These models can be divided into models for the RANS/URANS methodology and models for the LES Methodology. In this chapter, the zero-equation models for the RANS methodology will be presented. Sub-grid models for LES will be presented in Chap. 10. From a historical point of view, the first closure model was proposed by Prandtl [16] and Prandtl [17]. Despite its simplicity, this model is still used today and, furthermore, has great academic value. It will be presented below. Several examples of canonical turbulent flows will be solved. Illustrative results will be presented. It will be seen that for several types of canonical turbulent flows, it is possible to determine continuous solutions.

A. da Silveira Neto, *Turbulent Flows*, https://doi.org/10.1007/978-3-031-89222-6_7

7.2 Prandtl Mixing Length Model

The Prandtl Mixing Length model was proposed to model flows with a predominant physical behavior in a given flow direction, known as the main direction. For example: flows developed inside a flat channel; flows inside a circular conduit; jets, mixing layers and distant wakes. The self-similarity theory will be used, which requires that the flow must be spatially developed. This will be further discussed throughout this chapter. For these flows, an order of magnitude analysis of the terms of the filtered Continuity and Navier–Stokes Equations leads to a simplified model, for isothermal and incompressible flows:

$$\frac{\partial \overline{u}}{\partial x} + \frac{\partial \overline{\vartheta}}{\partial y} = 0, \tag{7.1}$$

$$\frac{\partial \overline{u}}{\partial t} + \overline{u}\frac{\partial \overline{u}}{\partial x} + \overline{\vartheta}\frac{\partial \overline{u}}{\partial y} = -\frac{1}{\rho}\frac{\partial \overline{p}}{\partial x} + \nu\frac{\partial^2 \overline{u}}{\partial y^2} - \frac{\partial\left(\overline{u'\vartheta'}\right)}{\partial y} \tag{7.2}$$

Using the Boussinesq hypothesis, the Boussinesq–Reynolds stresses are modeled using the concept of turbulent viscosity. For the group of canonical flows, mentioned above, and studied in this chapter, we have the following:

$$-\overline{u'\vartheta'} = \nu_t\frac{\partial \overline{u}}{\partial y}, \tag{7.3}$$

where $\overline{u}(y)$ is the filtered velocity, a function of the coordinate y, transversal to the flow. As previously presented, turbulent viscosity can be modeled, making an analogy with the modeling of molecular viscosity, using the kinetic theory of gases. Molecular kinematic viscosity can be modeled by the product of the velocity of sound in the gaseous medium and the molecular mean free path. This means that the turbulent kinematic viscosity of the fluid can be modeled as the product of a length and a velocity, characteristic of a turbulent flow. The characteristic velocity will be denoted here by $\hat{V}$ and the characteristic length comes from Prandtl's hypothesis for the mixing length, ℓ_m. This characteristic length represents all the turbulent structures that cohabit a given volume of fluid. Thus, the turbulent viscosity is calculated using the following equation:

$$\nu_t \approx \hat{V}\ell_m. \tag{7.4}$$

For flows with a predominant direction, a characteristic transport velocity of the turbulent structures can be obtained, as a function of the velocity gradient, normal to the predominant direction:

$$\hat{V} = \ell_m\left|\frac{\partial \overline{u}}{\partial y}\right| \tag{7.5}$$

Replacing Eq. (7.5) in Eq. (7.6), results:

$$\nu_t = \ell_m^2 \left| \frac{\partial \overline{u}}{\partial y} \right|. \tag{7.6}$$

The shear that is modeled through the derivative in Eq. (7.6) can be positive or negative, but it is desirable that the turbulent viscosity be always positive. This is why the strain rate is taken in modulus. Considering it positive, the process of direct transfer of information between the turbulent structures that make up the spectrum is modeled. If negative, the so-called reverse transfer of information is modeled. When negative, the diffusive term of the linear momentum balance equation makes numerical methods unstable. This is the reason for taking the modulus of the strain rate in Eq. (7.6).

The mixing length ℓ_m remains unknown. Taking as an example a mixing layer in spatial development, illustrated in Fig. 7.1, is a way to estimate the Prandtl mixing length. Analyzing this figure, a well-determined statistical behavior is inferred in terms of the thickness of the shear layer, a result of the dynamics of turbulent coherent structures. The characteristic length of this flow is the thickness of the mixing layer $\delta(x)$. So,

$$\ell_m \approx \delta(x) \Rightarrow \ell_m = \beta\delta(x) \tag{7.7}$$

The constant β is empirical and different for each type of flow, depending on its physical characteristics. In Table 7.1, different values of this constant for some classic flows, such as mixing layer, jets and wakes are presented.

The turbulent viscosity depends on x and on y, as presented in Eq. (7.8). The function $\delta(x)$ is empirical, determined from experiments, material or computational, for flows of interest:

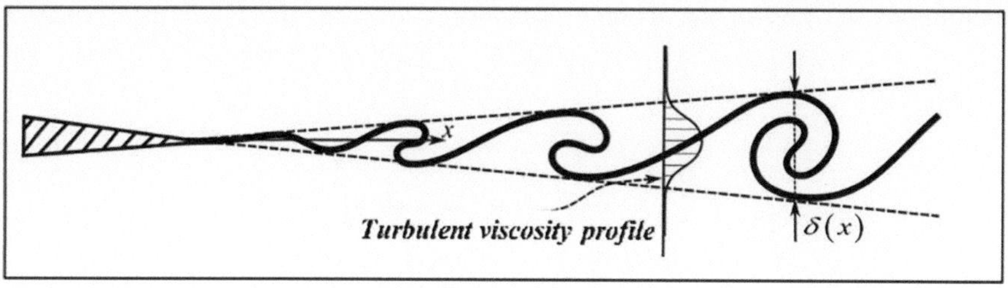

Fig. 7.1 Mixing layer in spatial development; ν_t: illustrative distribution of turbulent viscosity; $\delta(x)$: mixing layer thickness (characteristic length)

Table. 7.1 Constant β of Eq. (7.7), for four types of canonical flows

Flows	Spatial mixing layer	Plane jet	Round jet	Plane wake
$\frac{\ell_m}{\delta(x)}$	0.07	0.09	0.075	0.18

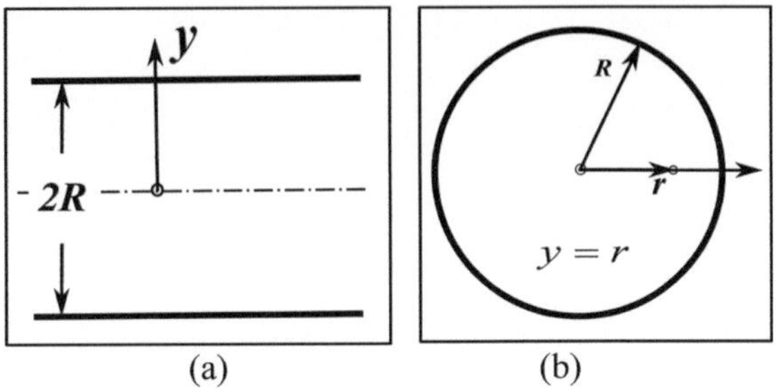

Fig. 7.2 Coordinate systems for: **a** a flat channel and **b** a circular duct

$$\nu_t(x, y) = \ell_m^2(x)\left|\frac{\partial \overline{u}(x, y)}{\partial y}\right|. \tag{7.8}$$

For fully developed flows in circular ducts or flat channels, Nikuradse, cited by Wilcox [26], proposed the following expression for the Prandtl mixing length:

$$\frac{\ell_m}{R} = 0.14 - 0.08\left(\frac{y}{R}\right)^2 - 0.06\left(\frac{y}{R}\right)^4, \tag{7.9}$$

where R is the radius for circular ducts, and the half-width for flat channels. For flat channels with a wide of $2R$, y is the Cartesian coordinate of the axis system with the origin placed at the center of the channel. For circular ducts, y is replaced by r, the radial coordinate of the cylindrical axis system. In Fig. 7.2, two coordinate systems are illustrated. Thus, for both cases: if $y = \pm R$ for flat channels, or if $r = R$ for the circular duct, $\ell_m = 0$, when Eq. (7.9) is used. On the other hand, if $y = 0$ for flat channels, or $r = 0$ for circular ducts, then $\ell_m = 0.14R$. As expected, turbulence has zero activity at the wall, since, due to the non-slip condition, the velocity and the velocity fluctuation are zero.

For layers close to walls, the Prandtl mixing length model is not sufficient to obtain consistent results. A damping function, known as the Van Driest Function [24], should be used in conjunction with the Prandtl mixing length model. The Van Driest function is presented in Eqs. (7.10) and (7.11):

$$\ell_m = \kappa(R - y)\left[1 - \exp\left(-\frac{R - y}{R}\frac{\mathrm{Re}_\tau}{26}\right)\right], \tag{7.10}$$

or

$$\tilde{\ell}_m = \kappa(1 - \tilde{y})\left[1 - \exp\left(-(1 - \tilde{y})\frac{\mathrm{Re}_\tau}{A}\right)\right], \tag{7.11}$$

where $A = 26$ is a dimensionless constant; $\kappa = 0.41$ is the Von Kármán Constant, also dimensionless; $\tilde{\ell}_m = (\ell_m/R), \tilde{y} = (y/R), \mathrm{Re}_\tau = (u_\tau R)/\nu, u_\tau = \sqrt{\frac{\|\tau_w\|}{\rho}}$ in (m/s), τ_w in $\left(\mathrm{N/m^2}\right)$ (viscous stress on the wall), $\ell_m^+ = \frac{\ell_m\sqrt{\tau_w/\rho}}{\nu}$ and $y^+ = \frac{y\sqrt{\tau_w/\rho}}{\nu}$ is the wall coordinate. The application domain of the Van Drist Function is such that $y^+ \leq 40$ or, equivalently, $\tilde{y}\mathrm{Re}_\tau \leq 40$. It is worth noting that for y tending to R, Eq. (7.10) can be replaced by $\ell_m = \kappa y$. It is also observed that the relationship $\ell_m = \kappa y$ does not result from Eq. (7.11), but these two equations give very similar results when approaching a wall.

Equation (7.10) is known as the Prandtl hypothesis [16], as discussed by Pope [15]. It is worth emphasizing that this model is used to treat areas close to walls. In order to have a general model, simultaneously evolving Nikuradse's proposal Eq. (7.9), and Prandtl's hypothesis Eq. (7.11), Cebecci and Bradshaw [5] proposed a single model, involving the two equations, as presented below:

$$L(\tilde{y}) = \frac{\ell_m}{R} = \left(0.14 - 0.08\tilde{y}^2 - 0.06\tilde{y}^4\right)\left\{1 - \mathrm{e}^{(\tilde{y}-1)\frac{\mathrm{Re}_\tau}{26}}\right\}. \tag{7.12}$$

In Eq. (7.12), for which $0 \leq \tilde{y} \leq 1$, Prandtl mixing length varies in the range $0 \leq L(\tilde{y}) \leq 0.14$. In Fig. 7.3, the function $L(\tilde{y})$ is illustrated, as proposed by Nikuradse, and with the modification proposed by Cebecci and Bradshaw. As can be seen in Fig. 7.3, the two models differ in proximity to the wall. The difference between the curves is apparently small, but as will be discussed later in Sect. 7.6, the consequences, in terms of velocity distribution, are substantial.

This damping near the walls is necessary since viscous effects predominate over turbulent effects. This topic is discussed in more detail in Sect. 7.7, where turbulent flow in a flat channel is mathematically modeled, using the Prandtl mixing length model, with Cebecci and Bradshaw damping model.

Examples of canonical turbulent flows are presented below, for which mathematical continuous solutions are developed and presented. These examples are very useful in understanding the importance of turbulence modeling. They also serve

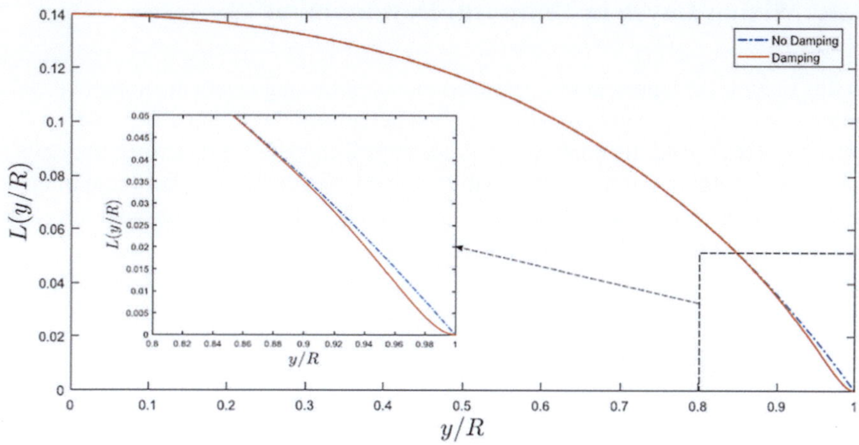

Fig. 7.3 Prandtl mixing length for a flat Channel: Nikuradse's proposal. Personal collection, Antonialli and Silveira-Neto [2]

Table. 7.2 Power law for thickness $\delta(x)$. increase, for centerline velocity, $\overline{u}(x)_{max}$, and for the velocity difference, $\overline{u}_1(x)$, also on the center line, Schlichting [21]

Regime	Lamar		Turbulent	
Type of flow	Thickness $\delta(x)$	Centerline velocity: $\overline{u}_{max}$ and $\overline{u}_1(x)$	Thickness $\delta(x)$	Centerline velocity: $\overline{u}_{max}$ and $\overline{u}_1(x)$
Spatial mixing layer (flat)	$x^{1/2}$	x^o	x	x^o
Two-dimensional flat jet	$x^{2/3}$	$x^{-1/3}$		$x^{-1/2}$
Circular jet	x	x^{-1}	x	x^{-1}
Two-dimensional flat wake	$x^{1/2}$	$x^{-1/2}$	$x^{1/2}$	$x^{-1/2}$
Circular wake	$x^{1/2}$	x^{-1}	$x^{1/3}$	$x^{-2/3}$

to see how limited continuous solutions are. However, following the simplifying hypotheses, the results obtained, when compared with the results of material experiments, lead to excellent agreement. The first example to be presented is one of the simplest canonical flows from a physical point of view. However, it is one of the most difficult to experience, from a material point of view. This is a mixing layer in temporal development, which was already presented and explored in Chap. 4 about the transition to turbulence.

Schlichting [21], presents a table with the so-called power laws for the most important information of some free, laminar and turbulent shear flows: thickness $\delta(x)$ and centerline velocity. This data is transcribed here, in Table 7.2.

7.3 Mixing Layer in Temporal Development

In this section the representative physical model of mixing layers in temporal development is shown, for the laminar and turbulent regimes. Next, the differential mathematical models and the continuous mathematical models are described separately for these two regimes. Finally, the results obtained with the application of continuous models are reported. The numerical-computational models used for the purpose of comparing results are also reported in this study.

7.3.1 Physical Model

A temporally developing mixing layer is one of the simplest canonical flows to be modeled. Two fluid layers move at velocities U_1 and U_2 defining and maintaining the flow. In an initial time, $t = 0$ (s), the velocity field will be considered discontinuous

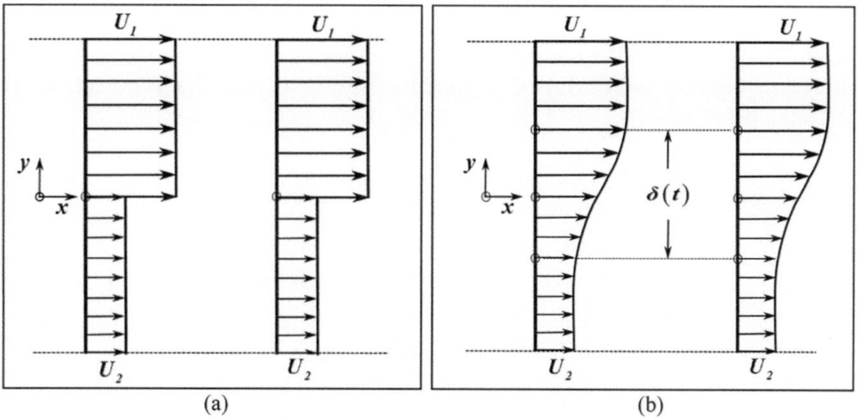

Fig. 7.4 Mixing layer in temporal development: **a** $t = 0$ (s) and **b** $t > 0$ (s)

at the center of the domain, with an infinite velocity gradient at that position. The thickness of the shear layer will be zero in the initial condition, $\delta(t = 0) = 0$ (m). This can be seen in Fig. 7.4a, in two positions x. As time passes, the diffusion process of linear momentum attenuates this gradient, establishing an inflectional velocity profile, as can be seen in Fig. 7.4b, for a generic time t. In a laminar regime, this enlargement of the mixing layer will be promoted only by molecular diffusion. In the turbulent regime, the turbulent structures will accelerate this molecular diffusion process by orders of magnitude. It is important to realize that turbulence accelerates the molecular diffusion transport, as it broadens local gradients and increases the number of gradients per unit of volume. It is observed that the thickness $\delta(x)$ evolves over time, but maintains similarity in space, which justifies the name of mixing layer in temporal development, for this type of flow.

In the present analysis, the objective will not be the modeling and the simulation of the process of formation of Kelvin–Helmholtz instabilities, but only to model the average flow behavior. It is observed that the velocity profiles, illustrated in Fig. 7.4, are self-similar, which allow the use of self-similarity theory to solve the differential mathematical model. Since only the average behavior will be determined, the physical model presented is valid for both the laminar and turbulent regimes. What would differentiate between the two regimes are the forms of temporal evolution.

7.3.2 Temporal Mixing Layer in Laminar Regime

7.3.2.1 Partial Differential Mathematical Model

Such a model can be derived from Eqs. (7.1) and (7.2). The Boussinesq–Reynolds tensor is set to zero, which is equivalent to not having filtered these equations, as is the case with the laminar regime. The advective term is simplified, in accordance with the physical nature of this problem, that is, the flow occurs only in the x-direction, and it only depends on the direction y and on the time t. This flow is modeled with the following equation:

$$\frac{\partial u}{\partial t} = \nu \frac{\partial^2 u}{\partial y^2} \tag{7.13}$$

being the initial and boundary conditions, illustrated in Fig. 7.4, given by:

$$\begin{aligned} &u(y, t = 0) = u_o(y), \\ &u\left(y = +\frac{\delta(t)}{2}\right) = 0.99U_1, \\ &u\left(y = -\frac{\delta(t)}{2}\right) = 1.01U_2. \end{aligned} \tag{7.14}$$

7.3.2.2 Ordinary Differential Mathematical Model

Equation (7.13) can be modified using the self-similarity theory, which makes it possible to transform the partial differential model into an ordinary differential model. So, a solution is proposed for the longitudinal velocity component, $u(y, t)$, as follows:

$$u(y, t) = \overline{U} + \tilde{U} \cdot f(\eta) \tag{7.15}$$

where $\tilde{U} = (U_1 + U_2)/2$ is the average of the current velocities and $\tilde{U} = (U_1 - U_2)/2$ is half difference in velocities. The similarity parameter, η, is given by:

$$\eta(y, t) = \frac{y}{\delta(t)} \tag{7.16}$$

Deriving Eq. (7.15) in relation to time, t, and in relation to space, y, taking into account Eq. (7.16), it follows that:

$$\frac{\partial u}{\partial t} = \tilde{U} \frac{\mathrm{d}f}{\mathrm{d}\eta} \frac{\partial \eta}{\partial \delta} \frac{\mathrm{d}\delta}{\mathrm{d}t} = \tilde{U} f' \left(-\frac{y}{\delta^2}\right) \frac{\mathrm{d}\delta}{\mathrm{d}t} = -\frac{\tilde{U}}{\delta} \frac{\mathrm{d}\delta}{\mathrm{d}t} \eta f' \tag{7.17}$$

$$\frac{\partial u}{\partial y} = \tilde{U} \frac{\mathrm{d}f}{\mathrm{d}\eta} \frac{\partial \eta}{\partial y} = \tilde{U} f' \frac{1}{\delta} \tag{7.18}$$

$$\frac{\partial^2 u}{\partial y^2} = \tilde{U} \frac{1}{\delta} \frac{\mathrm{d}^2 f}{\mathrm{d}\eta^2} \frac{\partial \eta}{\partial y} = \tilde{U} \frac{1}{\delta^2} f''. \tag{7.19}$$

Replacing Eqs. (7.17) and (7.19) in Eq. (7.19) yields:

$$f'' + \boxed{\frac{\delta}{\nu}\frac{\mathrm{d}\delta}{\mathrm{d}t}}\eta f' = 0. \tag{7.20}$$

The term inside the square in Eq. (7.20) must be a constant, since the solution of this differential equation must result in a function f' that depends only on η. Thus,

$$\frac{\delta(t)}{\nu}\frac{\mathrm{d}\delta}{\mathrm{d}t} = a_{o\ell} \Rightarrow \delta(t) = \sqrt{2\nu a_{o\ell}t} \Rightarrow \boxed{\delta(t) = K_\ell\sqrt{t}}, \tag{7.21}$$

where $K_\ell = \sqrt{2\nu a_{o\ell}}$. The index ℓ refers to the laminar regime. Obtaining Eq. (7.12) required determining the integration constant. For this purpose, the initial condition was used $\delta(t = 0) = 0$. This functional relationship $\delta(t)$ can also be determined using a power law, where $\delta(t) \sim t^p$. The two sides of Eq. (7.13) must follow the same power law. The left side, the rate of change of the linear momentum, is modeled and follows the temporal power law in t^{-1}, while the right side follows the law of t^{-2p}, which implies $p = 1/2$. Thus, we arrive at the same power for $\delta(t)$ obtained with Eq. (7.12).

Therefore, Eq. (7.20) can be rewritten as follows:

$$f'' + a_o\eta f' = 0. \tag{7.22}$$

To establish the boundary conditions for Eq. (7.22), one must understand the asymptotic behavior of the velocity distribution, $u(y, t)$, and of the similarity function, $f(\eta)$ and its derivative, $f'(\eta)$, as illustrated in Fig. 7.5. Inspecting these qualitative curves, presented in Fig. 7.5, the following equations were established:

$$\begin{aligned} f(\eta = 0.5) &= 0.99, \\ f(\eta = -0.5) &= -0.99, \\ f'(\eta = 0.5) &= 0.01. \end{aligned} \tag{7.23}$$

Thus, the ordinary differential mathematical model is established composing Eqs. (7.22) and (7.23). This model represents the physical problem of the mixing layer in temporal development and in a laminar regime. Next, the continuous mathematical model will be deduced and presented, which will result from the integration of the ordinary differential model. This is an approximate differential mathematical model for the present physical problem. Consequently, its solution will be called a continuous mathematical model and not an exact mathematical model. This is because it does not exactly represent physical problems.

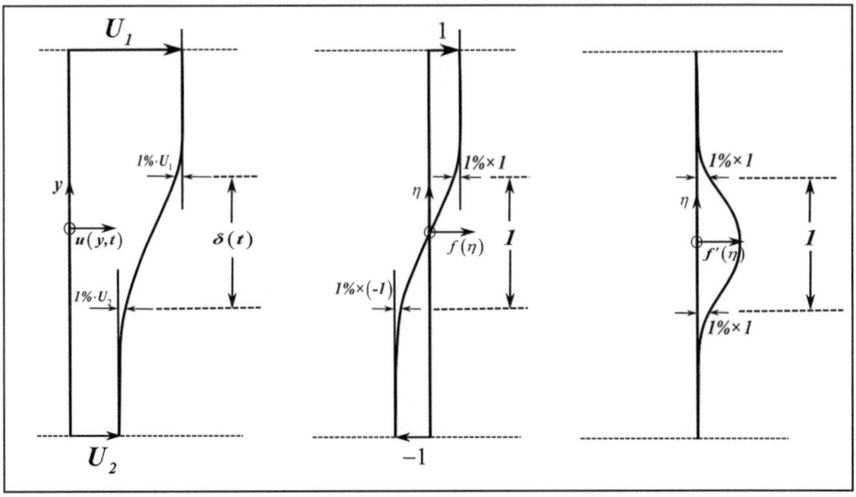

Fig. 7.5 Illustration of the thickness $\delta(t)$ of the shear layer; the velocity distribution $u(y, t)$; the similarity function $f(\eta)$ and its derivative $f'(\eta)$

7.3.2.3 Continuous Mathematical Model

The integration of Eq. (7.22) leads to the Gaussian error function, as presented with Eq. (7.24):

$$f(\eta) = c_1 \mathrm{erf}\left(\frac{\sqrt{2a_{o\ell}}}{2}\eta\right) + c_2, \tag{7.24}$$

where $\mathrm{erf}(\eta)$ is the Gauss error function, given by:

$$\mathrm{erf}(\eta) = \frac{2}{\sqrt{\pi}} \int_0^{\eta} \mathrm{e}^{-\tilde{\eta}^2} \mathrm{d}\tilde{\eta}. \tag{7.25}$$

The integration constants will be determined using two of the boundary conditions (7.23), which results in the following system of equations:

$$c_1 \mathrm{erf}\left(\frac{\sqrt{2a_{o\ell}}}{2}\frac{1}{2}\right) + c_2 = 0.99, \tag{7.26}$$

$$c_1 \mathrm{erf}\left(\frac{\sqrt{2a_{o\ell}}}{2}\left(-\frac{1}{2}\right)\right) + c_2 = -0.99 \tag{7.27}$$

The solution to this system of algebraic equations gives the constants, $c_1 = 1/\mathrm{erf}\left(\sqrt{2a_{o\ell}}/4\right)$ and $c_2 = 0$. Thus, we obtain $f(\eta)$:

$$f(\eta) = \text{erf}\left(\frac{\sqrt{2a_{o\ell}}}{2}\eta\right)/\text{erf}\left(\frac{\sqrt{2a_{o\ell}}}{4}\right). \tag{7.28}$$

It is recommended that the reader check whether the solution obtained for $f(\eta)$, Eq. (7.28), satisfies the boundary conditions (7.23). To determine the constant $a_{o\ell}$, which appears in Eq. (7.28), the last of the boundary conditions (7.23) will be used. The derivative $f'(\eta)$ is given by:

$$f'(\eta) = \frac{\text{d}f}{\text{d}\eta} = \sqrt{\frac{2a_{o\ell}}{\pi}}\left(\text{e}^{-\frac{a_{o\ell}\eta^2}{2}}/\text{erf}\frac{\sqrt{2a_{o\ell}}}{4}\right) \tag{7.29}$$

By also applying the asymptotic condition to the derivative, $f'(\eta = 1/2) = 0.01$, in Eq. (7.29), we obtain the results expressed in Eq. (7.30):

$$\sqrt{\frac{2a_{o\ell}}{\pi}}\text{e}^{-\frac{a_{o\ell}/4}{2}} - 0.01\text{erf}\frac{\sqrt{2a_{o\ell}}}{4} = 0. \tag{7.30}$$

By solving Eq. (7.30), the value of the constant is obtained, $a_{o\ell} = 50.745$. Thus, the constant $K_\ell = \sqrt{2\nu a_{o\ell}}$ is determined. Thus $\delta(t) = \sqrt{2 \times 50.745}\sqrt{\nu t} = 10.074\sqrt{\nu t}$.

Once the similarity function has been determined, Eq. (7.28), the function for velocity is determined from Eq. (7.15):

$$u(y,t) = \overline{U} + \tilde{U}\cdot\text{erf}\left(\frac{\sqrt{2a_{o\ell}}}{2}\frac{y}{\delta(t)}\right)/\text{erf}\left(\frac{\sqrt{2a_{o\ell}}}{4}\right) \Rightarrow$$

$$u(y,t) = \overline{U} + \tilde{U}\cdot\text{erf}\left(\frac{y}{2\sqrt{\nu t}}\right). \tag{7.31}$$

To arrive at Eq. (7.31), the results $\delta(t) = 10.074\sqrt{\nu t}$, $a_{o\ell} = 50.745$ and $\text{erf}\left(\frac{\sqrt{2a_{o\ell}}}{4}\right) \sim 1$ were used.

The EDP with which this problem is modeled is like the one-dimensional transient diffusion equation, for which the self-similarity solution can be found in the literature. However, the problem of determining coefficients and boundary conditions is not addressed in such resolutions. The solution presented for the mixing layer in temporal development in the laminar regime was not found in other bibliographic sources.

The solution for the turbulent temporal mixing layer will be developed in the following section, after which comparative results between the laminar and turbulent regimes will be presented.

7.3.3 Temporal Mixing Layer in Turbulent Regime

7.3.3.1 Partial Differential Mathematical Model

The differential mathematical model for the temporal mixing layer, in turbulent regime, can be extracted from Eqs. (7.1) and (7.2). It is considered that the diffusive transport process promoted by turbulence is orders of magnitude greater than the pure molecular diffusive process. Therefore, the molecular diffusion term will not be considered. The advective term was simplified, in accordance with the physical nature of this problem, that is, the average flow occurs only in the x direction and it just depends on the y direction and time t. Thus, the Filtered Navier–Stokes Equations, for the x direction result in:

$$\frac{\partial \overline{u}}{\partial t} = \frac{\partial \tau_{xy}}{\partial y}. \tag{7.32}$$

Using the turbulent viscosity hypothesis and the Boussinesq–Prandtl mixing length model, the component $\overline{u'\vartheta'}$ of the Boussinesq–Reynolds tensor can be modeled as follows:

$$\underbrace{\tau_{xy} = \nu_t \frac{\partial \overline{u}}{\partial y}}_{\text{Boussinesq}} = \underbrace{\ell_m^2 \left| \frac{\partial \overline{u}}{\partial y} \right| \frac{\partial \overline{u}}{\partial y}}_{\text{Boussinesq Prandtl}} \tag{7.33}$$

Considering that $\|(\partial \overline{u}/\partial y)\| > 0$ and that, for this flow, for all y, $(\partial \overline{u}/\partial y) \geq 0$, results that:

$$\left| \frac{\partial \overline{u}}{\partial y} \right| = \frac{\partial \overline{u}}{\partial y} \tag{7.34}$$

and

$$-\overline{u'\vartheta'} = \left(\ell_m \frac{\partial u}{\partial y} \right)^2. \tag{7.35}$$

Replacing Eq. (7.35) in Eq. (7.32), the following model is obtained:

$$\frac{\partial \overline{u}}{\partial t} = \ell_m^2 \frac{\partial}{\partial y} \left[\left(\frac{\partial \overline{u}}{\partial y} \right)^2 \right]. \tag{7.36}$$

As boundary and initial conditions, we have:

$$
\begin{aligned}
&\overline{u}(y, t = 0) = \overline{u}_o(y), \\
&\overline{u}\left(y = \frac{\delta(t)}{2}, t\right) = U_1, \\
&\frac{\partial \overline{u}}{\partial y}\left(y = \frac{\delta(t)}{2}, t\right) = 0, \\
&\overline{u}(y = 0) = \overline{U},
\end{aligned}
\tag{7.37}
$$

As previously established, Eq. (7.7), $\ell_m = \beta\delta(t)$. The coefficient β must be determined, either by material experimentation or by computational experimentation via DNS (Direct Numerical Simulation).

7.3.3.2 Ordinary Differential Mathematical Model

The self-similarity theory will be used to transform PDE (7.36) in an ODE, considering the continuous solution of the differential model. To this end, the similarity variable $\eta(y, t)$ is defined as follows:

$$
\eta(y, t) = \frac{y}{\delta(t)}
\tag{7.38}
$$

The velocity continuous solution can be proposed using a self-similarity function, $f(\eta)$, the average speed, $\overline{U} = (U_1 + U_2)/2$ and the half velocity difference, $\tilde{U} = (U_1 - U_2)/2$:

$$
\overline{u}(y, t) = \overline{U} + \tilde{U} \cdot f(\eta).
\tag{7.39}
$$

Deriving in time and space Eq. (7.39) results in the following equations:

$$
\frac{\partial \overline{u}}{\partial t} = \tilde{U} \frac{\mathrm{d}f}{\mathrm{d}\eta} \frac{\partial \eta}{\partial \delta} \frac{\mathrm{d}\delta}{\mathrm{d}t} = \tilde{U} f' \left(-\frac{y}{\delta^2} \frac{\mathrm{d}\delta}{\mathrm{d}t}\right) = -\frac{\tilde{U}}{\delta} \eta f' \frac{\mathrm{d}\delta}{\mathrm{d}t},
\tag{7.40}
$$

$$
\frac{\partial \overline{u}}{\partial y} = \frac{\tilde{U}}{\delta} f' \Rightarrow \frac{\partial}{\partial y}\left[\left(\frac{\partial \overline{u}}{\partial y}\right)^2\right] = \frac{\partial}{\partial y}\left[\left(\frac{\tilde{U}}{\delta} f'\right)^2\right] = \frac{2 \cdot \tilde{U}^2}{\delta^3} f' f''.
\tag{7.41}
$$

Replacing Eqs. (7.40) and (7.41) in Eq. (7.36) yields:

$$
f' f'' + \boxed{\frac{1}{2\tilde{U}\beta^2} \frac{\mathrm{d}\delta}{\mathrm{d}t}} \eta f' = 0.
\tag{7.42}
$$

The term inside the square, in Eq. (7.42), must be a constant, since the solution of this differential equation must result in a function $f(\eta)$. Thus:

$$\frac{1}{2\tilde{U}\beta^2}\frac{d\delta}{dt} = a_{ot} \Rightarrow \delta(t) = 2\tilde{U}\beta^2 a_{ot} t \Rightarrow \boxed{\delta(t) = K_t t}, \tag{7.43}$$

where $K_t = 2\tilde{U}\beta^2 a_{ot}$ is a constant related to turbulent flow, which will be determined using a boundary condition of the problem. Index "t" represents the turbulent regime. Obtaining Eq. (7.43) required the determination of a constant of integration. For this purpose, the initial condition $\delta(t = 0) = 0$ was used. The variation ratio of $\delta(t)$ can also be determined using a power law, where $\delta(t) \propto t^p$. The two sides of Eq. (7.13) must follow the same power law. The left side, the rate of change of linear momentum, follows the temporal power law in t^{-1}, while the right side follows the law of t^{-p}, which implies $p = 1$. Thus, we arrive at the same power law for $\delta(t)$, as the one in Eq. (7.43), which is rewritten with the constant a_{ot}:

$$f'f'' + a_{ot}\eta f' = 0 \tag{7.44}$$

To complement the ordinary model, it is necessary to add boundary conditions to Eq. (7.44):

$$f\left(\eta = \frac{1}{2}\right) = 1,$$
$$f'\left(\eta = \frac{1}{2}\right) = 0,$$
$$f(\eta = 0) = 0. \tag{7.45}$$

Similarly to the case of laminar flow, the ordinary differential model for the turbulent regime, established with Eqs. (7.44) and (7.45), also admits a continuous solution, as will be presented below.

7.3.3.3 Continuous Mathematical Model

For continuous integration of the differential model, Eqs. (7.44) and (7.45), Eq. (7.44) will be rewritten as follows:

$$f'\left(f'' + a_{ot}\eta\right) = 0 \tag{7.46}$$

which can be decomposed into two equations:

$$f' = 0,$$
$$f'' + a_{ot}\eta = 0. \tag{7.47}$$

The first of these leads to a trivial solution, $f(\eta) = \text{cte}$. The solution to the second equation is obtained by direct integration:

$$f(\eta) = -\frac{a_{ot}}{6}\eta^3 + a_1\eta + a_2. \tag{7.48}$$

The constant a_{ot} is given by $a_{ot} = K_t/\left(2\beta^2\tilde{U}\right)$. Using the third of the boundary conditions expressed by Eq. (7.45), it follows that $f(0) = 0 \Rightarrow a_2 = 0$. From the other two boundary conditions, $f(\eta = 1/2) = 1$ and $f'(\eta = 1/2) = 1$ it follows that $a_{ot} = 24$ and $a_1 = 3$. Thus, we have the function $f(\eta)$:

$$f(\eta) = -4\eta^3 + 3\eta \tag{7.49}$$

But,

$$a_{ot} = \frac{K_t}{2\beta^2\tilde{U}} = 24 \Rightarrow K_t = 48\beta^2\tilde{U} \tag{7.50}$$

Thus,

$$\delta(t) = K_t t = 48\beta^2\frac{1}{2}(U_1 - U_2)t = 24\beta^2(U_1 - U_2)t. \tag{7.51}$$

The constant β must be determined by material experimentation or by DNS. In the present text, in the absence of this information, we used $\beta = 0.07$, which corresponds to the value determined for the spatial mixing layer. Substituting Eqs. (7.49) and (7.50) in Eq. (7.39), results:

$$\overline{u}(\eta) = \overline{U} + \tilde{U}\cdot\left(3\eta - 4\eta^3\right). \tag{7.52}$$

or,

$$\overline{u}^*(y,t) = \frac{\overline{u}(y,t)}{\overline{U}} = 1 + \frac{\tilde{U}}{\overline{U}}\left[3\left(\frac{y}{\delta(t)}\right) - 4\left(\frac{y}{\delta(t)}\right)^3\right], \tag{7.53}$$

where $\delta(t) = 24\beta^2(U_1 - U_2)t$ and $\beta = 0.07$. Finally, the continuous model will be deduced for the component $\overline{u'\vartheta'}$ of the Boussinesq–Reynolds Tensor. The equations necessary for the deduction will be rewritten here. From Eq. (7.35), it follows:

$$-\tau_{xy} = -\overline{u'\vartheta'} = \left(\ell_m\frac{\partial u}{\partial y}\right)^2 \tag{7.54}$$

Equations (7.7), (7.41) and (7.49) result in:

$$\ell_m = \beta\delta(x), \tag{7.55}$$

$$\frac{\partial\overline{u}}{\partial y} = \frac{\tilde{U}}{\delta}f', \tag{7.56}$$

$$f'(\eta) = 3 - 12\eta^2. \tag{7.57}$$

Replacing Eqs. (7.55) and (7.57) in Eq. (7.54), the continuous model for the component $\overline{u'\vartheta'}$ of the Boussinesq–Reynolds tensor is obtained:

$$-\overline{u'\vartheta'} = \beta^2 \tilde{U}^2 \left(3 - 12\eta^2\right). \tag{7.58}$$

Bear in mind that β is the empirical constant of the Prandtl mixing length model and $\tilde{U} = (U_1 - U_2)/2$ is the semi-difference of the velocities.

7.3.4 Results for Temporal Mixing Layer in Laminar and Turbulent Regimes

The results obtained with the continuous models for the mixing layer in temporal development, for the laminar and turbulent regimes, are presented below. Initially, the temporal evolution of the thickness of the mixing layer for the laminar regime, Eq. (7.31), and for the turbulent regime, Eq. (7.53) is presented as the two curves of $\delta(t) \times t$, in Fig. 7.6. To this end, the kinematic viscosity of the air was used, at the temperature of 20 (°C): $\nu_{air} = 1.5 \times 10^{-5}\left(\mathrm{m}^2/\mathrm{s}\right)$. It is seen that, for the turbulent regime, the mixing layer becomes thicker in less time when compared to the laminar regime. This behavior was expected since, in the laminar regime, the transport of linear momentum in the direction normal to the streamwise direction is carried out only through the process of molecular diffusion. In fact, by Eq. (7.21), it is observed that the thickness $\delta(t)$ varies with the square root of the product of kinematic viscosity and time, while, according to Eq. (7.51), the thickness $\delta(t)$ varies linearly with time.

The velocity distributions as a function of time t and the coordinate y are presented in Fig. 7.7 for the laminar regime. Equation (7.31) was used to plot these curves. In Fig. 7.7, the comparison between the results obtained using the continuous solution (solid line) and the results obtained with computational simulation (symbols) is shown. The complete mathematical model for fluid dynamics was computationally solved. The MFSim Computational Code, developed at MFLab–FEMEC–UFU, was used. The results are visually coincident. This is not about validating either type of approach, but rather a matter of checking consistency (Fig. 7.8).

After presenting the illustrative results of the laminar mixing layer, we move on to the turbulent regime. The mixing layer was computationally simulated, using the MFSim Computational Code. The simulation domain has dimensions $8 \times 4 \times 8$ (m^3), considering a fluid with kinematic viscosity $\nu_{fluid} = 0.01$ (m^2/s) and the two streams with speeds $U_1 = 2.5$ (m/s) and $U_2 = 0.5$ (m/s). Molecular viscosity was only used for the purposes of computational simulation via MFSim. In fact, the continuous solution, by hypothesis, is independent of the viscosity of the fluid.

The domain was considered periodic in x and in z, which defines the horizontal planes. The velocities of the free currents, prescribed above, were imposed on the

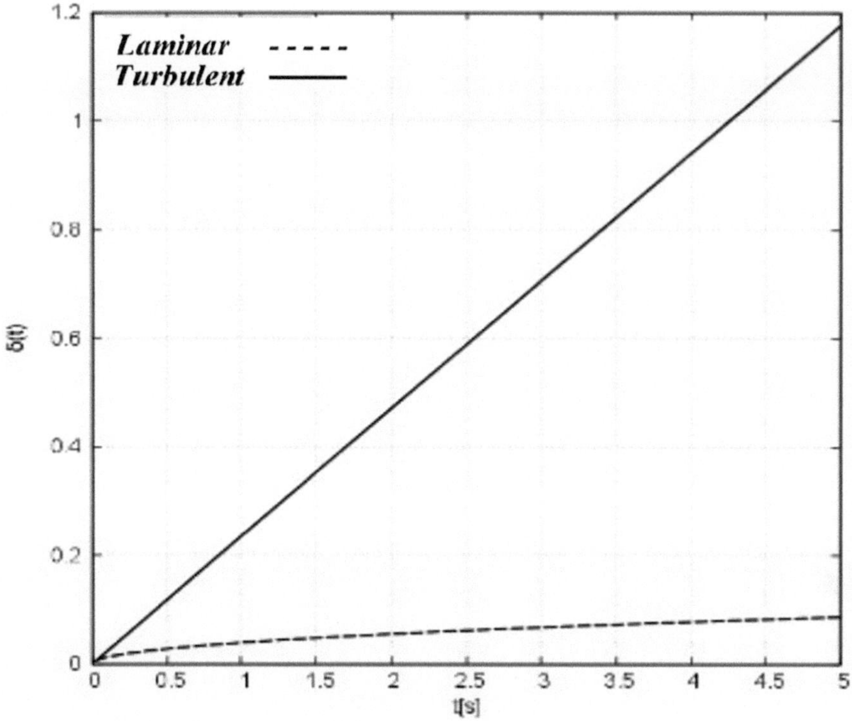

Fig. 7.6 Temporal evolution of the thickness of the temporal mixing layer for the laminar and turbulent regimes, considering $U_1 = 2.5\,(\mathrm{m/s})$, $U_2 = 0.5\,(\mathrm{m/s})$ and $\nu_{ar} = 1.5 \times 10^{-5}\,(\mathrm{m^2/s})$. Figure taken from Magalhães and Silveira-Neto [12], personal collection

horizontal planes that delimit the domain. The initial condition was defined by two uniform fields, namely: $U_1 = 2.5\,(\mathrm{m/s})$ in the upper half of the domain and $U_2 = 0.5\,(\mathrm{m/s})$ in the lower half of the domain, as illustrated in Fig. 7.4a. Initially, the flow is laminar, and it was necessary to impose a perturbation on the initial velocity field to trigger its transition to turbulence. White noise of intensity $10\% \times \overline{U}^2$ was superimposed on the three velocity components. As the flow is transitional and becomes turbulent further down, it was necessary to use a turbulence model. The Dynamic Sub-Mesh Model of Germano et al. [6] was used. In Fig. 7.9, visualizations are shown in a temporal sequence, starting from the initial noise, going through the formation of Kelvin–Helmholtz Instability, then the formation of longitudinal filaments and finally the flow three-dimensionalization.

The information inherent to this flow, that is, the velocity and pressure components, fluctuates in time and space. The statistics of this class of flow must be done in space, since the statistical moments are time dependent. For example, to determine the average velocity fields, for a given sampling time, the averages were made over horizontal planes (x, z). An average value was obtained for each coordinate value y. Thus, the average in each plane $y = cte$ was carried out using Eq. (7.59):

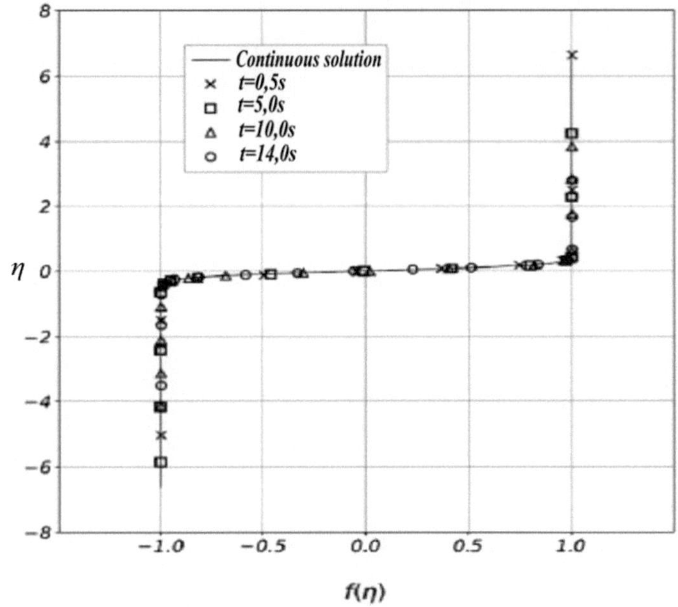

Fig. 7.7 Mixing layer developing over time, in a laminar regime. Spatio-temporal evolution of velocity distributions, considering $U_1 = 2.5\,(\mathrm{m/s})$, $U_2 = 0.5\,(\mathrm{m/s})$ and $\nu_{ar} = 1.5 \times 10^{-5}\left(\mathrm{m}^2/\mathrm{s}\right)$. Figure taken from Magalhães and Silveira-Neto [12], personal collection

$$\overline{u}(y) = \frac{\sum_{i=0}^{n} u(x_i, z_i, y = \mathrm{cte})\Delta x_i \Delta z_i}{\sum_{i=0}^{n} \Delta x_i \Delta z_i}. \tag{7.59}$$

Velocity distributions as a function of time t and of the coordinate y for the turbulent regime are presented in Fig. 7.10. Equation (7.53) was used to plot the curves.

In Fig. 7.11, the self-similarity function $f(\eta)$ is shown, plotted with Eq. (7.49) and with the results of the computer simulation, illustrated in Fig. 7.9. There is excellent agreement between the results obtained using the two methodologies for solving the problem, that is, the continuous solution via the self-similarity methodology and computer simulation.

It is instructive to compare the global behavior of the two flow regimes, laminar and turbulent. The velocity distributions $u(y, t = t_n)$ were plotted for both regimes. This has been done for a generic $t = t_n$, which is the same for both flow regimes. The comparison is presented in Fig. 7.12. For the turbulent regime, the temporal mixing layer thickens much more quickly than in the laminar regime. This was also evidenced in Fig. 7.10, comparing the temporal evolution of thicknesses $\delta(t)$ of mixing layers in laminar and turbulent regimes.

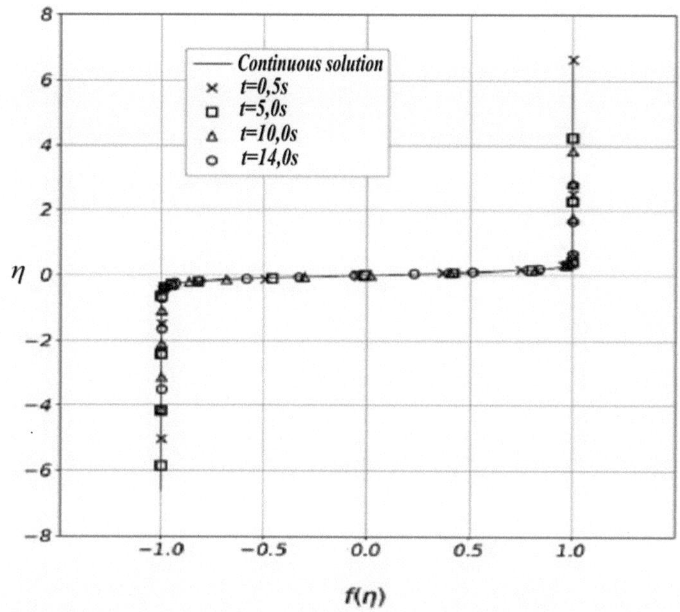

Fig. 7.8 Mixing layer developing over time, in a laminar regime. Exact similarity function (full line) and results obtained by computer simulation (symbols), considering $U_1 = 2.5\,(\text{m/s})$, $U_2 = 0.5\,(\text{m/s})$ and $\nu_{ar} = 1.5 \times 10^{-5}\,(\text{m}^2/\text{s})$. Figure taken from Magalhães and Silveira-Neto [12], personal collection

7.4 Mixing Layer in Spatial Development in Turbulent Regime

In this section, models for the mixing layer in spatial development, operating in the turbulent regime, will be presented. This analysis involves the establishment of the physical model, the simplified differential mathematical model and the corresponding continuous solution. Numerical and computational models will also be presented, starting with the physical model of the problem.

7.4.1 Physical Model

A spatially developing mixing layer is one of the simplest canonical flows to be solved. This flow is formed by two fluid streams with velocities U_1 and U_2, imposed at the entrance of a given physical domain. These two streams maintain the flow within the domain. Thus, at the center of the inlet, a strong gradient can be identified over the velocity field, which could theoretically be infinite. In this case, the thickness of the shear layer will be zero, $\delta(x = 0) = 0$. This can be viewed in Fig. 7.13.

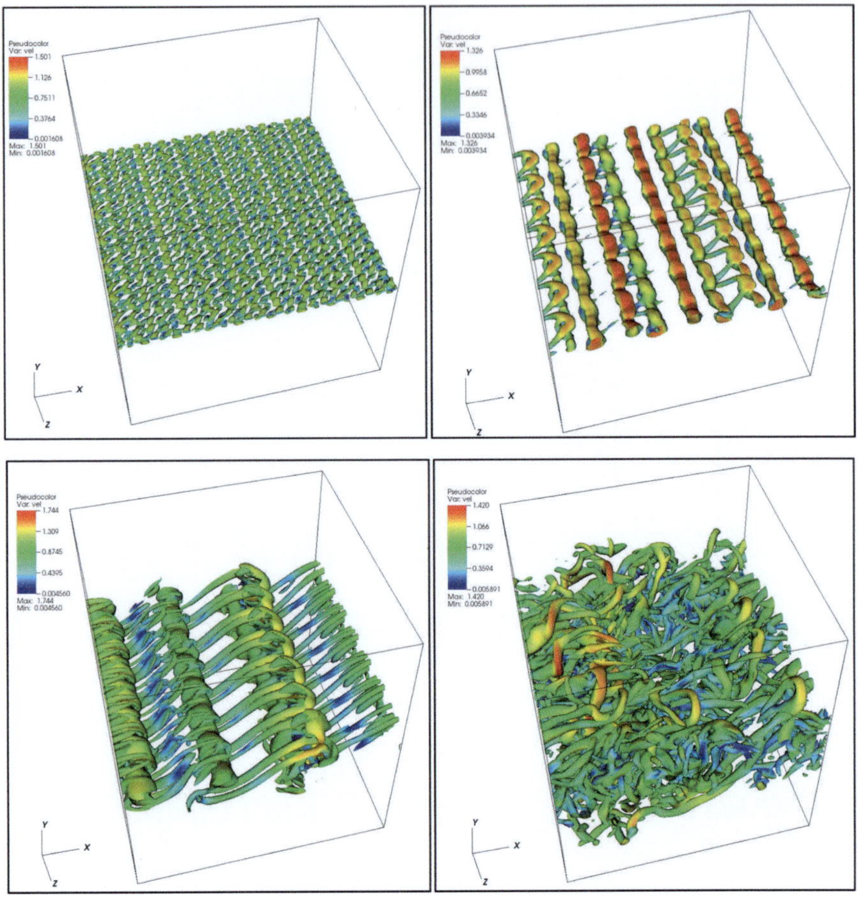

Fig. 7.9 Illustration of the transition to turbulence of a mixing layer in temporal development (personal collection, with thanks to MFLab researcher Millena M. Villar)

As an observer moves in the x direction, they observe that the diffusive transport of linear momentum attenuates this gradient, developing an inflectional velocity profile, as can be seen in Fig. 7.13. This diffusion will be promoted by molecular movement, in the laminar regime. In the turbulent regime, the turbulent structures will accelerate this molecular diffusion process by orders of magnitude. It is important to realize that this increase in molecular diffusion, in the turbulent regime, is due to the intensification of local velocity gradients and the increase in the number of gradients per unit volume of fluid. In addition to molecular diffusion, there will be turbulent advective transport that will be modeled via turbulent viscosity.

In the present analysis, the objective will not be to model and simulate the process of formation of Kelvin–Helmholtz Instabilities, but only to model the average flow behavior. As illustrated in Fig. 7.13, the evolution of the flow occurs in the plane (x, y). It is observed that the velocity profiles, illustrated in this figure, are self-similar,

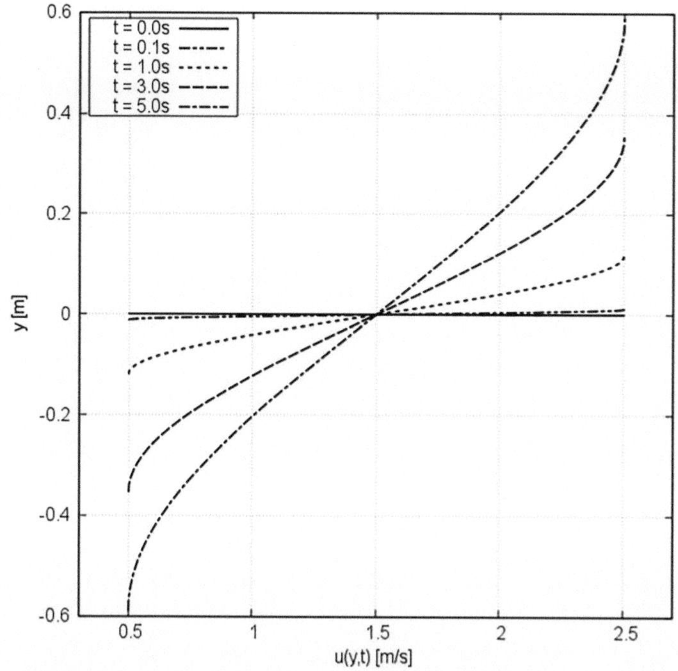

Fig. 7.10 Mixing layer in temporal development, in a turbulent regime. Spatio-temporal evolution of velocity distributions, considering $U_1 = 2.5\,(\text{m/s})$ and $U_2 = 0.5\,(\text{m/s})$. Figure taken from Magalhães and Silveira-Neto [12], personal collection

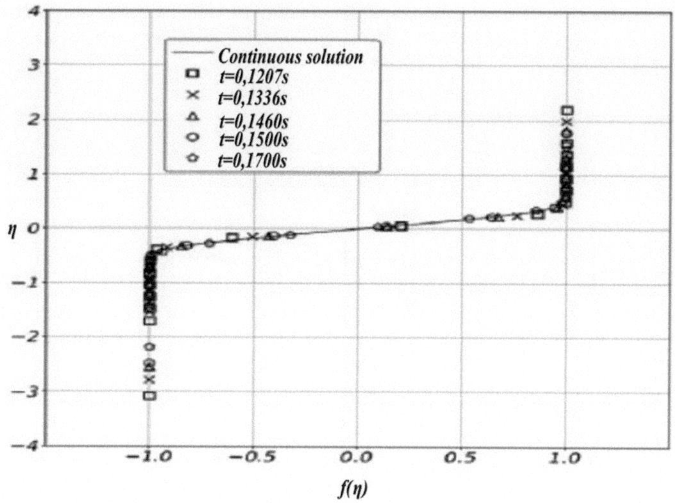

Fig. 7.11 Mixing layer in temporal development, in a turbulent regime. Self-similarity function $f(\eta)$ is plotted with Eq. (7.46) and with the results of the computer simulation, considering $U_1 = 2.5\,(\text{m/s})$ and $U_2 = 0.5\,(\text{m/s})$. Figure taken from Magalhães and Silveira-Neto [12], personal collection

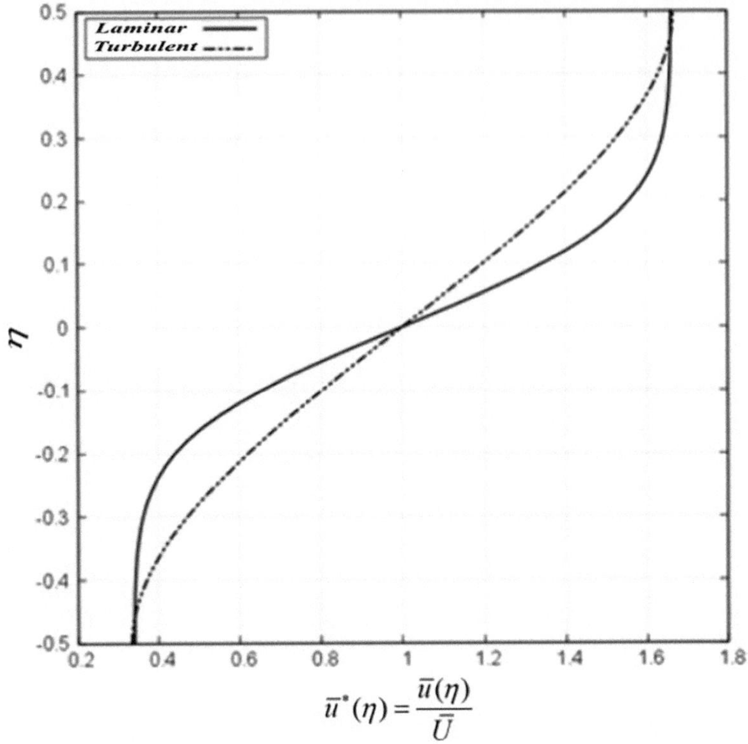

Fig. 7.12 Comparison between velocity profiles, in laminar and turbulent regimes, as a function of the similarity parameter η. The spatial–temporal evolution of velocity distributions is showcased, considering $U_1 = 2.5$ (m/s) and $U_2 = 0.5$ (m/s). Figure taken from Magalhães and Silveira-Neto [12], personal collection

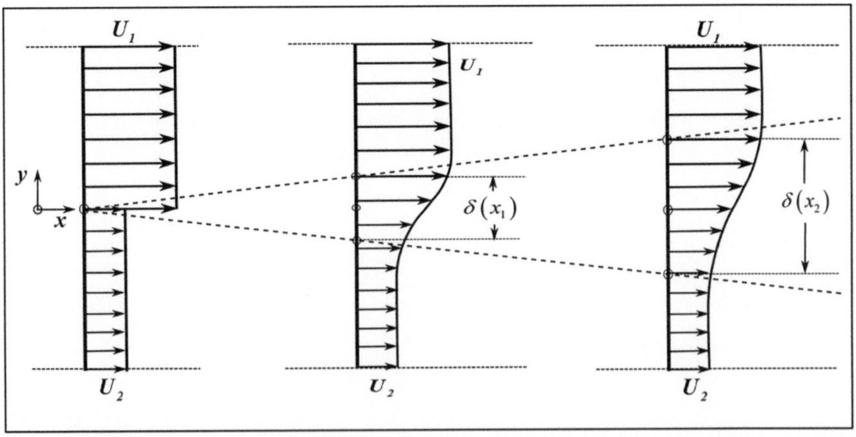

Fig. 7.13 Physical model for a spatially developing mixing layer: profiles for three values of x

which will allow the use of self-similarity theory to solve an ordinary differential model and determine a continuous solution. It is observed that the physical model presented here is qualitative and valid for both the laminar and turbulent regimes.

7.4.2 Partial Differential Mathematical Model

The differential mathematical model for the spatial mixing layer in the turbulent regime can be obtained from Eqs. (7.1) and (7.2). Turbulence accelerates the molecular diffusion process by orders of magnitude. Turbulent behavior will be modeled as a diffusive-like phenomena and separately from molecular diffusion. Therefore, the molecular diffusion term will not be considered. The advective term is simplified, in accordance with the physical nature of this problem, that is, the filtered flow occurs in the plane (x, y) and the average behavior is in a steady state. Thus, the continuity and the simplified and filtered Navier–Stokes equations are presented below:

$$\frac{\partial \overline{u}}{\partial x} + \frac{\partial \overline{\vartheta}}{\partial y} = 0, \tag{7.60}$$

$$\overline{u}\frac{\partial \overline{u}}{\partial x} + \overline{\vartheta}\frac{\partial \overline{u}}{\partial y} = \frac{\partial \tau_{xy}}{\partial y}. \tag{7.61}$$

The bar over the velocity components, $\overline{u}$ and $\overline{\vartheta}$ indicates that these are filtered quantities. Using the turbulent viscosity hypothesis and the Prandtl–Reichardt Mixing Length Model, the component τ_{xy} of the Boussinesq–Reynolds Tensor can be modeled as follows:

$$\underbrace{\tau_{xy} = \nu_t \frac{\partial \overline{u}}{\partial y}}_{\text{Boussinesq}} = \underbrace{\delta(x)\chi_1(U_1 - U_2)\frac{\partial \overline{u}}{\partial y}}_{\text{Prandtl - Reichardt}} \tag{7.62}$$

The Prandtl–Reichardt model is represented here:

$$\nu_t = cx\chi_1(U_1 - U_2) = cx\chi_1\frac{(U_1 - U_2)}{(U_1 + U_2)}\frac{(U_1 + U_2)}{2}2 = 2cx\chi_1\lambda\overline{U} = \frac{x\overline{U}}{2\sigma^2}, \tag{7.63}$$

where χ_1 and c are constant and

$$\lambda = \frac{U_1 - U_2}{U_1 + U_2} = \frac{\tilde{U}}{\overline{U}} \tag{7.64}$$

$$\overline{U} = \frac{(U_1 + U_2)}{2} \text{ and } \tilde{U} = \frac{(U_1 - U_2)}{2}, \tag{7.65}$$

$$\sigma = \frac{1}{2}(\chi_1 c\lambda)^{-1/2}. \tag{7.66}$$

Replacing Eq. (7.63) in Eq. (7.62), results:

$$\tau_{xy} = \frac{x\overline{U}}{2\sigma^2}\frac{\partial \overline{u}}{\partial y}. \tag{7.67}$$

Replacing Eq. (7.67) in Eq. (7.61) results:

$$\overline{u}\frac{\partial \overline{u}}{\partial x} + \overline{\vartheta}\frac{\partial \overline{u}}{\partial y} = \frac{x\overline{U}}{2\sigma^2}\frac{\partial^2 \overline{u}}{\partial y^2}. \tag{7.68}$$

The boundary conditions for this physical problem are:

$$\begin{aligned}
&\overline{u}(y \to +\infty) = U_1, \\
&\overline{u}(y \to -\infty) = U_2, \\
&\overline{u}(y = 0) = \overline{U}, \\
&\frac{\partial \overline{u}}{\partial y}(y = 0) = 0.
\end{aligned} \tag{7.69}$$

7.4.3 Ordinary Differential Mathematical Model

The solution of the partial differential model composed of Eqs. (7.68) and (7.69) can be obtained using similarity theory. Observing Fig. 7.13, the average velocity profiles, plotted at various x positions, are self-similar. In the search for a solution to this partial differential equation, this model will initially be transformed into an equivalent ordinary differential model. So, a self-similarity variable and a self-similarity function are defined:

$$\eta = \frac{\sigma y}{x}, \tag{7.70}$$

$$\psi(x, y) = x\overline{U}f(\eta), \tag{7.71}$$

where $\psi(x, y)$ is the stream function. Using the stream function definition, we obtain the components $\overline{u}(x, y)$ and $\overline{\vartheta}(x, y)$ and their respective derivatives:

$$\overline{u}(x, y) = \frac{\partial \psi}{\partial y} = x\overline{U}\frac{df}{d\eta}\frac{\partial \eta}{\partial y} = x\overline{U}f'(\eta)\frac{\sigma}{x} = \sigma\overline{U}f', \tag{7.72}$$

$$\overline{\vartheta}(x,y) = -\frac{\partial \psi}{\partial x} = -\left[\overline{U}f + x\overline{U}f'\left(-\frac{\sigma y}{x^2}\right)\right] = -\overline{U}\left[f - \eta f'\right]. \tag{7.73}$$

$$\frac{\partial \overline{u}}{\partial y} = \frac{\overline{U}\sigma^2 f''}{x}, \tag{7.74}$$

$$\frac{\partial^2 \overline{u}}{\partial y^2} = \frac{\overline{U}\sigma^3 f'''}{x^2}. \tag{7.75}$$

Replacing Eqs. (7.72) and (7.75) in Eq. (7.68), and after performing some algebraic manipulations, we obtain:

$$\sigma f''' + 2\sigma f f'' = 0 \Rightarrow f''' + 2ff'' = 0 \tag{7.76}$$

The boundary conditions for this equation are established using Eqs. (7.69), which leads to the following equations in the similarity variables:

$$\begin{aligned}
&\sigma f'(\eta \to +\infty) = \frac{U_1}{\overline{U}},\\
&\sigma f'(\eta \to -\infty) = \frac{U_2}{\overline{U}},\\
&\sigma f'(\eta = 0) = 1,\\
&\sigma f(\eta = 0) = 0.
\end{aligned} \tag{7.77}$$

The last equation in the set of Eqs. (7.77) was obtained considering that this equation must also be satisfied in $\eta = 0$. Therefore $f'''(\eta = 0) = 0$ and $f(\eta = 0) = 0$, given that $f''(\eta = 0) \neq 0$. The second derivative of the function f is the viscous tension, which cannot be zero at the wall. The ordinary model composed of Eqs. (7.76) and (7.77) must be integrated to obtain the corresponding continuous model.

7.4.4 Continuous Mathematical Model

The integration of the model composed of Eqs. (7.76) and (7.77) is not trivial, even using symbolic mathematics tools. To integrate it, the Görtler expansion will be used [21]:

$$\sigma f = \eta + \lambda f_1 + \lambda^2 f_2 + \cdots. \tag{7.78}$$

Schilichting [20] presents the approximate solution to the problem, considering only the first two terms of Eq. (7.78). In the present text, the author considered an additional term in the Görtler Expansion. Deriving Eq. (7.78) three times, in relation to the similarity variable, η, it follows:

$$\sigma f' = 1 + \lambda f_1 + \lambda^2 f_2 + \cdots, \tag{7.79}$$

$$\sigma f'' = \lambda f_1'' + \lambda^2 f_2'' + \cdots, \tag{7.80}$$

$$\sigma f''' = \lambda f_1''' + \lambda^2 f_2''' + \cdots, \tag{7.81}$$

Replacing Eqs. (7.79) and (7.81) in Eq. (7.76) yields:

$$\lambda\left[f_1' + 2\eta f_1''\right] + \lambda^2\left[f_2'' + 2\eta f_2'' + 2f_1 f_1'\right] + \cdots = 0 \tag{7.82}$$

Equation (7.82) can be decomposed into two others, whose solutions are used to compose a general solution for (7.82):

$$f_1 + 2\eta f_1'' = 0 \tag{7.83}$$

$$f_2 + 2\eta f_2'' + 2f_1 f_1'' = 0. \tag{7.84}$$

The boundary conditions for Eqs. (7.83) and (7.84) are set out below:

$$\begin{aligned}
&\overline{u}(y \to \infty) = \sigma \overline{U} f'(\eta \to \infty) = U_1 \Rightarrow \sigma f'(\eta \to \infty) = \frac{U_1}{\overline{U}},\\
&\overline{u}(y \to -\infty) = \sigma \overline{U} f'(\eta \to -\infty) = U_2 \Rightarrow \sigma f'(\eta \to -\infty) = \frac{U_2}{\overline{U}},\\
&\overline{u}(y = 0) = \sigma \overline{U} f'(\eta = 0) = \overline{U} \Rightarrow \sigma f'(\eta = 0) = 1.
\end{aligned} \tag{7.85}$$

But, from Eq. (7.69), it follows that:

$$\begin{aligned}
&\sigma f'(\infty) = 1 + \lambda f_1'(\infty) + \lambda^2 f_2'(\infty) = \frac{U_1}{\overline{U}},\\
&\sigma f'(-\infty) = 1 + \lambda f_1'(-\infty) + \lambda^2 f_2'(-\infty) = \frac{U_2}{\overline{U}},\\
&\sigma f'(0) = 1 + \lambda f_1'(0) + \lambda^2 f_2'(0) = 1.
\end{aligned} \tag{7.86}$$

By imposing homogeneous conditions for f_2', it follows that:

$$\begin{aligned}
&f_2'(\infty) = 0,\\
&f_2'(-\infty) = 0,\\
&f_2'(0) = 0.
\end{aligned} \tag{7.87}$$

Replacing (7.87) in (7.86), the boundary conditions for f_1' are:

$$\begin{aligned} 1+\lambda f_1'(\infty) &= \frac{U_1}{\overline{U}} \Rightarrow f_1'(\infty)=1, \\ 1+\lambda f_1'(-\infty) &= \frac{U_2}{\overline{U}} \Rightarrow f_1'(-\infty)=-1, \\ 1+\lambda f_1'(0) &= 1 \Rightarrow f_1'(0)=0. \end{aligned} \tag{7.88}$$

Thus, Eq. (7.87) in (7.88) are the boundary conditions for ordinary third-order Eq. (7.83) in (7.84). For the integration of Eq. (7.83), an order reduction is also carried out, defining a function $g(\eta)$, in such a way that:

$$g_1(\eta)=f_1'(\eta), \tag{7.89}$$

which allows us to rewrite Eq. (7.83):

$$g_1''+2\eta g_1'=0, \tag{7.90}$$

whose boundary conditions are derived from Eqs. (7.88):

$$\begin{aligned} g_1(\infty) &= f_1'(\infty)=1, \\ g_1(-\infty) &= f_1'(-\infty)=-1. \end{aligned} \tag{7.91}$$

Integrating Eq. (7.90), the following solution is obtained:

$$g_1(\eta)=c_1\sqrt{\pi}\,\mathrm{erf}(\eta)+c_2, \tag{7.92}$$

where the function erf(x) is given, in a generic way, by:

$$erf\,(x)=\frac{2}{\sqrt{\pi}}\int_0^x e^{-\tilde{x}^2}d\tilde{x}. \tag{7.93}$$

The constants c_1 and c_2 are determined using the boundary conditions given by Eqs. (7.91): $c_1=1/\sqrt{\pi}$ and $c_2=0$. Therefore:

$$g_1(\eta)=\mathrm{erf}(\eta) \tag{7.94}$$

Consequently:

$$f_1'(\eta)=\mathrm{erf}(\eta) \tag{7.95}$$

Integrating Eq. (7.89) results in:

$$f_1(\eta)=\eta\,\mathrm{erf}(\eta)+\frac{\mathrm{e}^{-\eta^2}}{\sqrt{\pi}}+c \tag{7.96}$$

and

$$f_1''(\eta) = \frac{2e^{-\eta^2}}{\sqrt{\pi}} \tag{7.97}$$

Replacing Eqs. (7.96) and (7.97) in Eq. (7.84) results in:

$$f_2 + 2\eta f_2'' = -2 \cdot \left(\eta \operatorname{erf}(\eta) + \frac{e^{-\eta^2}}{\sqrt{\pi}} + c \right) \cdot \left(\frac{2e^{-\eta^2}}{\sqrt{\pi}} \right). \tag{7.98}$$

Defining a function g_2 as below

$$g_2(\eta) = f_2'(\eta) \tag{7.99}$$

And replacing Eq. (7.98) in (7.99) results in:

$$g_2''(\eta) + 2\eta g_2'(\eta) = -2\left[\eta \operatorname{erf}(\eta) + \frac{e^{-\eta^2}}{\sqrt{\pi}} + c \right] \cdot \left(\frac{2e^{-\eta^2}}{\sqrt{\pi}} \right) \tag{7.100}$$

with the following boundary conditions:

$$\begin{aligned} g_2(\infty) &= f_2'(\infty) = 0, \\ g_2(-\infty) &= f_2'(-\infty) = 0, \\ g_2(0) &= f_2'(0) = 0 \end{aligned} \tag{7.101}$$

Integrating Equation (7.100), the following function is obtained:

$$\begin{aligned} g_2(\eta) = 2c\frac{e^{-\eta^2}}{\sqrt{\pi}} + \frac{1}{2}\sqrt{\pi}k_1 \operatorname{erf}(\eta) + \frac{\eta e^{-\eta^2} \operatorname{erf}(\eta)}{\sqrt{\pi}} \\ - \frac{1}{2}\operatorname{erf}^2(\eta) + k_2 + \frac{e^{-2\eta^2}}{\pi} \end{aligned} \tag{7.102}$$

where c, k_1 and k_2 are constants, which are determined using the boundary conditions (7.101):

$c = -\frac{\sqrt{\pi}(2+\pi)}{4\pi}$, $k_1 = 0$ and $k_2 = 1/2$. Thus, we have the solution:

$$f_2'(\eta) = g_2(\eta) = \frac{\eta e^{-\eta^2} \operatorname{erf}(\eta)}{\sqrt{\pi}} + \frac{1}{2}\left(1 - \operatorname{erf}^2(\eta) - e^{-\eta^2}\right). \tag{7.103}$$

Replacing (7.95) and (7.103) in (7.79), the following equation is obtained:

$$\sigma f' = 1 + \lambda\,\mathrm{erf}(\eta) + \lambda^2\left[\frac{\eta \mathrm{e}^{-\eta^2}\mathrm{erf}(\eta)}{\sqrt{\pi}} + \frac{1}{2}\left(1 - \mathrm{erf}^2(\eta) - \mathrm{e}^{-\eta^2}\right)\right] \tag{7.104}$$

Replacing Eq. (7.104) in Eq. (7.72) results in:

$$\overline{u}(\eta) = \overline{U}\left(1 + \lambda\,\mathrm{erf}(\eta) + \lambda^2\left[\frac{\eta \mathrm{e}^{-\eta^2}\mathrm{erf}(\eta)}{\sqrt{\pi}} + \frac{1}{2}\left(1 - \mathrm{erf}^2(\eta) - \mathrm{e}^{-\eta^2}\right)\right]\right). \tag{7.105}$$

Similarly, the component $\overline{\vartheta}(\eta)$ can be obtained by taking the above results to Eq. (7.73):

$$\overline{\vartheta}(\eta) = \frac{\overline{U}c}{\sigma}\left[\begin{array}{c}\lambda\left(\frac{\mathrm{e}^{-\eta^2}}{\sqrt{\pi}}\right) + \frac{\lambda^2}{2}\left(\frac{\eta^2\mathrm{e}^{-\eta^2}\mathrm{erf}(\eta)}{\sqrt{\pi}} - \frac{\sqrt{2}\mathrm{erf}(\sqrt{2}\eta)}{\sqrt{\pi}}\right. \\ \left. + \frac{3\mathrm{e}^{-\eta^2}\mathrm{erf}(\eta)}{2\sqrt{\pi}} - \frac{\eta\mathrm{e}^{-2\eta^2}}{\pi}\right)\end{array}\right] \tag{7.106}$$

These functions $\overline{u}(\eta)$ and $\overline{\vartheta}(\eta)$ satisfy Eq. (7.68) and the three boundary conditions expressed with Eq. (7.69). Similarly, the solution for the component $\overline{u'\vartheta'}$ of the Boussinesq-Reynolds tensor is:

$$-\overline{u'\vartheta'} = 2\tilde{U}^2\chi\sigma\lambda^2\left[\frac{2\mathrm{e}^{-\eta^2}}{\sqrt{\pi}} + \lambda\left(\frac{\mathrm{e}^{-\eta^2}\mathrm{erf}(\eta)}{\sqrt{\pi}}\right.\right.$$
$$\left.\left. + \frac{2\eta\mathrm{e}^{-2\eta^2}}{\pi} - \frac{2\mathrm{e}^{-\eta^2}\mathrm{erf}(\eta)}{\sqrt{\pi}}\right)\right]. \tag{7.107}$$

Therefore, solutions are presented for $\overline{u}(\eta)$, $\overline{\vartheta}(\eta)$ and $\overline{u'\vartheta'}(\eta)$ for mixing layers in spatial development in the turbulent regime.

7.4.5 Comparison Between Results Obtained with the Continuous Solution and with Material Experiment

In order to make a comparison, the approximate continuous solution of the physical problem, Eq. (7.105), was particularized, in the present work, to the case experienced by Reichardt [18]. This experiment was characterized by $U_2 = 0$, so that Eq. (7.105) can be rewritten as in Eq. (7.108):

$$\frac{\overline{u}(\eta)}{U_1} = \left(1 + \mathrm{erf}(\eta) + \left[\frac{\eta \mathrm{e}^{-\eta^2}\mathrm{erf}(\eta)}{\sqrt{\pi}} + \frac{1}{2}\left(1 - \mathrm{erf}^2(\eta) - \mathrm{e}^{-\eta^2}\right)\right]\right) \tag{7.108}$$

It is noted that $\lambda = (U_1 - U_2)/(U_1 + U_2) = (U_1 - 0)/(U_1 + 0) = 1.0$. The velocity distribution $\overline{u}(\eta)$, modeled with Eq. (7.108), is compared with the results of experimental measurements, obtained by Reichardt [18]. In Fig. 7.14, data from the material experiment and two curves referring to approximate continuous solutions are presented. One of the curves refers to the solution of the problem, when only the function $f_1(\eta)$ is considered in the Görtler Expansion, as presented by Schlichting [21]. The other curve was obtained with the solution of Eq. (7.108), particularized to Eq. (7.108), for which $f_1(\eta)$ and $f_2(\eta)$ were included. It is observed that the agreement between the approximate solutions and the experimental data is excellent, both for the solution based only on $f_1(\eta)$, as well as for the solution based on $f_1(\eta)$ and $f_2(\eta)$, proposed in this book. The latter presents a slight improvement in the overall behavior of the comparison. It should be noted that Schlichting [21] presents this solution with only the $f_1(\eta)$. He claims that $f_2(\eta)$ can be neglected.

In the present work, the expansion considering $f_2(\eta)$ was done to demonstrate mathematically that its influence is small. It is observed that in the modeling it was assumed that turbulent viscosity was orders of magnitude greater than the molecular viscosity. Thus, the present model becomes insensitive to the Reynolds Number. It is emphasized again that, to compare with the experiment, $U_2 = 0$ and U_1 was taken as a reference. To obtain the velocity $\overline{u}(x, y)$ (m/s), arbitrary values of U_1 (m/s) can be chosen, but $\sigma = 13.5$ must be used. This value, presented by Schlichting [21], was determined based on experimental data. With this, the norm

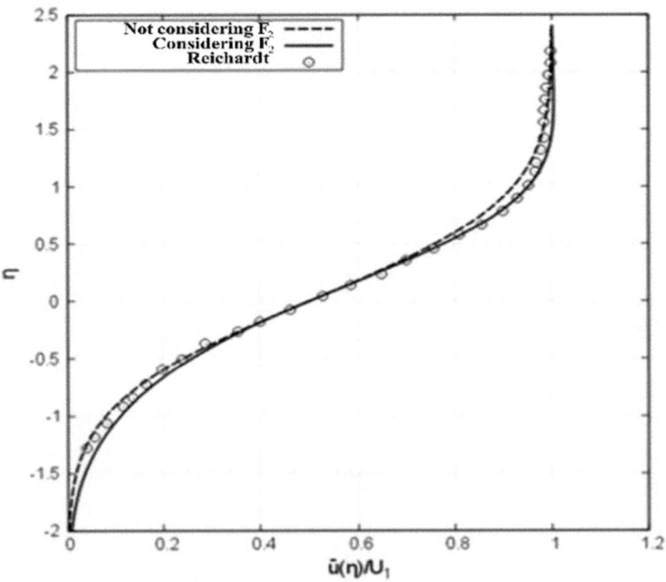

Fig. 7.14 Comparison between the approximate solution of a spatial mixing layer with experimental data from the literature, Reichardt [18]. Figure taken from Magalhães and Silveira-Neto [12], personal collection

$L_2 = \left[\sum_{i=1}^{N} \overline{u}_{ni} - \overline{u}_{ei}^2/N\right]^{1/2}$ was calculated, where N is the number of compared points, which make up the curves of Fig. 7.14; $\overline{u}_{ci}$ is the average velocity obtained by computer simulation; $\overline{u}_{ei}$ was obtained by material experimentation and the index "i" identifies the nodes at which the differences between $\overline{u}_{Ci}$ and $\overline{u}_{ei}$ are calculated.

With this standard, the statistical difference between theoretical values and experimental values is measured. It was calculated for the two curves: one obtained from the approximate solution considering only the function f_1 and the other for which the functions f_1 and f_2 were considered jointly. The values $L_{2(f_1)} = 0.0143$ and $L_{2(f_1 \text{ and } f_2)} = 0.0136$ were obtained. This shows an improvement of 5% in the overall difference in results when the function f_2 is added to the approximate theoretical procedure.

7.5 Distant Two-Dimensional Wake in Space Development

7.5.1 Physical Model

The flow illustrated in Fig. 7.15a is characterized by the approaching fluid from the left of a cylinder, placed transversally to the flow. The flow velocity upstream of the cylinder is uniform and equal to U_∞, and the downstream mean velocity distribution is $\overline{u}(x, y)$. The thickness of the shear layer is $2\delta(x)$. The difference between the free stream velocity and the local velocity is defined as $\overline{u}_1(x, y) = U_\infty - \overline{u}(x, y)$, and the maximum deficit, which occurs in $y = 0$, is $\overline{u}_o(x) = U_\infty - \overline{u}(x, 0)$. The coordinate axis system is positioned in the center of the cylinder.

In this flow, a wake is created downstream of the cylinder, which causes an obstruction within the flow, forming a region of lower velocity downstream. This allows the flow to recirculate, forming a region with a deficit in linear momentum. In the first velocity distribution, diagrammed in Fig. 7.15 b, a large gradient of linear momentum is perceived, which creates a strong diffusive transport process into the wake. Thus, as it moves away from the cylinder, in the positive direction of the x axis, flow develops. In the second velocity profile, drawn in Fig. 7.15 b, a reduction in the deficit can be observed. From this plan, the distant wake hypothesis can be used, where there is self-similarity between the speed profiles. Schlichting [21] presents this application as being a distant wake. This means that the region of analysis is far enough away from the source of wake formation to make use of the hypothesis of self-similarity. This is Schlichting's doctoral thesis, presented at Göttingen University, in Germany. Schlichting [21] resolved the distant wake promoted by a flat plate aligned with the flow. He proposed a suitable model for solving this specific problem.

In this book, an idealized model is presented for the wake formed downstream of any type of blunt body immersed in flows. This model is presented in Fig. 7.15b, which is similar to Schlichting's (1968) proposal. In this proposal, the genesis of the wake lies in $x = 0$, where $\delta(0) = 0$, $u_o(0) = U_\infty - \overline{u}(0, 0) = U_\infty - 0 = U_\infty$. Furthermore, $u_o(\infty) = 0$. It will be demonstrated that the influence of the type

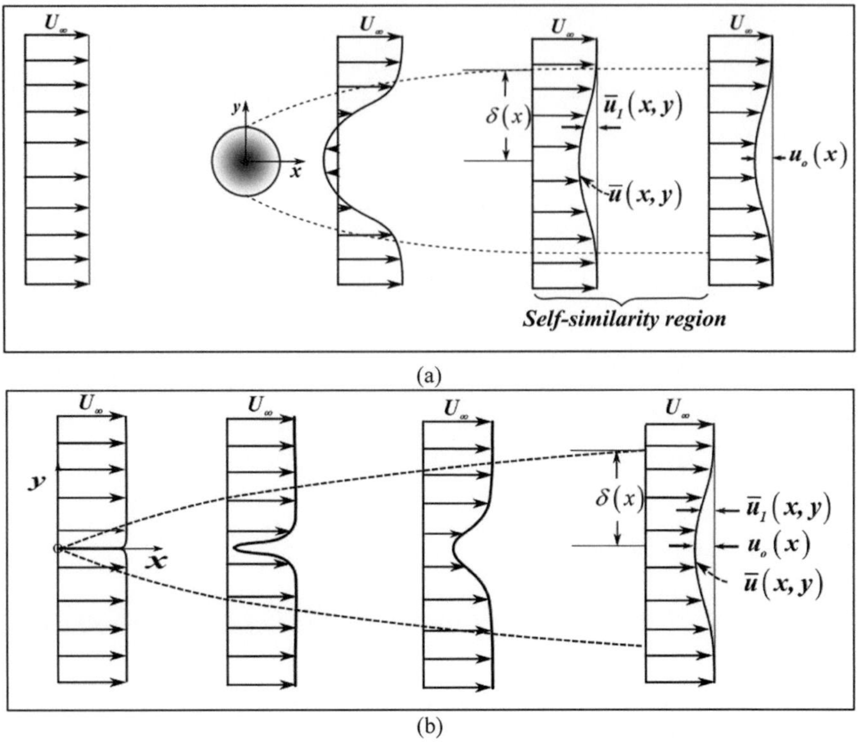

Fig. 7.15 Illustration of a wake downstream of a cylinder: region of self-similarity. **a**: wake downstream of a cylinder; **b** general model of a wake

of body that generates the wake will be considered through the drag coefficient experienced by the body.

It is observed in Fig. 7.15b that, in the proposed model, the entire field becomes self-similar, which makes it possible to obtain a boundary condition in $x = 0$, which would not be possible in the physical problems represented in Fig. 7.15a.

It should be noted that the procedure and solution presented are valid for various bodies immersed in the flow. For example, elliptical bodies, flat plates aligned to the flow and even airfoils. It is necessary to know the drag coefficient corresponding to the immersed body.

7.5.2 *Distant Wake in Laminar Regime*

7.5.2.1 Partial Differential Mathematical Model

The model for this flow is composed of the continuity equation and the Navier–Stokes equations, which were simplified considering the characteristics highlighted in the physical model. It was considered that the pressure field inside the belt is uniform and equal to the pressure outside the belt. Based on the order of magnitude analysis, it is concluded that the net advective flux of linear momentum, $\partial\left(\overline{u\vartheta}\right)/\partial x$, is negligible when compared to the net advective flow $\partial(\overline{uu})/\partial x$ in the x direction. It is also assumed that the regime is steady. The simplified model is presented below:

$$\frac{\partial \overline{u}}{\partial x}+\frac{\partial \overline{\vartheta}}{\partial y}=0, \tag{7.109}$$

$$\overline{u}\frac{\partial \overline{u}}{\partial x}=\nu\frac{\partial^2 \overline{u}}{\partial y^2}+\frac{\partial}{\partial y}\left(-\overline{u'\vartheta'}\right). \tag{7.110}$$

This model applies to both the turbulent regime and the laminar regime. For the latter case, the component of the Boussinesq–Reynolds Tensor, $\overline{u'\vartheta'}$, must be canceled and the filtering process must be disregarded in other terms. For the turbulent regime, the molecular viscous diffusion term can be disregarded when compared with the turbulent diffusion term. The closure for Eq. (7.110) will be done using the Prandtl Mixing Length Model. The boundary conditions will be established for both regimes in the following sections.

Equations (7.109) and (7.110) are rewritten and linearized. In Eq. (7.110), the filtering process, represented by the symbol $\left(^{-}\right)$, in laminar regime, is not necessary, consequently $\overline{u'\vartheta'}=0$. Using the transformation $u_1(x, y) = U_\infty - u(x, y)$ and $\vartheta_1(x, y) = \vartheta(x, y)$, as presented by Schilichting et al. [21] and considering the region distant downstream of the cylinder, also considering $u_1(x, y) \ll U_\infty$, the linearized model is obtained:

$$\frac{\partial u_1}{\partial x}+\frac{\partial \vartheta_1}{\partial y}=0, \tag{7.111}$$

$$U_\infty\frac{\partial u_1}{\partial x}=\nu\frac{\partial^2 u_1}{\partial y^2}, \tag{7.112}$$

The boundary conditions for the variable u_1 are given by:

$$\begin{aligned}
&u_1(x, y=\pm\delta(x))=0,\\
&\frac{\partial}{\partial x}\left[u_1(x, y=\pm\delta(x))\right]=0,\\
&u_1(x, y=0)=u_{1o}
\end{aligned} \tag{7.113}$$

7.5.2.2 Ordinary Differential Mathematical Model

Equations (7.111) and (7.113) make up the partial differential model for the problem under analysis. To proceed with the transformation of the partial differential model into an equivalent ordinary differential model, a variable and a similarity function will be defined:

$$\eta = \frac{y}{\delta(x)}, \tag{7.114}$$

$$u_1(x, y) = u_o(x) f(\eta), \tag{7.115}$$

The boundary conditions for the function $f(\eta)$ must be established, considering their asymptotic behavior at the extremes, and the fact that their maximum value is unity at the center of the flow. To transform Eq. (7.112), the derivatives will be evaluated in terms of the similarity variables:

$$\begin{aligned}\frac{\partial u_1}{\partial x} &= u_o(x)\frac{\mathrm{d}f}{\mathrm{d}\eta}\left(\frac{-y}{\delta^2(x)}\right)\frac{\mathrm{d}\delta}{\mathrm{d}x} + \frac{\mathrm{d}u_o}{\mathrm{d}x}f(\eta)\\ &= -\frac{u_o(x)}{\delta(x)}\eta f'(\eta) + \frac{\mathrm{d}u_o}{\mathrm{d}x}f(\eta),\end{aligned} \tag{7.116}$$

$$\frac{\partial u_1}{\partial y} = \frac{u_o(x)}{\delta(x)}f'(\eta), \tag{7.117}$$

$$\frac{\partial^2 u_1}{\partial y^2} = \frac{\partial \eta}{\partial y}\frac{\partial}{\partial \eta}\left[\frac{u_o(x)}{\delta(x)}f'(\eta)\right] = \frac{u_o(x)}{\delta^2(x)}f''(\eta). \tag{7.118}$$

Replacing Eqs. (7.116) and (7.118) in Eq. (7.112) yields:

$$f''(\eta) + \kappa\eta f'(\eta) + \beta f(\eta) = 0, \tag{7.119}$$

$$\kappa = \frac{U_\infty}{\nu}\delta(x)\frac{\mathrm{d}\delta}{\mathrm{d}x}, \tag{7.120}$$

$$\beta = -\frac{U_\infty}{\nu}\frac{\delta^2}{u_o(x)}\frac{\mathrm{d}u_o}{\mathrm{d}x}. \tag{7.121}$$

The quantities κ and β must be constant, since the solution of Eq. (7.119) must result in a function of the variable η only, in accordance with the proposal established with Eq. (7.115).

The boundary conditions for Eq. (7.119) are obtained by observing Eq. (7.113):

$$f(\eta = \pm 1) = 0.01, \tag{7.122}$$

$$f(\eta = 0) = 1.0, \tag{7.123}$$

$$f'(\eta = 0) = 0.0. \tag{7.124}$$

In the following sections, the continuous solution of the ordinary differential model established above will be determined. We will seek to incorporate the influence of the type of body that generates the wake through the drag coefficient of the body.

7.5.2.3 Relationship Between Drag Coefficient and Maximum Deficit $\overline{u}_o(x)$

- **Mass balance**

Applying the Reynolds–Leibniz Transport theorem, White [25], for the mass balance as in Fig. 7.16 results in:

$$U_\infty 2Lb = 2b\int_0^\delta \overline{u}(x, y)\mathrm{d}y \Rightarrow L = \frac{1}{U_\infty}\int_0^\delta \overline{u}(x, y)\mathrm{d}y \tag{7.125}$$

where b is the length of the cylinder and $2L$ is the distance between two streamlines upstream of the cylinder. These two streamlines delimit the control volume. At the outlet, downstream of the cylinder, the distance between these two streamlines is 2δ. The analysis that will be presented in this text refers to the case of a circular cylinder of diameter d and length b. However, this analysis can be extended to any type of immersed body, as long as the drag coefficient is known.

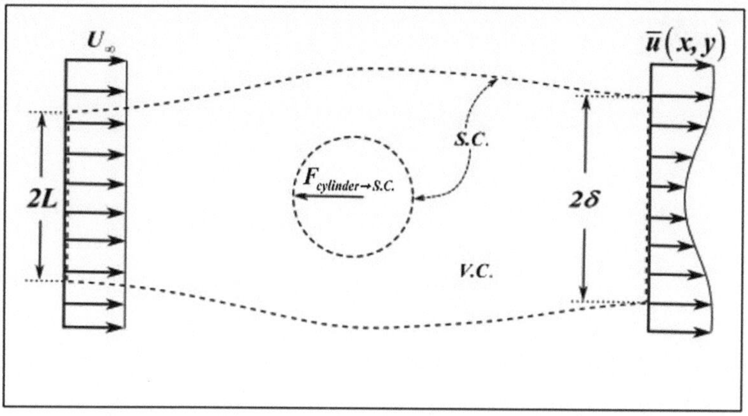

Fig. 7.16 Volume and control surface for calculating drag on a long cylinder

Linear Momentum Balance

Applying the Reynolds–Leibniz Transport theorem to the linear momentum balance, in the control volume illustrated in Fig. 7.16, results in:

$$F_{\text{cylinder}\to\text{S.C.}} = -2\rho b \int_0^\delta \overline{u}(U_\infty - \overline{u})\mathrm{d}y = -2\rho b \int_0^\delta \overline{u}_1(U_\infty - \overline{u}_1)\mathrm{d}y \qquad (7.126)$$

The drag force is the reaction to $F_{\text{cylinder}\to\text{S.C.}}$. Then,

$$F_\text{D} = -F_{\text{cylinder}\to\text{S.C.}} = 2\rho b \int_0^\delta \overline{u}(U_\infty - \overline{u})\mathrm{d}y$$

$$= 2\rho b \int_0^\delta \overline{u}_1(U_\infty - \overline{u}_1)\mathrm{d}y,$$

$$F_\text{D} = 2\rho b \int_0^\delta \left[\overline{u}_1(x, y)U_\infty - \overline{u}_1^2\right]\mathrm{d}y \qquad (7.127)$$

The solution being sought will only be valid for the region far from the obstacle that gives rise to this wake. In this region, it can be assumed that $u_1 \ll U_\infty$. Then, $\overline{u}_1^2 \ll \overline{u}_1 U_\infty$. It therefore follows that the drag force can be modeled with Eq. (7.128):

$$F_\text{D} = 2\rho b \int_0^\delta \overline{u}_1(x, y)U_\infty \mathrm{d}y \qquad (7.128)$$

Replacing Eqs. (7.114) and (7.115) in Eq. (7.128) yields:

$$F_\text{D} = 2\rho b U_\infty u_o(x)\delta(x) \int_0^\delta f(\eta)\mathrm{d}\eta. \qquad (7.129)$$

Using the definition of drag coefficient, C_D, the drag force is modeled by:

$$C_\text{D} = \frac{F_\text{D}}{\rho U_\infty^2 \mathrm{d}b/2} \Rightarrow F_\text{D} = C_\text{D}\rho U_\infty^2 \mathrm{d}b/2 \qquad (7.130)$$

From Eqs. (7.129) and (7.130), results:

$$u_o(x)\delta(x) = \frac{U_\infty C_\text{D} d}{4\int_0^1 f(\eta)\mathrm{d}\eta} \qquad (7.131)$$

The right side of Eq. (7.131) must be a constant since all terms, including the integral $\int_0^1 f(\eta)\mathrm{d}\eta$, are constant. Thus, a new constant is defined that will be used in the solution sequence:

$$\gamma = \frac{C_\mathrm{D} d}{2\int_0^1 f(\eta)\mathrm{d}\eta'} \tag{7.132}$$

which allows to rewrite Eq. (7.131) as

$$u_o(x)\delta(x) = \frac{U_\infty \gamma}{2}. \tag{7.133}$$

After presenting the hypotheses and considerations adopted for the case of the distant wake, the ordinary and continuous differential mathematical models will be presented below, for the laminar and turbulent regimes, respectively. Developments will be individualized with the aim of providing greater clarity in the process.

7.5.2.4 Continuous Mathematical Model

Equation (7.120) can be rewritten as follows:

$$\delta(x)\mathrm{d}\delta = \frac{\kappa \nu}{U_\infty}\mathrm{d}x \Rightarrow \delta(x) = \frac{2\kappa\nu}{U_\infty} x^{1/2}. \tag{7.134}$$

The integration of Eq. (7.134) gave rise to an integration constant, which was determined using the condition $\delta(x = 0) = 0$. This condition can be viewed in Fig. 7.15b. For the integration of Eq. (7.119), the coefficient β must depend only on $u_o(x)$ and some constants. To do so, we substitute Eq. (7.133) in Eq. (7.121), resulting in:

$$\beta = -\frac{U_\infty}{\nu}\frac{\left(\frac{U_\infty\gamma}{2u_o(x)}\right)^2}{u_o(x)}\frac{\mathrm{d}u_o}{\mathrm{d}x} \Rightarrow \beta = -\frac{U_\infty^3\gamma^2}{4\nu}\frac{1}{u_o^3(x)}\frac{\mathrm{d}u_o}{\mathrm{d}x}. \tag{7.135}$$

Integrating Eq. (7.135) results in a model for $u_o(x)$.:

$$u_o(x) = \left(\frac{8\nu\beta}{U_\infty^3\gamma^2}\right)^{-1/2} x^{-1/2}. \tag{7.136}$$

The constant of integration of Eq. (7.135) was determined using the condition $u_o(x \to \infty) \to 0$, resulting that the constant of integration can be taken asny real value, and that the best choice is $c = 0$. Replacing Eqs. (7.134) and (7.136) in Eq. (7.133) results in:

$$u_o(x)\delta(x) = \left(\frac{2\kappa\nu}{U_\infty}\right)^{1/2} x^{1/2}\left(\frac{8\nu\beta}{U_\infty^3\gamma^2}\right)^{-1/2} x^{-1/2} = \frac{U_\infty\gamma}{2} \Rightarrow k = \beta. \tag{7.137}$$

Substituting the result of Eq. (7.137), $\kappa = \beta$, in Eq. (7.119), results in a new ode:

$$f''(\eta) + \beta\eta f'(\eta) + \beta f(\eta) = 0. \tag{7.138}$$

Equation (7.138) can be rewritten using the product derivative rule, as follows:

$$\frac{d}{d\eta}\left[f'(\eta) + \beta\eta f(\eta)\right] = 0. \tag{7.139}$$

Integrating Eq. (7.139) and using the boundary condition expressed in Eq. (7.124), results in:

$$f'(\eta) + \beta\eta f(\eta) = 0. \tag{7.140}$$

Integrating Eq. (7.140) and using the boundary condition given by Eq. (7.132) results in:

$$f(\eta) = \exp\left(-\frac{\beta\eta^2}{2}\right), \tag{7.141}$$

where the value of β is determined using the asymptotic boundary condition, given by Eq. (7.122), applied to Eq. (7.141), resulting in:

$$\beta = 9.21034. \tag{7.142}$$

Replacing Eq. (7.142) in Eq. (7.143), the similarity function is completely determined:

$$\frac{u_1(x, y)}{u_o(x)} = f(\eta) = \exp\left(-4.60517\eta^2\right). \tag{7.143}$$

Upon determining the similarity function, $f(\eta)$, it is possible to calculate the function γ by realizing the integral present in Eq. (7.132):

$$\gamma = \frac{C_D d}{2\int_0^1 \exp\left(-4.60517\eta^2\right)d\eta} = 1.21365 C_D d. \tag{7.144}$$

The equalities $\kappa = \beta = 9.21034$ follow, respectively, from Eqs. (7.137) and (7.142). Substituting these equalities into Eq. (7.132), results in:

$$\delta(x) = 4.29193\left(\frac{\nu}{U_\infty}\right)^{1/2} x^{1/2}. \tag{7.145}$$

Equation (7.145) can be rewritten as a function of the Reynolds Number, based on the diameter of the cylinder, $\mathrm{Re}_d = \frac{U_\infty d}{\nu}$, as follows:

$$\delta(x) = 4.29193\left(\frac{1}{\mathrm{Re}_d}\right)^{1/2}\left(\frac{x}{d}\right)^{1/2}. \tag{7.146}$$

The maximum velocity deficit is also determined, $u_o(x)$, by substituting the values of $\beta = 9.21034$ and $\gamma = 1.21365 C_\mathrm{D} d$ in Eq. (7.136), resulting in:

$$\begin{aligned} u_o(x) &= 0.14139 U_\infty C_\mathrm{D}\left(\frac{U_\infty d^2}{\nu x}\right)^{1/2} \\ &= 0.14139 \frac{U_\infty C_\mathrm{D}}{\sqrt{\mathrm{Re}_d}}\left(\frac{x}{d}\right)^{-1/2}. \end{aligned} \tag{7.147}$$

Replacing the results for $\delta(x)$, Eq. (7.146), and for $u_o(x)$, Eq. (7.147), in Eq. (7.143), results in the following equation for the velocity deficit $u_1(x, y)$:

$$u_1(x, y) = 0.14139 \frac{U_\infty C_\mathrm{D}}{\sqrt{\mathrm{Re}_d}}\left(\frac{x}{d}\right)^{-1/2} \exp\left[-4.60517\left(\frac{y}{\delta(x)}\right)^2\right] \tag{7.148}$$

Remembering that $u_1(x, y) = U_\infty - u(x, y)$, from Eq. (7.148), the equation for the distribution of velocity $u(x, y)$ is obtained:

$$u(x, y) = U_\infty\left\{1 - 0.14139 \frac{C_\mathrm{D}}{\sqrt{\mathrm{Re}_d}}\left(\frac{x}{d}\right)^{-1/2} \exp\left[-4.60517\left(\frac{y}{\delta(x)}\right)^2\right]\right\}. \tag{7.149}$$

For illustration, in Fig. 7.17 a and b represent, respectively, the distributions of $u_1(\eta)/U_\infty$ and $u(\eta)/U_\infty$ which are plotted using Eqs. (7.148) and (7.149). Thus, the deduction of the continuous solution for distant wakes in the laminar regime created by circular cylinders is concluded. Next, the solution for the wakes in the turbulent regime will be presented.

7.5.3 Distant Wakes Under Turbulent Conditions

7.5.3.1 Partial Differential Mathematical Model

In this item, the simplified modeling will be presented for a distant, turbulent and two-dimensional wake, promoted by a cylinder of great length, compared to its diameter. It is worth remembering that a turbulent flow must necessarily be three-dimensional. However, it is pertinent to discuss the two-dimensional treatment when the interest is the average statistical behavior of the turbulent flow.

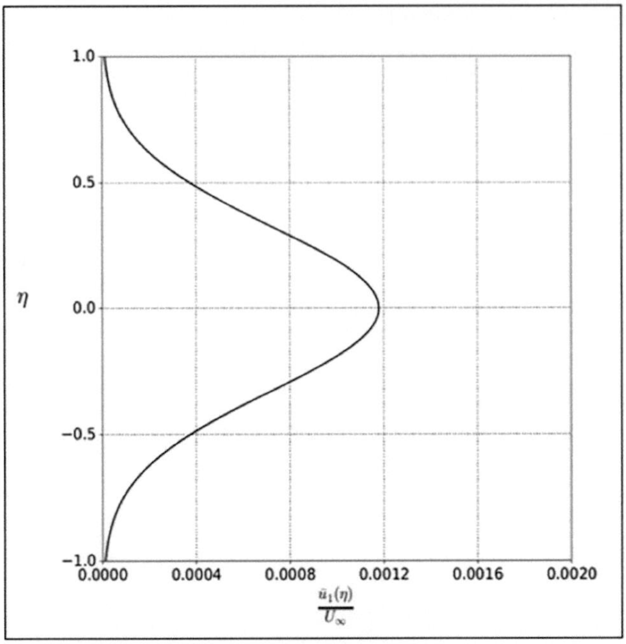

(a)

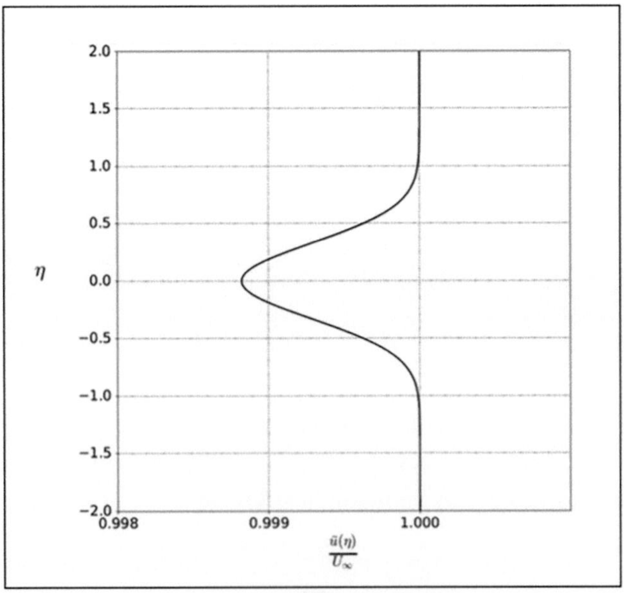

(b)

Fig. 7.17 Distribution of the velocity deficit $u_1(\eta)/U_\infty$ and the speed $u(\eta)/U_\infty$ as a function of the similarity coordinate. Figure taken from Magalhães and Silveira-Neto [12], personal collection

The basis of this example is the Prandtl Hypothesis for the mixing length model, given by Eqs. (7.150) and (7.151):

$$-\overline{u'\vartheta'} = \ell_m^2 \left|\frac{\partial \overline{u}}{\partial y}\right| \frac{\partial \overline{u}}{\partial y}, \tag{7.150}$$

$$\ell_m = \alpha \delta(x) \tag{7.151}$$

Equations (7.150) and (7.151) are used to model Eq. (7.2). With this, we have a closed PDE system, composed of Eqs. (7.152) and (7.153):

$$\frac{\partial \overline{u}}{\partial x} + \frac{\partial \overline{\vartheta}}{\partial y} = 0, \tag{7.152}$$

$$\overline{u}\frac{\partial \overline{u}}{\partial x} = \frac{\partial}{\partial y}\left(\alpha^2 \delta^2(x) \left|\frac{\partial \overline{u}}{\partial y}\right| \frac{\partial \overline{u}}{\partial y}\right) \tag{7.153}$$

The boundary conditions are represented by Eqs. (7.154) and (7.155):

$$\overline{u}(x, y = \pm\delta(x)) = U_\infty, \tag{7.154}$$

$$\frac{\partial}{\partial y}(\overline{u}(x, y = 0)) = 0 \tag{7.155}$$

In the derivation of Eqs. (7.152) and (7.153), the pressure inside the wake, in the region of similarity, is equal to the pressure of the free stream. Therefore, the pressure gradient in the direction x is considered null. It was also considered, based on the order of magnitude analysis, that the advection of linear momentum in the y direction, $\overline{\vartheta}(\partial \overline{u}/\partial y)$, is small when compared to the advection in the direction x, $\overline{u}(\partial \overline{u}/\partial x)$. Steady state flow was assumed, as well as that molecular viscous stresses are negligible when compared to turbulent stresses. Therefore, the term $\nu\left(\partial^2 \overline{u}/\partial y^2\right)$ was neglected.

To solve this system of equations, the self-similarity theory will be used, which requires that the velocity profiles be self-similar. To do so, one must focus on the region downstream and away from the cylinder. The starting point for finding the solution is based on the definition of the velocity difference:

$$\overline{u}_1(x, y) = U_\infty - \overline{u}(x, y), \tag{7.156}$$

where U_∞ is the velocity of the free stream, outside the shear layer, $\overline{u}(x, y)$ is the local velocity distribution and $\overline{u}(x, y)$ is the difference or local deficit of velocity. The hypothesis $\overline{u}_1(x, y) \ll U_\infty$ is adopted. These concepts and quantities are illustrated in 0. Replacing Eq. (7.156) in Eq. (7.153), results in:

$$-U_{\infty}\frac{\partial \overline{u}_1}{\partial x} = \frac{\partial}{\partial y}\left[\alpha^2\delta^2(x)\left\|-\frac{\partial \overline{u}_1}{\partial y}\right\|\left(-\frac{\partial \overline{u}_1}{\partial y}\right)\right]. \tag{7.157}$$

Since the term $\|-\partial u_1/\partial y\|$ is positive and knowing that the term $\partial u_1/\partial y$ is also always positive, at the bottom of the plane of symmetry as illustrated in Fig. 7.15, it can be said that:

$$\left\|-\frac{\partial u_1}{\partial y}\right\| = \frac{\partial u_1}{\partial y}. \tag{7.158}$$

This equality is valid for the specific case of Fig. 7.15. Therefore, the system of simplified equations for the distant wake is as follows:

$$\frac{\partial \overline{u}_1}{\partial x} + \frac{\partial \overline{\vartheta}_1}{\partial y} = 0, \tag{7.159}$$

$$U_{\infty}\frac{\partial \overline{u}_1}{\partial x} = 2\alpha^2\delta^2(x)\frac{\partial \overline{u}_1}{\partial y}\frac{\partial^2 \overline{u}_1}{\partial y^2}. \tag{7.160}$$

The boundary conditions are modeled with Eqs. (7.161) and (7.163):

$$\overline{u}_1(x, y = \pm\delta(x)) = 0, \tag{7.161}$$

$$\frac{\partial}{\partial y}(\overline{u}_1(x, y = 0)) = 0, \tag{7.162}$$

$$\overline{u}_1(x, y = 0) = u_o(x). \tag{7.163}$$

The reader should bear in mind that α is the empirical constant of the Prandtl mixing length model, $\delta(x)$ is half the thickness of the mat and $u_o(x, y)$ is the maximum velocity deficit. With this system of simplified equations, a distant turbulent wake is modeled, using the Boussinesq–Prandtl mixing length theory as a closure model for the turbulence equation.

Equations (7.159) and (7.160) can be resolved independently. First, solve Eq. (7.160), and obtain $\overline{u}_1(x, y)$ and then solve Eq. (7.159), and then solve Equation $\overline{\vartheta}_1(x, y)$. Thus, the fields of $\overline{u}_1(x, y)$ and $\overline{\vartheta}_1(x, y)$ will simultaneously satisfy the continuity equation and Newton's Second Law.

7.5.3.2 Ordinary Differential Mathematical Model

The partial differential Eq. (7.160) will be transformed into an ordinary differential equation, using the theory of self-similarity, for which purpose the variable η and the self-similarity function $f(\eta)$ are introduced:

$$\eta = \frac{y}{\delta(x)}, \tag{7.164}$$

$$\overline{u}_1(x, y) = u_o(x)f(\eta), \tag{7.165}$$

where $\delta(x)$ is half the thickness of the mat, as indicated in Fig. 7.15. By using the maximum deficit as a proportionality function in the self-similarity function, a new boundary condition can be obtained, since the largest value that $f(\eta)$ can assume is unity, and this occurs at $y = 0$. Thus, we have the boundary conditions for $f(\eta)$:

$$f(\eta = \pm 1) = 0, \tag{7.166}$$

$$f\prime(\eta = 0) = 0, \tag{7.167}$$

$$f(\eta = 0) = 1. \tag{7.168}$$

To transform Eq. (7.160) into an ODE, Eq. (7.165) will be derived as follows:

$$\begin{aligned}\frac{\partial u_1}{\partial x} = \frac{\partial}{\partial x}\left[u_o(x)f(\eta)\right] = u_o(x)f'(\eta)\left(\frac{-y}{\delta^2(x)}\right)\frac{\mathrm{d}\delta}{\mathrm{d}x}\\ +f(\eta)\frac{\mathrm{d}u_o}{\mathrm{d}x} = -\frac{u_o(x)}{\delta(x)}\frac{\mathrm{d}\delta}{\mathrm{d}x}\eta f'(\eta) + f(\eta)\frac{\mathrm{d}u_o}{\mathrm{d}x},\end{aligned} \tag{7.169}$$

$$\frac{\partial u_1}{\partial y} = \frac{u_o(x)}{\delta(x)}f'(\eta), \tag{7.170}$$

$$\frac{\partial^2 u_1}{\partial y^2} = \frac{u_o(x)}{\delta^2(x)}f''(\eta), \tag{7.171}$$

Replacing Eqs. (7.169)–(7.171) in Eq. (7.160) yields:

$$\begin{aligned}U_\infty\left[-\frac{u_o(x)}{\delta(x)}\frac{\mathrm{d}\delta}{\mathrm{d}x}\eta f'(\eta) + f(\eta)\frac{\mathrm{d}u_o}{\mathrm{d}x}\right] = 2\alpha^2\frac{u_o^2(x)}{\delta(x)}f'(\eta)f''(\eta) \Rightarrow\\ 2\alpha^2 f'(\eta)f''(\eta) + \frac{U_\infty}{u_o(x)}\frac{\mathrm{d}\delta}{\mathrm{d}x}\eta f'(\eta) - \frac{U_\infty\delta(x)}{u_o^2(x)}\frac{\mathrm{d}u_o}{\mathrm{d}x}f(\eta) = 0.\end{aligned} \tag{7.172}$$

Equation (7.172) is rewritten, defining the coefficients of the ODE:

$$2\alpha^2 f'(\eta)f''(\eta) + \kappa\eta f'(\eta) + \beta f(\eta) = 0, \tag{7.173}$$

Where

$$\kappa = U_\infty\frac{1}{u_o(x)}\frac{\mathrm{d}\delta}{\mathrm{d}x} \tag{7.174}$$

$$\beta = -U_\infty \frac{\delta(x)}{u_o^2(x)} \frac{du_o}{dx}. \tag{7.175}$$

Once we have the ordinary differential model, composed of Eq. (7.173) and by boundary conditions defined by Eqs. (7.166)–(7.168), we proceed to the integration phase and obtain the continuous model for the problem under analysis.

7.5.3.3 Continuous Mathematical Model

Before performing the integration of Eq. (7.173), Eqs. (7.174) and (7.175) will be integrated. Starting with Eq. (7.174), using Eq. (7.133), results in:

$$\begin{aligned} \kappa = U_\infty \frac{1}{u_o(x)} \frac{d\delta}{dx} &= \frac{2\delta(x)}{\gamma} \frac{d\delta}{dx} \Rightarrow \delta(x) d\delta \\ &= \frac{\kappa\gamma}{2} dx \Rightarrow \delta^2(x) = \kappa\gamma x + c. \end{aligned} \tag{7.176}$$

But $\delta(0) = 0$, thus $c = 0$. Then:

$$\delta(x) = (\kappa\gamma)^{1/2} x^{1/2} \tag{7.177}$$

Replacing Eq. (7.177) in Eq. (7.175), results in:

$$\beta = -U_\infty \frac{(\kappa\gamma)^{1/2} x^{1/2}}{u_o^2(x)} \frac{du_o}{dx} \Rightarrow \frac{du_o}{u_o^2} = -\frac{\beta}{U_\infty (\kappa\gamma)^{1/2}} x^{-1/2} dx. \tag{7.178}$$

Integrating Eq. (7.179), and using the boundary condition $u_o(x \to \infty) \to 0$ to determine the integration constant, we choose $c = 0$, since it can be any finite value. The following equation is then determined:

$$u_o(x) = \frac{U_\infty}{2\beta} (\kappa\gamma)^{1/2} x^{-1/2} \tag{7.179}$$

Once the equations for $\delta(x)$, Eq. (7.177), and for $u_o(x)$, Eq. (7.179), have been derived, a relationship between κ and γ is determined. It is now necessary to return to Eq. (7.173). Then, Eqs. (7.177) and (7.175) are used in Eq. (7.133), resulting in:

$$u_o(x)\delta(x) = \frac{U_\infty \gamma}{2} \Rightarrow \left[\frac{U_\infty}{2\beta} (\kappa\gamma)^{1/2} x^{-1/2} \right] . (\kappa\gamma)^{1/2} x^{1/2} = \frac{U_\infty \gamma}{2} \Rightarrow \beta = \kappa. \tag{7.180}$$

Substituting the result of Eq. (7.180) in Eq. (7.173) yields:

$$2\alpha^2 f'(\eta) f''(\eta) + \beta \eta f'(\eta) + \beta f(\eta) = 0. \tag{7.181}$$

Rewritten Eq. (7.181), using the derivation of the product rule, in a form convenient for integration, we have:

$$\frac{d}{d\eta}\left[\alpha^2 f'^2(\eta)\right] + \beta\frac{d}{d\eta}[\eta f(\eta)] = 0 \tag{7.182}$$

Integrating once, we obtain:

$$\alpha^2 f'^2(\eta) + \beta\eta f(\eta) = c_1. \tag{7.183}$$

Using the boundary conditions established by Eqs. (7.167) and (7.168), i.e., $f'(\eta = 0) = 0$ and $f(\eta = 0) = 1$, results in $c_1 = 0$ and, consequently:

$$\alpha^2 \left(f'\right)^2(\eta) + \beta\eta f(\eta) = 0 \Rightarrow \alpha\frac{\mathrm{d}f}{\mathrm{d}\eta} = \sqrt{-\beta\eta f}. \tag{7.184}$$

On the right side of Eq. (7.184), the positive sign was maintained, as a result of only the lower half of the wake being resolved, (7.184), for which the first derivative $\frac{\mathrm{d}f}{\mathrm{d}\eta} > 0$. It will be demonstrated that $\beta > 0$. By integrating Eq. (7.184), we obtain:

$$f(\eta) = \left(\frac{-\sqrt{\beta}}{3\alpha}\eta^{3/2} + c_2\right)^2. \tag{7.185}$$

The boundary condition $f(1) = 0$ leads to $c_2 = \left(\sqrt{\beta}\right)/(3\alpha)$. Thus, we obtain:

$$f(\eta) = \frac{\beta}{9\alpha^2}\left(1 - \eta^{3/2}\right)^2 = \frac{\beta}{9\alpha^2}\left(\eta^{3/2} - 1\right)^2. \tag{7.186}$$

Replacing Eq. (7.186) in Eq. (7.165) results in:

$$\frac{\overline{u}_1(x, y)}{\overline{u}_o(x)} = \frac{\beta}{9\alpha^2}\left(1 - \eta^{3/2}\right)^2. \tag{7.187}$$

It is known that in the center of the wake, $\eta = 0$, the velocity deficit $\overline{u}_1(x, 0) = \overline{u}_o(x)$.. Applying this equality in Eq. (7.187) results in:

$$\frac{\beta}{9\alpha^2} = 1 \Rightarrow \beta = 9\alpha^2 \tag{7.188}$$

and

$$\frac{\overline{u}_1(x, y)}{\overline{u}_o(x)} = \left[1 - \left(\frac{y}{\delta(x)}\right)^{3/2}\right]^2 \tag{7.189}$$

In terms of the similarity function, we have:

$$f(\eta) = \left(1 - \eta^{3/2}\right)^2. \tag{7.190}$$

Once the similarity function $f(\eta)$ is determined, Eq. (7.190), one can determine the integral present in the equation of the function γ, Eq. (7.132), transcribed for this point of the deduction:

$$\gamma = \frac{C_\mathrm{D} d}{2\int_0^1 f(\eta)\mathrm{d}\eta} \tag{7.191}$$

Carrying out the integral:

$$2\int_0^1 f(\eta)\mathrm{d}\eta = 2\int_0^1 \left(1 - \eta^{3/2}\right)^2 \mathrm{d}\eta = 2\frac{9}{20} = 9/10. \tag{7.192}$$

Replacing the given value using Eq. (7.192) in Eq. (7.191) results in:

$$\gamma = \frac{10 C_\mathrm{D} d}{9}. \tag{7.193}$$

Combining Eqs. (7.180) and (7.188) results in:

$$\kappa = \beta = 9\alpha^2. \tag{7.194}$$

Returning to Eq. (7.177), we obtain:

$$\begin{aligned} \delta(x) &= (\kappa\gamma)^{1/2} x^{1/2} = \left(9\alpha^2\right)^{1/2} \left(\frac{10 C_\mathrm{D} d}{9}\right)^{1/2} x^{1/2} \\ &= \sqrt{10}\alpha (C_\mathrm{D} \mathrm{d}x)^{1/2} \delta(x) = \sqrt{10}\alpha (C_\mathrm{D} \mathrm{d}x)^{1/2}. \end{aligned} \tag{7.195}$$

Also returning to Eq. (7.179), we obtain:

$$\begin{aligned} \overline{u}_o(x) = \frac{U_\infty}{2\beta} (\kappa\gamma)^{1/2} x^{-1/2} &= \frac{U_\infty}{18\alpha^2} \left(9\alpha^2 \frac{10 C_\mathrm{D} d}{9}\right)^{1/2} x^{-1/2} \\ &= U_\infty \frac{\sqrt{10 C_\mathrm{D}}}{18\alpha} \left(\frac{x}{d}\right)^{-1/2}. \end{aligned} \tag{7.196}$$

Replacing Eq. (7.196) in Eq. (7.189), we have the final equation for the velocity deficit:

$$\overline{u}_1(x, y) = U_\infty \frac{\sqrt{10 C_\mathrm{D}}}{18\alpha} \left(\frac{x}{d}\right)^{-1/2} \left[1 - \left(\frac{y}{\delta(x)}\right)^{3/2}\right]^2 \tag{7.197}$$

Being $\delta(x)$ modeled with Eq. (7.195), finally, the model for the velocity $\overline{u}(x, y)$ can be determined, substituting Eq. (7.197) in Eq. (7.156):

$$\overline{u}(x, y) = U_\infty \left\{ 1 - \frac{\sqrt{10C_D}}{18\alpha} \left(\frac{x}{d}\right)^{-1/2} \left[1 - \left(\frac{y}{\delta(x)}\right)^{3/2} \right]^2 \right\}. \tag{7.198}$$

It is of great importance to obtain the continuous model also for the component $\overline{u'\vartheta'}$ of the Boussinesq–Reynolds tensor. So, the model based on the Prandtl mixing length and the Boussinesq hypothesis is resumed:

$$\begin{aligned} -\overline{u'\vartheta'} &= \left(\ell_m \frac{\partial \overline{u}}{\partial y}\right)^2 = \left(\alpha\delta(x) \frac{\partial \overline{u}}{\partial y}\right)^2 \\ &= \left\{ \alpha\delta(x) \frac{U_\infty}{\delta(x)} \frac{\sqrt{10C_D}}{18\alpha} \left(\frac{x}{d}\right)^{-1/2} 3 \left[\left(\frac{y}{\delta(x)}\right)^{1/2} - \left(\frac{y}{\delta(x)}\right)^2 \right] \right\}^2, \end{aligned} \tag{7.199}$$

$$-\overline{u'\vartheta'} = U_\infty^2 \frac{10C_D}{36} \left(\frac{x}{d}\right)^{-1} \left[\left(\frac{y}{\delta(x)}\right)^{1/2} - \left(\frac{y}{\delta(x)}\right)^2 \right]^2. \tag{7.200}$$

The solutions for all variables of interest for an approximate, but continuous, mathematical characterization of the distant turbulent wake downstream of a body immersed in the flow were derived. Illustrative results will be presented below.

7.5.3.4 Comparison Between Continuous Solution and Material Experiment Results

In Fig. 7.18, the continuous solution, given by Eq. (7.189), is presented and compared with results from material experiments obtained by Schlichting [20] and Reichardt [18]. An excellent agreement is observed between the results obtained with the theoretical solution, via the Prandtl mixing length model, and the results of the material experiment.

The distribution of the component $\overline{u'\vartheta'}$ of the Boussinesq–Reynolds Tensor is presented in Fig. 7.19. This information cancels out in regions far from the center of the wake, and cancels out in the center, where the derivatives of the velocity components are zero. It attains a maximum at the inflection points or inflection lines of the velocity distribution, as expected.

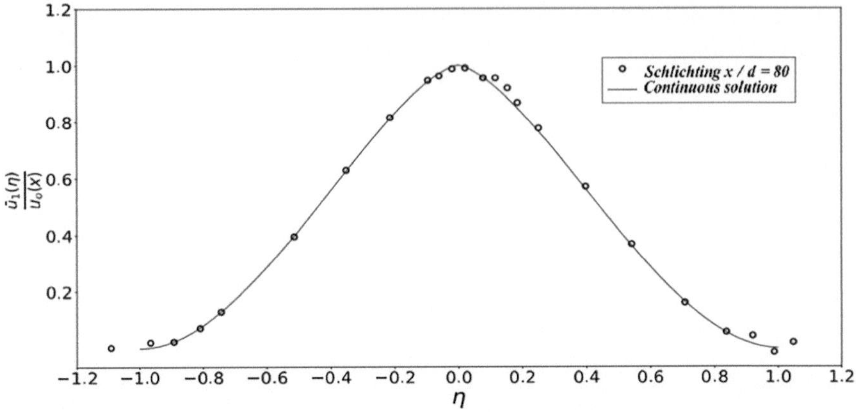

Fig. 7.18 Velocity deficit, $u_1(\eta)/u_o$, of a distant wake formed downstream of a long cylinder. Comparison between the results obtained using Eq. (7.189) and Schlichting's [20] and Reichardt [18] experimental results: (—) theoretical solution and (∘) experiment. Figure taken from Magalhães and Silveira-Neto [12], personal collection

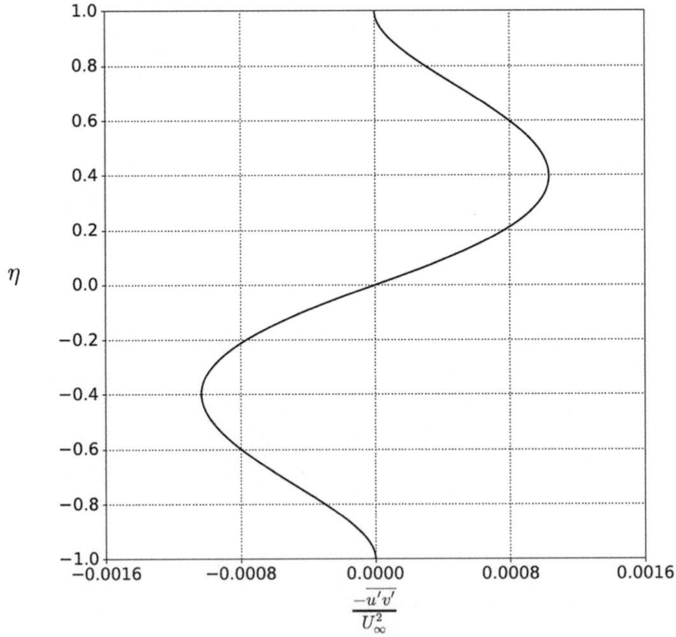

Fig. 7.19 Component $\overline{u'\vartheta'}/U_\infty^2$ of the Boussinesq–Reynolds Tensor, using Eq. (7.200). Figure taken from Magalhães and Silveira-Neto [12], personal collection

7.6 Plane Jet in Space Development

7.6.1 Physical Model for Plane Jets

Jets originate from the injection of a given fluid into another fluid, which can be quiescent or moving relative to the jet itself. The injection nozzle can be composed of any geometry. The fluid, leaving the nozzle, interacts with the environment under the effects of diffusive and advective transport processes. Plane jets originate from rectangular nozzles.

In Fig. 7.20, the physical model of a plane jet, developing spatially, is presented, where the thickness $\delta(x)$ and three regions of development can be identified. Region *A* refers to potential flow; in region *B*, the potential core is extinguished, giving rise to region *C*, where there is no longer a potential core. In regions *A* and *B*, the velocity profiles are not self-similar. Self-similarity theory can only be applied to region *C*. However, according to Tennekes and Lumley [23], the velocity distribution over the center line, determined by similarity theory, can also be used for region *B*. In the laminar regime, diffusion in the direction transverse to the main direction of fluid movement occurs for smaller values of *x*, when compared to the turbulent regime. Diffusion promotes a drop in centerline velocity. The distance for which the centerline velocity remains constant is shorter in the laminar regime than in the turbulent regime.

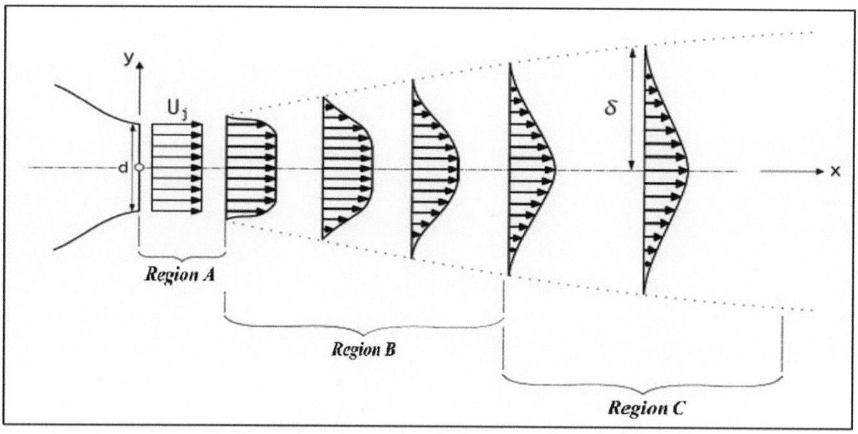

Fig. 7.20 Physical model of a flat jet in space development

7.6.2 *Plane Space Jet in Laminar Regime*

7.6.2.1 Partial Differential Mathematical Model

Plane jets of Newtonian fluids can be modeled using the continuity and Navier–Stokes equations:

$$\frac{\partial \overline{u}}{\partial x} + \frac{\partial \overline{\vartheta}}{\partial y} = 0, \tag{7.201}$$

$$\overline{u}\frac{\partial \overline{u}}{\partial x} + \overline{\vartheta}\frac{\partial \overline{u}}{\partial y} = \tilde{\nu}\frac{\partial^2 \overline{u}}{\partial y^2}, \tag{7.202}$$

where $\tilde{\nu}$ is the molecular kinematic viscosity. The possible boundary conditions for the problem are described below:

$$\overline{u}(x, y = 0) = U_c(x), \tag{7.203}$$

$$\overline{\vartheta}(x, y = 0) = 0, \tag{7.204}$$

$$\frac{\partial \overline{u}}{\partial y}(x, y = 0) = 0, \tag{7.205}$$

$$\overline{u}(x, y \to \pm\infty) = U_c(x), \tag{7.206}$$

Integrating Eq. (7.201) in the y direction yields:

$$\overline{\vartheta}(x, y) = -\frac{\partial}{\partial x}\int_0^y \overline{u}(x, y)\partial y. \tag{7.207}$$

The integration constant resulting from the application of the boundary condition (7.204) is equal to zero. Applying Eq. (7.207) in Eq. (7.202) results in an equation that only depends on $\overline{u}(x, y)$:

$$\overline{u}\frac{\partial \overline{u}}{\partial x} - \left(\frac{\partial}{\partial x}\int_0^y \overline{u}(x, y)\partial y\right)\frac{\partial \overline{u}}{\partial y} = \tilde{\nu}\frac{\partial^2 \overline{u}}{\partial y^2}. \tag{7.208}$$

7.6.2.2 Ordinary Differential Mathematical Model

For the transformation of the partial differential Eq. (7.208) into an ordinary differential equation, the similarity variable is defined:

$$\eta = \frac{y}{\delta(x)}. \tag{7.209}$$

The similarity function, for plane jets in a fluid at rest, is proposed as follows:

$$\frac{\overline{u}(x, y)}{U_C(x)} = f'(\eta), \tag{7.210}$$

where $U_C(x)$ is the jet velocity, at the center line, in $y = 0$. Developing the necessary derivatives to replace in Eq. (7.208), we have the ODE:

$$f''' - \kappa ff'' - \beta\left(f'^2 - ff''\right) = 0, \tag{7.211}$$

where

$$\kappa = \frac{U_C}{\tilde{\nu}} \frac{\partial^2\eta/\partial x\partial y}{(\partial\eta/\partial y)^3}, \tag{7.212}$$

$$\beta = \frac{1}{\tilde{\nu}} \frac{\mathrm{d}U_C}{\mathrm{d}x} \left(\frac{\partial\eta}{\partial y}\right)^{-2}, \tag{7.213}$$

The boundary conditions for Eq. (7.211) follow from the transformation of Eqs. (7.203)–(7.206):

$$\overline{\vartheta}(x, y = 0) = 0(\text{symmetry}) \Rightarrow f(\eta = 0) = 0, \tag{7.214}$$

$$\frac{\partial\overline{u}}{\partial y}(x, y = 0) = 0(\text{symmetry}) \Rightarrow f''(\eta = 0) = 0, . \tag{7.215}$$

$$\overline{u}(x, y = 0) = U_c(x) \Rightarrow f'(\eta = 0) = 1, \tag{7.216}$$

$$\overline{u}(x, y \to \infty) \to 0 \Rightarrow f'(\eta \to \infty) \to 0, \tag{7.217}$$

$$\overline{u}(x, y \to -\infty) \to 0 \Rightarrow f'(\eta \to -\infty) \to 0. \tag{7.218}$$

The model presented above can be used for both laminar and turbulent regimes.

7.6.2.3 Application of the Reynolds–Leibniz Transport Theorem (TTR) to the Linear Momentum Balance

View of determining ODE coefficients, the TTR will be applied to the control volume presented in Fig. 7.21, particularized to Newton's Second Law, which results in:

$$\int_{\text{entry}} \rho \overline{u}^2 b\text{dy} = \int_{\text{exit}} \rho \overline{u}^2 b\text{dy} = \text{constant.} \tag{7.219}$$

Replacing $\overline{u}(x, y) = U_C(x) f'(\eta)$ and $y = \eta\delta(x)$ in Eq. (7.219), results in:

$$\int_{-\infty}^{+\infty} \rho \overline{u}^2 b\text{dy} = J \Rightarrow J' = \frac{J}{b} = \int_{-\infty}^{+\infty} \rho \overline{u}^2 \text{dy} = \rho U_C^2(x)\delta(x) \int_{-\infty}^{+\infty} f'^2(\eta)\text{d}\eta \tag{7.220}$$

Thus,

$$U_c^2\delta = \frac{J'}{\rho \int_{-\infty}^{+\infty} f'^2(\eta)\text{d}\eta} = \text{constant} = \gamma \Rightarrow U_c^2\delta = \gamma, \tag{7.221}$$

It is worth noting that, even $U(x)$ and $\delta(x)$. are functions of the axis x, the product $U_c^2\delta$ is a constant, since $\int_{-\infty}^{+\infty} f'^2(\eta)\text{d}\eta$ is a constant. This result will be used later.

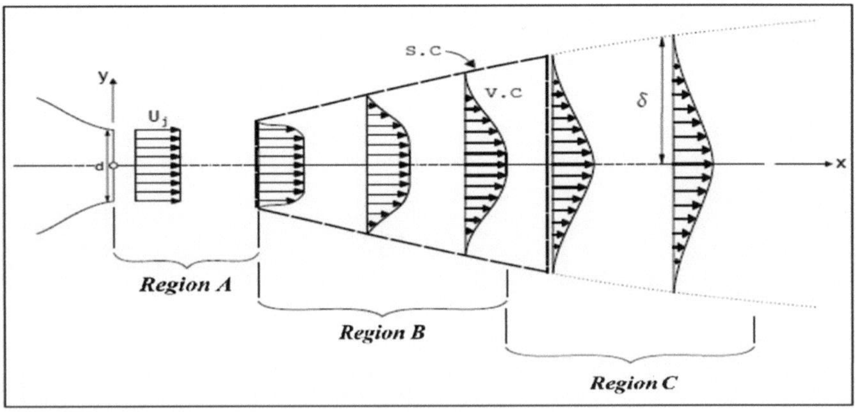

Fig. 7.21 Control volume over a plane jet in space development

7.6.2.4 Continuous Mathematical Model

The established ordinary differential model is composed of Eqs. (7.211)–(7.218). This model will be integrated, and the coefficients will be determined for the plane jet in the laminar regime.

To integrate Eq. (7.211), it is necessary to first establish a relationship between the constants κ and β, as will be done next. Replacing Eq. (7.209) in (7.212) yields:

$$\kappa = \frac{U_C}{\tilde{\nu}} \frac{\partial^2\eta/\partial x\partial y}{(\partial\eta/\partial y)^3} = \frac{U_C}{\tilde{\nu}} \frac{(-\mathrm{d}\delta/\mathrm{d}x)/\delta^2}{1/\delta^3} \Rightarrow \kappa = -\frac{U_C\delta}{\tilde{\nu}} \frac{\mathrm{d}\delta}{\mathrm{d}x}. \tag{7.222}$$

Taking Eq. (7.209) into Eq. (7.212) results in:

$$\beta = \frac{1}{\tilde{\nu}} \frac{\mathrm{d}U_C}{\mathrm{d}x} \left(\frac{\partial\eta}{\partial y} \right)^{-2} = \frac{1}{\tilde{\nu}\left(1/\delta^2\right)} \frac{\mathrm{d}U_C}{\mathrm{d}x} \Rightarrow \beta = \frac{\delta^2}{\tilde{\nu}} \frac{\mathrm{d}U_C}{\mathrm{d}x} \tag{7.223}$$

Dividing Eq. (7.223) by Eq. (7.222), a first relation is obtained involving β and κ:

$$\begin{aligned} \frac{\beta}{\kappa} &= \frac{\delta^2}{\tilde{\nu}} \frac{\mathrm{d}U_C}{\mathrm{d}x} \frac{\tilde{\nu}}{U_C\delta(\mathrm{d}\delta/\mathrm{d}x)} \Rightarrow \frac{\mathrm{d}U_C}{\mathrm{d}x} \frac{1}{U_C} \\ &= -\frac{\beta}{\kappa} \frac{1}{\delta} \frac{\mathrm{d}\delta}{\mathrm{d}x} \Rightarrow \frac{\mathrm{d}U_C}{U_C} = -\frac{\beta}{\kappa} \frac{\mathrm{d}\delta}{\delta} \end{aligned} \tag{7.224}$$

Integrating Eq. (7.224) results in:

$$U_C(x) = c_8 \delta^{-\beta/\kappa}(x), \tag{7.225}$$

Being c_8 a constant of integration to be determined. Substituting Eq. (7.221) into Eq. (7.223), and integrating it, we arrive at:

$$\frac{\mathrm{d}U_C}{U_C^4} = \frac{\beta\tilde{\nu}}{\gamma^2}\mathrm{d}x \Rightarrow U_C = -\left(\frac{1}{3} \frac{\gamma^2}{\beta\tilde{\nu}x} \right)^{1/3} + c_{10}, \tag{7.226}$$

where c_{10} is a constant of integration. $U_c(x)$ is the velocity at the jet center. When $x \to \infty$, $U_C \to 0$. This condition applied to Eq. (7.223) results in $c_{10} = 0$ and:

$$U_C = -\left(\frac{1}{3} \frac{\gamma^2}{\beta\tilde{\nu}} \right)^{1/3} x^{-1/3}, \tag{7.227}$$

Carrying out a similar procedure in relation to Eq. (7.222) and also using Eq. (7.227), results in:

$$\kappa = -\frac{U_C\delta}{\tilde{\nu}}\frac{\mathrm{d}\delta}{\mathrm{d}x} \Rightarrow \delta\frac{\mathrm{d}\delta}{\mathrm{d}x} = -\frac{\tilde{\nu}\kappa}{U_C} \Rightarrow \delta\frac{\mathrm{d}\delta}{\mathrm{d}x} = \left(\frac{\gamma^2}{3\beta\tilde{\nu}^4\kappa^3}\right)^{-1/3} x^{1/3}$$
$$\Rightarrow \frac{\delta^2}{2} = \left(\frac{\gamma^2}{3\beta\tilde{\nu}^4\kappa^3}\right)^{-1/3} \frac{3}{4}x^{4/3} + c_{\delta 1}, \tag{7.228}$$

where $\delta(x)$ is the spreading thickness of the jet. We have $\delta(0) = 0$. Consequently, $c_{\delta 1} = 0$ and:

$$\delta = \left(\frac{8\gamma^2}{81\beta\tilde{\nu}^4\kappa^3}\right)^{-1/6} x^{2/3}. \tag{7.229}$$

Substituting Eq. (7.229) in Eq. (7.225) results in:

$$U_C(x) = c_8\left(\frac{8\gamma^2}{81\beta\tilde{\nu}^4\kappa^3}\right)^{-\beta/6\kappa} x^{-2\beta/3\kappa}. \tag{7.230}$$

From Eq. (7.227) $U_c \sim x^{-1/3}$ and Eq. (7.230) $U_c \sim x^{-2\beta/3\kappa}$, then:

$$-\frac{1}{3} = -\frac{2}{3}\frac{\beta}{\kappa} \Rightarrow \kappa = 2\beta. \tag{7.231}$$

Taking this relationship between these two coefficients to Eq. (7.211) yields:

$$f''' - 2\beta ff'' - \beta\left(f'^2 - ff''\right) = 0 \Rightarrow f''' - \beta ff'' - \beta f'^2$$
$$= 0 \Rightarrow \frac{d}{\mathrm{d}\eta}\left[f'' + \tilde{\beta}ff'\right] = 0 \tag{7.232}$$

where $\tilde{\beta} = -\beta$. Integrating Eq. (7.232) results in:

$$f'' + \tilde{\beta}ff' = c_{11} \tag{7.233}$$

The constant of integration $c_{11} = 0$ results from the use of the conditions given by Eqs. (7.214)–(7.216). Then:

$$f'' + \tilde{\beta}ff' = 0. \tag{7.234}$$

The change of variable $g = \tilde{\beta}f$ is proposed, which is applied to Eq. (7.234), and we obtain:

$$\frac{g''}{\tilde{\beta}} + g\frac{g'}{\tilde{\beta}} = 0 \Rightarrow g'' + gg' = 0, \tag{7.235}$$

with the following boundary conditions:

$$g(\eta = 0) = 0, \tag{7.236}$$

$$g'(\eta \to \infty) = 0, \tag{7.237}$$

$$g'(\eta \to -\infty) = 0, \tag{7.238}$$

$$g''(\eta = 0) = 0. \tag{7.239}$$

The integration of Eq. (7.235) and the use of the above boundary conditions leads to:

$$g'(\eta) = \text{sech}^2\left(\frac{\eta}{\sqrt{2}}\right) \tag{7.240}$$

$$g(\eta) = \sqrt{2}\tanh\left(\frac{\eta}{\sqrt{2}}\right). \tag{7.241}$$

Using the boundary condition (7.216) results in:

$$g'(0) = \tilde{\beta} f'(0) = 1 \Rightarrow 1 = \tilde{\beta}.1 \Rightarrow \tilde{\beta} = 1 \Rightarrow \beta = -1. \tag{7.242}$$

Thus, considering that $g = \tilde{\beta} f$, the following is obtained:

$$f(\eta) = \sqrt{2}\tanh\left(\frac{\eta}{\sqrt{2}}\right). \tag{7.243}$$

and

$$f'(\eta) = \text{sech}^2\left(\frac{\eta}{\sqrt{2}}\right). \tag{7.244}$$

It is observed that these solutions satisfy the boundary conditions given by Eqs. (7.314)–(7.218). To determine the constant γ, we return to Eq. (7.221):

$$\gamma = \frac{J'}{\rho \int_{-\infty}^{+\infty} f'^2(\eta) \mathrm{d}\eta} = \frac{\rho U_j^2 d_j}{\rho \int_{-\infty}^{+\infty} \text{sech}^4\left(\frac{\eta}{\sqrt{2}}\right) \mathrm{d}\eta} \Rightarrow \gamma = 0.53033 U_j^2 d_j, \tag{7.245}$$

where U_j is the average velocity at the exit of the jet and d_j is the height of the flat nozzle, from which the jet originates. Substituting Eqs. (7.242) and (7.245) in Eq. (7.227) results in:

$$U_C(x) = \left(\frac{0.28125}{3} \frac{U_j^4 d_j^2}{\tilde{\nu}}\right)^{1/3} x^{-1/3}. \tag{7.246}$$

Rewriting this continuous solution in dimensionless form:

$$\frac{U_C(x)}{U_j} = 0.45428\,\mathrm{Re}_{d_j}^{1/3}\left(\frac{x}{d_j}\right)^{-1/3}. \tag{7.247}$$

To determine the continuous model for the thickness $\delta(x)$ of the jet, we insert Eqs. (7.231) and (7.245) into Eq. (7.227), yielding:

$$\delta(x) = \left(\frac{8\gamma^2}{81\beta\tilde{\nu}^4\kappa^3 x^4}\right)^{-1/6} = \left(\frac{8\left(0.53033U_j^2 d_j\right)^2}{81(-1)\tilde{\nu}^4(-2)^3 x^4}\right)^{-1/6}$$
$$= \left(\frac{0.28125U_j^4 d_j^2}{3\tilde{\nu}x}\frac{1}{27\tilde{\nu}^3 x^3}\right)^{-1/6}. \tag{7.248}$$

Replacing Eq. (7.246) in Eq. (7.248) results in:

$$\delta(x) = \left(U_C^3\frac{1}{27\tilde{\nu}^3 x^3}\right)^{-1/6} = \left(U_C\frac{1}{3\tilde{\nu}x}\right)^{-1/2} = \sqrt{\frac{3}{\mathrm{Re}_x}}x. \tag{7.249}$$

Therefore, substituting Eq. (7.247) in Eq. (7.210) and taking into account Eq. (7.249), we obtain:

$$u(x, y) = 0.45428U_j\mathrm{Re}_{d_j}^{1/3}\left(\frac{x}{d_j}\right)^{-1/2}\mathrm{sech}^2\left(\frac{1}{\sqrt{2}}\frac{y}{\delta(x)}\right). \tag{7.250}$$

To determine $\vartheta(x, y)$, use Eqs. (7.207) and (7.250):

$$\vartheta(x, y) = 0.389U_j\mathrm{Re}_{d_j}^{1/3}\left(\frac{x}{d_j}\right)^{-1/2}$$
$$\left[3\sqrt{2}\tanh\left(\frac{1}{\sqrt{2}}\frac{y}{\delta(x)}\right) + 2\frac{y}{\delta(x)}\mathrm{sech}^2\left(\frac{1}{\sqrt{2}}\frac{y}{\delta(x)}\right)\right]. \tag{7.251}$$

7.6.2.5 Comparison Between Continuous Model and Computational Experiment Results

After presenting the continuous solutions for the velocity components, the velocity at the center line of the jet and the thickness of the jet in a laminar regime, illustrative results will be presented. In Fig. 7.22, the distributions of the axial component of velocity, $u(\eta)$ are presented, for multiple stations x/d_j. Comparisons of the results from computer simulations are presented. This is a validation of the computational

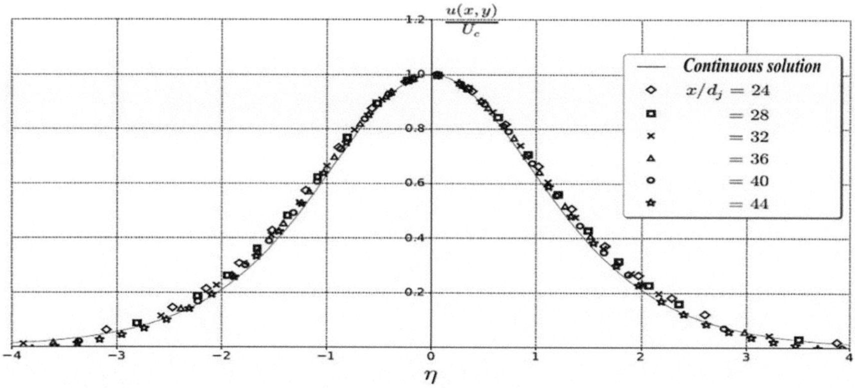

Fig. 7.22 Distributions of the axial component of velocity, $u(\eta)$, for multiple stations x/d_j of a plane jet in a laminar regime. Figure taken from Magalhães and Silveira-Neto [12]

results. The agreement is excellent. The simulation of a two-dimensional laminar plane jet was carried out, using the MFSim Computational Code. The following quantities were used: Reynolds Number, $Re_{dj} = 100$; jet diameter, $d_j = 0.075$ (m); jet velocity $U_j = 0.1$ (m/s).; the domain size is $50d_j$ in the x direction and $20d_j$ in the y direction. Six hundred (600) computational cells were used in the x direction and 200 cells in the y direction. The advective term was discretized with the CUBISTA scheme; a semi-implicit SBDF temporal discretization scheme was used; a Dirichlet boundary condition was used for the domain entrance, and a Neumann boundary condition was used for the other sides of the domain. The comparison is presented in Fig. 7.23, where an excellent agreement can be seen between the results obtained with the two methods.

7.6.3 Spatially Developing Plane Jet in the Turbulent Regime

7.6.3.1 Partial Differential Mathematical Model

The starting point for differential modeling of the turbulent regime will be Eqs. (7.60) and (7.61). It was assumed that the turbulent viscosity is much greater than the molecular viscosity, $\nu_t \gg \tilde{\nu}$. The filtered equations are rewritten below:

$$\frac{\partial \overline{u}}{\partial x} + \frac{\partial \overline{\vartheta}}{\partial y} = 0. \tag{7.252}$$

$$\overline{u}\frac{\partial \overline{u}}{\partial x} + \overline{\vartheta}\frac{\partial \overline{u}}{\partial y} = \frac{\partial \tau_{xy}}{\partial y}. \tag{7.253}$$

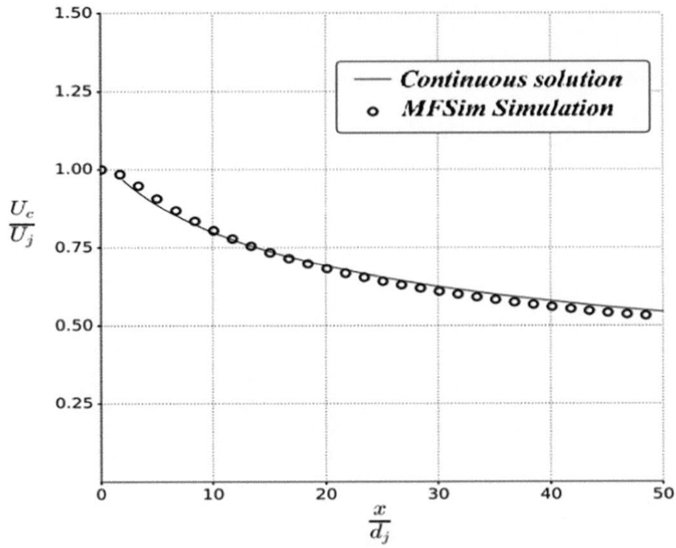

Fig. 7.23 Velocity distribution at the jet centerline, U_C/U_j, as a function of x/d_j; plane jet in laminar regime. Figure taken from Magalhães and Silveira-Neto [12]

The Prandtl–Reichardt turbulence closure model will be considered [18], as represented in Eq. (7.254), changing the characteristic velocity $(U_1 - U_2)$ by the speed at the centerline of the jet, i.e., U_C, and the constant χ_1 for χ:

$$\tau_{xy} = \nu_t \frac{\partial \overline{u}}{\partial y} = \delta(x) \chi U_c \frac{\partial \overline{u}}{\partial y}. \tag{7.254}$$

Substituting Eq. (7.254) into Eq. (7.253) results in:

$$\overline{u}\frac{\partial \overline{u}}{\partial x} + \overline{\vartheta}\frac{\partial \overline{u}}{\partial y} = \underbrace{\delta(x) \chi U_c}_{\nu_t} \frac{\partial^2 \overline{u}}{\partial y^2} = \nu_t \frac{\partial^2 \overline{u}}{\partial y^2}, \tag{7.255}$$

$$\overline{u}\frac{\partial \overline{u}}{\partial x} + \overline{\vartheta}\frac{\partial \overline{u}}{\partial y} = \nu_t \frac{\partial^2 \overline{u}}{\partial y^2}. \tag{7.256}$$

Equation (7.256) is identical to (7.202), except for the filtration and for the turbulent viscosity ν_t which replaces the molecular viscosity $\tilde{\nu}$.

7.6.3.2 Ordinary Differential Mathematical Model

Considering the similarity between Eqs. (7.255) and (7.202), the entire self-similarity transformation process carried out for the laminar regime will be used for the turbulent regime. Thus, the similarity variable η and the similarity function $f(\eta)$ will be considered as already defined for the laminar regime:

$$\eta = \frac{y}{\delta(x)}, \tag{7.257}$$

$$\frac{\overline{u}(x,y)}{U_C(x)} = f'(\eta) \tag{7.258}$$

After obtaining the derivatives and making the substitutions in Eq. (7.256), we arrive at the ODE for the turbulent regime, exactly like what was obtained for the laminar regime, i.e., simply replace $\tilde{\nu}$ by ν_t in the constants κ and β that appear in Eq. (7.211):

$$f''' - \kappa ff'' - \beta\left(f'^2 - ff''\right) = 0, \tag{7.259}$$

where

$$\kappa = \frac{U_c}{\nu_t}\frac{\partial^2\eta/\partial x\partial y}{(\partial\eta/\partial y)^3}, \tag{7.260}$$

$$\beta = \frac{1}{\nu_t}\frac{dU_c}{dx}\left(\frac{\partial\eta}{\partial y}\right)^{-2}. \tag{7.261}$$

Thus,

$$\kappa = \frac{U_c}{\nu_t}\frac{\partial^2\eta/\partial x\partial y}{(\partial\eta/\partial y)^3} = \frac{U_c(-\mathrm{d}\delta/\mathrm{d}x)/\delta^2}{\chi\delta U_c\left(1/\delta^3\right)} = -\frac{1}{\chi}\frac{\mathrm{d}\delta}{\mathrm{d}x}. \tag{7.262}$$

Integrating Eq. (7.262) results in:

$$\frac{\mathrm{d}\delta}{\mathrm{d}x} = -\kappa\chi \Rightarrow \delta(x) = -\kappa\chi x + c_{t1} = Ax + c_{t1}, \tag{7.263}$$

where $A = -\kappa\chi$ is the spreading rate of the mixing zone. Considering that in $x = 0$ the scattering of the jet is zero, i.e., $\delta(0) = 0$, it gives $c_{t1} = 0$. Thus,

$$\delta(x) = Ax, \tag{7.264}$$

and

$$\kappa = -\frac{A}{\chi}. \tag{7.265}$$

Replacing Eq. (7.264) in Eq. (7.261):

$$U_c^2\delta = \gamma \Rightarrow U_c^2(Ax) = \gamma \Rightarrow U_c = \sqrt{\frac{\gamma}{A}}x^{-1/2}. \tag{7.266}$$

The coefficient β will be determined based on Eq. (7.261)

$$\begin{aligned}\beta &= \frac{1}{\nu_t}\frac{\mathrm{d}U_c}{\mathrm{d}x}\left(\frac{\partial\eta}{\partial y}\right)^{-2} = \frac{1}{\chi\delta U_c\left(1/\delta^2\right)}\frac{\mathrm{d}U_c}{\mathrm{d}x} \Rightarrow \beta \\ &= \frac{\delta}{\chi U_c}\frac{\mathrm{d}U_c}{\mathrm{d}x} = \frac{-\kappa\chi x}{\chi U_c}\frac{\mathrm{d}U_c}{\mathrm{d}x} \Rightarrow \frac{\mathrm{d}U_c}{U_c} = -\frac{\beta}{\kappa}\frac{\mathrm{d}x}{x}.\end{aligned} \tag{7.267}$$

Integrating Eq. (7.267) yields:

$$U_c(x) = c_{t2}x^{-\beta/\kappa} \tag{7.268}$$

By comparing the exponents of the variable x that appear in Eqs. (7.266) and (7.268), it can be concluded that:

$$-\frac{\beta}{\kappa} = -\frac{1}{2} \Rightarrow \kappa = 2\beta. \tag{7.269}$$

Plugging Eq. (7.265) into Eq. (7.269), results in:

$$\beta = -\frac{A}{2\chi} \tag{7.270}$$

Substituting Eqs. (7.265) and (7.270) into Eq. (7.259) results in:

$$f''' + \frac{A}{\chi}ff'' + \frac{A}{2\chi}\left(f'^2 - ff''\right) = 0 \Rightarrow f''' + \frac{A}{2\chi}\left(f'^2 + ff''\right) = 0. \tag{7.271}$$

7.6.3.3 Continuous Mathematical Model

The task now is to integrate Eq. (7.271). Using the product derivative rule to rewrite Eq. (7.271), it follows that:

$$\frac{d}{\mathrm{d}\eta}\left(f'' + \frac{A}{2\chi}ff'\right) = 0. \tag{7.272}$$

Integrating Eq. (7.272), results in:

$$f'' + \frac{A}{2\chi} ff' = \text{constant}. \tag{7.273}$$

Using the boundary conditions expressed by Eqs. (7.214) and (7.215), the integration constant of Eq. (7.271) is null. Thus:

$$f'' + \frac{A}{2\chi} ff' = 0, \tag{7.274}$$

Equation (7.274) can also be rewritten, using the rule of the derivative of products:

$$\frac{d}{d\eta}\left(\chi f' + \frac{A}{4} f^2\right) = 0 \Rightarrow f' + \frac{A}{4\chi} f^2 = \text{constant}. \tag{7.275}$$

Using the boundary conditions expressed by Eqs. (7.214) and (7.216), it is concluded that the constant of integration of Eq. (7.275) is $c = 1$. Finally, by integrating Eq. (7.275), it follows that:

$$f(\eta) = 2\sqrt{\frac{\chi}{A}} \tanh\left(\sqrt{\frac{\chi}{A}} \frac{\eta}{2}\right), \tag{7.276}$$

and

$$f'(\eta) = \text{sech}^2\left(\sqrt{\frac{\chi}{A}} \frac{\eta}{2}\right). \tag{7.277}$$

It has been demonstrated, through material experiments on plane jet flows in the turbulent regime, that the spreading rate is, on average, $A = 0.1$, as stated by Wilcox [26]. The jet opening rate is defined as a function of the ratio y_a/x_a for which the jet velocity is equal to half the value of the jet center velocity for which $x = x_a$, i.e., $\overline{u}(x_a, y_a) = U_C(x)/2$. On the other hand, Bradbury [4] obtained, through material experimentation, that the constant A that appears in Eq. (7.264) takes on the value $A = 0.109$. Replacing the values of the opening rate and the constant A in $f'(\eta)$, there is a new restriction condition for Eq. (7.210), i.e.:

$$\frac{\overline{u}(x_a, y_a)}{U_C(x_a)} = f'(\eta_a) \Rightarrow \frac{(1/2)U_C}{U_C} = f'\left(\frac{1}{A}\frac{y_a}{x_a}\right) = f'\left(\frac{1}{0.109}0.1\right) = f'(0.9174) = 0.5. \tag{7.278}$$

Defining

$$\Gamma \equiv \frac{1}{2}\sqrt{\frac{A}{\chi}}. \tag{7.279}$$

and substituting Eq. (7.279) into Eq. (7.277), using the restriction given by Eq. (7.278), it follows that:

$$f'(0.9174) = \operatorname{sech}(0.91474\Gamma) = 0.5 \Rightarrow \Gamma = 0.96073 \tag{7.280}$$

With this result, Eq. (7.277) can be rewritten:

$$\begin{aligned}\frac{\overline{u}(x,y)}{U_c(x)} &= f'(\eta) = \sec^2\left(\frac{1}{2}\sqrt{\frac{A}{\chi}}\right) = f'(\eta)\\ &= \sec^2(\Gamma) = \operatorname{sech}^2(0.96073\eta)\\ \frac{\overline{u}(x,y)}{U_c(x)} &= \operatorname{sech}^2(0.96073\eta).\end{aligned} \tag{7.281}$$

The continuous model for the velocity over the centerline of the jet will also be obtained. To facilitate comparisons, a dimensionless form will be established, involving $U_c(x)$ and U_j, which is the average speed at the jet entrance. The width of the inlet channel, d_j, will also be involved. Therefore, Eq. (7.221) will be adjusted:

$$\begin{aligned}\gamma &= \frac{J'}{\rho\int_{-\infty}^{+\infty} f'^2(\eta)\mathrm{d}\eta} = \frac{\rho U_j^2 d_j}{\rho\int_{-\infty}^{+\infty} f'^2(\eta)d\eta} = \frac{U_j^2 d_j}{\int_{-\infty}^{+\infty} f'^2(\eta)\mathrm{d}\eta}\\ &= \frac{U_j^2 d_j}{\int_{-\infty}^{+\infty}\operatorname{sech}^4(0.9673\eta)\mathrm{d}\eta} \Rightarrow \gamma = 0.720549U_j^2 d_j\end{aligned} \tag{7.282}$$

Replacing Eq. (7.289) and the value $A = 0.109$ in Eq. (7.266), we have the velocity distribution at the center of the jet:

$$U_c = \sqrt{\frac{0.720549U_j^2 d_j}{Ax}} \Rightarrow \frac{U_C(x)}{U_j} = 2.5711\left(\frac{x}{d_j}\right)^{-1/2}. \tag{7.283}$$

In Eq. (7.283), it is assumed that the effective origin of the axis x is in the injection plane, as illustrated in Fig. 7.21. Andrade [1] suggests that the continuous solution fits better to data from material experiments when the solution is moved relative to the origin. Schlichting [21] suggests the following modification to Eq. (7.283):

$$\frac{U_C(x)}{U_j} = 2.5711\left(\frac{x+x_o}{d_j}\right)^{-1/2}, \tag{7.284}$$

where x_o is a positive constant, called the virtual origin of the jet. Tennekes and Lumley [23] suggest, based on material experiments, that at $x/d_j = 5$ the velocity distribution begins to decay. This value does not correspond to the end of the potential region of the jet. Thus,

$$1 = 2.5711\left(5 + \frac{x_o}{d_j}\right)^{-1/2} \Rightarrow \frac{x_o}{d_j} = 1.61054. \tag{7.285}$$

Thus,

$$U_C(x) = 2.5711U_j\left(1.61054 + \frac{x}{d_j}\right)^{-1/2}. \tag{7.286}$$

The model for the jet thickness, considering the value of the constant $A = 0.109$, is:

$$\delta(x) = 0.109x. \tag{7.287}$$

Substituting Eq. (7.286) in Eq. (7.281) results in:

$$\overline{u}(x, y) = 2.5711U_j\left(1.61054 + \frac{x}{d_j}\right)^{-1/2} \text{sech}^2\left(0.9673\frac{y}{\delta(x)}\right). \tag{7.288}$$

Finally, we can determine the component $\overline{u'\vartheta'}$ of the Boussinesq–Reynolds tensor. Returning to the Boussinesq hypothesis:

$$-\overline{u'\vartheta'} = \nu_t\frac{\partial \overline{u}}{\partial y} = \chi\delta(x)U_c(x)\frac{U_c(x)}{\delta(x)}f''(\eta) = \chi U_c^2(x)f''(\eta). \tag{7.289}$$

The second derivative $f''(\eta)$ reads:

$$f''(\eta) = -1.92146 \times \tanh(0.96073\eta) \times \text{sech}^2(0.96073\eta). \tag{7.290}$$

Thus, substituting Eqs. (7.286) and (7.290) into Eq. (7.289) results in:

$$\overline{u'\vartheta'} = 12.7019U_j^2\left(1.61054 + \frac{x}{d_j}\right)^{-1} \tanh(0.96073\eta)\text{sech}^2(0.9673\eta). \tag{7.291}$$

Solutions were presented for the component $\overline{u}(x, y)$ of velocity, as well as for the component $\overline{u'\vartheta'}$ of Boussinesq–Reynolds's Tensor. Illustrative results and comparisons are presented below.

7.6.3.4 Comparisons with Material Experiment Results

Figure 7.24 presents the comparison between the results obtained via continuous solution and the results obtained via Zinjnen's [27] Material experiment. Equation (7.286) was used. There is excellent agreement between the results from the two types of modeling. It is worth noting that, in the interval $0 \leq x/d_j \leq 5$, the similarity

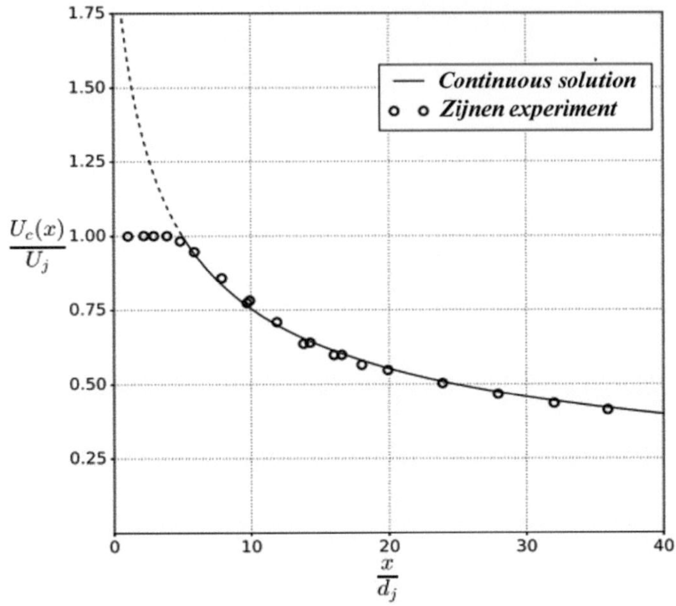

Fig. 7.24 Velocity distributions on the centerline of the plane jet in the turbulent regime, $U_c(x)/U_j$, function of x/d_j. Comparison with the material experiment results of Zijnen [27]

theory is not applicable. In this region, the curve relative to the continuous solution was drawn with a dotted line.

The continuous profile of the component $\overline{u}(x, y)/U_c(x)$, which is a function of x/y, is modeled with Eq. (7.288). In Fig. 7.24, the comparison of the continuous solution for the mean velocity field with the results obtained via material experimentation, obtained by Bradbury [4] and Heskestad [7], is presented. A departure from the results of the continuous solution is observed in relation to the results via material experimentation, especially when moving away from the center line, when $y/d_j \to \delta(x)/d_j$. This is due to the use of the asymptotic boundary condition for $y/d_j \to \infty$. This condition is not the best to represent the physical problem, which requires a finite distance to establish the jet limit. In other words, the asymptotic condition does not model well the physical problem (Fig. 7.25).

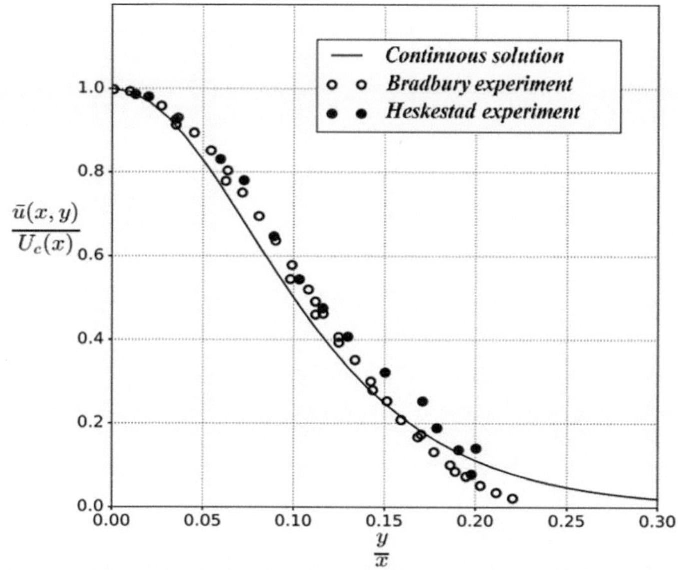

Fig. 7.25 Distributions of the average velocity of the plane jet in the turbulent regime, Z, as a function of y/x Comparison with results of the material experiments of Bradbury [4] and Heskestad [7]

7.7 Theoretical Analysis of Isothermal Plane Poiseuille Flows

7.7.1 Preliminary

Flows between two flat plates in a developed regime are known as Poiseuille flows, named after the French doctor and physicist, Jean-Léonard-Marie Poiseuille [14], who lived between the eighteenth and nineteenth centuries. These flows can be laminar or turbulent. They are driven by pressure gradients in the flow direction. The theoretical solution for the laminar regime is trivial. In this topic, the turbulent regime will be studied, also considering thermal effects.

7.7.2 Physical Model for Isothermal Planes Poiseuille Flows

A flat channel with a velocity distribution is illustrated in Fig. 7.26. The flat shape of the qualitative velocity distribution refers to the turbulent regime, where the molecular diffusion of linear momentum from the center of the channel to the walls is strongly accelerated by the turbulence. The advective transport of linear momentum, from the center of the channel to the vicinity of the walls, occurs with great intensity in

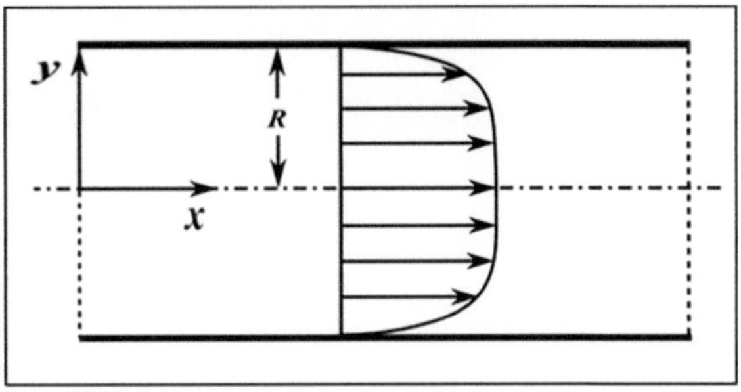

Fig. 7.26 Physical model of turbulent flows in flat channels

the turbulent regime. The flow will be considered two-dimensional, on average, by neglecting edge effects in the direction normal to the design. It is considered that the flow is developed in the direction x and that, on average, the velocity components in y as well as in z are null. With these simplifying assumptions, the differential model will be established for the mean flow.

Flows of this nature can be modeled with the average equations and the Mixing Length Model of Prandtl [17]. Nikuradse [13] suggested a polynomial for evaluating the mixing length in the form of a function of the coordinate y, normal to the walls. Due to the viscous effects being very strong in the vicinity of the walls, it is necessary to superimpose a damping function on the Polynomial of Nikuradse [13] when close to the walls. This model was presented in Eq. (7.12).

7.7.3 *Ordinary Differential Mathematical Model*

Starting with the average equations, (7.1) and (7.2), and simplifying them in accordance with the physical model presented above, we have:

$$\frac{d}{\mathrm{dy}}\left(\nu\frac{\mathrm{d}\overline{u}}{\mathrm{dy}}-\overline{u'v'}\right)=\frac{1}{\rho}\frac{\mathrm{d}\overline{p}}{\mathrm{dx}}. \tag{7.292}$$

To close this equation, we use the Hypothesis of Boussinesq [3] and the Prandtl mixing length model:

$$-\tau_{xy}=-\overline{u'v'}=\nu_t\frac{\partial\overline{u}}{\partial y}=\ell_m^2\left|\frac{\partial\overline{u}}{\partial y}\right|\frac{\partial\overline{u}}{\partial y}. \tag{7.293}$$

Substituting Eq. (7.293) into Eq. (7.292) results in:

$$\frac{d}{\mathrm{dy}}\left(\nu\frac{\mathrm{d}\overline{u}}{\mathrm{dy}}+l_m^2\left|\frac{\mathrm{d}\overline{u}}{\mathrm{dy}}\right|\frac{\mathrm{d}\overline{u}}{\mathrm{dy}}\right)=\frac{1}{\rho}\frac{\mathrm{d}\overline{p}}{\mathrm{dx}}. \tag{7.294}$$

Considering that the flow is symmetric with respect to y and that $(\mathrm{d}\bar{u}/\mathrm{d}y) < 0$ for $y > 0$, then $|\mathrm{d}\bar{u}/\mathrm{d}y| = -\mathrm{d}\bar{u}/\mathrm{d}y$ for $0 \leq y \leq R$. Thus, Eq. (7.294) is rewritten as follows:

$$\frac{d}{\mathrm{d}y}\left[\nu\frac{\mathrm{d}\bar{u}}{\mathrm{d}y} - l_m^2\left(\frac{\mathrm{d}\bar{u}}{\mathrm{d}y}\right)^2\right] = \frac{1}{\rho}\frac{\mathrm{d}\bar{p}}{\mathrm{d}x}. \tag{7.295}$$

The Prandtl Mixing Length model for turbulent flows in channels and ducts was proposed by Nikuradse [13], and is represented here by Eq. (7.296):

$$L(y) = \frac{l_m}{R} = 0.14 - 0.08\left(\frac{y}{R}\right)^2 - 0.06\left(\frac{y}{R}\right)^4. \tag{7.296}$$

To ensure adequate damping of $\ell_m(y)$ when $y = R$, Cebeci and Bradshaw [5], using the damping function of van Driest (1956)[24], proposed to modify the Nikuradse Model, as follows:

$$L(\tilde{y}) = \frac{l_m}{R} = \left(0.14 - 0.08\tilde{y}^2 - 0.06\tilde{y}^4\right)\left\{1 - \mathrm{e}^{\left((\tilde{y}-1)\frac{\mathrm{Re}_\tau}{A}\right)}\right\}, \tag{7.297}$$

where $A = 26$ is the constant of Cebeci, $0 \leq \tilde{y} = \frac{y}{R} \leq 1$. The parameters $\mathrm{Re}_\tau = (u_\tau R)/\nu$, $u_\tau = \sqrt{\|\tau_w\|/\rho}$ (m/s) and $\tau_w\left(\mathrm{N/m^2}\right)$ were defined at the beginning of this chapter and are reintroduced here. Equation (7.295) can be rewritten as follows:

$$\frac{d}{\mathrm{d}y}\left[\nu\frac{\mathrm{d}\bar{u}}{\mathrm{d}y} - L^2(\tilde{y})R^2\left(\frac{\mathrm{d}\bar{u}}{\mathrm{d}y}\right)^2\right] = \frac{1}{\rho}\frac{\mathrm{d}\bar{p}}{\mathrm{d}x}. \tag{7.298}$$

The boundary conditions for Eq. (7.298) are given by:

$$\begin{aligned}&\bar{u}(R) = 0,\\&\frac{\mathrm{d}\bar{u}}{\mathrm{d}y}(0) = 0.\end{aligned} \tag{7.299}$$

It should be noted that, unlike free shear flows, in the present problem, molecular viscous effects cannot be neglected when compared with turbulent effects, since in the vicinity of the walls, these molecular effects are predominant. Therefore, in the present model, there is the presence of molecular and turbulent effects simultaneously.

7.7.4 Numerical Model for Solving the Differential Model

Integrating Eq. (7.298) in the y direction yields:

$$\frac{1}{\rho}\frac{\mathrm{d}\overline{p}}{\mathrm{d}x}y = \nu\frac{\mathrm{d}\overline{u}}{\mathrm{d}y} - L^2R^2\left(\frac{\mathrm{d}\overline{u}}{\mathrm{d}y}\right)^2 + c_1. \tag{7.300}$$

Applying the second boundary condition of Eq. (7.299), $c_1 = 0$ is obtained. By isolating the term $\mathrm{d}\overline{u}/\mathrm{d}y$, the following equation is obtained:

$$\frac{d\overline{u}}{dy} = \frac{1-\sqrt{1-4\frac{L^2R^2}{\nu^2\rho}\frac{d\overline{p}}{dx}y}}{2\frac{L^2R^2}{\nu}}. \tag{7.301}$$

Equation (7.301) is characterized by a singularity, i.e., $L(\tilde{y}=0)=0$. With some algebraic operations, this singularity can be removed. Multiplying and dividing it by $\left\{1+\left[1-4\left(L^2R^2\right)/\left(\nu^2\rho\right)(\mathrm{d}p/\mathrm{d}x)y\right]^{1/2}\right\}$, results in the following equation, in which the singularity has been removed:

$$\frac{\mathrm{d}\overline{u}}{\mathrm{d}y} = \frac{2y\frac{1}{\rho}\frac{\mathrm{d}\overline{p}}{\mathrm{d}x}}{\nu\left(1+\sqrt{1-4\frac{L^2R^2}{\nu^2\rho}\frac{\mathrm{d}\overline{p}}{\mathrm{d}x}y}\right)}. \tag{7.302}$$

This is an ordinary, first-order, homogeneous, nonlinear differential equation. It was not possible to get its continuous solution. It can be integrated numerically, using, for example, a 4th Order Runge–Kutta Method. Before integrating it, in order to make comparisons with DNS results, the dimensional ODE (7.302) was transformed into a non-dimensionalized ODE. The following reference quantities were used: R (length), $u_{\text{average}} = (1/R)\int_0^R \overline{u}(y)\mathrm{d}y$ or $u_\tau = \sqrt{|\tau_w|/\rho}$ (velocity), where τ_W is the shear stress on the wall, which, for the steady state, can be obtained by making a force balance on a fluid element of length $\mathrm{d}x$, which results in $\tau_W = -R\frac{\mathrm{d}p}{\mathrm{d}x}$. From this last equation, we obtain $(\tau_w/R) = -(\mathrm{d}p/\mathrm{d}x) = \left(\rho u_\tau^2/R\right)$. Thus, $\tilde{y} = y/R$, $\tilde{\overline{u}} = \overline{u}/\overline{u}_\tau$, $\mathrm{Re}_\tau = (u_\tau R)/\nu$. The symbol "~" represents a dimensionless *information*. The velocity u_τ was used as a reference, which gave rise to the Reynolds number. After non-dimensionalization and algebra, Eq. (7.302) was transformed into:

$$\frac{\partial\tilde{\overline{u}}}{\partial\tilde{y}} = \frac{-2\tilde{y}\mathrm{Re}_\tau}{1+\sqrt{1+4L^2\mathrm{Re}_\tau^2\tilde{y}}}. \tag{7.303}$$

Equation (7.303) is accompanied by the boundary condition:

$$\tilde{\overline{u}}(\tilde{y}=1) = 0. \tag{7.304}$$

It is useful to write the global Reynolds Number, $\mathrm{Re_D}$, in function of Re_τ, as follows:

$$\frac{\mathrm{Re_D}}{\mathrm{Re}_\tau} = \frac{2R\overline{u}_{\mathrm{average}}}{\nu}\frac{1}{\mathrm{Re}_\tau} = \frac{2R\overline{u}_{\mathrm{average}}}{\nu}\frac{\nu}{u_\tau R} = \frac{2}{R}\int_0^R \frac{\overline{u}}{u_\tau}\mathrm{d}y$$
$$= 2\int_0^1 \tilde{\overline{u}}\mathrm{d}\tilde{y} \Rightarrow \mathrm{Re}_D = 2\mathrm{Re}_\tau \int_0^R \tilde{\overline{u}}\mathrm{d}\tilde{y}, \tag{7.305}$$

where the characteristic length for the global Reynolds Number is $D = 2R$. With Eq. (7.303), the viscous stress on the wall is continuously modeled. On the other hand, as already mentioned, it was not possible to obtain a continuous solution for the filtered velocity. Therefore, Eq. (7.303) was integrated using the four-step 5eme Runge–Kutta method, *RK4*. Results are presented below.

7.7.5 Continuous Mathematical Model for the Component τ_{xy} of the Boussinesq–Reynolds Tensor

The component $\tau_{xy}(y)$ of the Boussinesq–Reynolds stress tensor was modeled using the Prandtl mixture length theory:

$$\tau_{xy}(y) = \rho\overline{u'\vartheta'} = \rho\ell_m^2\left(\frac{\mathrm{d}\overline{u}}{\mathrm{d}y}\right)^2. \tag{7.306}$$

The wall stress is modeled using:

$$\tau_w = \rho u_\tau^2 \tag{7.307}$$

Dividing Eq. (7.306) by Eq. (7.307) results in:

$$\frac{\tau_{xy}}{\tau_w} = \frac{\rho\ell_m^2(\mathrm{d}\overline{u}/\mathrm{d}y)^2}{\rho u_\tau^2}\frac{R^2}{R^2} = L^2(\tilde{y})\left(\mathrm{d}\tilde{\overline{u}}/\mathrm{d}\tilde{y}\right)^2. \tag{7.308}$$

Since Eq. (7.303) represents the continuous solution for $\mathrm{d}\tilde{\overline{u}}/\mathrm{d}\tilde{y}$, Eq. (7.308) is the continuous solution for the Reynolds Stress, normalized by the wall stress. This stress ratio will vary from zero in the center of the channel to a maximum value close to the wall. This stress represents turbulence activity and, therefore, must be zero on the wall. Illustrative results will be presented below.

7.7.6 Proposed Model Results Compared with DNS Results

In Fig. 7.3, the Prandtl Mixing Length is presented with and without the damping function, showcasing that there is damping close to the wall, which causes the mixing length to tend to zero more quickly when compared to the model without damping. In other words, the damping function modifies the derivative $d\widetilde{\overline{u}}/d\tilde{y}$ in the proximity of the walls. The effect of the damp function might seem irrelevant, but results will be presented showing that it is, in fact, very important.

Solving the ordinary differential model, Eqs. (7.303) and (7.304), and taking into account Eq. (7.297) to calculate the length $L(\tilde{y})$, velocity profiles were obtained for four values of the Reynolds Number, Re_τ. The results obtained with and without the use of the damping function were compared with the results obtained using DNS, which were taken from Johns Hopkins Turbulence Database (JHTDB) [9] and ICES Turbulence Database [8]. Starting with values $Re_\tau = 180$ and $Re_\tau = 395$, that are equivalent, respectively, to $Re_D = 5660$ and $Re_D = 14{,}219$, the results are presented in Fig. 7.27. The great influence of the damping function can be observed. An excellent agreement is obtained between results of the Prandtl mixture length closure model's solution and the results obtained using DNS. This agreement is better for $Re_\tau = 395$.

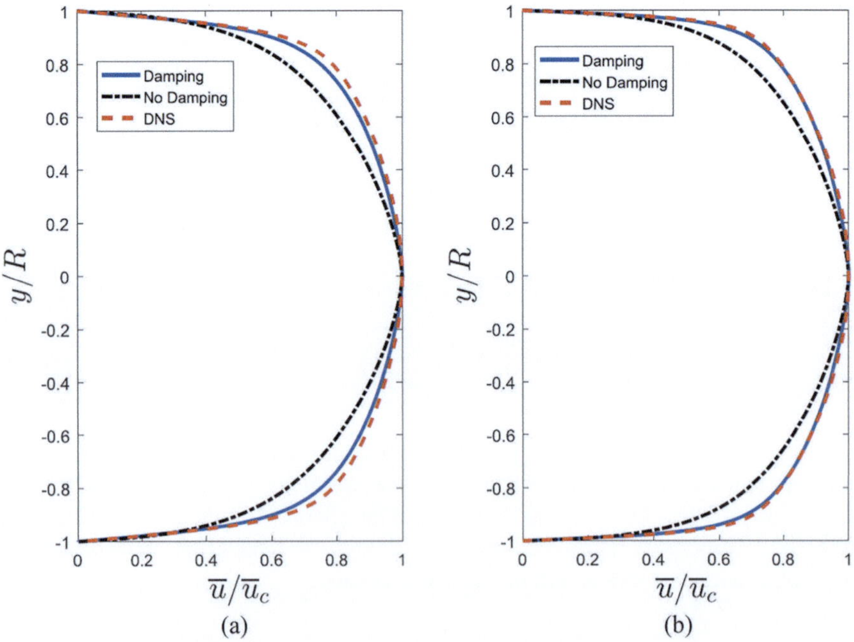

Fig. 7.27 Comparison $\overline{u}/\overline{u}_c$ profiles, obtained with and without the use of the damping function: **a** $Re_\tau = 180$ or $Re_D = 5660$ and **b** $Re_\tau = 395$ or $Re_D = 14{,}219$. Figure taken from Antonialli and Silveira-Neto [2], personal collection

In Fig. 7.28, results relating to $Re_\tau = 590$ and $Re_\tau = 1000$, which correspond, respectively, to $Re_D = 22{,}521$ and $Re_D = 40{,}950$. An excellent agreement between the results is observed and this agreement becomes visibly better for higher values of the Reynolds Number. This behavior is in line with expectations since, according to the theory on which the turbulence closure model that was used here is based, high values of the Reynolds Number are assumed. As this is a qualitative assessment, the norms L_2 and L_∞ were calculated, with which the differences between the results obtained with the Prandtl mixing length model and the DNS results are evaluated quantitatively. In Tables. 7.3 and 7.4, the obtained values are presented. It can now be observed quantitatively that the higher the Reynolds Number value, the smaller the differences between the results. The importance of the damping function is also clear. This can be seen from the two cases presented.

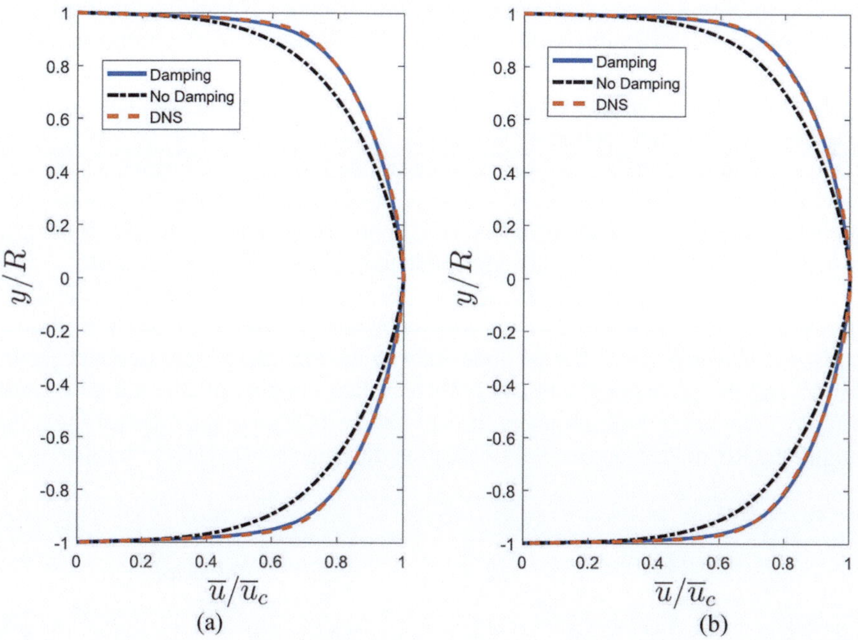

Fig. 7.28 Comparison of profiles $\overline{u}/\overline{u}_c$, obtained with and without the use of the damping function: **a** $Re_\tau = 590$ or $Re_D = 22{,}521$ and **b** $Re_\tau = 1000$ or $Re_D = 40{,}950$. Figure taken from Antonialli and Silveira-Neto [2], personal collection

Table. 7.3 Norm L_2 for the velocity profile, comparing the Prandtl mixing length model with DNS for various values of the Reynolds Number

Re_τ	180	395	590	1000
No damping function	0.0792	0.0684	0.0673	0.0567
With damping function	0.0200	0.0089	0.0063	0.0043

Table. 7.4 Norm L_∞ for the velocity profile, comparing the Prandtl Mixing Length Model with DNS for various values of the Reynolds Number

Re_τ	180	395	590	1000
No damping function	0.1511	0.1557	0.1566	0.1567
With damping function	0.0429	0.0295	0.0257	0.0212

The turbulent stresses, τ_{xy}/τ_w, are determined continuously using Eq. (7.308). The results are presented in Figs. 7.29 and 7.30. The Reynolds Number values are the same as those presented for the velocity distributions. An excellent agreement can be seen with the DNS results and again, the agreement becomes better as the values of the Reynolds Number increase. Another consistent finding obtained is that the higher the Reynolds Number value, the greater the maximum value of the turbulent stress distribution. Furthermore, the higher the Reynolds Number value, the closer to the wall is the maximum of the turbulent stress distribution. The turbulence itself transports all types of information in the direction transverse to the channel and towards the walls. This explains the fact that the velocity profiles become flatter as the flows become more and more turbulent.

It is important to understand how a simple modeling method can represent so well such a complex physical problem. However, it is also important to emphasize that, while accuracy is gained when this model is applied to a specific problem, generality is lost. That is, this model is only applicable to a specific class of problems.

This is, again, a qualitative assessment. To make it quantitative, the L_2 and L_∞ norms were calculated based on the difference between the distribution of viscous stresses transverse to the channel obtained with the Prandtl Mixture Length Closure Model and those obtained with DNS. These values are presented in Tables 7.5 and 7.6. By comparing these numbers, it can now be observed quantitatively that the higher the Reynolds Number, the smaller the differences between the results.

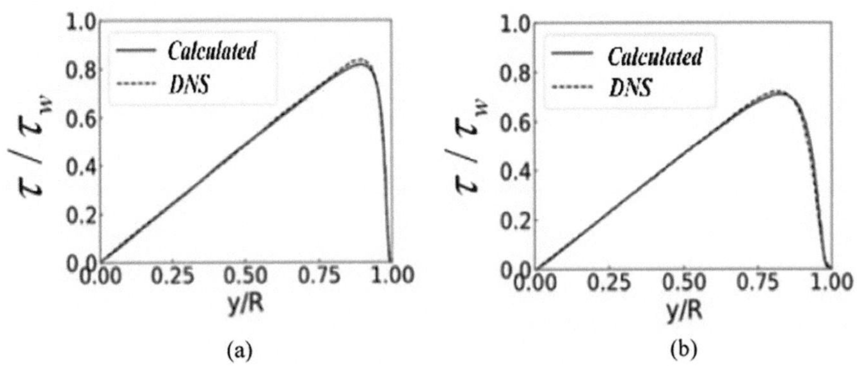

Fig. 7.29 τ_{xy}/τ_w—Profile comparison: **a** $Re_\tau = 180$ or $Re_D = 5660$ and **b** $Re_\tau = 395$ or $Re_D =$ 14,219. Figure taken from Antonialli and Silveira-Neto [2], personal collection

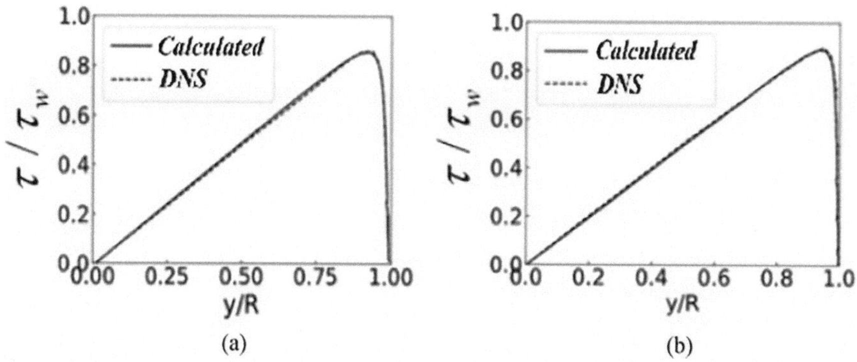

Fig. 7.30 τ_{xy}/τ_w—Profile comparison: **a** $Re_\tau = 590$ or $Re_D = 22{,}521$ and **b** $Re_\tau = 1000$ or $Re_D = 40{,}950$. Figure taken from Antonialli and Silveira-Neto [2], personal collection

Table. 7.5 Norm L_2 for the turbulent stress profile, comparing the Prandtl mixing length model and DNS, for various values of the Reynolds Number

Re_τ	180	395	590	1000
With damping function	0.0146	0.0095	0.0096	0.0075

Table. 7.6 Norm L_2 for the turbulent stress profile, comparing Prandtl mixing length model and DNS, for various values of the Reynolds Number

Re_τ	180	395	590	1000
With damping function	0.0644	0.0415	0.0380	0.0398

The problem that was proposed and solved above does not necessarily have any practical use since flows in channels and ducts have been studied experimentally for several decades, with several published results, including those that consider the effect of wall roughness. The main objective of the present study is educational, where we seek to demonstrate the resolution of one of the most used models for teaching turbulent flows. The present study can help students interested in this scientific area to visualize the complete solution of a turbulent flow, for which the solution in the laminar regime is known from the first contact with fluid mechanics courses. Another very important aspect is that the method presented can be used to validate numerical and computational models. The presented model can be used quickly as a validation reference. To date, the global scientific community has determined the rate of convergence of numerical methods using synthesized solutions for problems of a purely mathematical nature. With the model proposed in the present work, it is possible to determine the convergence rate of numerical and computational methods for a real turbulent flow. The exact solution for turbulent stress can be used for this purpose. Here is a suggestion for the student and/or advanced level reader: try developing this study for turbulent flows in circular conduits. The treatment is analogous, with the difference that the problem must be treated in a cylindrical coordinate system.

7.8 Theoretical Analysis of Plane Poiseuille Flows with Thermal Effects

7.8.1 Preliminary

The theoretical analysis of turbulent flows is a complex task, as explained at various points in this text. This complexity was described, for example, by Sommerer [22]. Simplifications in the full models can lead to important analytical capabilities. In this section, a one-dimensional RANS model (Boussinesq-Reynolds Averaged Navier–Stokes equations) is presented for analyzing turbulent flows with thermal effects in plane Poiseuille channels [14]. To model the temperature field, the thermal energy balance equation is used. The model for fluid dynamics, which has already been presented and validated in the previous item, is required.

7.8.2 Physical Model

Thermal effects were added to the physical model presented for fluid dynamics, Fig. 7.26. The same conditions adopted by Kawamura [10], to perform DNS of this problem were considered in the theoretical model to be presented. Kawamura's [10] results will be used as a reference for the development of this item. Three types of boundary conditions can be used to model this problem: the first type, or Dirichlet, for which the temperature is imposed on both walls; the second type, or Neumann, for which the thermal flow is imposed on the walls; and the third type, or Robin, for which an energy balance is carried out on the walls. In any of these three types of boundary conditions, the fluid will heat up or cool down as it flows within the channel. The flow regime depends on the operating conditions, for example, depending on the sign of the thermal energy flux.

It is very difficult to perform an experiment for which the fluid temperature remains constant along the axial direction of the channel. Therefore, it is not plausible to assume the condition of thermally developed flow, as was done for fluid dynamics. However, as proposed by Kawamura [10] and detailed by Ribeiro and Silveira-Neto [19], for the case of constant thermal flux imposed on the walls, it is observed that the temperature in the fluid and the wall will change linearly along the axial direction. This finding is analogous to fluid dynamics, for which, if the viscous stress on the wall is constant, the pressure in the fluid will vary linearly with the axial direction. This way, the temperature will increase linearly along the axial direction, but the temperature difference between the fluid and the wall will be constant with the axial direction.

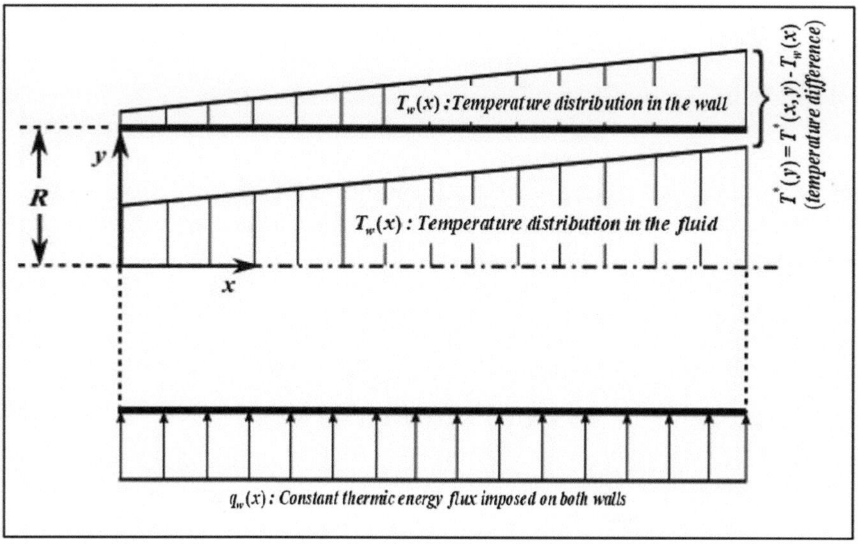

Fig. 7.31 Physical model of turbulent flow developed between two flat plates with thermal effects: temperature distributions

This temperature difference will be modeled here by: $T^*(y) = T(x, y) - T_w(x)$, where $T(x, y)$ is the temperature of the fluid, in two-dimensional flows, and $T_w(x)$ is the wall temperature. The temperature distributions are illustrated in Fig. 7.31. Through this physical analysis of the problem, the corresponding mathematical model can be established.

7.8.3 Differential Mathematical Model

The model for fluid dynamics was presented and validated in Sect. 7.7. Furthermore, the thermal model, showcased in Fig. 7.31, will be presented. The starting point is the thermal energy balance equation, written below, considering steady state and negligible specific viscosity transformation power. The temporal average and decomposition of the temperature field were taken, as presented in Chap. 6, Eq. (6.46):

$$\frac{\partial}{\partial x_j}(\overline{u}_j \overline{T}) = \frac{\partial}{\partial x_j}\left(\alpha \frac{\partial \overline{T}}{\partial x_j} - \overline{u_j' T'}\right), \tag{7.309}$$

where $q_j = \overline{u_j' T'}$ is the turbulent flux of thermal energy. Replacing $T^*(y) = T(x, y) - T_w(x)$ in Eq. (7.309), results in:

$$\frac{\partial}{\partial y}\left(\alpha\frac{\partial \overline{T^*}}{\partial y}-\overline{\vartheta' T^{*\prime}}\right)=\overline{u}\frac{\mathrm{d}\overline{T}_w}{\mathrm{d}x}. \tag{7.310}$$

The axial temperature gradient in the wall, $\mathrm{d}T_w/\mathrm{d}x$, is a constant, but is not known. Its modeling will be developed later. The turbulent flow of thermal energy, $\overline{\vartheta' T'}$, is part of the turbulence closure problem and must be modeled. It will be modeled via Boussinesq's hypothesis [3] and the concept of turbulent thermal diffusion coefficient will be used:

$$\begin{aligned} -\overline{\vartheta' T^{*\prime}} &= \alpha_t\frac{\partial \overline{T^*}}{\partial y},\\ -\overline{u' T^{*\prime}} &= \alpha_t\frac{\partial \overline{T^*}}{\partial x}=0. \end{aligned} \tag{7.311}$$

Replacing Eq. (7.311) in Eq. (7.310) yields:

$$\frac{d}{\mathrm{d}y}\left[(\alpha+\alpha_t)\frac{\mathrm{d}\overline{T^*}}{\mathrm{d}y}\right]=\overline{u}\frac{\mathrm{d}\overline{T}_w}{\mathrm{d}x}. \tag{7.312}$$

The turbulent diffusivity α_t can be evaluated with the concept of the Turbulent Prandtl Number, $\mathrm{Pr}_t=\nu_t/\alpha_t$. The turbulent viscosity ν_t was modeled in the previous section. The Turbulent Prandtl Number was adjusted by Ribeiro and Silveira-Neto [19]. Three correlations were proposed: the first establishes the turbulent Prandtl Number as a function of Re_τ, $\mathrm{Pr}_t=f(\mathrm{Re}_\tau)$; the second is the Cebeci function, $A_{\mathrm{din}}=g(\mathrm{Re}_\tau)$, which adds dynamic damping on the velocity field, and lastly, a thermal damping for the temperature field $A_{\mathrm{ter}}=h(\mathrm{Re}_\tau)$. The last two functions replace the constant A that appears in Eq. (7.297). These correlations are presented below:

$$\begin{aligned} \mathrm{Pr}_t(\mathrm{Re}_t) &= 0.7213-2.4892\times10^{-10}\times\mathrm{Re}_t^3+3.6036\times10^{-7}\\ &\quad\times\mathrm{Re}_t^2+3.9721\times10^{-5}\times\mathrm{Re}_t^3,\\ A_{\mathrm{ter}}(\mathrm{Re}_t) &= \frac{\mathrm{Re}_t^{0.0395\ln(\mathrm{Re}_t)+4.6637}}{e^{5.6703}},\\ A_{\mathrm{ter}}(\mathrm{Re}_t) &= \frac{\mathrm{Re}_t^{0.045\ln(\mathrm{Re}_t)+5.2753}}{e^{0.6094}}, \end{aligned} \tag{7.313}$$

Using Boussinesq's hypothesis, Eq. (7.6), and the Prandtl mixing length model with the damping function, Eq. (7.297), it follows that:

$$\frac{d}{\mathrm{d}y}\left[\left(\alpha-\frac{R^2L^2(y)}{\mathrm{Pr}_t(\mathrm{Re}_\tau)}\frac{\mathrm{d}\overline{u}}{\mathrm{d}y}\right)\frac{\mathrm{d}T^*}{\mathrm{d}y}\right]=\overline{u}\frac{\mathrm{d}\overline{T}_w}{\mathrm{d}x}. \tag{7.314}$$

To model the right side of Eq. (7.314), an energy balance will be made over the length of the domain:

$$2q_w b\Delta x = \dot{m}C_p\left[\overline{T}_m(x+\Delta x) - \overline{T}_m(x)\right] = \dot{m}C_p\frac{\mathrm{d}\overline{T}_w}{\mathrm{d}x}\Delta x$$
$$= 2\rho\overline{u}_m C_p Rb\frac{\mathrm{d}\overline{T}_w}{\mathrm{d}x}\Delta x \Rightarrow \frac{\mathrm{d}\overline{T}_w}{\mathrm{d}x} = \frac{q_w}{\rho\overline{u}_m C_p R} \tag{7.315}$$

The index m represents the average quantities calculated in the transverse plane of the duct, with area $2Rb$. Substituting Eq. (7.315) in (7.314) results in:

$$\frac{d}{\mathrm{d}y}\left[\left(\alpha - \frac{R^2L^2(y)}{\mathrm{Pr}_t(\mathrm{Re}_\tau)}\frac{\mathrm{d}\overline{u}}{\mathrm{d}y}\right)\frac{\mathrm{d}\overline{T}^*}{\mathrm{d}y}\right] = \frac{\overline{u}}{\overline{u}_m}\frac{q_w}{\rho C_p R}. \tag{7.316}$$

To adimensionalize Eq. (7.316), the following variables are used: $\tilde{y} = (y/R)\mathrm{Re}_\tau$, $\tilde{\overline{u}} = (\overline{u}/\overline{u}_\tau)$, $\tilde{\overline{T}} = \left(\overline{T}/\overline{T}_\tau\right)$, $\mathrm{Re}_\tau = (\overline{u}_\tau R)/\nu$, $\mathrm{Pr}_\tau = \nu_t/\alpha_t$, $\mathrm{Pr} = \nu/\alpha$ and $\overline{T}_\tau = q_w/\left(\rho C_p\overline{u}_\tau\right)$, resulting in the Eq. (7.317):

$$\frac{d}{\mathrm{d}\tilde{y}}\left[\left(\frac{\mathrm{Re}_\tau}{\mathrm{Pr}_t} - \frac{L^2(\tilde{y})\mathrm{Re}_t^3}{\mathrm{Pr}\mathrm{Re}_\tau}\frac{\mathrm{d}\tilde{\overline{u}}}{\mathrm{d}\tilde{y}}\frac{\mathrm{d}\tilde{\overline{T}}^*}{\mathrm{d}\tilde{y}}\right) = \frac{\tilde{\overline{u}}}{\tilde{\overline{u}}_m}\right]. \tag{7.317}$$

The term on the right-hand side of Eq. (7.317) must be obtained from the computational solution of the dynamic model. The term $\mathrm{d}\tilde{\overline{u}}/\mathrm{d}\tilde{y}$ is known in its continuous form, given, as already presented, by:

$$\frac{\mathrm{d}\tilde{\overline{u}}}{\mathrm{d}\tilde{y}} = \frac{2\tilde{y}\frac{1}{\mathrm{Re}_\tau}}{1+\sqrt{1+4L^2(\tilde{y})\mathrm{Re}_\tau^2\tilde{y}}}. \tag{7.318}$$

The boundary conditions for Eq. (7.317) are given by:

$$\frac{\mathrm{d}\tilde{\overline{T}}^*}{\mathrm{d}\tilde{y}}(\tilde{y}=0) = 0\,\tilde{\overline{T}}^*(\tilde{y}=1) = 0 \tag{7.319}$$

This differential model has no continuous solution. It can be solved computationally, using a CDS (Central Difference Scheme) method for the thermal model and the fourth-order Runge–Kutta Method for the dynamic model.

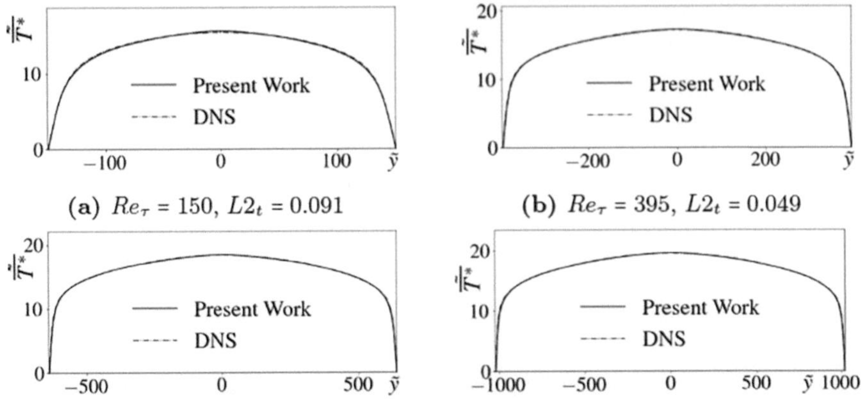

(a) $Re_\tau = 150$, $L2_t = 0.091$ (b) $Re_\tau = 395$, $L2_t = 0.049$

(c) $Re_\tau = 640$, $L2_t = 0.061$ (d) $Re_\tau = 1020$, $L2_t = 0.076$

Fig. 7.32 Comparison of the results obtained with the presented model and Kawamura's DNS results for four values of the Reynolds Number. DNS results taken from Kawamura's DNS results, Kawamura [10] and Kawamura [11]

7.8.4 Results for Temperature

Kawamura's DNS results, Kawamura [10] and Kawamura [11], were compared with the results of the approximate model, proposed by Ribeiro and Silveira-Neto [19]. The Turbulent prandtl number, Pr_t, was adjusted to various values of Re_t, as shown in Eq. (7.313). It was shown that the simulation results are very sensitive to small variations in this parameter. In this text, only the results for $\mathrm{Re}_\tau = 1020$, which correspond to $\mathrm{Re}_\mathrm{R} = 41{,}441$, are presented. In Fig. 7.32, the temperature distributions obtained with the presented model are shown, comparing them with the DNS results.

The error made when using the present model, when compared with DNS, was evaluated by calculating the norm L_2, whose valueobtained are of the order of 0.1%.

It is not viable to use the Prandtl Mixing Length Model for solving turbulent flows that are more complex than those presented in this chapter, due to difficulties in evaluating the mixing length in complex flows. Therefore, following Prandtl's Mixing Length Theory, Kolmogorov and Prandtl proposed the model of a balance equation, based on the concept of turbulent kinetic energy. Other more complete models were also developed. They will be presented in the next chapter.

7.9 Exercises

1. Once more, integrate Eq. 7.22 subjected to boundary conditions given by Eq. 7.23 to obtain the solution expressed in Eq. 7.31. Determine the constant $a_{o\ell}$. Using a CFD platform, model this problem computationally, simulate the laminar regime and compare it with the continuous solution for two different times.
2. Use Eqs. 7.31 and 7.53 to plot the molecular viscous stress between two layers of fluid around $y = 0$. Additionally, plot the component of the Reynolds tensor $\overline{u'\vartheta'}$ using Eq. 7.58. Compare these stresses in laminar and turbulent regimes.
3. Develop the simplified equation for a flat jet in laminar regime and in temporal development. Use the self-similarity theory. Determine the continuous solution and compare it with computational data available in the literature.
4. Develop the simplified equation for a turbulent flat jet in temporal development. Use the Prandtl Mixing Length theory to model the closure of turbulence. Use the self-similarity theory for the equations. Determine the continuous solution and compare it with computational data available in the literature.
5. Develop a computer simulation of a flat jet in the laminar regime and compare its results with Eqs. 7.250 and 7.251.
6. Repeat the study presented in Sect. 7.7 for circular ducts.
7. Repeat the study presented in Sect. 7.8 for circular ducts.

References

1. Andrade, E.D.C.: The velocity distribution in a liquid-into-liquid jet. part 2: the plane jet. Proc. Phys. Soc. **51**, 784–793 (1939)
2. Antonialli, L.A., Silveira-Neto.: Theoretical study of fully developed turbulent flow in a channel, using Prandtl mixing length model. J. Appl. Math. Phys. **06**, 677–692 (2018)
3. Boussinesq, J.:, Theorie de l´coulement tourbillonnant et tumultueux des liquides dans les lits rectilignes a grande section. Fautiers-Villars, Paris, 2nd edn (1877)
4. Bradbury, L.: The structure of a self-preserving turbulent plane jet. In: Journal of Fluid Mechanics, vol. 23, n° 1, pp. 31–64. Cambridge University Press (1965)
5. Cebeci, T., Bradshaw, P.: Physical and Computational Aspects of Convective Heat Transfer. Springer-Verlag, Berlin-Heidelberg-New York-Tokyo (1984)
6. Germano, M.: A dynamic sub grid-scale eddy viscosity model. Phys. Fluids A Fluid Dyn. AIP **3**(7), 1760–1765 (1991)
7. Heskestad, G.: Hot-wire measurements in a plane turbulent jet. J. Appl. Mech. **32**(4), 721–750 (1965)
8. ICES Turbulence Database: DNS Database of the Institute for Computational Engineering and Sciences da University of Texas (2007). http://turbulence.ices.utexas.edu/MKM%20-%201999.html
9. Johns Hopkins Turbulence Database (JHTDB): *DNS* results of Johns Hopkins University (2016). http://turbulence.pha.jhu.edu/
10. Kawamura, A.H.: DNS of turbulence and heat transport in a channel flow with Different Reynolds and Prandtl numbers and boundary conditions. In: Turbulence, Heat and Mass Transfer 3, (Proceedings of the 3rd International Symposium on Turbulence, Heat and Mass Transfer). Yamazaki, Tokyo (2000)

11. Kawamura, H.: Direct numerical simulation database for turbulent channel flow with heat transfer. http://www.rs.tus.ac.jp/t2lab/db/index.html, Laboratory of Thermo-fluid dynamics, Department of Mechanical Engineering, Faculty of Science and Technology, Tokyo University of Science, Noda-Shi, Chiba-ken, Japan (2007)
12. Magalhães. G.M., e Silveira-Neto, A.: Soluções contínuas e discretas de escoamentos cisalhantes livres, dissertação de mestrado, FEMEC-UFU (2018)
13. Nikuradse, J.: Laws of turbulent flow in smooth pipes (English translation). NASA TT F-10: p. 359 (1966)
14. Poiseuille, J.: Recherches experimentales sur le mouvement des liquides dans des tubes de tres-petits diametres, a l´Academie Royale des Sciences de l´Institut de France, pp. 433–544 (1846)
15. Pope, S.B.: Turbulent Flows. Cambridge University Press (2000)
16. Prandtl, Z.A.: Bericht über Untersuchugen zur Ausgebildeten Turbulenz. As Angew. Math. Mech. **5**, 136–169 (1925)
17. Prandtl, Z.A.: Über ein neues Formelsystem für die ausgebildete Turbulenz. Nachr. Akad. Wiss. Göttingen Math-Phys. **K1**, 6–19 (1945)
18. Reichardt, H.: Gesetzmässigkeiten der frein turbulenz, VDI-Forschungsheft 414, 2nd edn (1951)
19. Ribeiro, F.J.O., Silveira-Neto: Theoretical analyze of turbulent channel flow with thermal effects—the influence of the turbulent Prandtl number. In: 17th Brazilian Congress of Thermal Sciences and Engineering, November 25th–28th, Águas de Lindoia, SP-Brazil (2018)
20. Schilichting, H.: Über das ebene Windschattenproblem. Diis. Göttingen 1930. Ing.-Arch. **1**, 533–571 (1930)
21. Schlichting, H.: Boundary-Layer Theory. McGraw-Hill Inc., New York (1968)
22. Sommerer, J.C., Ott, E.: Modeling two-dimensional flows chaos theory. J. Hopkins APL Tech. Dig. **2**, 193–202 (1997)
23. Tennekes, H., Lumley, J.L.: A first course in turbulence. [S.I.]: MIT Press (1972)
24. van Driest, E.R.: On turbulent flow near a wall. J. Aerospace Sci. **23**, 1007–1011 (1956)
25. White, F.M.: Mecânica dos Fluidos. McGraw Hill, Rio de Janeiro, Brasil (1999)
26. Wilcox, C.W.: Turbulence modeling for CFD, La Cañada, Ca: DCW Industries (1998)
27. Zijnen, H., Van der. B.: Measurements of the velocity distribution in a plane turbulent jet of air. Appl. Sci. Res. **7**(4), 256–276 (1958)

Chapter 8
URANS Closure Models Based on One or More Balance Equations

In this chapter, the URANS closure modeling is presented using one or more balance equations for the characteristic quantities of turbulence. In Sect. 8.1, an introduction is presented. In Sect. 8.2, a model composed of one balance equation is presented, called the $\boldsymbol{k-\ell}$ model. Still in this item, the $\boldsymbol{S-A}$ model is presented, also with a balance equation. In Sect. 8.3, some closure models for two balance equations are presented, including the $\boldsymbol{k-\varepsilon}$ model and the $\boldsymbol{k-\omega-SST}$ model. In Sect. 8.4, the ***RSM*** model is presented, which is composed of seventeen balance equations, including thermal effects. In Sect. 8.5, the ***IASM*** model is presented as well as the ***EASM*** model. In Sect. 8.6, a performance analysis of the ***URANS*** closure models is carried out. This chapter ends with a list of exercises.

8.1 Introduction

The closure models used for the ***URANS*** Methodology may or may not be directly based on the calculation of turbulent viscosity. All of them, however, serve to model the Boussinesq–Reynolds Tensor, which comes from the filtering process of the Navier–Stokes Equations. The same happens with turbulent flows, arising from the filtering of energy balance equations and concentrations of chemical species, or even concentrations of contaminants. The Boussinesq–Reynolds tensor and turbulent flux must be closed. When closure is based on the concept of turbulent viscosity, it is necessary to evaluate at least two characteristic quantities of turbulence, such as length and velocity. In Chap. 7, the Prandtl mixing length model was presented and exemplified with applications. This model, although working exceptionally well for specific flows, is not applicable to complex turbulent flows, due to the difficulty of empirically evaluating the characteristic velocity and length of turbulence. Following the proposal of Prandtl's mixing length model, Kolmogorov proposed a model based

A. da Silveira Neto, *Turbulent Flows*, https://doi.org/10.1007/978-3-031-89222-6_8

on the concept of turbulent kinetic energy, from which the characteristic velocity of turbulence could be obtained.

To solve increasingly complex turbulent flows, equally more complex and more complete models were developed. With these developments, researchers translated more and more physical aspects into mathematical models, enriching closure modeling. Thus, a whole family of models will be presented in this chapter, involving algebraic-differential models, composed of one or more balance equations, as well as purely algebraic models, which do not involve balance equations.

8.2 Closure Models Based on One Balance Equation

8.2.1 $k-\ell$ Closure Model

The authors of this proposal developed the first closure model for which the spatio-temporal evolution of turbulent flows is considered in a mathematically structured way. The boundary conditions that make up the mathematical models for turbulent flows are also considered. This model is known as a balance equation model, that is, the balance equation for turbulent kinetic energy k.

In the closure models of one balance equation, the aim is to calculate the characteristic quantities of turbulence, used to calculate turbulent viscosity, for example, turbulent kinetic energy, k, with which the characteristic velocity is calculated: $\hat{V} = \sqrt{k}$, where the turbulent kinetic energy is defined as $k = \left(\overline{u'^2} + \overline{\vartheta'^2} + \overline{w'^2}\right)/2$.

Remembering that $\nu_t = C\hat{V}\ell$, thus: $\nu_t = C_\mu\sqrt{k}\ell$, with $C_\mu = 0.09$, as reported by Davidson [7], among several other authors. This relationship was proposed by Kolmogorov and Prandtl. Prandtl suggested deriving a balance equation for k. The derivation of this equation will be done in detail in this text. To do so, the Navier–Stokes equations are rewritten for incompressible flows:

$$\frac{\partial u_i}{\partial t} + u_j\frac{\partial u_i}{\partial x_j} = -\frac{1}{\rho}\frac{\partial p}{\partial x_i} + \nu\frac{\partial^2 u_i}{\partial x_j \partial x_j} - \beta g_i(T - T_o), \tag{8.1}$$

where β is the volumetric expansion coefficient. Carrying out the decompositions $u_i = \overline{u}_i + u'_i$, $T = \overline{T} + T'$ and $p = \overline{p} + p'$, multiplying all terms by u'_i, and performing spatio-temporal filtering, we obtain:

$$\underbrace{\overline{u_i'\frac{\partial\left(\overline{u}_i+u_i'\right)}{\partial t}}}_{1}+\underbrace{\overline{u_i'\left(\overline{u}_j+u_j'\right)\frac{\partial}{\partial x_j}\left(\overline{u}_i+u_i'\right)}}_{3}=-\frac{1}{\rho}\underbrace{\overline{u_i'\frac{\partial}{\partial x_i}(\overline{p}+p')}}_{2}$$
$$+\underbrace{\overline{\nu u_i'\frac{\partial^2}{\partial x_j\partial x_j}\left(\overline{u}_i+u_i'\right)}}_{4}-\underbrace{g_i\beta\overline{u_i'\left(\overline{T}+T'-T_o\right)}}_{5}. \tag{8.2}$$

Manipulating each term of Eq. (8.2), using the filtering process properties $\overline{u_i'}=0$, $\overline{u}_i$ and $\overline{T}$ considered constant in relation to the filtering process, we have:

$$\overline{u_i'\frac{\partial\left(\overline{u}_i+u_i'\right)}{\partial t}}=\overline{u_i'}\frac{\partial\overline{u}_i}{\partial t}+\overline{u_i'\frac{\partial u_i'}{\partial t}}=0+\frac{1}{2}\frac{\partial\overline{u_i'u_i'}}{\partial t}=\frac{\partial k}{\partial t}, \tag{8.3}$$

$$\begin{aligned}\overline{u_i'\left(\overline{u}_j+u_j'\right)\frac{\partial}{\partial x_j}\left(\overline{u}_i+u_i'\right)}&=\overline{u_i'\overline{u}_j\frac{\partial}{\partial x_j}\left(\overline{u}_i+u_i'\right)}+\overline{u_i'u_j'\frac{\partial}{\partial x_j}\left(\overline{u}_i+u_i'\right)}\\&=\overline{u_i'\overline{u}_j\frac{\partial\overline{u}_i}{\partial x_j}}+\overline{u_i'\overline{u}_j\frac{\partial u_i'}{\partial x_j}}+\overline{u_i'u_j'\frac{\partial\overline{u}_i}{\partial x_j}}+\overline{u_i'u_j'\frac{\partial u_i'}{\partial x_j}}\\&=0+\overline{u}_j\frac{1}{2}\frac{\partial\overline{\left(u_i'u_i'\right)}}{\partial x_j}+\overline{u_i'u_j'\frac{\partial\overline{u}_i}{\partial x_j}}+\frac{1}{2}\overline{\frac{\partial\left(u_i'u_i'u_j'\right)}{\partial x_j}}\\&=\overline{u}_j\frac{\partial k}{\partial x_j}+\frac{\partial\overline{u}_i}{\partial x_j}\overline{u_i'u_j'}+\frac{1}{2}\frac{\partial\overline{\left(u_i'u_i'u_j'\right)}}{\partial x_j},\end{aligned} \tag{8.4}$$

$$\overline{u_i'\frac{\partial}{\partial x_i}(\overline{p}+p')}=\overline{u_i'}\frac{\partial\overline{p}}{\partial x_i}+\frac{\partial\left(\overline{u_i'p'}\right)}{\partial x_i}=0+\frac{\partial\left(\overline{u_i'p'}\right)}{\partial x_i}, \tag{8.5}$$

$$\begin{aligned}\overline{u_i'\frac{\partial^2}{\partial x_j\partial x_j}\left(\overline{u}_i+u_i'\right)}&=\overline{u_i'\frac{\partial^2\overline{u}_i}{\partial x_j\partial x_j}}+\overline{u_i'\frac{\partial^2u_i'}{\partial x_j\partial x_j}}\\&=0+\overline{u_i'}\frac{\partial^2\overline{u}_i}{\partial x_j\partial x_j}+\overline{\frac{\partial}{\partial x_j}\left(u_i'\frac{\partial u_i'}{\partial x_j}\right)}-\overline{\frac{\partial u_i'}{\partial x_j}\frac{\partial u_i'}{\partial x_j}}\\&=\frac{\partial}{\partial x_j}\left(\frac{\partial\left(\overline{u_i'u_i'}/2\right)}{\partial x_j}\right)-\overline{\frac{\partial u_i'}{\partial x_j}\frac{\partial u_i'}{\partial x_j}}=\frac{\partial^2k}{\partial x_j\partial x_j}-\overline{\frac{\partial u_i'}{\partial x_j}\frac{\partial u_i'}{\partial x_j}},\end{aligned} \tag{8.6}$$

$$\begin{aligned}g_i\beta\overline{u_i'\left(\overline{T}+T'-T_o\right)}&=g_i\beta\left[\overline{u_i'\overline{T}}+\overline{u_i'T'}-\overline{u_i'T_o}\right]\\&=g_i\beta\left[\overline{u_i'}\,\overline{T}+\overline{u_i'T'}-\overline{u_i'}T_o\right]=g_i\beta\overline{u_i'T'}.\end{aligned} \tag{8.7}$$

Replacing Eqs. (8.4) to (8.7) in Eq. (8.2), results in:

$$\underbrace{\underbrace{\frac{\partial k}{\partial t}}_{\substack{\text{Rate of change of turbulent kinetic energy} \\ k\text{, contained in a fluid particle}}}}_{\substack{\text{Process of temporal accumulation/} \\ \text{decay of the information}}}$$

$$= \underbrace{\underbrace{-\overline{u}_j \frac{\partial k}{\partial x_j}}_{\substack{\text{Advective net flux of turbulent kinetic} \\ \text{energy}, k\text{, through the fluid particle} \\ \text{surface}}} + \underbrace{\nu \frac{\partial^2 k}{\partial x_j \partial x_j}}_{\substack{\text{Molecular diffusive net} \\ \text{flux of } k \text{ across the fluid} \\ \text{particle surface}}} \underbrace{- \frac{\partial}{\partial x_j}\left(\frac{1}{2}\overline{u_i' u_i' u_j'} + \frac{1}{\rho}\overline{u_i' p'}\delta_{ij}\right)}_{\substack{\text{Turbulent diffusive net flux} \\ \text{of } k\text{, across the surface} \\ \text{of the fluid particle}}}}_{\text{Advective and diffusive transport processes of } k}$$

$$\underbrace{\underbrace{- \overline{u_i' u_j'}\frac{\partial \overline{u}_i}{\partial x_j}}_{\substack{\text{Transformation of mean} \\ \text{flow kinetic energy into} \\ \text{turbulent kinetic energy } k}} \underbrace{- \nu \overline{\frac{\partial u_i'}{\partial x_j}\frac{\partial u_i'}{\partial x_j}}}_{\substack{\text{Transformation of turbulent} \\ \text{kinetic energy } k \text{ into} \\ \text{thermal energy}}} \underbrace{- g_i \beta \overline{u_i' T'}}_{\substack{\text{Transformation of gravitational} \\ \text{potential energy into turbulent} \\ \text{kinetic energy } k \text{ and vice versa}}}}_{\text{Transformation processes involving turbulent kinetic energy } k \text{ into other information and vice versa}} \tag{8.8}$$

This is a balance equation, not a transport equation as we usually find in the literature, let alone a conservation equation! In fact, we find other terms in it besides transportation. Accumulation rate term and transformative terms. All these processes: accumulation rate, transport and transformations, make up the balance of information under analysis, in this case, turbulent specific kinetic energy.

Thus, this equation can be rewritten in a general way as being composed of only three types of physical processes: the rate of variation of the information contained in a fluid particle; the process of transporting information and the process of transforming information into other information. The transport process is divided into two in the present model: diffusive and advective. The transformation process is divided into the transformation of average kinetic energy into turbulent kinetic energy, the transformation of turbulent kinetic energy into thermal energy and the process of transforming potential energy into turbulent kinetic energy, and vice versa. To physically interpret each of the terms in Eq. (8.8), it is rewritten below, naming each term:

$$\underbrace{\frac{\partial k}{\partial t}}_{1\Leftrightarrow T^k} + \underbrace{\overline{u}_j \frac{\partial k}{\partial x_j}}_{2\Leftrightarrow A^k} = \underbrace{-\overline{u_i' u_j'}\frac{\partial \overline{u}_i}{\partial x_j}}_{3\Leftrightarrow P^k} \underbrace{- \frac{\partial}{\partial x_j}\left(\frac{1}{2}\overline{u_i' u_i' u_j'} + \frac{1}{\rho}\overline{u_i' p'}\delta_{ij}\right)}_{4\Leftrightarrow D^{tk}}$$
$$\underbrace{- \nu \overline{\frac{\partial u_i'}{\partial x_j}\frac{\partial u_i'}{\partial x_j}}}_{3\Leftrightarrow P^k} + \underbrace{\nu \frac{\partial^2 k}{\partial x_j \partial x_j}}_{6\Leftrightarrow D^{mk}} \underbrace{- g_i \beta \overline{u_i' T'}}_{7\Leftrightarrow G^k}. \tag{8.9}$$

This is the turbulent kinetic energy balance equation, in which each of the seven terms represents a different physical effect, as specified below:

T^k: Rate of change of turbulent kinetic energy contained in a fluid particle;

A^k: Net advective flux of turbulent kinetic energy across the control surface surrounding the fluid particle;

P^k: Transformation of average kinetic energy into turbulent kinetic energy. This term can be positive or negative, representing both the energy that leaves the turbulence for the mean field, and that leaving the mean field for the turbulence. This term will be referred to in this text as "kinematic transformation", instead of "production", as used in the literature in general, due to the physical inconsistency of the term "production". Energy cannot be "produced";

D^{tk}: Turbulent diffusive net flux of turbulent kinetic energy through the control surface surrounding a fluid particle. This transport process results from fluctuations in speed and pressure;

ε^k: Specific power of viscous transformation of turbulent kinetic energy into thermal energy, inside a fluid particle. This physical effect has unity [W/kg]. In this text, it will be called the specific power of viscous transformation or, simply, viscous transformation instead of "viscous dissipation", due to the multiple meanings of the word "dissipation". Energy cannot be "dissipated", in the sense of disappearing, like a cloud that "dissipates". Energy can only be accumulated, transformed or transported;

G^k: Specific power of transformation [W/kg] of gravitational potential energy into turbulent kinetic energy and vice versa. This term will be called "gravitational transformation" instead of "generation", used in the literature. Energy cannot be "generated".

Note, for example, that when deducing a balance equation for a second-order statistical moment, as is the case with turbulent kinetic energy, third-order statistical moments appear, which are terms 4 of Eq. (8.9), which are part of the turbulence closure problem. To complete this equation, it is necessary to model these third-order correlations. It is worth noting that term (3) is already modeled, since it contains the Boussinesq–Reynolds tensor, which is modeled with the Boussinesq hypothesis. The viscous transformation, ε^k, must also be modeled.

The first to propose a closure model for this equation was Prandtl [40]. The natural tendency indicates that, by analogy with the molecular diffusive process, any correlation of the type $-\overline{u_j'\phi'} \approx \nu_t \partial\overline{\phi}/\partial x_j$, where $\overline{\phi}$ represents the filtered part of the function ϕ, whose fluctuation ϕ', is being transported by the flow. In the absence, for example, of material experiments, Direct Computational Simulation [26] has been used as a tool to indicate that the moment involving pressure fluctuation can also be modeled in a similar way. Researchers have pointed out, through vast literature, that:

$$-\left(\frac{1}{2}\overline{u_i'u_i'u_j'} + \frac{1}{\rho}\overline{p'u_j'}\right) = \frac{\nu_t}{\sigma_k}\frac{\partial k}{\partial x_j},$$

$$
-\overline{u_i'u_j'} = \nu_t\left(\frac{\partial u_i}{\partial x_j} + \frac{\partial u_j}{\partial x_i}\right) - \frac{2}{3}k\delta_{ij},
$$

$$
-\overline{u_i'T'} = \alpha_t \frac{\partial \overline{T}}{\partial x_i}. \tag{8.10}
$$

where σ_k is a constant to be adjusted, depending on the method of solving the equations and the type of flow to be simulated. The function α_t is calculated with the assumption of a Turbulent Prandtl Number: $\mathrm{Pr}_{\overline{u_j'T'}} = \nu_t/\alpha_t \Rightarrow \alpha_t = \nu_t/\mathrm{Pr}_{\overline{u_j'T'}}$. The function $\mathrm{Pr}_{\overline{u_i'T'}}$, in accordance with the literature, varies in the range $0.7 \leq \mathrm{Pr}_{\overline{u'T'}} \leq 0.9$. Ribeiro and Silveira-Neto [41] concluded that the best adjusted value for this parameter is $\mathrm{Pr}_{\overline{u'T'}} = 0.91$ for turbulent flows in a flat channel with thermal energy transfer. However, these authors, Ribeiro and Silveira-Neto [41], extended this study and proposed correlations for the Turbulent Prandtl Number $\mathrm{Pr}_{\overline{u_i'T'}}$, for the dynamic and thermal Cebeci functions, as presented in Chap. 7, Eq. (7.313).

The term with which the specific power of viscous transformation of turbulent kinetic energy is modeled, ε^k, must also be closed. It is noted that, overall, there are still two unknowns: ε^k and the characteristic length ℓ. Taylor [52], using Kolmogorov dimensional analysis, proposed the relationship $\varepsilon^k = c_D k^{3/2}/\ell$. The constant c_D must be determined empirically. Note that in this proposal, it is assumed that ε^k and ℓ do not depend on any physical property of the fluid. It remains to be determined what the characteristic length of turbulent flows.

Rewriting the balance equation for k, including closure terms, we have:

$$
\frac{\partial k}{\partial t} + \frac{\partial}{\partial x_j}(\overline{u}_j k) = \left[\nu_t\left(\frac{\partial \overline{u}_i}{\partial x_j} + \frac{\partial \overline{u}_j}{\partial x_i}\right) - \frac{2}{3}k\delta_{ij}\right]\frac{\partial \overline{u}_i}{\partial x_j} + \frac{\partial}{\partial x_j}\left[\left(\nu + \frac{\nu_t}{\sigma_k}\right)\frac{\partial k}{\partial x_j}\right]
$$

$$
- c_D\frac{k^{3/2}}{\ell} + g_i\beta\alpha_t\frac{\partial \overline{T}}{\partial x_i} \Leftrightarrow
$$

$$
\frac{\partial k}{\partial t} + \frac{\partial}{\partial x_j}(\overline{u}_j k) = \nu_t\left(\frac{\partial \overline{u}_i}{\partial x_j} + \frac{\partial \overline{u}_j}{\partial x_i}\right)\frac{\partial \overline{u}_i}{\partial x_j} + \frac{\partial}{\partial x_j}\left[\left(\nu + \frac{\nu_t}{\sigma_k}\right)\frac{\partial k}{\partial x_j}\right]
$$

$$
- c_D\frac{k^{3/2}}{\ell} + g_i\beta\alpha_t\frac{\partial \overline{T}}{\partial x_i}. \tag{8.11}
$$

In establishing Eq. (8.11), the incompressibility hypothesis was used:

$$
\frac{2}{3}k\delta_{ij}\frac{\partial \overline{u}_i}{\partial x_j} = \frac{2}{3}k\frac{\partial \overline{u}_i}{\partial x_i} = 0. \tag{8.12}
$$

The Boussinesq–Reynolds tensor $-\overline{u_i'u_j'}$ is already modeled, apart from the calculation of ℓ. The turbulent viscosity can be rewritten as follows: $\nu_t = k^{1/2}\ell = c_D k^2/\varepsilon$. This proposal was also made by Emmons [8] and, together with Glushko [13], proposed adjustments for the constants, $0.07 \leq c_D \leq 0.09$ and $\sigma_k = 1.0$. The mixing length was used by both.

Bobyleva et al. [2], following Von Kármán's ideas, they proposed the following model for calculating ℓ:

$$\ell = \sqrt{k^{3/2}/\left(\frac{\partial \psi}{\partial y}\right)}$$
$$\psi = \frac{k^{1/2}}{\ell}. \tag{8.13}$$

It is observed that a new variable was introduced, ψ, and one more equation. With this model, the auxiliary function ψ and the length ℓ are updated, as well as the k field. This can be done iteratively: given k, calculate ψ and ℓ, then recalculate k using the solution to its balance equation, iterations continue until convergence. A long discussion on the values of constants and about ℓ can also be found in Wilcox [54].

It can be observed that the filtered temperature has not yet been modeled. This requires filtering the balance equation for the thermal energy:

$$\overline{\frac{\partial T}{\partial t} + \frac{\partial \left(u_j T\right)}{\partial x_j}} = \alpha \frac{\partial^2 \overline{T}}{\partial x_j \partial x_j} \Rightarrow \frac{\partial \overline{T}}{\partial t} + \frac{\partial \left(\overline{u_j T}\right)}{\partial x_j}$$
$$= \alpha \frac{\partial^2 \overline{T}}{\partial x_j \partial x_j} \Rightarrow \frac{\partial \overline{T}}{\partial t} + \frac{\partial \left[\overline{\left(\overline{u}_j + u_j'\right)(\overline{T} + T')}\right]}{\partial x_j}$$
$$= \alpha \frac{\partial^2 \overline{T}}{\partial x_j \partial x_j} \Rightarrow \frac{\partial \overline{T}}{\partial t} + \frac{\partial \left(\overline{\overline{u}_j \overline{T}} + \overline{\overline{u}_j T'} + \overline{u_j' \overline{T}} + \overline{u_j' T'}\right)}{\partial x_j}$$
$$= \alpha \frac{\partial^2 \overline{T}}{\partial x_j \partial x_j} \Rightarrow \frac{\partial \overline{T}}{\partial t} + \frac{\partial \left(\overline{u}_j \overline{T}\right)}{\partial x_j} = \alpha \frac{\partial^2 \overline{T}}{\partial x_j \partial x_j} - \frac{\partial \left(\overline{u_j' T'}\right)}{\partial x_j}. \tag{8.14}$$

Thus, we have a balance equation for the filtered temperature. However, a new unknown vector variable emerged, the turbulent flux of thermal energy, $\overline{u_j'T'}$, which must be modeled. The model, based on the first gradient hypothesis and the definition of a turbulent thermal diffusion coefficient, α_t, was presented in Eq. (8.10), i.e., $-\overline{u_i'T'} = \alpha_t\left(\partial \overline{T}/\partial x_i\right)$. The modeled balance equation for thermal energy, in a turbulent regime, can then be rewritten as follows:

$$\frac{\partial \overline{T}}{\partial t} + \frac{\partial \left(\overline{u}_j \overline{T}\right)}{\partial x_j} = \alpha \frac{\partial^2 \overline{T}}{\partial x_j \partial x_j} + \frac{\partial}{\partial x_j}\left(\alpha_t \frac{\partial \overline{T}}{\partial x_i}\right)$$
$$= \frac{\partial}{\partial x_j}\left[(\alpha + \alpha_t)\frac{\partial \overline{T}}{\partial x_i}\right] \Leftrightarrow \frac{\partial \overline{T}}{\partial t} + \frac{\partial \left(\overline{u}_j \overline{T}\right)}{\partial x_j}$$
$$= \frac{\partial}{\partial x_j}\left[\alpha_{ef}\frac{\partial \overline{T}}{\partial x_i}\right], \tag{8.15}$$

where $\alpha_{ef} = \alpha + \alpha_t$.

Aiming for a better understanding of the model presented, a summary is made and a sequence of steps for solving the equations is presented. Further on, Sect. 8.4, a more complete closure model will be deduced, where balance equations for turbulent flux $\overline{u_i' u_j'}$ and $\overline{u_j' T'}$ will be presented, eliminating the direct use of the second and third lines of Eq. (8.10).

The system of differential equations that make up the model for non-isothermal turbulent flows is given by:

$$\frac{\partial \overline{u}_i}{\partial x_i} = 0, \tag{8.16}$$

$$\frac{\partial \overline{u}_i}{\partial t} + \frac{\partial}{\partial x_j}(\overline{u}_i \overline{u}_j) = -\frac{1}{\rho}\frac{\partial \overline{p^*}}{\partial x_i} + \frac{\partial}{\partial x_j}\left[\nu_{ef}\left(\frac{\partial \overline{u}_i}{\partial x_j} + \frac{\partial \overline{u}_j}{\partial x_i}\right)\right] - g_i \beta\left(\overline{T} - T_o\right), \tag{8.17}$$

$$\frac{\partial k}{\partial t} + \frac{\partial}{\partial x_j}\left(\overline{u}_j k\right) = \nu_t\left(\frac{\partial \overline{u}_i}{\partial x_j} + \frac{\partial \overline{u}_j}{\partial x_i}\right)\frac{\partial \overline{u}_i}{\partial x_j} + \frac{\partial}{\partial x_j}\left\{\left(\nu + \frac{\nu_t}{\sigma_k}\right)\frac{\partial k}{\partial x_j}\right\}$$
$$- c_D \frac{k^{3/2}}{\ell} + g_i \beta \alpha_t \frac{\partial \overline{T}}{\partial x_i}, \tag{8.18}$$

$$\frac{\partial \overline{T}}{\partial t} + \frac{\partial \left(\overline{u}_j \overline{T}\right)}{\partial x_j} = \frac{\partial}{x_j}\left[\alpha_{ef}\frac{\partial \overline{T}}{\partial x_j}\right]. \tag{8.19}$$

To complete the model, Eq. (8.13) is added. Below, we briefly propose a calculation sequence:

1. Initialize all variables;
2. With ν_t and $\overline{T}$ from the previous time, the equations for the filtered velocities and pressure are solved;
3. With ν_t in the previous time, the equation for k is solved;
4. The length ℓ is then calculated;
5. With k and ℓ recalculate ν_t and then calculate α_t;
6. The thermal energy equation is solved, and the filtered temperature is determined;
7. Return to step 2 and recalculate the velocity and pressure;
8. Verify the convergence of all variables;
9. Time is advanced.

Although this model has been important for the progress of modeling turbulence closure, many limitations are inherent to it for simulations of complex flows. Adjustment of two proportionality constants is required, c_D and σ_k. In the literature, a variation range is presented $0.3 \leq c_D \leq 0.8$ [31] and $\sigma_k = 1.0$ [7]. It also requires determination of the characteristic length ℓ of turbulence, which must be a function of time and space. Hence, the motivation to propose other models is to contemplate the evaluation of this characteristic length function or another equivalent function such as ε^k. Before presenting the model based on ε^k balance equation a new model based on one balance equation for $\tilde{\nu}$ will be presented in the next section.

8.2.2 *Closure Model Based on Turbulent Kinematic Viscosity,* $\tilde{\nu}$

Models with one balance equation have been proposed for different types of information, such as turbulent kinetic energy and turbulent kinematic viscosity itself. Researchers Nee and Kovasznay [33] proposed a model for a balance equation for turbulent viscosity, ν_t. It is a balance equation composed of conventional terms. The model also involves four constants to be adjusted. The major problem with this proposal is that it requires the evaluation of the characteristic length function of turbulence, which returns to the main problem of the previous model, based on the balance equation for turbulent kinetic energy. Following these developments, Sekundov [46] proposed another model, also based on a balance equation for ν_t. This model was widely used in Russian schools but had no repercussions in other countries.

These models served as the basis for other developments, such as the Baldwin and Barth Proposal [1]. They proposed a new version of the closure model, based on a balance equation for the Turbulent Reynolds Number. In this proposal, the authors made the model independent of the evaluation of a characteristic turbulence length. This was the greatest contribution of these authors. However, to achieve that, they had to write the open term of the viscous transformation of the related transported *information*, that is, the Turbulent Reynolds Number, as a function of the turbulent viscosity gradient, $\partial \nu_t / \partial x_i$. This fact caused inconsistencies in this model and also a diffusion of a purely numerical nature, which limits its application. Furthermore, this model requires the adjustment of seven constants and three damping functions.

Spalart and Allmaras [48] proposed a new version of the balance equation for $\tilde{\nu}$, whose closure does not require the evaluation of the characteristic velocity and characteristic length functions of the turbulence. This model was tested and is widely used by the scientific community. This model will be presented as follows in this book. Calculation of viscosity ν_t is done with the following equation:

$$\nu_t = \tilde{\nu} f_{\vartheta 1}. \tag{8.20}$$

The function $\tilde{\nu}$ is determined by solving the balance equation proposed by the authors:

$$\frac{\partial \tilde{\nu}}{\partial t}+U_j\frac{\partial \tilde{\nu}}{\partial x_j}=c_{b1}(1-f_{t2})\tilde{S}\tilde{\nu}-\left(c_{w1}f_w-\frac{c_{b1}}{\kappa^2}f_{t2}\right)\left(\frac{\tilde{\nu}}{d}\right)^2$$
$$+\frac{1}{\sigma}\frac{\partial}{x_k}\left[(\nu+\tilde{\nu})\frac{\partial \tilde{\nu}}{\partial x_k}\right]+\frac{c_{b2}}{\sigma}\frac{\partial \tilde{\nu}}{\partial x_k}\frac{\partial \tilde{\nu}}{\partial x_k}+f_{t1}\Delta U^2, \tag{8.21}$$

where

$$c_{w1}=\frac{c_{b1}}{\kappa^2}+\frac{(1+c_{b2})}{\sigma}, \tag{8.22}$$

$$f_{\vartheta 1}=\frac{\chi^3}{\chi^3+c_{\vartheta 1}^3}, \tag{8.23}$$

$$f_{\vartheta 2}=1-\frac{\chi}{1+\chi f_{\vartheta 1}}, \tag{8.24}$$

$$f_w=g\left[\frac{1+c_{w3}^6}{g^6+c_{w3}^6}\right]^{1/6}, \tag{8.25}$$

$$\chi=\frac{\tilde{\nu}}{\nu}, \tag{8.26}$$

$$g=r+c_{w2}\left(r^6-r\right), \tag{8.27}$$

$$r=\frac{\tilde{\nu}}{\tilde{S}\kappa^2 d^2}, \tag{8.28}$$

$$\tilde{S}=S+\frac{\tilde{\nu}}{\kappa^2 d^2}f_{\vartheta 2},\ \ S=\sqrt{2S_{ij}S_{ij}} \tag{8.29}$$

$$S_{ij}=\frac{1}{2}\left(\frac{\partial U_i}{\partial x_j}+\frac{\partial U_j}{\partial x_i}\right), \tag{8.30}$$

$$f_{t2}=c_{t3}\exp\left(-c_{t4}\chi^2\right),\ f_{t1}=c_{t1}g_t\exp\left[-c_{t2}\frac{w_t^2}{\Delta U^2}\left(d^2+g_t^2 d_t^2\right)\right],$$
$$g_t\equiv\min\left(0.1;\ \frac{\Delta U}{w_t\Delta x_t}\right), \tag{8.31}$$

The constants are:

$$c_{b1}=0.1355,\ c_{b2}=0.622,\ \sigma=2/3;\ \kappa=0.41;\ c_{w2}=0.3;\ c_{w3}=2;\ c_{\vartheta 1}=7.1;$$
$$c_{t1}=1.0;\ c_{t2}=2.0;\ c_{t3}=1.1;\ c_{t4}=2.0. \tag{8.32}$$

S_{ij} is the strain rate tensor and d being the distance to the nearest surface. The variable d_t represents the distance from a given cell to the transition position, from the laminar regime to the turbulent regime. The Spalart and Allmaras Model, presented above, makes it possible to introduce a correction to consider the boundary layer transition, which requires the use of four coefficients and two additional empirical functions. It is observed that the term $c_{w1}f_w(\tilde{\nu}/d)^2$ tends to zero as d increases. The term $(\partial\tilde{\nu}/\partial x_k)(\partial\tilde{\nu}/\partial x_k)$, function of the gradient $\tilde{\nu}$, cancels out at points far from the wall. The reader must bear in mind that the rule for the sum of repeated indices applies. For the region of potential flow, away from walls, it is also not possible to predict the viscosity $\tilde{\nu}$ decay. The authors of this proposal showed how the model works, comparing computational results with results from material experiments for distant wakes, spatial mixing layers, plane jets and circular jets.

It is worth noting the following: once the turbulent viscosity field has been calculated, the Boussinesq–Reynolds Tensor has been modeled without calculating the characteristic velocity and length of the turbulence. Hence, the turbulent kinetic energy can be determined using the trace of the tensor τ_{ij}: $k = (1/2)\tau_{ii}$. Having ν_t and k, the viscous transformation power ε^k is determined.

To demonstrate the functionalities of these models, the comparison of statistical quantities of several free shear flows, obtained computationally, using the two models presented in this item, is presented below. In particular, the spreading rates of free shear flows in spatial development will be presented.

For the case of a round jet, the concepts that lead to the definition of scattering rate will be presented here. A round jet issues from a nozzle, which injects the fluid into a free medium. Along the centerline of the jet the flow begins to develop in the form of a characteristic Gaussian. Finally, at about 20 diameters from the nozzle exit, the velocity profile reaches and maintains the self-similar shape. The value of this distance depends on the type of jet, whether it is plane or axisymmetric. The self-similar region begins at a "virtual" origin (z_o), as outlined in Fig. 8.1. A characteristic of circular jets is that the flux of linear momentum, in steady state, must remain constant on the planes normal to the axis of development. The same does not happen with the mass flow, due to the process of drawing mass from the environment into the jet [39]. This mechanism is known as entrainment and is represented in Fig. 8.1 by curved arrows.

Jet-type flows can be divided into three regions, taking the radial direction of the jet as reference: potential core, shear layer and region of purely shear flow. The axis system origin was placed as shown in Fig. 8.1. Along the jet axis of symmetry, i.e., the center line, the average (temporal) velocity is given by:

$$\overline{w}_c(z) = \overline{w}_c(0, 0, z). \tag{8.33}$$

The thickness of the shear layer, $\delta(z)$, can be viewed in Fig. 8.2. The half-width of the jet is also defined as:

$$\overline{w}\big(r = \delta_{1/2}(z), z\big) = \frac{1}{2}\overline{w}_c(z). \tag{8.34}$$

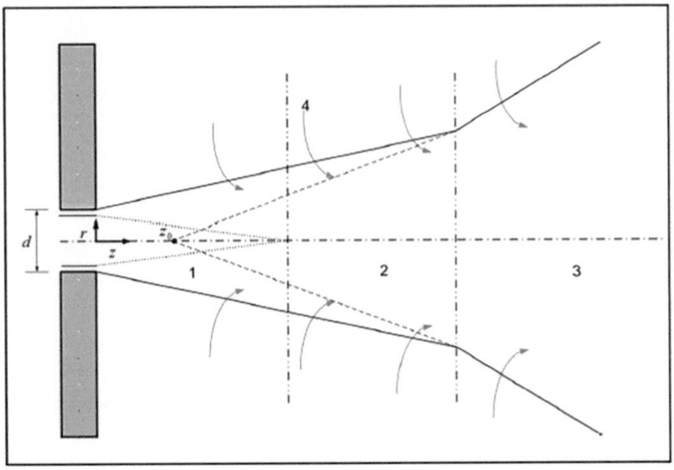

Fig. 8.1 Sketch of the development of a jet near the nozzle: (1) region containing a shear subregion and an irrotational subregion; (2) transition region and (3) purely shear region, without potential core

The virtual origin z_o is the point at which the self-similarity of the velocity profiles begins. This point, $(0, z_o)$, is illustrated in Fig. 8.1. The velocity at the center of the jet is inversely proportional to the axial position z:

$$\overline{w}_c(z) \sim z^{-1}. \tag{8.35}$$

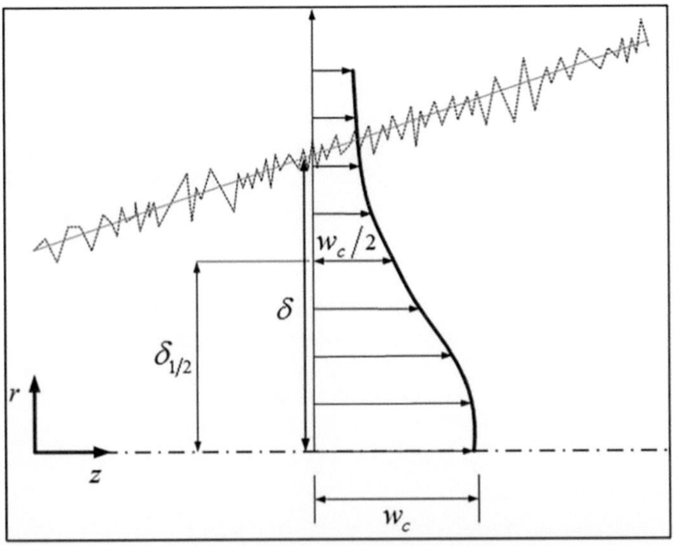

Fig. 8.2 Illustrative scheme of the shear layer and half-width in a turbulent round jet

Thus, the inverse of this velocity depends linearly on z. The inverse of this dimensionless velocity will also depend linearly on z:

$$\frac{\overline{w}_{\mathrm{J}}}{\overline{w}_{\mathrm{c}}(z)} = \frac{1}{a}\left(\frac{z}{d} - \frac{z_{\mathrm{o}}}{d}\right), \tag{8.36}$$

where the constant a is the velocity decay constant at the center of the jet, $\overline{w}_{\mathrm{c}}(z)$, d is the diameter of the nozzle and z_{o} is the coordinate of the virtual origin. Knowing the distribution $\overline{w}_{\mathrm{c}}(z)/\overline{w}_{\mathrm{J}}$, either by material experiment or by computational experimentation, the constants a and z_{o}/d can be determined.

From empiricism or computational experiments, we are aware that the jet expands with the axial position z. This expansion follows a linear law, so that a spatial expansion rate of the jets can be defined as follows:

$$s \equiv \frac{\mathrm{d}}{\mathrm{d}z}\left[\delta_{1/2}(z)\right]. \tag{8.37}$$

It can be concluded, experimentally, that s is a constant. Integrating Eq. (8.37), and assuming an asymptotic behavior of the self-similar part of the jet, it is assumed that at $z = z_{\mathrm{o}}$ the half width $\delta_{1/2}(z) = 0$, the equation for the half-width $\delta_{1/2}(z)$ is obtained:

$$\delta_{1/2}(z) = s(z - z_{\mathrm{o}}). \tag{8.38}$$

In a similar way, the expansion rate of other free shear flows, i.e., distant wake, plane jet and radial jet, can be defined. This latter flow can be created through a collision between two round jets, after which a radial jet appears. For a spatially developing mixing layer, Wilcox [54] suggests that the expansion rate can be calculated based on the difference in velocity values of y/z, in which the information $(U - U_2)^2/(U_1 - U_2)^2$ assumes the values 9/10 and 1/10. This is a definition that must be adopted equally for two databases, which you want to compare with each other. Several authors present the following comparative results, showing how the two models, Baldwin–Barth [1] and Spalart–Allmaras [48], respond when compared with experimental data. In Table 8.1, comparisons are presented.

Table 8.1 Application of the one balance equation models, proposed by Baldwin–Barth [1] and Spalart–Allmaras [48], for the solution of free shear flows

Type of flow	Model of Baldwin–Barth [1]	Model of Spalart–Allmaras [48]	Experiments [54]
Distant wake	0.315	0.341	0.365
Spatial mixing layer	–	0.108	0.115
Flat jet	–	0.156	0.1–0.11
Round jet	–	0.246	0.086–0.096

Comparison with experimental data (presented by Wilcox [54]). Spread rates are compared

The results for the Spalart–Allmaras Model were obtained by Wilcox [54] and the results for the distant wake, using the Baldwin–Barth Model, were obtained by Baldwin–Barth [1]. Wilcox [54] argues that the lack of results with the Baldwin–Barth model, for other flows, is due to the poorly conditioned matrices that bring about numerical instability issues.

As can be seen in Table 8.1, good results were obtained for the distant wake, with errors in 14% and 6%, respectively, with the Baldwin–Barth and Spalart–Allmaras models. For the mixed layer in spatial development, an error of 6% is achieved with the Spalart–Allmaras model. For plane and radial jets, the errors obtained with the Spalart–Allmaras model are very high, showing the great limitation of this model for the analysis of free shear flows. This is due to these models being adjusted for applications in aerodynamic problems, that is, for flows with walls. Thus, the lack of generality of these models is clear. However, the Spalart–Allmaras model has great potential for aeronautical applications. These results are not the only ones presented in the literature. Serious failures occur with these models when applied to complex flows, such as those that occur in a sudden expansion.

In search of more general models that enable the prediction of complex flows that present boundary layer detachment and reattachment, the global scientific community has invested efforts in developing models for the separate evaluation of characteristic quantities of turbulence, such as length and velocity. In this sense, some types of closure models are presented below, based on two balance equations.

8.3 Closure Models Based on Two Balance Equations

8.3.1 k−ω Closure Model

Turbulence closure models based on two balance equations have served as the basis for the vast majority of developments in this scientific area. The characteristics and main limitations of the closure models based on zero and one balance equations were presented in the previous sections. Kolmogorov [17] presented his ideas and concerns in proposing a new balance equation for the specific power of viscous transformation, ε [W/kg], or for the Kolmogorov Frequency, $\omega = \varepsilon/k\left[\mathrm{s}^{-1}\right]$, which can also be seen as a specific power of viscous transformation, that is, per unit of specific turbulent kinetic energy, $\omega\left[(\mathrm{W/kg})/(\mathrm{J/kg}) = \mathrm{s}^{-1}\right]$. This information has a frequency unit; thus, it can be interpreted as a characteristic frequency of turbulence, as presented by Safman [45]. The basis of all developments in closure models is the Boussinesq Hypothesis. Kolmogorov [17] also presented, based on dimensional analysis, the following proportionality relationships:

$$\varepsilon \sim \omega k = \frac{k^{3/2}}{\ell}, \ell \sim \frac{k^{1/2}}{\omega} = \frac{k^{3/2}}{\varepsilon} = k^{1/2}\tau \textit{ and } \nu_t \sim \frac{k}{\omega} = \frac{k^2}{\varepsilon} = k^{1/2}\ell = k\tau. \tag{8.39}$$

In Eq. (8.39), the characteristic time of turbulence was also introduced, τ [s]. Closure models must, in short, enable the calculation of turbulent viscosity, $\nu_t\left[\mathrm{m}^2/\mathrm{s}\right]$. It can be seen, from Eq. (8.39), that once the turbulent kinetic energy has been calculated, k, the calculation of ν_t still requires the evaluation of ω or ε, even ℓ. For this reason, in the following items, the transport equations for the Kolmogorov frequency will be presented, ω, and for the specific power of viscous transformation ε, which added to the transport equation for turbulent kinetic energy, k, will compose the closure models to two balance equations.

Kolmogorov [17] proposed a balance equation for the characteristic frequency ω. He sought to incorporate into this equation, in a phenomenological way, the two main processes of transport of any information, that is, the advection process and the diffusion process. It also incorporated the process of viscous transformation of the information being transported, that is, the characteristic frequency ω. Therefore, when taking a balance of ω in an infinitesimal particle of fluid and by differentially translating the advective and diffusive liquid flows of ω, Kolmogorov arrived at the following equation:

$$\underbrace{\frac{\partial \omega}{\partial t}}_{1} + \underbrace{\overline{u}_j \frac{\partial \omega}{\partial x_j}}_{2} = \underbrace{\frac{\partial}{\partial x_j}\left(\sigma \nu_t \frac{\partial \omega}{\partial x_j}\right)}_{3} - \underbrace{\beta \omega^2}_{4}. \tag{8.40}$$

The physical interpretation of the terms that make up Eq. (8.40) is presented below: term (1) represents the rate of variation of the *information* ω contained in a fluid particle; term (2) represents the net advective flux of ω which occurs across the particle control surface; term (3) represents the turbulent net diffusive flux of ω that occurs through the particle control surface and term (4) represents the specific power of viscous transformation of ω. Kolmogorov possibly concluded that molecular diffusion would not be relevant when compared to turbulent diffusion. Therefore, in term (3) the influence of molecular viscous diffusion does not appear. Thermal effects were not included in this closure model.

In this text, the way Kolmogorov adopted to evaluate the closure constants that appear in the proposed model is not presented, that is, σ and β, as well as the values he proposed. Some important observations made by Wilcox [54] on Kolmogorov's Proposal are listed below: (1) in this equation the interaction term of quantities with the filtered part of the flow does not appear, which translates into the fact that, for Kolmogorov, this frequency is characteristic of the so-called turbulent structures of Kolmogorov, whose interactions with large turbulence structures are negligible; this would be modeled as a function of the strain rate of the filtered field $c_{\omega 1}(\omega/k)\tau_{ij}\left(\partial \overline{u}_i/\partial x_j\right)$; therefore, they do not present interactions with the filtered velocity field and (2) this model did not incorporate molecular diffusion into this equation, which restricts its use to flows at high values of Reynolds Number, not including regions of flows in viscous sublayers, in the vicinity of walls, where molecular viscous effects are important.

By analogy with the balance equation for turbulent kinetic energy, $k = \left(\overline{u'^2} + \overline{\vartheta'^2} + \overline{w'^2}\right)/2$ one wonders why Kolmogorov did not propose a balance equation for enstrophy, $\Im = \omega^2/2$! In this equation, ω^2 represents the *variance* of the fluctuations of the vorticity vector ω_i. The enstrophy magnitude can be seen as a kind of energy, calculated using vorticity. Kolmogorov's Choice was made by ω, instead of enstrophy $\Im$, which affected all the developments that followed his proposal.

With this additional balance equation, the problem of closing the turbulence equations is solved. Its main weakness is the restriction of application only to flows at high Reynolds Numbers, which prevents its application to simulate problems involving walls. This proposal served, however, as the basis for subsequent developments, giving rise to more complete models. These models will be presented at the appropriate time in this text, following a historical chronological sequence. Next, other developments related to closure modeling based on two balance equations will be reported.

8.3.2 $k-\varepsilon$ Closure Model

Following Kolmogorov's Ideas, Chou [3] proposed a balance equation, in addition to that for k, the specific power of viscous transformation, $\varepsilon\ [\mathrm{W/kg}]$. Improvements were implemented by Davidov [6] and Harlow and Nakayama [14]. This development served as inspiration for the classic closing models, known as the Model $k-\varepsilon$. Its most consolidated version was proposed by Jones and Launder [16]. This model also gave rise to several others, based on the same balance equations, however, with variations in closure constants and variants in the closure modeling of these equations themselves. Thus, the well-known Jones and Lauder Model is more commonly referred to as the standard $k-\varepsilon$. Launder and Sharma [18] adjusted the closure constants and, therefore, this version of the model $k-\varepsilon$ is the most used by researchers in turbulence modeling. The complete deduction process of the balance equation will not be presented in this text. The balance equation, however, in its final form, proposed by Jones and Launder [16], before being modeled, the closure, it is transcribed below, which contains all the possible terms that should compose a balance equation:

$$\frac{\partial \varepsilon}{\partial t} + \overline{u}_j \frac{\partial \varepsilon}{\partial x_j} = P^{\varepsilon} + D^{m\varepsilon} + D^{t\varepsilon} + G^{\varepsilon} - \psi^{\varepsilon}. \tag{8.41}$$

Seeking again to give physical interpretations to each term of this equation, it will be rewritten. The purpose of the author of this text is to show that balance equations can always be written based on only three fundamental processes: accumulation/decay of information contained in a material particle; transport processes and processes of transforming the *information* (object of the balance) into other correlated information and vice versa. Thus:

$$
\underbrace{\underbrace{\frac{\partial \varepsilon}{\partial t}}_{\substack{\text{Rate of change of } \varepsilon \\ \text{contained in a particle of fluid}}}}_{\text{Accumulation/decay process}}
= \underbrace{\underbrace{-\bar{u}_j \frac{\partial \varepsilon}{\partial x_j}}_{\substack{\text{Advective net flux of } \varepsilon \text{ by the} \\ \text{surface of a fluid particle}}} + \underbrace{D^{m\varepsilon}}_{\substack{\text{Molecular diffusive net} \\ \text{flux of } \varepsilon \text{ by the surface} \\ \text{of a fluid particle}}} + \underbrace{D^{t\varepsilon}}_{\substack{\text{Turbulent diffusive} \\ \text{net flux of } \varepsilon \text{ by the} \\ \text{surface of a fluid particle}}}}_{\text{Transport processes}}
= \underbrace{\underbrace{P^{\varepsilon}}_{\substack{\text{Transformation of } \varepsilon\text{, from the} \\ \text{mean flow, to the spectrum of} \\ \text{turbulence structures}}} + \underbrace{G^{\varepsilon}}_{\substack{\text{Transformation of others} \\ \text{quantities into } \varepsilon\text{, due to the} \\ \text{gravitational effects}}} - \underbrace{\psi^{\varepsilon}}_{\substack{\text{Viscous transformation of} \\ \varepsilon \text{ into others quantities}}}}_{\text{Transformation processes}} \tag{8.42}
$$

The physical meanings of the terms that make up Eq. (8.42) are presented in the keys beneath each term in Eq. (8.42). Some emphasis will be given below to some of these terms. For example, the term $D^{m\varepsilon}$ represents the net molecular diffusive flux of ε, while $D^{t\varepsilon}$ represents the net turbulent diffusive flux of ε, both on the control surface surrounding the fluid particle; the term P^{ε} represents the transformation of the *information* ε, contained in the mean flow, for the spectrum of turbulent structures that make up the turbulent flow, which is called by the global scientific community by the inappropriate term "production"; term G^{ε} represents a process of transforming other correlated information into ε, due to the gravitational effects; term ψ^{ε} represents the viscous transformation of ε in another correlated kind of information. This term is also known by the inappropriate name "destruction of ε". It is observed that these denominations, "production" and "destruction" are not suitable to translate any physical aspect, since in nature "nothing is generated and nothing is destroyed, everything is transformed", according to the famous French Chemist Antoine Lavoisier, eighteenth century. One of Lavoisier's major concerns was nomenclature. It is a shame that the global scientific community pays little attention to this important concern of this thinker. As a rule, when the community names a scientific term, little effort is made to bring into this naming the physical nature of what it represents.

These terms can be modeled using what has already been done for the balance equation for turbulent kinetic energy k. To do so, the two balance equations are written together for comparative purposes:

$$
\begin{aligned}
\frac{Dk}{Dt} &= \frac{\partial k}{\partial t} + \bar{u}_j \frac{\partial k}{\partial x_j} = P^k + D^k + G^k - \varepsilon \\
\frac{D\varepsilon}{Dt} &= \frac{\partial \varepsilon}{\partial t} + \bar{u}_j \frac{\partial \varepsilon}{\partial x_j} = P^{\varepsilon} + D^{\varepsilon} + G^{\varepsilon} - \psi^{\varepsilon},
\end{aligned} \tag{8.43}
$$

where $D^{\varepsilon} = D^{m\varepsilon} + D^{t\varepsilon}$, $D^k = D^{mk} + D^{tk}$. To model these terms that appear in the second member of the balance equation for ε, a dimensional analysis of the terms will be presented. The unity of *information* k is energy per unit of mass, i.e., specific energy: $[k] = \left[\mathrm{J/kg}\right]$. The unit of ε is specific power, i.e., $[\varepsilon] = \left[\mathrm{W/kg}\right]$. Thus, the unit of $[\varepsilon/k] = [1/\mathrm{s}]$. The unity of the term $\partial k/\partial t$ is $\left[\mathrm{W/kg}\right]$, while the unit $\partial\varepsilon/\partial t$ is $\left[(\mathrm{W/kg})/\mathrm{s}\right]$. These units must be the same for the other terms of the respective equations. In other words, a balance equation must be dimensionally homogeneous. Thus, there is a difference of $[1/\mathrm{s}]$ between the terms of the balance equation for k and the terms of the balance equation for ε. A possible modeling of the open terms of the balance equation for ε can be proposed as follows:

$$P^{\varepsilon} + G^{\varepsilon} - \psi^{\varepsilon} = \frac{\varepsilon}{k}\left(c_{\varepsilon 1}P^k + c_{\varepsilon 1}G^k - c_{\varepsilon 2}\varepsilon\right), \tag{8.44}$$

$$D^{\varepsilon} = \frac{\partial}{\partial x_j}\left[\left(\nu + \frac{\nu_t}{\sigma_{\varepsilon}}\right)\frac{\partial \varepsilon}{\partial x_j}\right]. \tag{8.45}$$

Thus, the final and modeled form of the balance equation for ε is:

$$\frac{\partial \varepsilon}{\partial t} + \overline{u}_j\frac{\partial \varepsilon}{\partial x_j} = \frac{\varepsilon}{k}\left(c_{\varepsilon 1}P^k + c_{\varepsilon 1}G^k - c_{\varepsilon 2}\varepsilon\right) + \frac{\partial}{\partial x_j}\left[\left(\nu + \frac{\nu_t}{\sigma_{\varepsilon}}\right)\frac{\partial \varepsilon}{\partial x_j}\right]. \tag{8.46}$$

The modeling of this equation is thus concluded, since the terms P^k and G^k were modeled for the equation k. Substituting the terms in the right sides of the equations to k and ε results in:

$$\frac{\partial k}{\partial t} + \overline{u}_j\frac{\partial k}{\partial x_j} = \tau_{ij}\frac{\partial \overline{u}_i}{\partial x_j} - \varepsilon + \frac{\partial}{\partial x_j}\left[\left(\nu + \frac{\nu_t}{\sigma_k}\right)\frac{\partial k}{\partial x_j}\right] + g_i\beta\alpha_t\frac{\partial \overline{T}}{\partial x_i}, \tag{8.47}$$

$$\frac{\partial \varepsilon}{\partial t} + \overline{u}_j\frac{\partial \varepsilon}{\partial x_j} = \frac{\varepsilon}{k}c_{\varepsilon 1}\tau_{ij}\frac{\partial \overline{u}_i}{\partial x_j} - c_{\varepsilon 2}\frac{\varepsilon^2}{k} + \frac{\partial}{\partial x_j}\left[\left(\nu + \frac{\nu_t}{\sigma_{\varepsilon}}\right)\frac{\partial \varepsilon}{\partial x_j}\right] + c_{\varepsilon 1}g_i\beta\alpha_t\frac{\partial \overline{T}}{\partial x_i}. \tag{8.48}$$

Recap: the Boussinesq's hypothesis is:

$$-\tau_{ij} = 2\nu_t S_{ij} - \frac{2}{3}k\delta_{ij}, \tag{8.49}$$

and the turbulent viscosity can be calculated using k and ε:

$$\nu_t = c_{\mu}\frac{k^2}{\varepsilon}. \tag{8.50}$$

The filtered temperature $\overline{T}$ is modeled with the balance Eq. (8.18). Thus, to finalize the establishment of the closure model, the constants that appear in the equations above must be established. Lauder and Sharma [18] and Wilcox [54] suggested the following values for these constants: $c_\mu = 0.09$, $\sigma_k = 1.0$, $\sigma_\varepsilon = 1.3$, $c_{\varepsilon 1} = 1.44$, $c_{\varepsilon 2} = 1.92$. It can also be calculated:

$$\omega = \frac{\varepsilon}{c_\mu k}, \tag{8.51}$$

$$\ell = c_\mu k^{3/2}/\varepsilon. \tag{8.52}$$

This is the Standard Closing Model $k-\varepsilon$ of Launder and Sharma [18]. Other versions of this model have been proposed. The variants between them are characterized by different values of the constants and by the addition of damping functions or wall laws for regions neighboring the walls.

8.3.3 k−ε−RNG Closure Model

Yhakhot and Orszag [56] and Yhakhot et al. [56] developed one of the most promising versions of the model $k-\varepsilon$, based on Renormalization Group Theory. This version is known as $k-\varepsilon-$RNG. This model is composed of Eqs. (8.47) to (8.50), with important modifications regarding the viscous transformation term of the ε balance equation, that is, in the term $c_{\varepsilon 2}\frac{\varepsilon^2}{k}$. The coefficient that was a constant becomes a function $c_{\varepsilon 2}(\vec{x}, t)$. This gives a dynamic nature, since this function must be evaluated during the solution of the problem. The proposing authors established the following model for the function $c_{\varepsilon 2}(\vec{x}, t)$:

$$c_{\varepsilon 2}(\vec{x}, t) = \tilde{c}_{\varepsilon 2} + \frac{c_\mu \lambda^3\left(1 - \frac{\lambda}{\lambda_o}\right)}{1 + \beta\lambda^3}, \tag{8.53}$$

$$\lambda = \frac{k}{\varepsilon}\sqrt{2 S_{ij} S_{ij}}. \tag{8.54}$$

The following values were adjusted to the closure model coefficients: $c_\mu = 0.085$, $\sigma_k = 0.72$, $\sigma_\varepsilon = 0.72$, $c_{\varepsilon 1} = 1.44$, $\tilde{c}_{\varepsilon 2} = 1.68$, $\beta = 0.012$ and $\lambda_o = 4.38$. Returning to the two-equation model based on the balance equation for ω, the following item will summarize the evolution over the last fifty years of Kolmogorov's original proposal.

8.3.4 k−ω Closure Model—Revisiting Kolmogorov's Proposal

The balance equation for ω, proposed by Kolmogorov [17], was taken up by Saffman [45], who proposed modifications and improvements. Important work in this area was developed by Launder and Spalding [19]. Wilcox [53], Speziale et al. [50] and Peng et al. [36] contributed greatly to the evolution of the $k-\omega$ closure. The most advanced version of this closure model was presented by Wilcox [53, 54].

The derivation of the equation for ω can be done using the equations for k and ω, as presented by Davidson [7]. To make this text self-sufficient, this deduction will be presented here, adding more details and seeking to give more consistency to the notation. The relationship between ω, k and ε is given by: $\omega = \frac{\varepsilon}{c_\mu k}$. Thus,

$$\begin{aligned}
\frac{D\omega}{Dt} &= \frac{D}{Dt}\left(\frac{\varepsilon}{c_\mu k}\right) = \frac{1}{c_\mu k}\frac{D\varepsilon}{Dt} - \frac{\varepsilon}{c_\mu k^2}\frac{Dk}{Dt} = \frac{1}{c_\mu k}\frac{D\varepsilon}{Dt} - \frac{\omega}{k}\frac{Dk}{Dt} \Rightarrow \\
\frac{D\omega}{Dt} &= P^\omega - \psi^\omega + D^{t\omega} + D^{m\omega} + G^\omega.
\end{aligned} \tag{8.55}$$

The last equality of Eq. (8.55) was proposed, using a phenomenological notion. Replacing Eq. (8.43) in Eq. (8.55) results in:

$$\begin{aligned}
\frac{D\omega}{Dt} &= \frac{1}{c_\mu k}(P^\varepsilon - \psi^\varepsilon + D^{t\varepsilon} + D^{m\varepsilon} + G^\varepsilon) - \frac{\omega}{k}(P^k - \psi^k + D^{tk} + D^{mk} + G^k) \\
&= \left(\frac{1}{c_\mu k}P^\varepsilon - \frac{\omega}{k}P^k\right) - \left(\frac{1}{c_\mu k}\psi^\varepsilon - \frac{\omega}{k}\psi^k\right) + \left(\frac{1}{c_\mu k}D^{t\varepsilon} - \frac{\omega}{k}D^{tk}\right) \\
&+ \left(\frac{1}{c_\mu k}D^{m\varepsilon} - \frac{\omega}{k}D^{mk}\right) + \left(\frac{1}{c_\mu k}G^\varepsilon - \frac{\omega}{k}G^k\right) \\
&= P^\omega - \psi^\omega + D^{t\omega} + D^{m\omega} + G^\omega.
\end{aligned} \tag{8.56}$$

Following the nomenclature proposal of this text, the terms P^ω, ψ^ω, $D^{t\omega}$, $D^{m\omega}$ and G^ω will be called as follows: P^ω, transformation of the *information* ω contained in the mean flow into ω contained in the turbulence; Ψ^ω, viscous transformation of *information* ω into another type of information; $D^{t\omega}$, net turbulent diffusive flux of the *information* ω through the control surface of the fluid particle; $D^{m\omega}$, net molecular diffusive flux of the *information* ω through the control surface of the fluid particle and G^ω, conversion of other *information* into *information* ω through thermal and gravitational effects. These terms will be modeled below:

G^ω: transformation of ω from the mean flow to ω of turbulence:

$$P^\omega = \frac{1}{c_\mu k}P^\varepsilon - \frac{\omega}{k}P^k = c_{\varepsilon 1}\frac{\varepsilon}{c_\mu k^2}P^k - \frac{\omega}{k}P^k = (c_{\varepsilon 1} - 1)\frac{\omega}{k}P^k. \tag{8.57}$$

To derive Eq. (8.57), the following equalities were used $P^\varepsilon = c_{\varepsilon 1}\frac{\varepsilon}{k}P^k$ and $\omega = \frac{\varepsilon}{c_\mu k}$.

ψ^ω: viscous transformation of information ω in another type of information:

$$
\begin{aligned}
\psi^{\omega} &= \frac{1}{c_{\mu}k}\psi^{\varepsilon} - \frac{\omega}{k}\psi^{k} = \frac{1}{c_{\mu}k}c_{\varepsilon 2}\frac{\varepsilon^2}{k} - \frac{\omega}{k}\varepsilon \\
&= c_{\varepsilon 2}\frac{\varepsilon^2}{c_{\mu}k^2}\frac{c_{\mu}}{c_{\mu}} - \frac{\omega}{k}\varepsilon\frac{c_{\mu}}{c_{\mu}} \\
&= c_{\varepsilon 2}c_{\mu}\omega^2 - c_{\mu}\omega^2 = (c_{\varepsilon 2} - 1)c_{\mu}\omega^2.
\end{aligned}
\tag{8.58}
$$

$D^{m\omega}$: molecular diffusive net flux of ω between a fluid particle and its surroundings:

$$
\begin{aligned}
D^{m\omega} &= \frac{1}{c_{\mu}k}D^{m\varepsilon} - \frac{\omega}{k}D^{mk} = \nu\frac{1}{c_{\mu}k}\frac{\partial^2\varepsilon}{\partial x_j\partial x_j} - \nu\frac{\omega}{k}\frac{\partial^2 k}{\partial x_j\partial x_j} \\
&= \frac{\nu}{k}\frac{\partial}{\partial x_j}\left[\frac{\partial}{\partial x_j}\left(\frac{\varepsilon}{c_{\mu}}\frac{k}{k}\right)\right] - \nu\frac{\omega}{k}\frac{\partial^2 k}{\partial x_j\partial x_j} \\
&= \frac{\nu}{k}\frac{\partial}{\partial x_j}\left[\frac{\partial}{\partial x_j}(\omega k)\right] - \nu\frac{\omega}{k}\frac{\partial^2 k}{\partial x_j\partial x_j} \\
&= \frac{\nu}{k}\frac{\partial}{\partial x_j}\left[\omega\frac{\partial k}{\partial x_j} + k\frac{\partial\omega}{\partial x_j}\right] - \nu\frac{\omega}{k}\frac{\partial^2 k}{\partial x_j\partial x_j} \\
&= \frac{\nu}{k}\left[\frac{\partial\omega}{\partial x_j}\frac{\partial k}{\partial x_j} + \omega\frac{\partial^2 k}{\partial x_j\partial x_j} + \frac{\partial k}{\partial x_j}\frac{\partial\omega}{\partial x_j} + k\frac{\partial^2\omega}{\partial x_j\partial x_j}\right] \\
&\quad - \nu\frac{\omega}{k}\frac{\partial^2 k}{\partial x_j\partial x_j} = 2\frac{\nu}{k}\frac{\partial\omega}{\partial x_j}\frac{\partial k}{\partial x_j} + \nu\frac{\partial^2\omega}{\partial x_j\partial x_j}.
\end{aligned}
\tag{8.59}
$$

$D^{t\omega}$: turbulent diffusive net flux of ω between a fluid particle and its surroundings:

$$
\begin{aligned}
D^{t\omega} &= \frac{1}{c_{\mu}k}D^{t\varepsilon} - \frac{\omega}{k}D^{tk} = \frac{2\nu_t}{\sigma_{\varepsilon}k}\frac{\partial k}{\partial x_j}\frac{\partial\omega}{\partial x_j} \\
&+ \frac{\omega}{k}\left(\frac{\nu_t}{\sigma_{\varepsilon}} - \frac{\nu_t}{\sigma_k}\right)\left(\frac{\partial^2 k}{\partial x_j\partial x_j} + \frac{\partial\nu_t}{\partial x_j}\frac{\partial k}{\partial x_j}\right) + \frac{\partial}{\partial x_j}\left(\frac{\nu_t}{\sigma_{\varepsilon}}\frac{\partial\omega}{\partial x_j}\right).
\end{aligned}
\tag{8.60}
$$

It is worth remembering that for the standard model $k-\varepsilon$, $\sigma_{\varepsilon} = 1.3$ and $\sigma_k = 1.0$. For the $k-\omega$ model, the following approach is made $\sigma_{\varepsilon} = \sigma_k = 1.0$. Thus, Eq. (8.60) reduces to:

$$
D^{t\omega} = \frac{1}{c_{\mu}k}D^{t\varepsilon} - \frac{\omega}{k}D^{tk} = \frac{2\nu_t}{\sigma_{\varepsilon}k}\frac{\partial k}{\partial x_j}\frac{\partial\omega}{\partial x_j} + \frac{\partial}{\partial x_j}\left(\frac{\nu_t}{\sigma_{\varepsilon}}\frac{\partial\omega}{\partial x_j}\right).
\tag{8.61}
$$

G^{ω}: conversion of some correlated *information* into the ω *information* through thermal-gravitational effects:

$$G^{\omega} = \frac{1}{c_{\mu}k}G^{\varepsilon} - \frac{\omega}{k}G^{k} = \frac{1}{c_{\mu}k}c_{\varepsilon 1}g_i\beta\alpha_{\mathrm{t}}\frac{\partial \overline{T}}{\partial x_i} - \frac{\omega}{k}g_i\beta\alpha_{\mathrm{t}}\frac{\partial \overline{T}}{\partial x_i}$$
$$= \left(\frac{c_{\varepsilon 1}}{c_{\mu}k} - \frac{\omega}{k}\right)g_i\beta\alpha_t\frac{\partial \overline{T}}{\partial x_i}. \tag{8.62}$$

Thus, using Eqs. (8.57) to (8.62), we have the balance equation for the Kolmogorov frequency, ω:

$$\frac{D\omega}{Dt} = (c_{\varepsilon 1} - 1)\frac{\omega}{k}P^k - (c_{\varepsilon 2} - 1)c_{\mu}\omega^2 + 2\frac{\nu}{k}\frac{\partial \omega}{\partial x_j}\frac{\partial k}{\partial x_j} + \nu\frac{\partial^2 \omega}{\partial x_j \partial x_j}$$
$$+ \frac{2\nu_{\mathrm{t}}}{\sigma_{\varepsilon}k}\frac{\partial k}{\partial x_j}\frac{\partial \omega}{\partial x_j} + \frac{\partial}{\partial x_j}\left(\frac{\nu_{\mathrm{t}}}{\sigma_{\varepsilon}}\frac{\partial \omega}{\partial x_j}\right) + \left(\frac{c_{\varepsilon 1}}{c_{\mu}k} - \frac{\omega}{k}\right)g_i\beta\alpha_{\mathrm{t}}\frac{\partial \overline{T}}{\partial x_i}$$
$$= (c_{\varepsilon 1} - 1)\frac{\omega}{k}P^k - (c_{\varepsilon 2} - 1)c_{\mu}\omega^2 + \frac{\partial}{\partial x_j}\left[\left(\nu + \frac{\nu_{\mathrm{t}}}{\sigma_{\varepsilon}}\right)\frac{\partial \omega}{\partial x_j}\right]$$
$$+ \frac{2}{k}\left(\nu + \frac{\nu_{\mathrm{t}}}{\sigma_{\varepsilon}}\right)\frac{\partial \omega}{\partial x_j}\frac{\partial k}{\partial x_j} + \left(\frac{c_{\varepsilon 1}}{c_{\mu}k} - \frac{\omega}{k}\right)g_i\beta\alpha_{\mathrm{t}}\frac{\partial \overline{T}}{\partial x_i}. \tag{8.63}$$

In the derivation by Wilcox [54], a mathematical approach was used rather than the phenomenological approach presented above. As a result, the cross term, $(2/k)(\nu + \nu_{\mathrm{t}}/\sigma_{\varepsilon})\left(\partial\omega/\partial x_j\right)\left(\partial k/\partial x_j\right)$, which appears in Eq. (8.63), does not appear in the Wilcox model. The balance equations that make up the $k{-}\omega$ model, Wilcox [54], are presented below, including the values of the coefficients:

$$\frac{\partial k}{\partial t} + \frac{\partial}{\partial x_j}\left(\overline{u}_j k\right) = -\overline{u_i' u_j'}\frac{\partial \overline{u}_i}{\partial x_j} + \frac{\partial}{\partial x_j}\left(\left(\nu + \sigma^k \nu_{\mathrm{t}}\right)\frac{\partial k}{\partial x_j}\right) - \beta^k k\omega + g_i\beta\alpha_{\mathrm{t}}\frac{\partial \overline{T}}{\partial x_i'} \tag{8.64}$$

$$\frac{\partial \omega}{\partial t} + \frac{\partial}{\partial x_j}\left(\overline{u}_j \omega\right) = -\overline{u_i' u_j'}\alpha^{\omega}\frac{\omega}{k}\frac{\partial \overline{u}_i}{\partial x_j} + \frac{\partial}{\partial x_j}\left(\left(\nu + \sigma^{\omega}\nu_{\mathrm{t}}\right)\frac{\partial \omega}{\partial x_j}\right)$$
$$- \beta^{\omega}\omega^2 + \left(\frac{c_{\varepsilon 1}}{c_{\mu}k} - \frac{\omega}{k}\right)g_i\beta\alpha_{\mathrm{t}}\frac{\partial \overline{T}}{\partial x_i}. \tag{8.65}$$

Equations (8.64) and (8.65) were written from Eqs. (8.18) and (8.63). The following redefinitions, in relation to Eq. (8.18), were made: $c_{\mathrm{D}}k^{3/2}/\ell = c_{\mathrm{D}}^2 c_{\mu}\left[\varepsilon/\left(c_{\mathrm{D}}c_{\mu}k\right)\right]k = \beta^k k\omega$, $\beta^k = c_{\mathrm{D}}^2 c_{\mu}$, $\varepsilon = k^{3/2}/\ell$ and $\omega = \varepsilon/\left(kc_{\mathrm{D}}c_{\mu}\right)$. It was also defined that $\sigma^k = 1/\sigma_k$. Regarding Eq. (8.60) the following redefinition was made: $\alpha^{\omega} = c_{\varepsilon 1} - 1$, $\beta^{\omega} = (c_{\varepsilon 2} - 1)c_{\mu}$ and $\sigma^{\omega} = 1/\sigma_{\omega}$. The following equality was also used $P^k = -\overline{u_i' u_j'}\left(\partial \overline{u}_i/\partial x_j\right)$. An important feature of the model presented by Wilcox [54] is the fact that, to establish the closure of Eqs. (8.64) and (8.65), dynamic proportionality functions were established, that is, in these functions, there is a dependence on the flow itself. Thus, part of the coefficients that were considered constants became functions of time and space. These functions and constant coefficients are presented by Wilcox [54] and are presented here:

$$\alpha^{\omega} = \frac{13}{25}; \quad \beta^{\omega} = \beta_{\text{o}}^{\omega} f_{\beta}^{\omega}; \quad \beta^{k} = \beta_{0}^{k} f_{\beta}^{k};$$
$$\sigma^{\omega} = \sigma^{k} = \frac{1}{2}; \quad \beta_{\text{o}}^{\omega} = \frac{9}{125}; \quad \beta_{\text{o}}^{k} = \frac{9}{100} \tag{8.66}$$

$$f_{\beta}^{k} = \begin{cases} 1 & \text{if } \chi^{k} \leq 0 \\ \frac{1+680(\chi^{k})^{2}}{1+400(\chi^{k})^{2}} & \text{if } \chi^{k} > 0 \end{cases} \quad \text{and} \quad f_{\beta}^{\omega} = \frac{1+70\chi^{\omega}}{1+80\chi^{\omega}}, \tag{8.67}$$

$$\chi^{k} = \frac{1}{\omega^{3}} \frac{\partial k}{\partial x_j} \frac{\partial \omega}{\partial x_j} \quad \text{and} \quad \chi^{\omega} = \left| \frac{\Omega_{ij}\Omega_{jk}S_{ki}}{(\beta_{\text{o}}^{k}\omega)^{3}} \right|, \tag{8.68}$$

$$\nu_{\text{t}} = k/\omega, \tag{8.69}$$

$$\Omega_{ij} = \frac{1}{2}\left(\frac{\partial \overline{u}_i}{\partial x_j} - \frac{\partial \overline{u}_j}{\partial x_i}\right) \quad \text{and} \quad S_{ij} = \frac{1}{2}\left(\frac{\partial \overline{u}_i}{\partial x_j} + \frac{\partial \overline{u}_j}{\partial x_i}\right). \tag{8.70}$$

In Eq. (8.67) Ω_{ij} is the absolute vorticity tensor and S_{ij} is the strain rate tensor. Thus, the closure model of the type $k\!-\!\omega$, which, together with the filtered equations, makes up the model for turbulence, is completely established. With this model, the problem of modeling turbulence closure within the internal layer of a boundary layer or in the vicinity of walls is presented. However, its application in the outer layer of a boundary layer causes problems due to its high sensitivity to boundary conditions at free boundaries. On the other hand, for all models presented so far in this book, approximations and hypotheses have been made with serious negative consequences when modeling flows with the presence of walls and adverse pressure gradients. Seeking to solve these problems, Menter [27] developed, based on the Models $k\!-\!\varepsilon$ and $k\!-\!\omega$, the model *SST* (*Shear Stress Transport*), which will be the subject of study in the next item.

8.3.5 Closure SST Model (Shear Stress Transport)

The Closure Model SST (*Shear Stress Transport*) was proposed by Menter [27], featuring the search to overcome the deficiencies carried by the two Models $k\!-\!\varepsilon$ and $k\!-\!\omega$, which are based on two balance equations and the concept of turbulent viscosity. Thus, the hybridization of these two closure models was proposed. One of the shortcomings of the $k\!-\!\varepsilon$ model is related to the fact that the shear stresses in the external region of a boundary layer are over predicted, which implies an under prediction of the specific power of viscous transformation, ε. This implies over-prediction of the characteristic turbulence length, which affects the calculation of turbulent viscosity. A second deficiency of the $k\!-\!\varepsilon$ model is the need for additional modeling of turbulent viscosity, ν_t, in regions close to walls. To remedy this deficiency, damping functions are used for turbulent viscosity, evaluated in flow regions, at low values of the Reynolds Number, especially in the vicinity of walls.

This second deficiency of the Standard Model was the main reason for the resumption of developments of the $k-\omega$ model. It is emphasized that with the $k-\omega$ model it is possible to solve flows in the vicinity of walls without the use of damping functions. However, the $k-\omega$ model presents a problem with high sensitivity to the presence of free boundaries [7, 27, 53], i.e., the sensitivity to the value of ω_f, to model the boundary condition on a free boundary of the calculation domain.

In order to overcome the shortcomings of both models, Menter [27] proposed the fusion of the two models so that the regions close to walls are modeled with the closure model $k-\varepsilon$ while regions far from the walls and characterized by high values of Reynolds Number are modeled with the $k-\varepsilon$ closure model, which does not present excessive sensitivity to the boundary condition at the free boundary.

Another important point that must be presented is the issue of turbulent flows with adverse pressure gradients. This characteristic is difficult to model mathematically. Rodi and Scheurer [42] and Wilcox [54] showed the shortcomings of the $k-\varepsilon$ model to simulate flows that present adverse pressure gradients. It has been demonstrated over time by the closure model development community that, to model well and consider the effects of adverse pressure gradients in a flow, the most important factor is the level of values assumed by the turbulent viscosity, ν_t, in the logarithmic region of flows in the vicinity of walls. This is what determines the quality of closure models based on the concept of turbulent viscosity in modeling and predicting turbulent flows characterized by strong adverse pressure gradients.

Johnson and King [15] proposed an algebraic model, mathematically simple, that allows good representation for flows with adverse pressure gradients. Johnson and King [15] in this proposal, they sought to ensure Bradshaw's finding, in which he states that "the main component of the turbulent shear stress, that is, in the main flow direction, is always proportional to the turbulent kinetic energy present in the external region of the boundary layer". By forcing this proportionality in the closure model, a balance effect of shear stresses (Shear Stress Transport—SST) is ensured. This question, of a physical nature, gives rise to the name of the model proposed by Menter [27], which is the object of this text. Turbulence models based on the concept of turbulent viscosity, ν_t, that do not incorporate this finding by Bradshaw, fail to model flows with adverse pressure gradients.

As for free shear flows, that is, without walls, the standard $k-\varepsilon$ model presented above can be used. These are flows of the mixing layer, jet and wake types. However, with the URANS modeling family, there are difficulties in successfully modeling all flows in that family.

Flows in the vicinity of walls can be understood as composed of two large regions: the internal region, composed of the viscous sublayer of a linear nature, the logarithmic layer of a nonlinear nature and the region between the two, called the buffer region; the second is the external region, that is, the region of turbulent nature and, normally, characterized by high values of the local Reynolds Number. It is emphasized that it is not possible to represent the physical nature of these two regions well, simultaneously, using turbulence closure models. With the standard $k-\varepsilon$ model, the external region is well modeled, but the internal region is not very well modeled, thus compromising the modeling of the whole. Similarly, with the $k-\omega$ model, the internal region is well modeled, but not the external region, leaving the whole also compromised. With the *SST* model, we seek a solution to this problem, as well as to the problem of the presence of adverse pressure gradients.

For the SST model proposal, the standard $k-\varepsilon$ model was transformed into the $k-\varepsilon$ model, so that the latter retained the characteristics of the former. Therefore, there are two versions of the $k-\omega$ model, i.e., the original $k-\omega$ model, proposed by Wilcox and the $k-\omega$ model, that resulted from the transformed $k-\varepsilon$ model. Hybridization is carried out using a function F_1 that varies between 0 and 1, being 1 in the internal region, that is, in the viscous sublayer, logarithmic layer and buffer layer, and assumes 0 in the external region, which covers the entire flow beyond the logarithmic layer. Between these two values, the function takes values that vary continuously and gradually between 0 and 1. The $k-\omega$ model, derived from the standard $k-\varepsilon$ model, is multiplied by $(1 - F_1)$ and the original $k-\omega$ model is multiplied by F_1. The two models are then added together to obtain a third model, which retains the characteristics of the two base models, discussed above.

It should be stressed that the new model is based on the $k-\omega$ proposed by Wilcox, which is active in the internal region of the boundary layer and also on the $k-\omega$ model, derived from the standard $k-\varepsilon$ model, which is active in the outer region of the boundary layer. With the coupling between the two models, the objective to have a model with the characteristics of the two original models is reached. However, even in this new model, which results from this composition, the effects of adverse pressure gradients are not included. Therefore, the calculation of turbulent viscosity was also modified, so that Bradshaw's finding was considered.

Before presenting the deduction of the SST Model, the $k-\omega$ model, proposed by Wilcox, and the $k-\omega$ model, derived from the standard $k-\varepsilon$ model will be presented. Indices will be introduced in the coefficients, to better differentiate between the two models. Index "1" for $k-\omega$, proposed by Wilcox, and the index "2" for the $k-\omega$, derived from the standard $k-\varepsilon$ model.

Wilcox [54] $k-\omega$ model:

$$\frac{\partial k}{\partial t} + \frac{\partial}{\partial x_j}(\overline{u}_j k) = -\overline{u_i' u_j'}\frac{\partial \overline{u}_i}{\partial x_j} + \frac{\partial}{\partial x_j}\left(\left(\nu + \sigma_1^k \nu_t\right)\frac{\partial k}{\partial x_j}\right) - \beta_1^k k\omega + g_i \beta \alpha_t \frac{\partial \overline{T}}{\partial x_i}. \quad (8.71)$$

$$\frac{\partial \omega}{\partial t} + \frac{\partial}{\partial x_j}(\overline{u}_j \omega) = -\overline{u_i' u_j'}\alpha_1^{\omega}\frac{\omega}{k}\frac{\partial \overline{u}_i}{\partial x_j} + \frac{\partial}{\partial x_j}\left(\left(\nu + \sigma_1^{\omega} \nu_t\right)\frac{\partial \omega}{\partial x_j}\right) - \beta_1^{\omega}\omega^2$$
$$+ \left(\frac{c_{q11}}{c_{\mu_1 k}} - \frac{\omega}{k}\right) g_i \beta \alpha_t \frac{\partial \overline{T}}{\partial x_i}. \quad (8.72)$$

$k-\omega$ model, derived from standard $k-\varepsilon$ model:

$$\frac{\partial k}{\partial t}+\frac{\partial}{\partial x_j}(\overline{u}_j k)=-\overline{u_i' u_j'}\frac{\partial \overline{u}_i}{\partial x_j}+\frac{\partial}{\partial x_j}\left(\left(\nu+\sigma_2^k \nu_t\right)\frac{\partial k}{\partial x_j}\right)-\beta_2^k k\omega+g_i\beta\alpha_t\frac{\partial \overline{T}}{\partial x_i'} \tag{8.73}$$

$$\frac{\partial \omega}{\partial t}+\frac{\partial}{\partial x_j}(\overline{u}_j \omega)=-\overline{u_i' u_j'}\alpha_2^\omega\frac{\omega}{k}\frac{\partial \overline{u}_i}{\partial x_j}+\frac{\partial}{\partial x_j}\left(\left(\nu+\sigma_2^\omega \nu_t\right)\frac{\partial \omega}{\partial x_j}\right)-\beta_2^\omega\omega^2$$
$$+\frac{2}{k}\left(\nu+\frac{\nu_t}{\sigma_{\varepsilon 2}}\right)\frac{\partial \omega}{\partial x_j}\frac{\partial k}{\partial x_j}+\left(\frac{c_{\varepsilon 12}}{c_{\mu 2}k}-\frac{\omega}{k}\right)g_i\beta\alpha_t\frac{\partial \overline{T}}{\partial x_i}. \tag{8.74}$$

The two models multiplied by F_1 and by $1-F_1$ are rewritten below.
Wilcox $k-\omega$ original model:

$$F_1\left[\frac{\partial k}{\partial t}+\frac{\partial}{\partial x_j}(\overline{u}_j k)\right]=F_1\left\{-\overline{u_i' u_j'}\frac{\partial \overline{u}_i}{\partial x_j}+\frac{\partial}{\partial x_j}\left[\left(\nu+\sigma_1^k \nu_t\right)\frac{\partial k}{\partial x_j}\right]\right.$$
$$\left.-\beta_1^k k\omega+g_i\beta\alpha_t\frac{\partial \overline{T}}{\partial x_i}\right\}, \tag{8.75}$$

$$F_1\left[\frac{\partial \omega}{\partial t}+\frac{\partial}{\partial x_j}(\overline{u}_j \omega)\right]=F_1\left\{-\overline{u_i' u_j'}\alpha_1^\omega\frac{\omega}{k}\frac{\partial \overline{u}_i}{\partial x_j}+\frac{\partial}{\partial x_j}\left[\left(\nu+\sigma_1^\omega \nu_t\right)\frac{\partial \omega}{\partial x_j}\right]\right.$$
$$\left.-\beta_1^\omega\omega^2+\left(\frac{c_{\varepsilon 11}}{c_{\mu 1}k}-\frac{\omega}{k}\right)g_i\beta\alpha_t\frac{\partial \overline{T}}{\partial x_i}\right\}. \tag{8.76}$$

Transformed $k-\omega$ model:

$$(1-F_1)\left[\frac{\partial k}{\partial t}+\frac{\partial}{\partial x_j}(\overline{u}_j k)\right]=(1-F_1)\left\{-\overline{u_i' u_j'}\frac{\partial \overline{u}_i}{\partial x_j}+\frac{\partial}{\partial x_j}\left[\left(\nu+\sigma_2^k \nu_t\right)\frac{\partial k}{\partial x_j}\right]\right.$$
$$\left.-\beta_2^k k\omega+g_i\beta\alpha_t\frac{\partial \overline{T}}{\partial x_i}\right\}, \tag{8.77}$$

$$(1-F_1)\left[\frac{\partial \omega}{\partial t}+\frac{\partial}{\partial x_j}(\bar{u}_j \omega)\right]=(1-F_1)\left\{-\overline{u_i' u_j'}\alpha_2^\omega\frac{\omega}{k}\frac{\partial \bar{u}_i}{\partial x_j}+\frac{\partial}{\partial x_j}\left[\left(\nu+\sigma_2^\omega \nu_t\right)\frac{\partial \omega}{\partial x_j}\right]\right.$$
$$+\frac{2}{k}\left(\nu+\frac{\nu_t}{\sigma_2^\varepsilon}\right)\frac{\partial \omega}{\partial x_j}\frac{\partial k}{\partial x_j}$$
$$\left.-\beta_2^\omega\omega^2+\left(\frac{c_{\varepsilon 12}}{c_{\mu 2}k}-\frac{\omega}{k}\right)g_i\beta\alpha_t\frac{\partial \bar{T}}{\partial x_i}\right\}. \tag{8.78}$$

Adding Eq. (8.75) with Eq. (8.77), results in:

$$
\begin{aligned}
\frac{\partial k}{\partial t}+\frac{\partial}{\partial x_j}\left(\overline{u}_j k\right)=F_1&\left\{-\overline{u_i'u_j'}\frac{\partial \overline{u}_i}{\partial x_j}+\frac{\partial}{\partial x_j}\left[\left(\underbrace{\nu+\sigma_1^k\,\nu_{\rm t}}_{\text{constant}}\right)\frac{\partial k}{\partial x_j}\right]\right.\\
&\left.-\beta_1^k k\omega+g_i\beta\alpha_{\rm t}\frac{\partial \overline{T}}{\partial x_i}\right\}+(1-F_1)\\
&\left\{-\overline{u_i'u_j'}\frac{\partial \overline{u}_i}{\partial x_j}+\frac{\partial}{\partial x_j}\left[\left(\underbrace{\nu+\sigma_2^k\,\nu_{\rm t}}_{\text{constant}}\right)\frac{\partial k}{\partial x_j}\right]-\beta_2^k k\omega\right.\\
&\left.+g_i\beta\alpha_{\rm t}\frac{\partial \overline{T}}{\partial x_i}\right\}=-\overline{u_i'u_j'}\frac{\partial \overline{u}_i}{\partial x_j}+\frac{\partial}{\partial x_j}\left(\nu\frac{\partial k}{\partial x_j}\right)\\
&+\left[\underbrace{F_1\sigma_1^k+(1-F_1)\sigma_2^k}_{\sigma^k=\text{ constant}}\right]\frac{\partial}{\partial x_j}\left(\nu_{\rm t}\frac{\partial k}{\partial x_j}\right)\\
&-\left[\frac{F_1\beta_1^k+(1-F_1)\beta_2^k}{\beta^k\ =\ \text{constant}}\right]k\omega+g_i\beta\alpha_{\rm t}\frac{\partial \overline{T}}{\partial x_i}.
\end{aligned}
\tag{8.79}
$$

Using the definitions of σ^k and β^k, results in:

$$
\frac{\partial k}{\partial t}+\frac{\partial}{\partial x_j}\left(\overline{u}_j k\right)=-\overline{u_i'u_j'}\frac{\partial \overline{u}_i}{\partial x_j}+\frac{\partial}{\partial x_j}\left(\left(\nu+\sigma^k\nu_{\rm t}\right)\frac{\partial k}{\partial x_j}\right)-\beta^k k\omega+g_i\beta\alpha_{\rm t}\frac{\partial \overline{T}}{\partial x_i}.
\tag{8.80}
$$

Adding Eq. (8.76) to Eq. (8.78) results in:

$$
\begin{aligned}
\frac{\partial \omega}{\partial t}+\frac{\partial}{\partial x_j}\left(\overline{u}_j\omega\right)=F_1&\left\{-\overline{u_i'u_j'}\alpha_1^\omega\frac{\omega}{k}\frac{\partial \overline{u}_i}{\partial x_j}+\frac{\partial}{\partial x_j}\left[\left(\nu+\sigma_1^\omega\nu_{\rm t}\right)\frac{\partial \omega}{\partial x_j}\right]\right.\\
&\left.-\beta_1^\omega\omega^2+\left(\frac{c_{\varepsilon 11}}{c_{\mu 1}k}-\frac{\omega}{k}\right)g_i\beta\alpha_{\rm t}\frac{\partial \overline{T}}{\partial x_i}\right\}\\
&+(1-F_1)\left\{-\overline{u_i'u_j'}\alpha_2^\omega\frac{\omega}{k}\frac{\partial \overline{u}_i}{\partial x_j}-\beta_2^\omega\omega^2+\frac{\partial}{\partial x_j}\left[\left(\nu+\sigma_2^\omega\nu_{\rm t}\right)\frac{\partial \omega}{\partial x_j}\right]\right.\\
&\left.+\frac{2}{k}\left(\nu+\sigma_2^\omega\nu_{\rm t}\right)\frac{\partial \omega}{\partial x_j}\frac{\partial k}{\partial x_j}+\left(\frac{c_{\varepsilon 12}}{c_{\mu 2}k}-\frac{\omega}{k}\right)g_i\beta\alpha_{\rm t}\frac{\partial \overline{T}}{\partial x_i}\right\}\\
=&-\left[\underbrace{F_1\alpha_1^\omega+(1-F_1)\alpha_2^\omega}_{\alpha^\omega}\right]\overline{u_i'u_j'}\frac{\omega}{k}\frac{\partial \overline{u}_i}{\partial x_j}\\
&-\left[\underbrace{F_1\beta_1^\omega+(1-F_1)\beta_2^\omega}_{\beta^\omega}\right]\omega^2+\frac{\partial}{\partial x_j}\left(\nu\frac{\partial \omega}{\partial x_j}\right)
\end{aligned}
$$

$$
\begin{aligned}
&+ \Bigg[\underbrace{F_1\sigma_1^{\omega} + (1-F_1)\sigma_2^{\omega}}_{\sigma^{\omega}=\text{constant}} \Bigg] \frac{\partial}{\partial x_j}\left(\nu_t \frac{\partial \omega}{\partial x_j}\right) \\
&+ (1-F_1)\frac{2}{k}\left(\nu + \sigma_2^{\omega}\nu_t\right)\frac{\partial \omega}{\partial x_j}\frac{\partial k}{\partial x_j} \\
&+ \Bigg[\underbrace{F_1\left(\frac{c_{\varepsilon 11}}{c_{\mu 1}k} - \frac{\omega}{k}\right) + (1-F_1)\left(\frac{c_{\varepsilon 12}}{c_{\mu 2}k} - \frac{\omega}{k}\right)}_{\gamma^{\omega}} \Bigg] g_i \beta \alpha_t \frac{\partial \overline{T}}{\partial x_i} \\
= &-\alpha^{\omega}\overline{u_i' u_j'}\frac{\omega}{k}\frac{\partial \overline{u}_i}{\partial x_j} - \beta^{\omega}\omega^2 + \frac{\partial}{\partial x_j}\left[\left(\nu + \nu_t\sigma^{\omega}\right)\frac{\partial \omega}{\partial x_j}\right] \\
&+ 2(1-F_1)\frac{1}{k}\left(\nu + \sigma_2^{\omega}\nu_t\right)\frac{\partial \omega}{\partial x_j}\frac{\partial k}{\partial x_j} + \gamma^{\omega} g_i \beta \alpha_t \frac{\partial \overline{T}}{\partial x_i}.
\end{aligned} \tag{8.81}
$$

Simplifying the terms, we have:

$$
\begin{aligned}
\frac{\partial \omega}{\partial t} + \frac{\partial}{\partial x_j}\left(\overline{u}_j \omega\right) = &-\frac{\alpha^{\omega}}{\nu_t}\overline{u_i' u_j'}\frac{\partial \overline{u}_i}{\partial x_j} - \beta^{\omega}\omega^2 \\
&+ \frac{\partial}{\partial x_j}\left[\left(\nu + \nu_t\sigma^{\omega}\right)\frac{\partial \omega}{\partial x_j}\right] + 2(1-F_1) \\
&\frac{1}{k}\left(\nu + \sigma_2^{\omega}\nu_t\right)\frac{\partial \omega}{\partial x_j}\frac{\partial k}{\partial x_j} + \gamma^{\omega} g_i \beta \alpha_t \frac{\partial \overline{T}}{\partial x_i}.
\end{aligned} \tag{8.82}
$$

To establish Eqs. (8.80) and (8.82), the following mathematical relations were used:

$$
\begin{aligned}
&\nu_t = \frac{k}{\omega}; \quad \sigma^k = F_1\sigma_1^k + (1-F_1)\sigma_2^k; \quad \beta^k = F_1\beta_1^k + (1-F_1)\beta_2^k; \\
&\sigma^{\omega} = F_1\sigma_1^{\omega} + (1-F_1)\sigma_2^{\omega}; \quad \alpha^{\omega} = F_1\alpha_1^{\omega} + (1-F_1)\alpha_2^{\omega}; \\
&\beta^{\omega} = F_1\beta_1^{\omega} + (1-F_1)\beta_2^{\omega}; \quad \gamma^{\omega} = F_1\left(\frac{c_{\varepsilon 11}}{c_{\mu 1}k} - \frac{\omega}{k}\right) + (1-F_1)\left(\frac{c_{\varepsilon 12}}{c_{\mu 2}k} - \frac{\omega}{k}\right)
\end{aligned} \tag{8.83}
$$

The coefficients defined above were presented by Menter [27]: $\sigma_1^k = 0.85$, $\sigma_1^{\omega} = 0.5$, $\beta_1^k = 0.09$, $\beta_1^{\omega} = 0.075$, $\alpha_1^{\omega} = \frac{\beta_1^{\omega}}{\beta_1^k} - \frac{\sigma_1^{\omega}\kappa^2}{\sqrt{\beta_1^k}}$, $\kappa = 0.41$, $\sigma_2^k = 1.5$, $\sigma_2^{\omega} = 0.856$, $\beta_2^k = 0.09$, $\beta_2^{\omega} = 0.0828$ and $\alpha_2^{\omega} = \left(\beta_2^{\omega}/\beta_2^k\right) - \left(\sigma_2^{\omega}\kappa^2/\sqrt{\beta_2^k}\right)$. In the model proposed by Menter [27], the thermal part was not included. In the present text, it was considered, which gave rise to the term γ^{ω}. To determine this constant, it is necessary to adjust $c_{\varepsilon 11}$, $c_{\varepsilon 12}$, $c_{\mu 1}$ and $c_{\mu 2}$. A simplifying option would be to consider that these constants assume the same values inside and outside the boundary layer: $c_{\varepsilon 11} = c_{\varepsilon 12} = c_{\varepsilon 1} = 1.44$, $c_{\mu 1} = c_{\mu 2} = c_{\mu} = 0.09$ and the function $\gamma^{\omega} = F_1\left[\left(c_{\varepsilon 1}/c_{\mu}k\right) - (\omega/k)\right]$.

The hybridization function F_1 is given by:

$$F_1 = \tanh\left(\arg_1^4\right), \tag{8.84}$$

where

$$\arg_1 = \min\left\{\max\left(\frac{\sqrt{k}}{0.09\omega y}; \frac{500\nu}{\omega y^2}\right); \frac{4\rho\sigma_2^{\omega}k}{CD^{k\omega}y^2}\right\}. \tag{8.85}$$

The coefficient CD^{kw} is the positive part of the cross-diffusion term:

$$CD^{kw} = \max\left(2\sigma_2^{\omega}\frac{1}{\omega}\frac{\partial k}{\partial x_j}\frac{\partial \omega}{\partial x_j}, 10^{-20}\right), \tag{8.86}$$

being y the distance from a given point to the wall. The hyperbolic tangent argument, $\arg_1$, has the role of carrying the function $F_1 = 0$ in the outer zone of the boundary layer. This will happen because in Eq. (8.85) the three terms tend to zero, as y grows sufficiently, once the terms $1/y$ and $1/y^2$, appear in the three terms of $\arg_1$. Within a boundary layer, these three terms have well-defined roles. The first, $\sqrt{k}/(0.09\omega y) = \left(\sqrt{k}/\omega\right)/(0.09y) = \ell_{\text{turb}}/(0.09y)$, according to Menter [27], tends to 2.5 in the logarithmic region of the boundary layer and tends to zero, when k tends to zero, since the product ωy tends to unity and turbulent kinetic energy, k, tends to zero. The second term, $(500\nu)/\left(\omega y^2\right)$, tends to be a constant, close to the solid surface and tends to zero in the log region of the boundary layer. The third term, $\left(4\rho\sigma_2^{\omega}k\right)/\left(CD^{kw}y^2\right)$, serves to ensure that the solution is independent of the free boundary condition. Menter [27] ensures that this term tends to zero when approaching the boundary layer, ensuring that the turbulence closure model assumes the characteristics of the Boussinesq–Reynolds Stress Tensor. This is because the function $\arg_1 \to 0$, the same happens with the hybridization function F_1. The Boussinesq–Reynolds Tensor is modeled as proposed by Boussinesq:

$$-\overline{u_i'u_j'} = \nu_t\left(\frac{\partial u_i}{\partial x_j} + \frac{\partial u_j}{\partial x_i}\right) - \frac{2}{3}k\delta_{ij}, \tag{8.87}$$

where in the *SST* Closure Model proposal, the turbulent viscosity is modeled with the following equation:

$$\nu_t = \frac{a_1 k}{\max(a_1\omega; \Omega F_2)}, \tag{8.88}$$

with $a_1 = 0.31$ and Ω is the absolute value of vorticity. The function $F_2 = \tanh\left(\arg_2^2\right)$, and its argument is $\arg_2 = \max\left(2\frac{\sqrt{k}}{0.09\omega y}; \frac{500\nu}{y^2\omega}\right)$. Regarding the term specific power of viscous transformation, $P_\omega = \alpha^\omega \frac{\omega}{k}\tau_{ij}\frac{\partial u_i}{\partial x_j}$, Menter [27] highlights that the dimensionless group $\nu_t\frac{\omega}{k}$ is unity for the $k-\omega$ original model. Regarding boundary conditions, the author of this model recommends the following conditions:

For a free stream:

$$\omega_\infty = (1 \rightarrow 10)\frac{U_\infty}{L},\ \nu_{t\infty} = 10^{-(2\rightarrow 5)}\nu_\infty,\ k_\infty = \nu_{t\infty}\omega_\infty, \tag{8.89}$$

where L is the characteristic length of the problem to be analyzed. For a solid wall:

$$\omega(y=0) = \frac{60\nu}{\beta_1^\omega(\Delta y_1)^2},\ k = 0,\ \nu_t = 0, \tag{8.90}$$

where Δy_1 is the distance between the wall and the nearest node, in the direction normal to the wall. The mesh refinement condition for using this model is that $\Delta y_1^+ < 3$.

8.3.6 SAS Closure Model

The acronym *SAS* means "Scale Adaptive Simulation". With this type of modeling, we sought to better adapt the characteristic length of the turbulence to the physical nature of different regions of the turbulent flows. For example, the characteristic length of turbulence near walls differs from the characteristic length of turbulence in regions away from walls.

Currently, there are several models of this type. The first model in this family was developed by Menter et al. [28]. These authors based this proposal on the Rotta Model [44], which is a closure model based on two equations, which is presented below:

$$\frac{\partial k}{\partial t} + \frac{\partial}{\partial x_j}\left(\overline{u}_j k\right) = -\overline{u_i' u_j'}\frac{\partial \overline{u}_i}{\partial x_j} + \frac{\partial}{\partial x_j}\left(\left(\nu + \sigma^k \nu_t\right)\frac{\partial k}{\partial x_j}\right) - c_\mu^{3/4}\frac{k^{3/2}}{L}, \tag{8.91}$$

$$\frac{\partial \Psi}{\partial t} + \frac{\partial}{\partial x_j}\left(\overline{u}_j \Psi\right) = -\overline{u_i' u_j'}\left(\hat{\xi}_1 L\frac{\partial \overline{u}}{\partial y} + \hat{\xi}_2 L^3\frac{\partial^3 \overline{u}}{\partial y^3}\right) - \hat{\xi}_3 k^{3/2} + \frac{\partial}{\partial x_j}\left(\left(\nu + \sigma^k \nu_t\right)\frac{\partial \Psi}{\partial x_j}\right), \tag{8.92}$$

where

$$\Psi = kL \text{ and } \nu_t = c_\mu^{1/4}\frac{\Psi}{\sqrt{k}}. \tag{8.93}$$

Two-equation turbulence closure models are based on the determination of two characteristic turbulence quantities. In the present case, these quantities are turbulent kinetic energy k and the function Ψ. The main difference and the main difficulty in implementing this two-equation model is the presence of a third-order derivative, which appears in Eq. (8.92). According to Menter and Egorov [30], this model presents mathematical inconsistency in this third-order derivative term, and its implementation is costly and unstable. However, according to Menter and Egorov [30], the use of this model, without this term, leads to worse results than the other models with two balance equations.

Reviewing the Rotta Model [44], Menter, Kuntz and Bender [28] found that the argument that led to the third-order derivative term was inconsistent. After reevaluating the hypotheses, a new model was proposed for the balance of variables k and a function $\Phi = \sqrt{kL}$. This model is named *KSKL* (*K* − *Square* − *root KL*) model:

$$\frac{\partial \Phi}{\partial t} + \frac{\partial}{\partial x_j}(\overline{u}_j \Phi) = \frac{\Phi}{k}\overline{u_i' u_j'}\frac{\partial \overline{u}_i}{\partial x_j}\left(\xi_1 - \xi_2\left(\frac{L}{L_{\nu K}}\right)^2\right) - \xi_3 k + \frac{\partial}{\partial x_j}\left(\left(\nu + \sigma^k \nu_t\right)\frac{\partial \Phi}{\partial x_j}\right), \tag{8.94}$$

where

$$\nu_t = c_\mu^{1/4}\Phi,\, L_{\vartheta K} = \kappa\left(\frac{U'}{U''}\right),\, U' = S = \sqrt{2S_{ij}S_{ij}},\, U'' = \sqrt{\frac{\partial^2 U_i}{\partial x_k \partial x_k}\frac{\partial^2 U_i}{\partial x_j \partial x_j}}. \tag{8.95}$$

$$S_{ij} = \frac{1}{2}\left(\frac{\partial U_i}{\partial x_j} + \frac{\partial U_j}{\partial x_i}\right). \tag{8.96}$$

The term $L = \Phi^2/k$ is calculated once k and Φ are determined. In the process of reformulating the model proposed by Rotta [44], the derivative of order three gave way to the second-order derivatives of velocity. They are used to evaluate the von Kármán Length, $L_{\vartheta K}$. This term emerges as a term of accumulation of information ϕ, enabling the adaptation of the characteristic length of the turbulence to sudden changes in information in the flow. Classical models of turbulence, such as $k-\varepsilon$ and $k-\omega$, require characteristic turbulence lengths proportional to the thickness of the boundary layer, which does not require the adaptivity to the local nature of turbulent flows [30].

8.3.7 SST–SAS Closure Model

Aiming to include the adaptivity of the characteristic magnitudes of turbulence to classical closure models, relationships are established between the variables ϕ, ε and ω, as presented by Menter and Egorov [29]:

$$k = c_\mu^{1/4}\Phi\omega \text{ and } \varepsilon = c_\mu \omega k. \tag{8.97}$$

Menter and Egorov [29] proposed an extension of the $k-\omega-$SST model for a version $k-\omega$SST$-$SAS. A transformation term was added, q_{SAS}, in the balance equation of ω, with the role of modeling the adaptivity of the characteristic quantities of turbulence. Thus, the equations for k and ω are presented, as follows:

$$\frac{\partial k}{\partial t} + \frac{\partial}{\partial x_j}(\overline{u}_j k) = -\overline{u_i' u_j'}\frac{\partial \overline{u}_i}{\partial x_j} + \frac{\partial}{\partial x_j}\left((\nu + \sigma^k \nu_t)\frac{\partial k}{\partial x_j}\right) - \beta^k k\omega + g_i \beta \alpha_t \frac{\partial \overline{T}}{\partial x_i'}, \tag{8.98}$$

$$\frac{\partial \omega}{\partial t} + \frac{\partial}{\partial x_j}(\overline{u}_j \omega) = -\frac{\alpha^\omega}{\nu_t}\overline{u_i' u_j'}\frac{\partial \overline{u}_i}{\partial x_j} - \beta^\omega \omega^2 + \frac{\partial}{\partial x_j}\left[(\nu + \nu_t \sigma^\omega)\frac{\partial \omega}{\partial x_j}\right]$$
$$+ 2(1 - F_1)\frac{1}{k}(\nu + \sigma_2^\omega \nu_t)\frac{\partial \omega}{\partial x_j}\frac{\partial k}{\partial x_j} + \gamma^\omega g_i \beta \alpha_t \frac{\partial \overline{T}}{\partial x_i} + q_{SAS}, \tag{8.99}$$

where

$$q_{SAS} = \max\left[\rho \eta_2 \kappa S^2 \left(\frac{L}{L_{\vartheta K}}\right)^2 - C_{SAS}\frac{2\rho k}{\sigma_\phi}\max\left(\frac{1}{\omega^2}\frac{\partial \omega}{\partial x_j}\frac{\partial \omega}{\partial x_j}, \frac{1}{k^2}\frac{\partial k}{\partial x_j}\frac{\partial k}{\partial x_j}\right), 0\right]. \tag{8.100}$$

The characteristic lengths are given by:

$$L = \frac{\sqrt{k}}{c_\mu^{1/4}\omega}, \tag{8.101}$$

$$L_{\vartheta K} = \max\left[\kappa\left(\frac{U'}{U''}\right), C_s\sqrt{\frac{\kappa \eta_2}{(\beta/c_\mu) - \alpha}}\Delta\right], \tag{8.102}$$

where $\Delta = \sqrt[3]{\Delta x \Delta y \Delta z}$, Δx, Δy and Δz are the sizes of the edges of the discretization cells. The main feature incorporated in this proposal is that the original formulation is preserved near walls. In regions away from the walls, the von Kármán Characteristic Length, $L_{\vartheta K}$, Eq. (8.102), becomes dominant, putting into action the adaptivity of the magnitude of turbulence. It is also observed that the use of the characteristic cell size, Δ, in Eq. (8.102), limits the length $L_{\vartheta K}$. This treatment allows the damping of high frequencies or high wave numbers.

8.3.8 SA–SAS Closure Model

Coder [5] proposed, through the derivation of a model for one balance equation, starting from the two-equation model presented by Menter et al. [28], the following variation with adaptive characteristic quantities of the Spalart and Allmaras [48]. Starting from the model established with Eqs. (8.20) and (8.32), results in:

$$\frac{\partial \tilde{\nu}}{\partial t} + U_j \frac{\partial \tilde{\nu}}{\partial x_j} = c_{b1}\tilde{S}\tilde{\nu} - c_{w1}f_w\left(\frac{\tilde{\nu}}{d}\right)^2 + \frac{1}{\sigma}\frac{\partial}{x_k}\left[(\nu+\tilde{\nu})\frac{\partial \tilde{\nu}}{\partial x_k}\right] + \frac{c_{b2}}{\sigma}\frac{\partial \tilde{\nu}}{\partial x_k}\frac{\partial \tilde{\nu}}{\partial x_k} - q_{SAS}, \tag{8.103}$$

where

$$q_{SAS} = c_{s1}\max\left\{\left(\frac{\kappa^2}{L^2} - \frac{c_{s2}}{d^2}\right), 0\right\}\tilde{\nu}, \tag{8.104}$$

$$L = \max(L_{\theta K}, \kappa C_{SA-SAS}\Delta) \text{ and } L_{\theta K} = \kappa S\Big/\sqrt{\frac{\partial^2 u_i}{\partial x_n \partial x_n}\frac{\partial^2 u_i}{\partial x_n \partial x_n}}. \tag{8.105}$$

The constants are the following: $c_{s1} = 4.9$, $c_{s2} = 2.0$ and $C_{SA-SAS} = C_{DES} = 0.65$. Again, the size of the mesh cells, Δ, was used as a limit for the von Kármán Characteristic Length. In a similar way to that presented in the model *SST–SAS*, the von Kármán Length was introduced independently of the original Spalart and Allmaras model [48], through the transformation term q_{SAS}. The resulting model tends to be the behavior of the Classic *URANS* model close to walls, while far from walls, it works like a *SAS*.

8.4 Boussinesq–Reynolds Stress Model or RSM Model (Reynolds Stress Model)

The model based on Boussinesq–Reynolds Stresses indirectly relies on the concept of Boussinesq Turbulent Viscosity, contrary to what is stated by several authors. As will be seen, to close the balance equations of the components of the Boussinesq–Reynolds Tensor, $\overline{u_i'u_j'}$, and other information, such as specific viscosity transformation power, ε, the concept of turbulent viscosity, ν_t, will be used. Another important observation concerns the order of this model, which is classified by many authors as a second-order model, which creates confusion, since all models with two balance equations are also second order. This is because the information that is transported, such as: turbulent kinetic energy, k, the specific power of viscous transformation, ε, and turbulent thermal energy flux, $\overline{u_j'T'}$, they are all second-order correlations.

In the previous items, the Boussinesq–Reynolds tensions were modeled, via turbulent viscosity, and an attempt was made to calculate, in different ways, the characteristic quantities of turbulence, with which the turbulent viscosity itself was determined. The Boussinesq–Reynolds stress Model is based on deriving balance equations for each component of the Boussinesq–Reynolds Stress tensor. However, to close the third-order statistical moments which appear in the balance equations for the second-order moments, that is, the components of the Boussinesq–Reynolds Tensor, the concepts of turbulent kinetic energy k and the viscous transformation function, ε, are used. A balance equation for this last *information* is required and used. As we already know, the closure of the balance equation for ε requires the use of turbulent viscosity, ν_t. Turbulent kinetic energy, k, is obtained from the trace of the Boussinesq–Reynolds tensor $\overline{u_i'u_j'}$, i.e., $k = \overline{u_i'u_i'}/2$. Therefore, it is not necessary to solve a balance equation for this *information.*

8.4.1 *Balance Equations for the Boussinesq–Reynolds Tensor—$\overline{u_i'u_j'}$*

Using a procedure similar to that used to obtain the equations for the turbulent kinetic energy, k, and for the specific power of viscous transformation ε, a set of six equations for the components of the Boussinesq–Reynolds Tensor $\overline{u_i'u_j'}$ can be obtained, starting from the Navier–Stokes equations. Hence, the resulting equations will be called exact differential equations, since they will not have approximations throughout the deduction. Incompressible flows of Newtonian fluids with constant physical properties are considered. The equations are written below, in index form:

$$\frac{\partial u_i}{\partial t} + \frac{\partial}{\partial x_k}(u_i u_k) = -\frac{1}{\rho}\frac{\partial p}{\partial x_i} + \nu\frac{\partial^2 u_i}{\partial x_k \partial x_k} - g_i\beta(T - T_o). \tag{8.106}$$

Applying the filtering operator (—) to Eq. (8.106), results:

$$\frac{\partial \overline{u}_i}{\partial t} + \frac{\partial}{\partial x_k}(\overline{u}_i\overline{u}_k) = -\frac{1}{\rho}\frac{\partial \overline{p}}{\partial x_i} + \nu\frac{\partial^2 \overline{u}_i}{\partial x_k \partial x_k} - \frac{\partial}{\partial x_k}\left(\overline{u_i'u_k'}\right) - g_i\beta\left(\overline{T} - T_o\right). \tag{8.107}$$

Subtracting Eq. (8.107) from Eq. (8.106), introducing the concept of temperature fluctuation, T', such that $T = \overline{T} + T'$, and observing that $(T - T_0) - \left(\overline{T} - T_0\right) = T - \overline{T} = T'$, an equation for the velocity fluctuation u_i' is obtained:

$$\frac{\partial u_i'}{\partial t} + \frac{\partial}{\partial x_k}\left(u_i u_k - \overline{u}_i \overline{u}_k\right) = -\frac{1}{\rho}\frac{\partial p'}{\partial x_i} + \nu \frac{\partial^2 u_i'}{\partial x_k \partial x_k} + \frac{\partial}{\partial x_k}\left(\overline{u_i' u_k'}\right) - g_i \beta T'. \quad (8.108)$$

Rewriting Eq. (8.108) in the form of a balance equation for u_j' follows:

$$\frac{\partial u_j'}{\partial t} + \frac{\partial}{\partial x_k}\left(u_j u_k - \overline{u}_j \overline{u}_k\right) = -\frac{1}{\rho}\frac{\partial p'}{\partial x_j} + \nu \frac{\partial^2 u_j'}{\partial x_k \partial x_k} + \frac{\partial}{\partial x_k}\left(\overline{u_j' u_k'}\right) - g_j \beta T'. \quad (8.109)$$

Multiplying Eq. (8.108) by u_j' and Eq. (8.109) by u_i', adding the two resulting equations, grouping the terms, using the rule for derivation of products of functions, applying the filtering operator, together with its properties, and using the continuity equation for incompressible flows, $\partial \overline{u}_k / \partial x_k = 0$, results in the following equations:

$$\begin{aligned}
&\frac{\partial\left(\overline{u_i' u_j'}\right)}{\partial t} + \overline{u_j' \frac{\partial}{\partial x_k}\left(u_i u_k - \overline{u}_i \overline{u}_k\right)} + \overline{u_i' \frac{\partial}{\partial x_k}\left(u_j u_k - \overline{u}_j \overline{u}_k\right)} \\
&= -\frac{1}{\rho}\left(\overline{u_j' \frac{\partial p'}{\partial x_i}} + \overline{u_i' \frac{\partial p'}{\partial x_j}}\right) + \nu\left(\overline{u_j' \frac{\partial^2 u_i'}{\partial x_k \partial x_k}} + \overline{u_i' \frac{\partial^2 u_j'}{\partial x_k \partial x_k}}\right) \\
&\quad + \overline{u_j' \frac{\partial}{\partial x_k}\left(\overline{u_i' u_k'}\right)} + \overline{u_i' \frac{\partial}{\partial x_k}\left(\overline{u_j' u_k'}\right)} - \left(g_i \beta \overline{u_j' T'} + g_j \beta \overline{u_i' T'}\right),
\end{aligned} \quad (8.110)$$

$$\begin{aligned}
&\frac{\partial\left(\overline{u_i' u_j'}\right)}{\partial t} + \overline{u_j' \frac{\partial}{\partial x_k}\left[\left(\overline{u}_i + u_i'\right)\left(\overline{u}_k + u_k'\right) - \overline{u}_i \overline{u}_k\right]} \\
&\quad + \overline{u_i' \frac{\partial}{\partial x_k}\left[\left(\overline{u}_j + u_j'\right)\left(\overline{u}_k + u_k'\right) - \overline{u}_j \overline{u}_k\right]} \\
&= -\frac{1}{\rho}\left(\overline{u_j' \frac{\partial p'}{\partial x_i}} + \overline{u_i' \frac{\partial p'}{\partial x_j}}\right) + \nu\left(\overline{u_j' \frac{\partial^2 u_i'}{\partial x_k \partial x_k}} + \overline{u_i' \frac{\partial^2 u_j'}{\partial x_k \partial x_k}}\right) \\
&\quad + \boxed{\overline{u_j'}} \frac{\partial}{\partial x_k}\left(\overline{u_i' u_k'}\right) + \boxed{\overline{u_i'}} \frac{\partial}{\partial x_k}\left(\overline{u_j' u_k'}\right) - \left(g_i \beta \overline{u_j' T'} + g_j \beta \overline{u_i' T'}\right).
\end{aligned} \quad (8.111)$$

The two terms inside the rectangles are zero, due to:

$$\overline{u'_j \frac{\partial}{\partial x_k}\left(\overline{u'_i u'_k}\right)} + \overline{u'_i \frac{\partial}{\partial x_k}\left(\overline{u'_j u'_k}\right)} = \boxed{\overline{u'_j}}\frac{\partial}{\partial x_k}\left(\overline{u'_i u'_k}\right) + \boxed{\overline{u'_i}}\frac{\partial}{\partial x_k}\left(\overline{u'_j u'_k}\right) = 0, \text{ and}$$

$$\begin{aligned}
&\frac{\partial\left(\overline{u'_i u'_j}\right)}{\partial t} + \overline{u'_j \frac{\partial}{\partial x_k}\left(\boxed{\overline{u}_i\overline{u}_k} + \overline{u}_i u'_k + \overline{u}_k u'_i + u'_i u'_k - \boxed{\overline{u}_i\overline{u}_k}\right)} \\
&+ \overline{u'_i \frac{\partial}{\partial x_k}\left(\boxed{\overline{u}_j\overline{u}_k} + \overline{u}_j u'_k + u'_j\overline{u}_k + u'_j u'_k - \boxed{\overline{u}_j\overline{u}_k}\right)} \\
&= -\frac{1}{\rho}\left[\frac{\partial\left(\overline{u'_j p'}\right)}{\partial x_i} + \frac{\partial\left(\overline{u'_i p'}\right)}{\partial x_j} - \overline{p'\frac{\partial u'_j}{\partial x_i}} - \overline{p'\frac{\partial u'_i}{\partial x_j}}\right] \\
&+ \nu\left[\frac{\partial}{\partial x_k}\left(\overline{u'_j\frac{\partial u'_i}{\partial x_k}}\right) + \frac{\partial}{\partial x_k}\left(\overline{u'_i\frac{\partial u'_j}{\partial x_k}}\right) - 2\overline{\frac{\partial u'_i}{\partial x_k}\frac{\partial u'_j}{\partial x_k}}\right] \\
&- \left(g_i\beta\overline{u'_j T'} + g_j\beta\overline{u'_i T'}\right),
\end{aligned} \tag{8.112}$$

$$\begin{aligned}
&\frac{\partial\left(\overline{u'_i u'_j}\right)}{\partial t} + \overline{u'_j u'_k}\frac{\partial\overline{u}_i}{\partial x_k} + \overline{u'_i u'_k}\frac{\partial\overline{u}_j}{\partial x_k} + \overline{u}_k\left[\overline{u'_j\frac{\partial u'_i}{\partial x_k}} + \overline{u'_i\frac{\partial u'_j}{\partial x_k}}\right] + \frac{\partial}{\partial x_k}\left(\overline{u'_i u'_j u'_k}\right) \\
&= -\frac{1}{\rho}\left[\delta_{ik}\frac{\partial\left(\overline{u'_j p'}\right)}{\partial x_k}\delta_{jk}\frac{\partial\left(\overline{u'_i p'}\right)}{\partial x_k} - \overline{p'\left(\frac{\partial u'_j}{\partial x_i} + \frac{\partial u'_i}{\partial x_j}\right)}\right] \\
&+ \nu\frac{\partial^2\left(\overline{u'_i u'_j}\right)}{\partial x_k \partial x_k} - 2\nu\overline{\frac{\partial u'_i}{\partial x_k}\frac{\partial u'_j}{\partial x_k}} - \left(g_i\beta\overline{u_j T'} + g_j\beta\overline{u_i T'}\right),
\end{aligned} \tag{8.113}$$

$$\begin{aligned}
&\frac{\partial\left(\overline{u'_i u'_j}\right)}{\partial t} + \overline{u}_k\frac{\partial\left(\overline{u'_i u'_j}\right)}{\partial x_k} = -\overline{u'_i u'_k}\frac{\partial\overline{u}_j}{\partial x_k} - \overline{u'_j u'_k}\frac{\partial\overline{u}_i}{\partial x_k} \\
&- \frac{\partial}{\partial x_k}\left[\frac{1}{\rho}\delta_{ik}\left(\overline{u'_j p'}\right) + \frac{1}{\rho}\delta_{jk}\left(\overline{u'_i p'}\right) + \left(\overline{u'_i u'_j u'_k}\right)\right] \\
&+ \nu\frac{\partial^2\left(\overline{u'_i u'_j}\right)}{\partial x_k \partial x_k} + \overline{\frac{p'}{\rho}\left(\frac{\partial u'_j}{\partial x_i} + \frac{\partial u'_i}{\partial x_j}\right)} - 2\nu\overline{\frac{\partial u'_i}{\partial x_k}\frac{\partial u'_j}{\partial x_k}} \\
&- \left(g_i\beta\overline{u'_j T'} + g_j\beta\overline{u'_i T'}\right).
\end{aligned} \tag{8.114}$$

Equation (8.114) is the exact form of the balance equation of the components of the Boussinesq–Reynolds tensor, $\left(\overline{u'_i u'_j}\right)$. It will be rewritten here in the general form proposed by the author of this book, which can be used for any type of balance equation in fluid mechanics:

$$\left\{\begin{array}{c}\text{Rate of change}\\ \text{of quantity } \overline{u_i'u_j'}\end{array}\right\} = \left\{\begin{array}{c}\text{Transport}\\ \text{processes of}\\ \overline{u_i'u_j'}\end{array}\right\} + \left\{\begin{array}{c}\text{Transformation}\\ \text{processes involving}\\ \overline{u_i'u_j'} \text{ and other}\\ \text{informations}\end{array}\right\}, \tag{8.115}$$

$$\underbrace{\frac{\partial\left(\overline{u_i'u_j'}\right)}{\partial t}}_{1\Leftrightarrow T_{ij}:\ \text{Rate of change of information } \overline{u_i'u_j'} \text{ contained in the fluid particle}}$$
$$= \underbrace{\left\{\begin{array}{l}\underbrace{-\bar{u}_k\frac{\partial\left(\overline{u_i'u_j'}\right)}{\partial x_k}}_{2\Leftrightarrow A_{ij}} + \underbrace{\nu\frac{\partial^2\left(\overline{u_i'u_j'}\right)}{\partial x_k\partial x_k}}_{5\Leftrightarrow D_{ij}^m} - \underbrace{\frac{\partial}{\partial x_k}\left[\frac{1}{\rho}\delta_{ik}\left(\overline{u_j'p'}\right) + \frac{1}{\rho}\delta_{jk}\left(\overline{u_i'p'}\right) + \left(\overline{u_i'u_j'u_k'}\right)\right]}_{4\Leftrightarrow D_{ij}^t}\\ \qquad + \underbrace{\overline{p'\left(\frac{\partial u_j'}{\partial x_i} + \frac{\partial u_i'}{\partial x_j}\right)}}_{6\Leftrightarrow \Pi_{ij}}\end{array}\right\}}_{\text{Transport terms: advective and diffusive (molecular and turbulent) net flux of the information } \overline{u_i'u_j'}}$$
$$\underbrace{-\left\{\underbrace{\overline{u_i'u_k'}\frac{\partial \bar{u}_j}{\partial x_k} + \overline{u_j'u_k'}\frac{\partial \bar{u}_i}{\partial x_k}}_{3\Leftrightarrow P_{ij}} + \underbrace{2\nu\overline{\frac{\partial u_i'}{\partial x_k}\frac{\partial u_j'}{\partial x_k}}}_{7\Leftrightarrow \varepsilon_{ij}} + \underbrace{\left(g_i\beta\overline{u_j'T'} + g_j\beta\overline{u_i'T'}\right)}_{8\Leftrightarrow G_{ij}}\right\}}_{\text{Transformation terms involving information } \overline{u_i'u_j'} \text{ and other related informations}} \tag{8.116}$$

This equation has nine components and is written in index form. As the tensor $\overline{u_i'u_j'}$ is symmetrical, only six of its nine components remain. The physical interpretation of each term of Eq. (8.116) is given below:

- Term 1, T_{ij}: represents the rate of variation of the information transported, that is, the components of the Boussinesq–Reynolds Tensor, contained in a differential fluid particle.
- Term 2, A_{ij}: represents the net advective flux of the Boussinesq–Reynolds Tensor Components across the boundary of a fluid particle.
- Term 5, D_{ij}^m: represents the net molecular diffusive flux of the *information* being transported, the components of the Boussinesq–Reynolds tensor.
- Term 4, D_{ij}^t: is composed of three other terms, which represent third-order correlations, between pressure fluctuations and velocity fluctuations, $\partial\left[\delta_{jk}\left(\overline{u_i'p'}\right)/\rho + \delta_{ik}\left(\overline{u_j'p'}\right)/\rho\right]/\partial x_k$, and correlations of velocity fluctuations with themselves, $\partial\left(\overline{u_i'u_j'u_k'}\right)/\partial x_k$. Pressure fluctuations are proportional to second-order correlations of velocity fluctuations., $p' \propto \left(u_i'u_i'\right)$. Therefore, it is considered

that a correlation of a pressure fluctuation by a velocity fluctuation, $\overline{u_j'p'}$, is a third-order correlation. The divergence of these correlations, $\partial\left[\left(\overline{u_j'p'}\right)\delta_{ik}/\rho\right]/\partial x_k$, $\partial\left[\left(\overline{u_i'p'}\right)\delta_{jk}/\rho\right]/\partial x_k$ and $\partial\left(\overline{u_i'u_j'u_k'}\right)/\partial x_k$, represent the net flux of *information*, components of the Boussinesq–Reynolds stress tensor, between the different lengths and times of the turbulence. These are nonlinear interactions. These terms, as they are third order while the transported *information* is second order, constitute the problem of closing this type of turbulence modeling. To prepare for the closure modeling of these balance equations, these terms also receive the interpretation of turbulent diffusion of the transported information. Thus, the sum of these terms can be called the turbulent net diffusive flux of the components of the Boussinesq–Reynolds tensor, through the control surface of the fluid particle.

- Term 6, Π_{ij}: represents the net flux of the *information* object of the balance, i.e., the components of the Boussinesq–Reynolds Tensor, promoted by pressure fluctuations. It is also known as an *information* redistribution term. For this reason, it cannot be seen as a transformative term, but rather as a transport term. It differs from term (4) in the sense that pressure fluctuations correlate with the gradient of velocity fluctuations and with the transposed gradient of the same velocity fluctuations. It also has a diffusive nature.
- Term 3, P_{ij}: represents the transformation of the *information* in question, i.e., the components of the Boussinesq–Reynolds Tensor, from the filtered velocity field into the modeled turbulence, not directly resolved. This term is called "production" in the literature, which is naturally not an appropriate name, since, in nature, quantities are not produced! Only transport or convert that *information* into something else! In this text, it will be called "transformation P_{ij}".
- Term 7, ε_{ij}: represents the specific power of viscous transformation of the components of the Boussinesq–Reynolds tensor in other related quantities.
- Term 8, G_{ij}: represents the process of transforming gravitational potential energy into components of the Boussinesq–Reynolds tensor, and vice versa. It will be closed by deducing a balance equation for the turbulent flux of thermal energy, $\overline{u_k'T'}$.

Terms (1), (2) and (3) are exact terms and must be calculated by solving the PDEs. Terms (4), (5), (6) and (7) must be modeled, as they constitute additional unknowns to the components of the Boussinesq–Reynolds tensor. Additional equations must be proposed for these terms, which are presented below:

- Term 4, D_{ij}^t, modeling: this term represents the turbulent diffusive net flux, which can be modeled, as recommended by Davidson [7], using the first gradient hypothesis. In this case, the gradient of the balance object *information* is the gradient of the components of the Boussinesq–Reynolds tensor:

$$D_{ij}^{t} = -\frac{\partial}{\partial x_k}\left[\frac{1}{\rho}\delta_{ik}\left(\overline{u_j'p'}\right) + \frac{1}{\rho}\delta_{jk}\left(\overline{u_i'p'}\right) + \left(\overline{u_i'u_j'u_k'}\right)\right] = \frac{\partial}{\partial x_m}\left[\frac{\nu_t}{\sigma_k}\frac{\partial\left(\overline{u_i'u_j'}\right)}{\partial x_m}\right], \tag{8.117}$$

where $\sigma_k = 1.0$.

- Term $\Pi_{ij} = \overline{p'\left(\partial u_i'/\partial x_j + \partial u_j'/\partial x_i\right)/\rho}$ modeling: as highlighted by Davidson [7], this term plays a very important role in the balance equation of the tensor components $\overline{u_i'u_j'}$. It models a transport term that occurs through the *information* redistribution mechanism, which can be positive or negative. For turbulent flux in a channel, for example, this term will be negative for the balance equation for $\overline{u_1'u_1'}$, that is, towards the streamwise direction of the flow. On the other hand, it will be positive in the other directions: normal to the horizontal walls, $\overline{u_2'u_2'}$, and for the direction transverse to the flow and parallel to the vertical walls, $\overline{u_3'u_3'}$. So, this term acts to make the flow isotropic, removing quantities from the main direction and injecting the same quantities in the direction transverse to the flow. As explained by Rotta [43] and Davidson [7], the tensor Π_{ij} is decomposed into: Π_{ij}^1, quick part; Π_{ij}^2, slow part; Π_{ij}^{1w}, quick part on the wall and Π_{ij}^{2w} slow part on the wall. This effect is also known as the "Robin Wood Effect". It is often referred to in the literature as a "source term". Clearly, these same authors allude to the redistribution effect, which leads us to conclude that it is a transport effect and not a transformative term (source). To facilitate its modeling, this term will be written in decomposed form, as follows:

$$\Pi_{ij} = \Pi_{ij}^1 + \Pi_{ij}^2 + \Pi_{ij}^{1w} + \Pi_{ij}^{2w}. \tag{8.118}$$

Rota [43], Gibson and Launder [11] and Davidson [7] proposed modeling all of these terms. This modeling is transcribed into this text in a summarized form:

$$\Pi_{ij}^1 = -c_1\frac{\varepsilon}{k}\left(\overline{u_i'u_j'} - \frac{2}{3}k\delta_{ij}\right), \tag{8.119}$$

$$\Pi_{ij}^2 = -c_2\left(P_{ij} - \frac{2}{3}P^k\delta_{ij}\right), \tag{8.120}$$

$$\Pi_{ij}^{1w} = c_1\frac{\varepsilon}{k}\left[\overline{u_k'u_m'}n_k n_m\delta_{ij} - \frac{3}{2}\left(\overline{u_i'u_k'}n_k n_j + \overline{u_j'u_k'}n_k n_i\right)\right]f, \tag{8.121}$$

$$\Pi_{ij}^{2w} = c_2\frac{\varepsilon}{k}\left[\Pi_{km,2}n_k n_m\delta_{ij} - \frac{3}{2}\left(\Pi_{ik,2}n_k n_j + \Pi_{jk,2}n_k n_i\right)\right]f, \tag{8.122}$$

where $f = k^{3/2}/(2.55\|\vec{n}\cdot(\vec{x}-\vec{x}_w)\|\varepsilon)$ and $\|\vec{n}\cdot(\vec{x}-\vec{x}_w)\|$ is the distance from a point $\vec{x}$ to the nearest wall, described by $\vec{x}_w$; the constants $c_1 = 1.5$ and $c_2 = 0.6$ was adjusted empirically.

- The term $\varepsilon_{ij} = 2\nu\overline{\left(\partial u_i'/\partial x_k\right)\left(\partial u_j'/\partial x_k\right)}$, which represents the specific power of viscous transformation, can be modeled using the hypothesis of isotropy of small turbulent structures, for which this term becomes relevant. For large turbulent structures, the viscous transformation process is not relevant. Thus,

$$\overline{\frac{\partial u_1'}{\partial x_k}\frac{\partial u_1'}{\partial x_k}} = \overline{\frac{\partial u_2'}{\partial x_k}\frac{\partial u_2'}{\partial x_k}} = \overline{\frac{\partial u_3'}{\partial x_k}\frac{\partial u_3'}{\partial x_k}} \text{ and } \overline{\frac{\partial u_i'}{\partial x_k}\frac{\partial u_j'}{\partial x_k}} = 0 \text{ if } i \neq j. \tag{8.123}$$

This leads to the following model for ε_{ij}, as a function of ε:

$$\varepsilon_{ij} = 2\nu\overline{\frac{\partial u_i'}{\partial x_k}\frac{\partial u_j'}{\partial x_k}} = \frac{2}{3}\varepsilon\delta_{ij}, \tag{8.124}$$

where δ_{ij} is the Kronecker Delta. The coefficient 2/3 is for restoring the equality $\varepsilon = \frac{1}{2}\varepsilon_{ii}$ effectively, $\varepsilon_{ii} = \frac{2}{3}\varepsilon\delta_{ii} \Rightarrow \varepsilon_{11} + \varepsilon_{22} + \varepsilon_{33} = \frac{2}{3}\varepsilon(1+1+1) \Rightarrow \varepsilon = \frac{1}{2}\varepsilon_{ii}$.

Substituting these models into Eq. (8.116), we arrive at the balance equation, partially modeled, for $\overline{u_i'u_j'}$.

$$\begin{aligned}
\frac{\partial\left(\overline{u_i'u_j'}\right)}{\partial t} + \overline{u}_k\frac{\partial\overline{u_i'u_j'}}{\partial x_k} &= -\overline{u_i'u_k'}\frac{\partial\overline{u}_j}{\partial x_k} - \overline{u_j'u_k'}\frac{\partial\overline{u}_i}{\partial x_k} - c_1\frac{\varepsilon}{k}\left(\overline{u_i'u_j'} - \frac{2}{3}k\delta_{ij}\right) + \\
&- c_2\left(P_{ij} - \frac{2}{3}P^k\delta_{ij}\right) + c_1\rho\frac{\varepsilon}{k}\left[\overline{u_k'u_m'}n_kn_m\delta_{ij} - \frac{3}{2}\left(\overline{u_i'u_k'}n_kn_j + \overline{u_j'u_k'}n_kn_i\right)\right]f \\
&+ c_2\rho\frac{\varepsilon}{k}\left[\Pi_{km,2}n_kn_m\delta_{ij} - \frac{3}{2}\left(\Pi_{ik,2}n_kn_j + \Pi_{jk,2}n_kn_i\right)\right]f + \frac{2}{3}\varepsilon\delta_{ij} \\
&+ \frac{\partial}{\partial x_m}\left[\left(\nu + \frac{\nu_t}{\sigma_k}\right)\frac{\partial\left(\overline{u_i'u_j'}\right)}{\partial x_m}\right] - \left(g_i\beta\overline{u_j'T'} + g_j\beta\overline{u_i'T'}\right).
\end{aligned} \tag{8.125}$$

There is still a lack of modeling for the viscous transformation power of turbulent kinetic energy, ε, which appears in Eq. (8.125). The balance equation for ε, was derived previously—Eq. (8.48). It is worth highlighting the fact that this equation relies on turbulent viscosity, ν_t, whose calculation depends on the turbulent kinetic energy, k, which in turn can be obtained by knowing the diagonal of the tensor $\overline{u_i'u_j'}$, already modeled. Upon having k and ε available, ν_t is calculated. It can be seen, therefore, that it is not pertinent to say that the Boussinesq–Reynolds tensor model, the *RSM*, is independent of turbulent viscosity. It does depend on ν_t.

Equation (8.125) is almost completely modeled, except for the term:

$$G_{ij} = \left(g_i \beta \overline{u_j' T'} + g_j \beta \overline{u_i' T'}\right). \tag{8.126}$$

To model this term, a balance equation for the turbulent flux of thermal energy, $\overline{u_j' T'}$, must be established.

8.4.2 *Balance Equation for Turbulent Thermal Energy Flux:* $\overline{u_j' T'}$

The balance equation for thermal energy is rewritten below:

$$\frac{\partial T}{\partial t} + \frac{\partial (u_k T)}{\partial x_k} = \alpha \frac{\partial^2 T}{\partial x_k \partial x_k}. \tag{8.127}$$

Filtering Eq. (8.127), results in the following equation:

$$\frac{\partial \overline{T}}{\partial t} + \frac{\partial \left(\overline{u}_k \overline{T}\right)}{\partial x_k} = \alpha \frac{\partial^2 \overline{T}}{\partial x_k \partial x_k} - \frac{\partial \left(\overline{u_k' T'}\right)}{\partial x_k}. \tag{8.128}$$

This equation represents the balance of filtered thermal energy contained in a fluid particle. It is constituted by the terms: the rate of change of filtered thermal energy, $\partial \overline{T}/\partial t$, contained in the fluid particle; the net advective flux of thermal energy across the particle surface, $\partial \left(\overline{u}_k \overline{T}\right)/\partial x_k$; the net molecular diffusive flux of thermal energy across the particle surface, $\partial \left(\alpha \partial \overline{T}/\partial x_k\right)/\partial x_k = \alpha \partial^2 \overline{T}/\partial x_k \partial x_k$ and the turbulent net diffusive flux of thermal energy across the particle surface, $\partial \left(\overline{u_k' T'}\right)/\partial x_k$. This last term, composed of the correlations between speed and temperature fluctuations, $\overline{u_j' T'}$, is unknown and must be modeled. Subtracting Eq. (8.127) from Eq. (8.128) yields:

$$\frac{\partial T'}{\partial t} + \frac{\partial \left(u_k T - \overline{u}_k \overline{T}\right)}{\partial x_k} = \alpha \frac{\partial^2 T'}{\partial x_k \partial x_k} + \frac{\partial \left(\overline{u_k' T'}\right)}{\partial x_k}. \tag{8.129}$$

Decomposing temperature and velocity as $T = \overline{T} + T'$ and $u_k = \overline{u}_k + u_k'$, results in $u_k T - \overline{u}_k \overline{T} = \overline{u}_k T' + u_k' \overline{T} + u_k' T'$. Thus, Eq. (8.129) can be rewritten as:

$$\frac{\partial T'}{\partial t} + \frac{\partial \left(u_k T - \overline{u}_k \overline{T}\right)}{\partial x_k} = \alpha \frac{\partial^2 T'}{\partial x_k \partial x_k} + \frac{\partial \left(\overline{u_k' T'}\right)}{\partial x_k}. \tag{8.130}$$

Equation (8.108) for velocity fluctuation, u_i', is rewritten as follows: $u_iu_k - \overline{u}_i\overline{u}_k = (\overline{u}_i + u_i')(\overline{u}_k + u_k') = \boxed{\overline{u}_i\overline{u}_k} + \overline{u}_iu_k' + u_i'\overline{u}_k + u_i'u_k' \boxed{-\overline{u}_i\overline{u}_k} = \overline{u}_iu_k' + u_i'\overline{u}_k + u_i'u_k'$:

$$\frac{\partial u_i'}{\partial t} + \frac{\partial}{\partial x_k}\left(\overline{u}_iu_k' + u_i'\overline{u}_k + u_i'u_k'\right) = -\frac{1}{\rho}\frac{\partial p'}{\partial x_i} + \nu\frac{\partial^2 u_i'}{\partial x_k \partial x_k} + \frac{\partial}{\partial x_k}\left(\overline{u_i'u_k'}\right) - g_i\beta T'. \tag{8.131}$$

Multiplying Eq. (8.130) by u_i' and Eq. (8.131) by T', and adding the two resulting equations and filtering, we reach:

$$\begin{aligned}
&\overline{u_i'\frac{\partial T'}{\partial t} + T'\frac{\partial u_i'}{\partial t}} + \overline{u_i'\frac{\partial}{\partial x_k}\left(\overline{u}_kT' + u_k'\overline{T} + u_k'T'\right) + T'\frac{\partial}{\partial x_k}\left(\overline{u}_iu_k' + u_i'\overline{u}_k + u_i'u_k'\right)} \\
&= -\frac{1}{\rho_0}\overline{T'\frac{\partial p'}{\partial x_i}} + \nu\overline{T'\frac{\partial^2 u_i'}{\partial x_k\partial x_k}} + \alpha\overline{u_i'\frac{\partial^2 T'}{\partial x_k\partial x_k}} + \boxed{\overline{u_i'\frac{\partial\left(\overline{u_k'T'}\right)}{\partial x_k}}} + \boxed{\overline{T'\frac{\partial\left(\overline{u_i'u_k'}\right)}{\partial x_k}}} \\
&- g_i\beta\overline{T'T'} \\
&\Leftrightarrow \overline{u_i\frac{\partial T'}{\partial t} + T'\frac{\partial u_i'}{\partial t}} + \overline{u_i'\frac{\partial(\overline{u}_kT')}{\partial x_k} + T'\frac{\partial\left(u_i'\overline{u}_k\right)}{\partial x_k}} \\
&+ \overline{u_i'\frac{\partial\left(u_k'\overline{T}\right)}{\partial x_k} + T'\frac{\partial}{\partial x_k}\left(\overline{u}_iu_k'\right)} + \overline{u_i'\frac{\partial}{\partial x_k}\left(u_k'T'\right)} + \overline{T'\frac{\partial}{\partial x_k}\left(u_i'u_k'\right)} \\
&= -\frac{1}{\rho_0}\overline{T'\frac{\partial p'}{\partial x_i}} + \nu\overline{T'\frac{\partial^2 u_i'}{\partial x_k\partial x_k}} + \alpha\overline{u_i'\frac{\partial^2 T'}{\partial x_k\partial x_k}} - g_i\beta\overline{T'T'}.
\end{aligned} \tag{8.132}$$

Considering that the following terms are zero, $\overline{u_i'\partial\left(\overline{u_k'T'}\right)/\partial x_k} + \overline{T'\partial\left(\overline{u_i'u_k'}\right)/\partial x_k} = \overline{u_i'}\partial\left(\overline{u_k'T'}\right)/\partial x_k + \overline{T'}\partial\left(\overline{u_i'u_k'}\right)/\partial x_k = 0$, as $\overline{u_i'} = \overline{T'} = 0$, and $\partial\overline{u}_k/\partial x_k = \partial u_k'/\partial x_k = 0$, for incompressible flows, rearranging the terms of Eq. (8.132), results in:

$$\Rightarrow \overline{u_i'\frac{\partial T'}{\partial t} + T'\frac{\partial u_i'}{\partial t}} = \frac{\partial\left(\overline{u_i'T'}\right)}{\partial t}$$

$$\Rightarrow \overline{u_i'\frac{\partial(\overline{u}_kT')}{\partial x_k} + T'\frac{\partial\left(u_i'\overline{u}_k\right)}{\partial x_k}} = \overline{u}_k\overline{\left[u_i'\frac{\partial(T')}{\partial x_k} + T'\frac{\partial\left(u_i'\right)}{\partial x_k}\right]} = \overline{u}_k\frac{\partial\left(\overline{u_i'T'}\right)}{\partial x_k}$$

$$\Rightarrow \overline{u_i'\frac{\partial\left(u_k'\overline{T}\right)}{\partial x_k} + T'\frac{\partial}{\partial x_k}\left(\overline{u}_iu_k'\right)} = \overline{u_i'u_k'}\frac{\partial\overline{T}}{\partial x_k} + \overline{u_k'T'}\frac{\partial\overline{u}_i}{\partial x_k}$$

$$\Rightarrow \overline{u_i'\frac{\partial\left(u_k'T'\right)}{\partial x_k} + T'\frac{\partial\left(u_i'u_k'\right)}{\partial x_k}} = \overline{\left(u_k'u_i'\frac{\partial T'}{\partial x_k} + T'\frac{\partial\left(u_i'u_k'\right)}{\partial x_k}\right)} = \frac{\partial\left(\overline{u_i'u_k'T'}\right)}{\partial x_k}$$

$$\Rightarrow \alpha\overline{u_i'\frac{\partial^2 T'}{\partial x_k \partial x_k}} + \nu\overline{T'\frac{\partial^2 u_i'}{\partial x_k \partial x_k}} = \alpha\overline{\frac{\partial}{\partial x_k}\left(u_i'\frac{\partial T'}{\partial x_k}\right)}$$
$$+ \nu\overline{\frac{\partial}{\partial x_k}\left(T'\frac{\partial u_i'}{\partial x_k}\right)} - (\alpha+\nu)\overline{\frac{\partial u_i'}{\partial x_k}\frac{\partial T'}{\partial x_k}}$$
$$= (\alpha+\nu)\overline{\frac{\partial^2 (u_i'T')}{\partial x_k \partial x_k}} - \alpha\overline{\frac{\partial}{\partial x_k}\left(T'\frac{\partial u_i'}{\partial x_k}\right)} - \nu\overline{\frac{\partial}{\partial x_k}\left(u_i'\frac{\partial T'}{\partial x_k}\right)}$$
$$- (\alpha+\nu)\overline{\frac{\partial u_i'}{\partial x_k}\frac{\partial T'}{\partial x_k}} \Rightarrow -\frac{1}{\rho_o}\overline{T'\frac{\partial p'}{\partial x_i}}$$
$$= -\frac{1}{\rho_o}\left(\overline{\frac{\partial}{\partial x_i}(T'p') - p'\frac{\partial T'}{\partial x_i}}\right) = -\frac{1}{\rho_o}\frac{\partial}{\partial x_j}\overline{(T'p')}\delta_{ij} + \frac{1}{\rho_o}\overline{p'\frac{\partial T'}{\partial x_i}}. \quad (8.133)$$

Substituting Eq. (8.133) in Eq. (8.132) results in:

$$\frac{\partial\left(\overline{u_i'T'}\right)}{\partial t} + \bar{u}_k\frac{\partial\left(\overline{u_i'T'}\right)}{\partial x_k} = -\frac{1}{\rho_o}\frac{\partial}{\partial x_j}\overline{(T'p')}\delta_{ij} + \frac{1}{\rho_o}\overline{p'\frac{\partial T'}{\partial x_i}} - \overline{u_i'u_k'}\frac{\partial \overline{T}}{\partial x_k}$$
$$- \overline{u_k'T'}\frac{\partial \bar{u}_i}{\partial x_k} - \frac{\partial\left(\overline{u_i'u_k'T'}\right)}{\partial x_k} + (\alpha+\nu)\overline{\frac{\partial^2 (u_i'T')}{\partial x_k \partial x_k}}$$
$$- \alpha\overline{\frac{\partial}{\partial x_k}\left(T'\frac{\partial u_i'}{\partial x_k}\right)} - \nu\overline{\frac{\partial}{\partial x_k}\left(u_i'\frac{\partial T'}{\partial x_k}\right)}$$
$$- (\alpha+\nu)\overline{\frac{\partial u_i'}{\partial x_k}\frac{\partial T'}{\partial x_k}} - g_i\beta\overline{T'T'}$$
$$\Leftrightarrow \frac{\partial\left(\overline{u_i'T'}\right)}{\partial t} + \bar{u}_k\frac{\partial\left(\overline{u_i'T'}\right)}{\partial x_k}$$
$$= -\frac{\partial}{\partial x_k}\left(\alpha\overline{T_i'\frac{\partial u'}{\partial x_k}} + \nu\overline{u_i'\frac{\partial T'}{\partial x_k}} + \overline{u_i'u_k'T'} + \frac{1}{\rho_o}\overline{T'p'\delta_{ij}}\right)$$
$$- \overline{u_i'u'}_k\frac{\partial \overline{T}}{\partial x_k} - \overline{u_k'T'}\frac{\partial \bar{u}_i}{\partial x_k} - g_i\beta\overline{T'T'} + (\alpha+\nu)$$
$$\frac{\partial^2\left(\overline{u_i'T'}\right)}{\partial x_k \partial x_k} - (\alpha+\nu)\overline{\frac{\partial u_i'}{\partial x_k}\frac{\partial T'}{\partial x_k}} + \frac{1}{\rho_o}\overline{p'\frac{\partial T'}{\partial x_i}}. \quad (8.134)$$

Equation (8.134) is rewritten below, in the general format of balance equations, seeking to use a nomenclature consistent with the physical meaning that each of its terms represents, as presented in Eq. (8.135):

$$
\underbrace{\frac{\partial\left(\overline{u_i'T'}\right)}{\partial t}}_{1\Leftrightarrow T_{\overline{u_i'T'}}:\ \text{Rate of change of information } \overline{u_i'T'} \text{ contained in the fluid particle}} =
$$

$$
\underbrace{\underbrace{-\bar{u}_k\frac{\partial\left(\overline{u_i'T'}\right)}{\partial x_k}}_{2\Leftrightarrow A_{\overline{u_i'T'}}} + \underbrace{(\alpha+\nu)\frac{\partial^2\left(\overline{u_i'T'}\right)}{\partial x_k\partial x_k}}_{3\Leftrightarrow D^m_{\overline{u_i'T'}}} \underbrace{-\frac{\partial}{\partial x_k}\left(\alpha\overline{u_i'\frac{\partial T'}{\partial x_k}} + \nu\overline{T'\frac{\partial u_i'}{\partial x_k}} + \overline{u_i'u_k'T'} + \frac{1}{\rho_o}\overline{T'p'}\delta_{ij}\right)}_{4\Leftrightarrow D^t_{\overline{u_i'T'}}} \underbrace{+\frac{1}{\rho_o}\overline{p'\frac{\partial T'}{\partial x_i}}}_{5\Leftrightarrow \Pi_{\overline{u_i'T'}}}}_{\text{Transport processes of the information } \overline{u_i'T'} \text{ between a fluid particle and its surroundings}}
$$

$$
\underbrace{-\left(\underbrace{\overline{u_i'u'_k}\frac{\partial \bar{T}}{\partial x_k} + \overline{u_k'T'}\frac{\partial \bar{u}_i}{\partial x_k}}_{6\Leftrightarrow P_{\overline{u_i'T'}}}\right) \underbrace{-(\alpha+\nu)\frac{\overline{\partial u_i'\frac{\partial T'}{\partial x_k}}}{\partial x_k}}_{7\Leftrightarrow \varepsilon_{\overline{u_i'T'}}} \underbrace{-g_i\beta\overline{T'T'}}_{8\Leftrightarrow G_{\overline{u_i'T'}}}}_{\text{Transformation processes of the information } \overline{u_i'T'} \text{ in another related property and vice versa}} \tag{8.135}
$$

The terms indicated in Eq. (8.135) have the following meanings:

- Term 1, $T_{\overline{u_i'T'}}$: rate of change of *information* $\overline{u_i'T'}$ contained in the fluid particle.
- Term 2, $A_{\overline{u_i'T'}}$: represents the net advective flux of *information* $\overline{u_i'T'}$, through the surface of the fluid particle.
- Term 3, $D^m_{\overline{u_i'T'}}$: represents the net molecular diffusive flux of *information* $\overline{u_i'T'}$, through the surface of the fluid particle.
- Term 4, $D^t_{\overline{u_i'T'}}$: represents the turbulent net diffusive flux of *information* $\overline{u_i'T'}$, through the surface of the fluid particle.
- Term 5, $\Pi^i_{\overline{u_i'T'}}$: Represents the dispersion of *information* $\overline{u_i'T'}$, by the action of turbulence. This is a term for transporting *information* that is the subject of the balance equation.
- Term 6, $P_{\overline{u_i'T'}}$: represents the transformation of *information* $\overline{u_i'T'}$, between the mean flow and the entire spectrum of turbulent structures that make up the turbulent flow, promoted by the gradients of the mean velocity and temperature fields.
- Term 7, $\varepsilon_{\overline{u_i'T'}}$: represents the power of molecular viscous transformation of *information* $\overline{u_i'T'}$, in other similar *information.* This term is broken down into two

parts $\varepsilon_{\overline{u_i'T'}} = \varepsilon^{\alpha}_{\overline{u_i'T'}} + \varepsilon^{\nu}_{\overline{u_i'T'}}$, where $\varepsilon^{\alpha}_{\overline{u_i'T'}} = \alpha\overline{(\partial u_i'/\partial x_k)(\partial T'/\partial x_k)}$ and $\varepsilon^{\nu}_{\overline{u_i'T'}} = \nu\overline{(\partial u_i'/\partial x_k)(\partial T'/\partial x_k)}$. These portions represent, respectively, the thermal and viscous transformation power of the *information* $\overline{u_i'T'}$, in another type of related *information*. Here is a question: what does "another type of related *information*" mean? In the thermal energy balance equation, this "viscous transformation" transforms one type of energy into another type of energy! In the present case, this "viscous transformation" transforms the correlation $\overline{u_i'T'}$ in another similar correlation. The question remains unanswered! What is important to highlight here is that using the correct nomenclature leads us to reflect and search for answers! If we said that $\varepsilon_{\overline{u_i'T'}}$ represents a "dissipation", we wouldn't need to look for any answers! "Dissipate" is to transform something into "nothing"! And that doesn't require any further investigation! It is left to the reader's reflection.
- Term 8, $G_{\overline{u_i'T'}}$: represents the transformation of temperature fluctuations into *information* $\overline{u_i'T'}$, due to the gravitational field.

Terms (1), (2), (3) and (6) are calculated via the solution of the respective PDEs already presented. For term (8) in particular, a balance equation for the temperature variance $\overline{T'T'}$ will be derived. Terms (4), (5) and (7), must be modeled, as they constitute additional unknowns to the components of the turbulent thermal energy flux vector. However, before modeling all these terms, an order of magnitude analysis allows us to disregard the influence of some of them, when compared with the others. Following the ideas presented by Pimont [37] and Mompean [31], all the terms that make up Eq. (8.135) are of the same order of magnitude, with the exception of the terms $\alpha\overline{u_i'(\partial T'/\partial x_k)} \sim \mathrm{Re}_\ell^{-1/2}$, $\nu\overline{T'(\partial u_i'/\partial x_k)} \sim \mathrm{Pe}_\ell^{-1/2}$ and $(\alpha+\nu)\overline{(\partial u_i'/\partial x_k)(\partial T'/\partial x_k)} \sim \mathrm{Re}_\ell^{-1/2}$, where $\mathrm{Re}_\ell = (U\ell)/\nu$, $\mathrm{Pe}_\ell = (U\ell)/\alpha$ and ℓ is the integral length of the turbulence. Consequently, for flows at high values of these two dimensionless parameters, these terms are negligible when compared to others. Therefore, the following additional equations are proposed for the appropriate closures:

- Modeling of term (5) $\Pi_{\overline{u_i'T'}}$: Launder [31] proposed the modeling below, decomposing this term into three parts:

$$\Pi_{\overline{u_i'T'}} = \Pi_{1\overline{u_i'T'}} + \Pi_{2\overline{u_i'T'}} + \Pi_{3\overline{u_i'T'}}. \tag{8.136}$$

For the first part, $\Pi^1_{\overline{u_i'T'}}$, the most used model was proposed by Monin [32]:

$$\Pi^1_{\overline{u_i'T'}} = -c_{1\overline{u_i'T}}\overline{u_i'T'}\frac{\varepsilon}{k}. \tag{8.137}$$

Various values for the constant $c^1_{\overline{u_i'T'}}$ were proposed by different authors: Owen [35] proposed $c^1_{\overline{u_i'T'}} = 4.1$ for pipe flows; Wyngaard and Cote [55] proposed $c^1_{\overline{u_i'T'}} = 4.4$ for the component $\overline{u'T'}$ and $c^1_{\overline{u_i'T'}} = 9.7$ for the component $\overline{w'T'}$, to simulate an atmospheric boundary layer and finally Launder [21] proposed $c^1_{\overline{u_i'T'}} = 3.2$ to model and simulate turbulence downstream of a grid.

The other two terms, $\Pi^2_{\overline{u_i'T'}}$ and $\Pi^3_{\overline{u_i'T'}}$, are modeled as presented by Mompean [31]:

$$\Pi^2_{\overline{u_i'T'}} = c^2_{\overline{u_i'T'}}\overline{u_j'T'}\frac{\partial \overline{u}_i}{\partial x_j}, \tag{8.138}$$

$$\Pi^3_{\overline{u_i'T'}} = c^3_{\overline{u_i'T'}}\beta g_i \overline{T'T'}. \tag{8.139}$$

Regarding the constants, Lauder [22] used free shear flow simulations and proposed $c^2_{\overline{u_i'T'}} = 0.5$ and $c^3_{\overline{u_i'T'}} = 0.333$. Owen [35] also proposed $c^2_{\overline{u_i'T'}} = 0.5$, but did not propose any value for $c^3_{\overline{u_i'T'}}$.

- Term (4) modeling: Mompean [31] presents a model proposal for the triple correlation or the third-order moment, $\overline{u_i'u_k'T'}$. Regarding the third-order moment, $-\overline{T'p'}\delta_{ij}/\rho_o$, was disregarded, as proposed by Mompean [31]. However, the sum of these two terms that make up term 4 of Eq. (8.135) has a turbulent diffusion nature and can be modeled jointly, as proposed in the present work:

$$-\overline{u_i'u_k'T'} - \frac{1}{\rho_o}\overline{T'p'}\delta_{ij} = \frac{\nu_t}{\mathrm{Pr}_{\overline{u_i'T'}}}\left[\frac{\partial\left(\overline{u_i'T'}\right)}{\partial x_k} + \frac{\partial\left(\overline{u_k'T'}\right)}{\partial x_i}\right]. \tag{8.140}$$

The term $\mathrm{Pr}_{\overline{u_i'T'}} = 1$ can be used. So, substituting the terms modeled in Eq. (8.135) produces the equation below:

$$\begin{aligned}\frac{\partial\left(\overline{u_i'T'}\right)}{\partial t} + \overline{u}_k\frac{\partial\left(\overline{u_i'T'}\right)}{\partial x_k} = {} & \frac{\partial}{\partial x_k}\left\{\frac{\nu_t}{\mathrm{Pr}_{\overline{u_i'T'}}}\left[\frac{\partial\left(\overline{u_i'T'}\right)}{\partial x_k} + \frac{\partial\left(\overline{u_k'T'}\right)}{\partial x_i}\right]\right\} - \overline{u_i'u_k'}\frac{\partial \overline{T}}{\partial x_k} \\ & - c^1_{\overline{u_i'T'}}\overline{u_i'T'}\frac{\varepsilon}{k} - \left(1 - c^2_{\overline{u_i'T'}}\right)\overline{u_j'T'}\frac{\partial \overline{u}_i}{\partial x_j} \\ & - \left(1 - c^3_{\overline{u_i'T'}}\right)g_i\beta\overline{T'T'}.\end{aligned} \tag{8.141}$$

All terms of Eq. (8.141) are properly modeled, except for the second-order moment $\overline{T'T'}$. This term closure can be done with an additional balance equation, which is presented below.

8.4.3 *Balance Equation for Temperature Fluctuation Variance—$\overline{T'T'}$*

Starting from Eq. (8.130), multiplying it by $2T'$ and using the incompressibility hypothesis, the following equation is obtained:

$$
\begin{aligned}
&2T'\frac{\partial T'}{\partial t}+2T'\frac{\partial\left(\bar{u}_k T'\right)}{\partial x_k}+2T'\frac{\partial\left(u'_k T\right)}{\partial x_k}+2T'\frac{\partial\left(u'_k T'\right)}{\partial x_k}\\
&=2\alpha T'\frac{\partial^2 T'}{\partial x_k\partial x_k}-2T'\frac{\partial\left(\overline{u'_k T'}\right)}{\partial x_k}\Leftrightarrow\frac{\partial\left(T'T'\right)}{\partial t}\\
&+\bar{u}_k\frac{\partial\left(T'T'\right)}{\partial x_k}+2T'u'_k\frac{\partial\bar{T}}{\partial x_k}+\frac{\partial\left(u'_k T'T'\right)}{\partial x_k}\\
&=\alpha\frac{\partial^2\left(T'T'\right)}{\partial x_k\partial x_k}-2\alpha\frac{\partial T'}{\partial x_k}\frac{\partial T'}{\partial x_k}-2T'\frac{\partial\left(\overline{u'_k T'}\right)}{\partial x_k}\Leftrightarrow\cdot\frac{\partial\left(\overline{T'T'}\right)}{\partial t}\\
&+\bar{u}_k\frac{\partial\left(\overline{T'T'}\right)}{\partial x_k}+2\overline{T'u'_k}\frac{\partial\bar{T}}{\partial x_k}+\frac{\partial\left(\overline{u'_k T'T'}\right)}{\partial x_k}\\
&=\alpha\frac{\partial^2\left(\overline{T'T'}\right)}{\partial x_k\partial x_k}-2\alpha\overline{\frac{\partial T'}{\partial x_k}\frac{\partial T'}{\partial x_k}}-2\boxed{\overline{T'}}\frac{\partial\left(\overline{u'_k T'}\right)}{\partial x_k}.
\end{aligned}
\tag{8.142}
$$

The interpretation of each term of Eq. (8.142) is presented below, in the standard format of this text:

$$
\underbrace{\underbrace{\frac{\partial\left(\overline{T'T'}\right)}{\partial t}}_{1\Leftrightarrow T_{\overline{T'T'}}}}_{\substack{\text{Rate of change of information }\overline{T'T'}\\ \text{contained in the fluid particle}}}
$$

$$
=\underbrace{\underbrace{-\bar{u}_k\frac{\partial\left(\overline{T'T'}\right)}{\partial x_k}}_{2\Leftrightarrow A_{\overline{T'T'}}}\underbrace{+\alpha\frac{\partial^2\left(\overline{T'T'}\right)}{\partial x_k\partial x_k}}_{3\Leftrightarrow D^m_{\overline{T'T'}}}\underbrace{-\frac{\partial^2\left(\overline{u'_k T'T'}\right)}{\partial x_k}}_{4\Leftrightarrow D^t_{\overline{T'T'}}}}_{\substack{\text{Transport processes of the information }\overline{T'T'}\\ \text{between a fluid particle and its surroundings}}}\quad\underbrace{\underbrace{-2\overline{T'u'_k}\frac{\partial\bar{T}}{\partial x_k}}_{5\Leftrightarrow P_{\overline{T'T'}}}\underbrace{-2\alpha\overline{\frac{\partial T'}{\partial x_k}\frac{\partial T'}{\partial x_k}}}_{6\Leftrightarrow\varepsilon_{\overline{T'T'}}}}_{\substack{\text{Processes of transforming information }\overline{T'T'}\\ \text{into another related quantity and vice versa}}}.
\tag{8.143}
$$

The terms marked in Eq. (8.143) can be interpreted as:

- Term 1, $T_{\overline{T'T'}}$: rate of change of *information* $\left(\overline{T'T'}\right)$ which is contained in a differential fluid particle.

- Term 2, $A_{\overline{T'T'}}$: advective net flux of *information* $\left(\overline{T'T'}\right)$ through the surface of the fluid particle.
- Term 3, $D^m_{\overline{T'T'}}$: molecular diffusive net flux of *information* $\left(\overline{T'T'}\right)$ through the control surface surrounding the fluid particle.
- Term 4, $D^t_{\overline{T'T'}}$: turbulent diffusive net flux of *information* $\left(\overline{T'T'}\right)$ through by the control surface surrounding the fluid particle.
- Term 5, $P_{\overline{T'T'}}$: transformation of *information* $\left(\overline{T'T'}\right)$ contained in the filtered flow into the same *information* contained in the entire spectrum of turbulent structures. This transformation is promoted by filtered temperature gradients.
- Term 6, $\varepsilon_{\overline{T'T'}}$: specific molecular viscous transformation power of temperature fluctuation variance $\left(\overline{T'T'}\right)$, in another related *information.*

Terms (1), (2) and (3) are objects of the balance Eq. (8.143) itself. Term (5), $2\overline{T'u'_k}\partial \overline{T}/\partial x_k$, is modeled through the modeling of the term $\overline{T'u'_k}$, which has already been done with Eq. (8.141). The term $\alpha\partial^2\left(\overline{T'T'}\right)/\partial x_k \partial x_k$, is of the order of magnitude given by the reciprocal of the Peclet Number, Pe_ℓ^{-1}, and, by the order of magnitude of the other significant terms in Eq. (8.143), this term can be neglected.

Terms (4) and (6) must be modeled. Term (4) can be modeled using the linear or first gradient modeling approach, commonly used to model the turbulent diffusive flux of a given *information.* In this case, as presented by Mompean [31], the following model proposal is transcribed:

$$-\overline{u'_k T'T'} = c_{\overline{T'T'}} \frac{k^2}{\varepsilon} \frac{\partial\left(\overline{T'T'}\right)}{\partial x_k}. \tag{8.144}$$

Spalding [49] proposed $c_{\overline{T'T'}} = 0.13$, while Lemos and Sensonske [24] proposed $c_{\overline{T'T'}} = 0.1$. Another way to model this term is to use the Prandtl Number for the temperature variance $P_{\overline{T'T'}}$ and remembering that $\nu_t = c_\mu k^2/\varepsilon$:

$$-\overline{u'_k T'T'} = c_{\overline{T'T'}} \frac{\nu_t}{c_\mu} \frac{\partial\left(\overline{T'T'}\right)}{\partial x_k} \equiv \frac{\nu_t}{\mathrm{Pr}_{\overline{T'T'}}} \frac{\partial\left(\overline{T'T'}\right)}{\partial x_k}, \tag{8.145}$$

where $\mathrm{Pr}_{\overline{T'T'}} = C_\mu / C_{\overline{T'T'}}$. As $C_\mu = 0.09$ and $c_{\overline{T'T'}} = 0.13$, we have $\mathrm{Pr}_{\overline{T'T'}} = 0.7$. Substituting these closure models in Eq. (8.143), results in:

$$\frac{\partial\left(\overline{T'T'}\right)}{\partial t} + \overline{u}_k \frac{\partial\left(\overline{T'T'}\right)}{\partial x_k} = -2\overline{T'u'_k}\frac{\partial \overline{T}}{\partial x_k} + \frac{\partial}{\partial x_k}\left[\frac{\nu_t}{\mathrm{Pr}_{\overline{T'T'}t}} \frac{\partial}{\partial x_k}\left(\overline{T'T'}\right)\right] - 2\varepsilon_{\overline{T'T'}}. \tag{8.146}$$

Modeling of term (6), $\varepsilon_{\overline{T'T'}}$, in Eq. (8.143), can be done by deducing a balance equation for this information, which is presented below.

8.4.4 Balance Equation for the Molecular Viscous Transformation—$\varepsilon_{\overline{T'T'}}$

Starting from Eq. (8.130) for T', rewritten below:

$$\frac{\partial T'}{\partial t} + u'_j \frac{\partial \overline{T}}{\partial x_j} + \overline{u}_j \frac{\partial T'}{\partial x_j} + \frac{\partial}{\partial x_j}\left(u'_j T' - \overline{u'_j T'}\right) = \alpha \frac{\partial^2 T'}{\partial x_j \partial x_j}. \tag{8.147}$$

After applying the gradient operator, $\partial()/\partial x_k$, to all terms of this Equation, we have Eq. (8.148):

$$\begin{aligned}&\frac{\partial}{\partial t}\left(\frac{\partial T'}{\partial x_k}\right) + \frac{\partial}{\partial x_k}\left(u'_j \frac{\partial \overline{T}}{\partial x_j}\right) + \frac{\partial}{\partial x_k}\left(\overline{u}_j \frac{\partial T'}{\partial x_j}\right)\\ &+ \frac{\partial}{\partial x_k}\left[\frac{\partial}{\partial x_j}\left(u'_j T' - \overline{u'_j T'}\right)\right] = \alpha \frac{\partial}{\partial x_k}\left(\frac{\partial^2 T'}{\partial x_j \partial x_j}\right).\end{aligned} \tag{8.148}$$

Taking the dot product between the vector $\alpha \partial T'/\partial x_k$ and all terms of Eq. (8.148) yields:

$$\begin{aligned}&\underbrace{\alpha \frac{\partial T'}{\partial x_k}\frac{\partial}{\partial t}\left(\frac{\partial T'}{\partial x_k}\right)}_{1} + \underbrace{\alpha \frac{\partial T'}{\partial x_k}\frac{\partial}{\partial x_k}\left(u'_j \frac{\partial \overline{T}}{\partial x_j}\right)}_{4} + \underbrace{\alpha \frac{\partial T'}{\partial x_k}\frac{\partial}{\partial x_k}\left(\overline{u}_j \frac{\partial T'}{\partial x_j}\right)}_{3}\\ &+ \underbrace{\alpha \frac{\partial T'}{\partial x_k}\frac{\partial}{\partial x_k}\left[\frac{\partial}{\partial x_j}\left(u'_j T' - \overline{u'_j T'}\right)\right]}_{5} = \underbrace{\alpha^2 \frac{\partial T'}{\partial x_k}\frac{\partial}{\partial x_k}\left(\frac{\partial^2 T'}{\partial x_j \partial x_j}\right)}_{4}.\end{aligned} \tag{8.149}$$

Applying the filtering operator to each term in Eq. (8.149) and dealing with each one separately, we arrive at:

$$\text{Term 1:} \left\{ \overline{\alpha \frac{\partial T'}{\partial x_k}\frac{\partial}{\partial t}\left(\frac{\partial T'}{\partial x_k}\right)} = \frac{\partial}{\partial t}\left(\frac{1}{2}\alpha \overline{\frac{\partial T'}{\partial x_k}\frac{\partial T'}{\partial x_k}}\right) = \boxed{\frac{1}{2}\frac{\partial \varepsilon_{\overline{T'T'}}}{\partial t}} \right.. \tag{8.150}$$

$$\text{Term 2:} \left\{ \begin{aligned}&\overline{\alpha \frac{\partial T'}{\partial x_k}\frac{\partial}{\partial x_k}\left(u'_j \frac{\partial \overline{T}}{\partial x_j}\right)} = \overline{\alpha \frac{\partial T'}{\partial x_k}\left[\frac{\partial \overline{T}}{\partial x_j}\frac{\partial u'_j}{\partial x_k} + u'_j \frac{\partial^2 \overline{T}}{\partial x_k \partial x_j}\right]}\\ &= \boxed{\alpha \overline{\frac{\partial T'}{\partial x_k}\frac{\partial u'_j}{\partial x_k}}\frac{\partial \overline{T}}{\partial x_j} + \alpha \overline{u'_j \frac{\partial T'}{\partial x_k}}\frac{\partial^2 \overline{T}}{\partial x_k \partial x_j}}\end{aligned} \right.. \tag{8.151}$$

$$\text{Term 3:}\left\{\begin{aligned}
&\alpha\overline{\frac{\partial T'}{\partial x_k}\frac{\partial}{\partial x_k}\left(\bar{u}_j\frac{\partial T'}{\partial x_j}\right)}=\alpha\overline{\frac{\partial T'}{\partial x_k}\left[\bar{u}_j\frac{\partial}{\partial x_j}\left(\frac{\partial T'}{\partial x_k}\right)+\frac{\partial T'}{\partial x_j}\frac{\partial \bar{u}_j}{\partial x_k}\right]}\\
&=\alpha\bar{u}_j\overline{\frac{\partial T'}{\partial x_k}\frac{\partial}{\partial x_j}\left(\frac{\partial T'}{\partial x_k}\right)}+\alpha\overline{\frac{\partial T'}{\partial x_k}\frac{\partial T'}{\partial x_j}}\frac{\partial \bar{u}_j}{\partial x_k}=\frac{\bar{u}_j}{2}\frac{\partial}{\partial x_j}\left(\alpha\overline{\frac{\partial T'}{\partial x_k}\frac{\partial T'}{\partial x_k}}\right)\\
&+\alpha\overline{\frac{\partial T'}{\partial x_k}\frac{\partial T'}{\partial x_j}}\frac{\partial \bar{u}_j}{\partial x_k}=\boxed{\frac{\bar{u}_j}{2}\frac{\partial \varepsilon_{\overline{T'T'}}}{\partial x_j}+\alpha\overline{\frac{\partial T'}{\partial x_k}\frac{\partial T'}{\partial x_j}}\frac{\partial \bar{u}_j}{\partial x_k}}
\end{aligned}\right. . \tag{8.152}$$

$$\text{Term 4:}\left\{\begin{aligned}
&\alpha\overline{\frac{\partial T'}{\partial x_k}\frac{\partial}{\partial x_k}\left[\frac{\partial}{\partial x_j}\left(u'_jT'-\overline{u'_jT'}\right)\right]}=\alpha\overline{\frac{\partial T'}{\partial x_k}\frac{\partial}{\partial x_k}\frac{\partial}{\partial x_j}\left(u'_jT'\right)}\\
&-\alpha\overline{\frac{\partial T'}{\partial x_k}\frac{\partial}{\partial x_k}\left[\frac{\partial}{\partial x_j}\left(\overline{u'_jT'}\right)\right]}=\alpha\overline{\frac{\partial T'}{\partial x_k}\frac{\partial}{\partial x_k}\left(u'_j\frac{\partial T'}{\partial x_j}\right)}-\alpha\boxed{\overline{\frac{\partial T'}{\partial x_k}}}\frac{\partial}{\partial x_k}\left[\frac{\partial}{\partial x_j}\left(\overline{u'_jT'}\right)\right]\\
&=\alpha\overline{\frac{\partial T'}{\partial x_k}\frac{\partial u'_j}{\partial x_k}\frac{\partial T'}{\partial x_j}}+\alpha\overline{u'_j\frac{\partial T'}{\partial x_k}\frac{\partial}{\partial x_k}\left(\frac{\partial T'}{\partial x_j}\right)}=\alpha\overline{\frac{\partial T'}{\partial x_k}\frac{\partial u'_j}{\partial x_k}\frac{\partial T'}{\partial x_j}}\\
&+\alpha\overline{u'_j\frac{\partial T'}{\partial x_k}\frac{\partial}{\partial x_j}\left(\frac{\partial T'}{\partial x_k}\right)}=\boxed{\alpha\overline{\frac{\partial T'}{\partial x_k}\frac{\partial u'_j}{\partial x_k}\frac{\partial T'}{\partial x_j}}+\frac{\alpha}{2}\overline{u'_j\frac{\partial}{\partial x_j}\left(\frac{\partial T'}{\partial x_k}\frac{\partial T'}{\partial x_k}\right)}}
\end{aligned}\right. . \tag{8.153}$$

$$\text{Term 5:}\left\{\begin{aligned}
&\alpha^2\overline{\frac{\partial T'}{\partial x_k}\frac{\partial}{\partial x_k}\left(\frac{\partial^2 T'}{\partial x_j\partial x_j}\right)}=\alpha^2\frac{\partial}{\partial x_k}\overline{\left(\frac{\partial T'}{\partial x_k}\frac{\partial^2 T'}{\partial x_j\partial x_j}\right)}-\overline{\left(\alpha\frac{\partial^2 T'}{\partial x_j\partial x_j}\right)^2}\\
&=\alpha^2\frac{\partial}{\partial x_k}\left[\overline{\frac{\partial}{\partial x_j}\left(\frac{\partial T'}{\partial x_k}\frac{\partial T'}{\partial x_j}\right)}-\overline{\frac{\partial T'}{\partial x_j}\frac{\partial}{\partial x_j}\left(\frac{\partial T'}{\partial x_k}\right)}\right]-\overline{\left(\alpha\frac{\partial^2 T'}{\partial x_j\partial x_j}\right)^2}\\
&=\alpha^2\frac{\partial}{\partial x_k}\left[\overline{\frac{\partial}{\partial x_j}\left(\frac{\partial T'}{\partial x_k}\frac{\partial T'}{\partial x_j}\right)}-\overline{\frac{\partial T'}{\partial x_j}\frac{\partial}{\partial x_k}\left(\frac{\partial T'}{\partial x_j}\right)}\right]-\overline{\left(\alpha\frac{\partial^2 T'}{\partial x_j\partial x_j}\right)^2}\\
&=\alpha^2\frac{\partial}{\partial x_k}\left[\frac{\partial}{\partial x_j}\left(\overline{\frac{\partial T'}{\partial x_k}\frac{\partial T'}{\partial x_j}}\right)\right]-\frac{1}{2}\alpha\frac{\partial}{\partial x_k}\left[\frac{\partial}{\partial x_k}\left(\alpha\overline{\frac{\partial T'}{\partial x_j}\frac{\partial T'}{\partial x_j}}\right)\right]-\overline{\left(\alpha\frac{\partial^2 T'}{\partial x_j\partial x_j}\right)^2}
\end{aligned}\right. . \tag{8.154}$$

Returning these manipulated terms to Eq. (8.149) produces:

$$\begin{aligned}
&\frac{1}{2}\frac{\partial \varepsilon_{\overline{T'T'}}}{\partial t}+\alpha\overline{\frac{\partial T'}{\partial x_k}\frac{\partial u'_j}{\partial x_k}}\frac{\partial \overline{T}}{\partial x_j}+\alpha\overline{u'_j\frac{\partial T'}{\partial x_k}}\frac{\partial^2 \overline{T}}{\partial x_k\partial x_j}+\frac{\bar{u}_j}{2}\frac{\partial \varepsilon_{\overline{T'T'}}}{\partial x_j}+\alpha\overline{\frac{\partial T'}{\partial x_k}\frac{\partial T'}{\partial x_j}}\frac{\partial \bar{u}_j}{\partial x_k}\\
&+\alpha\overline{\frac{\partial T'}{\partial x_k}\frac{\partial u'_j}{\partial x_k}\frac{\partial T'}{\partial x_j}}+\frac{\alpha}{2}\overline{u'_j\frac{\partial}{\partial x_j}\left(\frac{\partial T'}{\partial x_k}\frac{\partial T'}{\partial x_k}\right)}=\alpha^2\frac{\partial}{\partial x_k}\overline{\left(\frac{\partial T'}{\partial x_k}\frac{\partial^2 T'}{\partial x_j\partial x_j}\right)}\\
&-\overline{\left(\alpha\frac{\partial^2 T'}{\partial x_j\partial x_j}\right)^2}\Leftrightarrow\frac{\partial \varepsilon_{\overline{T'T'}}}{\partial t}+\bar{u}_j\frac{\partial \varepsilon_{\overline{T'T'}}}{\partial x_j}=-2\alpha\overline{\frac{\partial T'}{\partial x_k}\frac{\partial u'_j}{\partial x_k}}\frac{\partial T}{\partial x_j}\\
&-2\alpha\overline{u'_j\frac{\partial T'}{\partial x_k}}\frac{\partial^2 \overline{T}}{\partial x_k\partial x_j}-2\alpha\overline{\frac{\partial T'}{\partial x_k}\frac{\partial T'}{\partial x_j}}\frac{\partial \bar{u}_j}{\partial x_k}-2\overline{\left(\alpha\frac{\partial^2 T'}{\partial x_j\partial x_j}\right)^2}\\
&-2\alpha\overline{\frac{\partial T'}{\partial x_k}\frac{\partial u'_j}{\partial x_k}\frac{\partial T'}{\partial x_j}}-\frac{2\alpha}{2}\overline{u'_j\frac{\partial}{\partial x_j}\left(\frac{\partial T'}{\partial x_k}\frac{\partial T'}{\partial x_k}\right)}-\frac{1}{2}2\alpha\frac{\partial}{\partial x_j}\left[\frac{\partial}{\partial x_j}\left(\alpha\overline{\frac{\partial T'}{\partial x_k}\frac{\partial T'}{\partial x_k}}\right)\right]\\
&+2\alpha^2\frac{\partial}{\partial x_k}\left[\overline{\frac{\partial}{\partial x_j}\left(\frac{\partial T'}{\partial x_k}\frac{\partial T'}{\partial x_j}\right)}\right].
\end{aligned}\tag{8.155}$$

The term $2\alpha^2\partial\left[\partial\left(\overline{(\partial T'/\partial x_k)(\partial T'/\partial x_j)}\right)/\partial x_j\right]/\partial x_k$ is not considered for thermally isotropic turbulence, in accordance with Mompean [31] and Newman et al. [34]. Using the following settings $\varepsilon^{\alpha}_{\overline{u'_i T'},j} = \overline{\left(\partial u'_j/\partial x_k\right)(\partial T'/\partial x_k)}$ and $\varepsilon'_{\overline{T'T'}} = \overline{\alpha\left(\partial u'_i/\partial x_k\right)(\partial T'/\partial x_k)}$, in Eq. (8.155), it can be rewritten as Eq. (8.156):

$$\frac{\partial\varepsilon_{\overline{T'T'}}}{\partial t}+\bar{u}_j\frac{\partial\varepsilon_{\overline{T'T'}}}{\partial x_j}=-\left(2\varepsilon\frac{\alpha}{\overline{u'_iT'}}\frac{\partial\overline{T}}{\partial x_j}+2\alpha\overline{u'_j\frac{\partial T'}{\partial x_k}}\frac{\partial^2\overline{T}}{\partial x_k\partial x_j}\right)-2\alpha\overline{\frac{\partial T'}{\partial x_k}\frac{\partial T'}{\partial x_j}}\frac{\partial\bar{u}_j}{\partial x_k}$$
$$-2\alpha\overline{\frac{\partial T'}{\partial x_k}\frac{\partial u'_j}{\partial x_k}\frac{\partial T'}{\partial x_j}}-2\overline{\left(\alpha\frac{\partial^2 T'}{\partial x_j\partial x_j}\right)^2}-\frac{\partial}{\partial x_j}\left(\alpha\frac{\partial\varepsilon_{\overline{T'T'}}}{\partial x_j}\right)-\frac{\partial}{\partial x_j}\left(\overline{u'_j\varepsilon_{\overline{T'T'}}}\right). \quad (8.156)$$

Equation (8.156) is rewritten below, highlighting the fundamental physical processes involved in the balance of *information* $\varepsilon_{\overline{T'T'}}$, that is, the rate of variation, the transport processes and the transformation processes of this *information*:

$$\underbrace{\underbrace{\frac{\partial\varepsilon_{\overline{T'T'}}}{\partial t}}_{1\Leftrightarrow T\varepsilon_{\overline{T'T'}}}}_{\text{Rate of change of the information }\varepsilon_{\overline{T'T'}}\text{ contained in the fluid particle}}=$$

$$\underbrace{\underbrace{-\bar{u}_j\frac{\partial\varepsilon_{\overline{T'T'}}}{\partial x_j}}_{2\Leftrightarrow A\varepsilon_{\overline{T'T'}}}\underbrace{-\frac{\partial}{\partial x_j}\left(\alpha\frac{\partial\varepsilon_{\overline{T'T'}}}{\partial x_j}\right)}_{3\Leftrightarrow D^m_{\varepsilon\,\overline{T'T'}}}\underbrace{-\frac{\partial}{\partial x_j}\left(\overline{u'_j\varepsilon'_{\overline{T'T'}}}\right)}_{4\Leftrightarrow D^t_{\varepsilon\overline{T'T'}}}\underbrace{-2\alpha\overline{\frac{\partial T'\partial u'_j\partial T'}{\partial x_k\partial x_k\partial x_j}}}_{5\Leftrightarrow\Re_{\varepsilon\,\overline{T'T'}}}}_{\text{Transport processes of the information }\varepsilon_{\overline{T'T'}}\text{ between a fluid particle and its surroundings}}$$

$$\underbrace{\underbrace{-\left(2\varepsilon\frac{\alpha}{\overline{u'_iT'}_{,j}}\frac{\partial\bar{T}}{\partial x_j}+2\alpha\overline{u'_j\frac{\partial T'}{\partial x_k}}\frac{\partial^2\bar{T}}{\partial x_k\partial x_j}\right)}_{6\Leftrightarrow P^{\bar{T}}_{\varepsilon\,\overline{T'T'}}}\underbrace{-2\alpha\overline{\frac{\partial T'\partial T'}{\partial x_k\partial x_j}}\frac{\partial\bar{u}_j}{\partial x_k}}_{7\Leftrightarrow P^{\bar{u}}_{\varepsilon\,\overline{T'T'}}}\underbrace{-2\alpha\overline{\left(\frac{\partial^2 T'}{\partial x_j\partial x_j}\right)^2}}_{8\Leftrightarrow\Im\varepsilon\,\overline{T'T'}}}_{\text{Processes of transforming information }\varepsilon_{\overline{T'T'}}\text{ into other related informations and vice versa}} \quad (8.157)$$

The terms that appear in Eq. (8.157) are described below:

- Term 1, $T_{\varepsilon_{\overline{T'T'}}}$: represents the rate of change in the amount of *information* $\varepsilon_{\overline{T'T'}}$, contained in the fluid particle.
- Term 2, $A_{\varepsilon_{\overline{T'T'}}}$: represents the net advective flux of *information* $\varepsilon_{\overline{T'T'}}$, through the surface surrounding the fluid particle.
- Term 3, $D^m_{\varepsilon_{\overline{T'T'}}}$: the vector *information* $\alpha\partial\varepsilon_{\overline{T'T'}}/\partial x_j$ represents the molecular diffusive flux of *information* $\varepsilon_{\overline{T'T'}}$. The divergence of this vector, $\partial\left(\alpha\partial\varepsilon_{\overline{T'T'}}/\partial x_j\right)/\partial x_j$, represents the net molecular diffusive flux across the boundary of a fluid particle.

- Term 4, $D^{t}_{\varepsilon_{\overline{T'T'}}}$: the vector *information* $\overline{u'_j\varepsilon'_{\overline{T'T'}}}$ has effectively an advective nature. However, preparing it for its closure modeling, it comes to be interpreted as a turbulent diffusive flux of *information* $\varepsilon_{\overline{T'T'}}$. The divergence of this flux, $\partial\left(\overline{u'_j\varepsilon'_{\overline{T'T'}}}\right)/\partial x_j$, represents the turbulent net diffusive flux of information $\varepsilon_{\overline{T'T'}}$, through the surface of the fluid particle.
- Term 5, $\Re_{\varepsilon_{\overline{T'T'}}}$: represents the redistribution of information $\varepsilon_{\overline{T'T'}}$ due to the action of turbulence.
- Term 6, $P^{\overline{T}}_{\varepsilon_{\overline{T'T'}}}$: represents how much of the information $\varepsilon_{\overline{T'T'}}$ is being exchanged between the filtered temperature field and turbulence. This is a local, volumetric process, and not a transport process, which would occur if it were from one point to another. Therefore, it is characterized as a transformation process, which is promoted by filtered temperature gradients.
- Term 7, $P^{\overline{u}}_{\varepsilon_{\overline{T'T'}}}$: represents how much of the *information* $\varepsilon_{\overline{T'T'}}$ is being removed from the filtered field. This is also a local, volumetric process, and not transport from one point to another. Therefore, it is also characterized as a transformation process, promoted by velocity gradients and filtered temperature gradients.
- Term 8, $\Im_{\varepsilon_{\overline{T'T'}}}$: represents the power of molecular transformation of *information* $\varepsilon_{\overline{T'T'}}$ in another type of correlated information. This term is strictly positive, but with a negative sign, coming from the balance. Thus, it always represents a negative portion of the rate of change, $\partial\varepsilon_{\overline{T'T'}}/\partial t$.

By means of an order of magnitude analysis of all terms in Eq. (8.157) it is shown that $\alpha\left(\overline{u'_j\partial T'/\partial x_k}\right)\left(\partial^2\overline{T}/\partial x_k\partial x_j\right)\sim \mathrm{Re}_\ell^{-1/2}$ and $\partial\left(\alpha\partial\varepsilon_{\overline{T'T'}}/\partial x_j\right)/\partial x_j\sim \mathrm{Pe}_\ell^{-1}$. Comparing with the order of magnitude of the other terms, it is shown that these two terms are negligible. After eliminating these terms from Eq. (8.157), this equation can be simplified, giving rise to Eq. (8.158), in which its terms were renumbered:

$$\underbrace{\frac{\partial\varepsilon_{\overline{T'T'}}}{\partial t}}_{1\Leftrightarrow T_{\varepsilon_{\overline{T'T'}}}}+\underbrace{\frac{\partial\varepsilon_{\overline{T'T'}}}{\partial x_j}}_{2\Leftrightarrow A_{\varepsilon_{\overline{T'T'}}}}=\underbrace{-\frac{\partial}{\partial x_j}\left(\overline{u'_j\varepsilon'_{\overline{T'T'}}}\right)}_{3\Leftrightarrow D^{t}_{\varepsilon_{\overline{T'T'}}}}\underbrace{-2\varepsilon^{\alpha}_{\overline{T'_iT'}}\frac{\partial\overline{T}}{\partial x_j}}_{4\Leftrightarrow P^{\overline{T}}_{\varepsilon_{\overline{T'T'}}}}\underbrace{-2\alpha\overline{\frac{\partial T'}{\partial T'}\frac{\partial T'}{\partial x_j}\frac{\partial\overline{u}_j}{\partial x_k}}}_{5\Leftrightarrow P^{\overline{u}}_{\varepsilon_{\overline{T'T'}}}}$$
$$\underbrace{-2\alpha\overline{\frac{\partial u'_j}{\partial x_k}\frac{\partial T'}{\partial x_k}\frac{\partial T'}{\partial x_j}}}_{6\Leftrightarrow\Re_{\varepsilon\overline{T'T'}}}\underbrace{-2\overline{\left(\alpha\frac{\partial^2T'}{\partial x_j\partial x_j}\right)^2}}_{7\Leftrightarrow\Im_{\overline{T'T'}}}. \tag{8.158}$$

All terms on the right side of Eq. (8.158) must be modeled. Term 3 can be modeled, as presented by Mompean [31], using the first gradient rule as the basis:

$$-\overline{u'_j\varepsilon'_{\overline{T'T'}}}=\frac{\nu_t}{\mathrm{Pr}_{\varepsilon_{\overline{T'T'}}}}\frac{\partial\varepsilon_{\overline{T'T'}}}{\partial x_j}, \tag{8.159}$$

where $\mathrm{Pr}_{\varepsilon\overline{T'T'}}=1.0$ is used in the literature. Modeling terms 4 and 5 follow the proposal of Chung and Sung [4]:

$$-2\varepsilon^{\alpha}_{u_i'T'}\frac{\partial \overline{T}}{\partial x_j}-2\alpha\overline{\frac{\partial T'}{\partial x_k}\frac{\partial T'}{\partial x_j}\frac{\partial \overline{u}_j}{\partial x_k}}=-c_c\frac{\varepsilon_{\overline{T'T'}}}{\overline{T'T'}}\overline{u_j'T'}\frac{\partial \overline{T}}{\partial x_j}+c_d\left(P^k+G^k\right)\frac{\varepsilon_{\overline{T'T'}}}{k}. \quad (8.160)$$

Based on the works of Newman et al. [34], Mompean [31] presents the modeling for terms 6 and 7. This modeling is presented again below:

$$2\left\{\alpha\overline{\frac{\partial T'}{\partial x_k}\frac{\partial u_j'}{\partial x_k}\frac{\partial T'}{\partial x_j}}+\overline{\left(\alpha\frac{\partial^2 T'}{\partial x_j\partial x_j}\right)^2}\right\}=\left\{c_a\varepsilon\overline{T'T'}\frac{\varepsilon}{k}+c_b\frac{\left(\varepsilon_{\overline{T'T'}}\right)^2}{\overline{T'T'}}\right\}. \quad (8.161)$$

Newman et al. [33] determined the first three constants using grid turbulence experiments, $c_a = 0.8$, $2.0 \leq c_b \leq 3.3$ and $c_c = 1.96$. Chung and Sung [4], to simulate boundary layer flows, used $c_b = 2.0$ and proposed $c_d = 0.8$. Returning these modeled terms to Eq. (8.158), the following balance equation was obtained, completely modeled:

$$\frac{\partial \varepsilon\overline{T'T'}}{\partial t}+\overline{u}_j\frac{\partial \varepsilon\overline{T'T'}}{\partial x_j}=\frac{\partial}{\partial x_j}\left(\frac{\nu_t}{\mathrm{Pr}_\varepsilon}\frac{\partial \varepsilon\overline{T'T'}}{\partial x_j}\right)-\left(c_a\varepsilon\overline{T'T'}\frac{\varepsilon}{k}+c_b\frac{\left(\varepsilon\overline{T'T'}\right)^2}{\overline{T'T'}}\right)$$
$$-c_c\frac{\varepsilon\overline{T'T'}}{\overline{T'T'}}\overline{u_j'T'}\frac{\partial \overline{T}}{\partial x_j}+c_d\left(P^k+G^k\right)\frac{\varepsilon\overline{T'T'}}{k}. \quad (8.162)$$

This completes the full RSM Model, including all fluid dynamics modeling as well as thermal modeling. The objective of this book was to organize the derivation and present the complete model in a single text. Despite all the development in more modern turbulence modeling methodologies, such as DNS, LES and hybrid methodologies, the author's point of view on the importance of URANS methodologies for engineering applications is recorded here. Undoubtedly, the RSM Methodology is one of the most representative mathematical models of the physical nature of turbulence. However, the degree of difficulty for its computational implementation is also noted and, mainly, the difficult computational convergence, as it is a model composed of 17 balance equations. The composition of this mathematical model is summarized below:

- Continuity equation: 1
- Filtered Navier–Stokes equations: 3 ($\overline{u}_i$, $i = 1, 2, 3$)
- Filtered thermal energy equation: 1
- Components of the Boussinesq–Reynolds tensor: 6 ($\overline{u_i'u_j'}$, $i = 1, 2, 3$, $j = 1, 2, 3$). This is a symmetric tensor
- Specific power of viscous transformation of turbulent kinetic energy: 1 (ε)
- Turbulent thermal flux: 3 ($\overline{u_j'T'}, j = 1, 2, 3$)
- Temperature variance: 1 ($\overline{T'T'}, j = 1, 2, 3$) and
- Specific power of molecular viscous transformation of temperature variance: 1 ($\varepsilon_{\overline{T'T'}}$).

8.5 Boussinesq–Reynolds Algebraic-Differential Stress Model (ASM—Algebraic Stress Model)

8.5.1 *Implicit Algebraic-Differential Shear Model (IASM)*

Turbulent flows are characterized by a broad band of quantities, including characteristic lengths and times. The integral quantities, which globally characterize turbulent flows, change dramatically from one flow to another.

Among the two-equation closure models, the $k-\varepsilon$ Model [20] stands out. Unlike Prandtl's mixing length model, for the $k-\varepsilon$ model, it is not necessary, a priori, to specify empirical characteristic lengths and times. However, despite this great advantage of the $k-\varepsilon$ closure model, it continues to be directly dependent on Boussinesq's Turbulent Viscosity concept, which characterizes serious limitations, like all closure models that are based on this concept. Among these limitations, the difficulty of modeling the effects of body forces caused by curvature or rotation stands out. Lumley [25], Lauder [23] and Speziale et al. [51] identified difficulties in simulating secondary recirculation in turbulent flows inside rectangular conduits. The model based on balance equations for the components of the Boussinesq–Reynolds Tensor was proposed to overcome these difficulties, as presented in the previous item. With this type of model, it was possible to treat the so-called non-local effects of turbulence, which are anisotropic in nature.

The Algebraic Stresses Model (ASM) for the Boussinesq–Reynolds tensor, the object of this item, is a simplification of the Boussinesq–Reynolds Stress Model (RSM), with the concomitant use of the $k-\varepsilon$. These two models are rewritten here, using the notation adopted in this book:

$$\underbrace{\frac{\partial\left(\overline{u_i'u_j'}\right)}{\partial t}}_{1\Leftrightarrow\overline{T_{ij}}}+\underbrace{\bar{u}_k\frac{\partial\left(\overline{u_i'u_j'}\right)}{\partial x_k}}_{2\Leftrightarrow A_{ij}}=\underbrace{-\overline{u_i'u_k'}\frac{\partial\bar{u}_j}{\partial x_k}-\overline{u_j'u_k'\frac{\partial\bar{u}_i}{\partial x_k}}}_{3\Leftrightarrow P_{ij}}$$

$$-\underbrace{\frac{\partial}{\partial x_k}\left[\frac{1}{\rho}\delta_{ik}\left(\overline{u_j'p'}\right)+\frac{1}{\rho}\delta_{jk}\left(\overline{u_i'p'}\right)+\left(\overline{u_i'u_j'u_k'}\right)\right]}_{4\Leftrightarrow D_{ij}^t}$$

$$+\underbrace{\nu\frac{\partial^2\left(\overline{u_i'u_j'}\right)}{\partial x_k\partial x_k}}_{5\Leftrightarrow D_{ij}^m}+\underbrace{\overline{\frac{p'}{\rho}\left(\frac{\partial u_j'}{\partial x_i}+\frac{\partial u_i'}{\partial x_j}\right)}}_{6\Leftrightarrow\Pi_{ij}}$$

$$-\underbrace{2\nu\overline{\frac{\partial u_i'}{\partial x_k}\frac{\partial u_j'}{\partial x_k}}}_{7\Leftrightarrow\overline{\varepsilon_{ij}}}-\underbrace{\left(g_i\beta\overline{u_j'T'}+g_j\beta\overline{u_i'T'}\right)}_{8\Leftrightarrow G_{ij}}\Leftrightarrow$$

$$A_{ij} - D_{ij} = P_{ij} + \Pi_{ij} - \varepsilon_{ij}. \tag{8.163}$$

where $D_{ij} = D_{ij}^t + D_{ij}^m$. The terms T_{ij} and G^k were not considered. Concerning the balance equation for turbulent kinetic energy

$$\underbrace{\frac{\partial k}{\partial t} + \overline{u}_j \frac{\partial k}{\partial x_j}}_{A^k} = P^k + D^k + G^k - \varepsilon \Leftrightarrow A^k - D^k = P^k - \varepsilon, \tag{8.164}$$

where $D^k = D^{mk} + D^{tk}$. To derive the ASM Model, it is proposed that the difference between advection and diffusion components of the Boussinesq–Reynolds tensor $\overline{u_i'u_j'}$, i.e., $A_{ij} - D_{ij}$, be modeled as being proportional to the difference between advection and diffusion of turbulent kinetic energy, k, i.e.:

$$A_{ij} - D_{ij} = \frac{\overline{u_i'u_j'}}{k}\left(A^k - D^k\right) = \frac{\overline{u_i'u_j'}}{k}\left(P^k - \varepsilon\right). \tag{8.165}$$

Substituting Eq. (8.165) in Eq. (8.163) yields:

$$\frac{\overline{u_i'u_j'}}{k}\left(P^k - \varepsilon\right) = P_{ij} + \Pi_{ij} - \varepsilon_{ij}. \tag{8.166}$$

Thus, the differential model for the components of the Boussinesq–Reynolds tensor was transformed into a purely algebraic relationship. Since the terms $\overline{P^k}$, P_{ij} and Π_{ij} depend on the components of the Boussinesq–Reynolds Tensor, $\overline{u_i'u_j'}$, Eq. (8.166) becomes a nonlinear algebraic equation for the components $\overline{u_i'u_j'}$. It is worth remembering that the pressure-strain function was modeled by decomposition into 4 terms, $\Pi_{ij} = \Pi_{ij}^1 + \Pi_{ij}^2 + \Pi_{ij}^{1w} + \Pi_{ij}^{2w}$. For didactic reasons, the proposed models are rewritten here for each term, i.e.:

$$\Pi_{ij}^1 = \overline{p'\left(\frac{\partial u_i'}{\partial x_j} + \frac{\partial u_j'}{\partial x_i}\right)} = -c_1 \frac{\varepsilon}{k}\left(\overline{u_i'u_j'} - \frac{2}{3}k\delta_{ij}\right), \tag{8.167}$$

$$\Pi_{ij}^2 = -c_2\left(P_{ij} - \frac{2}{3}P^k\delta_{ij}\right), \tag{8.168}$$

$$\Pi_{ij}^{1w} = c_1 \frac{\varepsilon}{k}\left[\overline{u_k'u_m'}n_k n_m \delta_{ij} - \frac{3}{2}\left(\overline{u_i'u_k'}n_k n_j + \overline{u_j'u_k'}n_k n_i\right)\right] f, \tag{8.169}$$

$$\Pi_{ij}^{2w} = c_2 \frac{\varepsilon}{k}\left[\Pi_{km}^2 n_k n_m \delta_{ij} - \frac{3}{2}\left(\Pi_{ik}^2 n_k n_j + \Pi_{jk}^2 n_k n_i\right)\right] f, \tag{8.170}$$

$$f = \frac{k^{3/2}}{2.55|\vec{n} \cdot (\vec{x} - \vec{x}_w)|\varepsilon}. \tag{8.171}$$

Also remembering the isotropic model for the tensor ε_{ij}:

$$\varepsilon_{ij} = \frac{2}{3}\varepsilon\delta_{ij}, \tag{8.172}$$

where ε is determined with the solution of the respective partial differential balance equation. It is also known that:

$$P^k = -\overline{u_i'u_j'}\frac{\partial\overline{u}_i}{\partial x_j'}, \tag{8.173}$$

$$P_{ij} = -\overline{u_i'u_k'}\frac{\partial\overline{u}_j}{\partial x_k} - \overline{u_j'u_k'}\frac{\partial\overline{u}_i}{\partial x_k}. \tag{8.174}$$

Equation (8.166) is an algebraic relation for the components of the Boussinesq–Reynolds Tensor, $\overline{u_i'u_j'}$. This term is present in all variables in this equation, except in ε and in ε_{ij}, under the hypothesis of isotropy.

As can be observed, Eq. (8.166) is algebraic and implicit in its main variable, i.e., $\overline{u_i'u_j'}$. It can be rewritten in a more convenient way to be solved by replacing the decomposition by $\Pi_{ij} = \Pi_{ij}^1 + \Pi_{ij}^2 + \Pi_{ij}^{1w} + \Pi_{ij}^{2w}$ and models for Π_{ij}^1 and Π_{ij}^2 (Eqs. 8.167 and 8.168) and the isotropic model for ε_{ij}, Eq. (8.172), as follows:

$$\begin{aligned}
&\frac{\overline{u_i'u_j'}}{k}\left(P^k - \varepsilon\right) = P_{ij} + \Pi_{ij}^1 + \Pi_{ij}^2 + \Pi_{ij}^{1w} + \Pi_{ij}^{2w} - \varepsilon_{ij}\\
&\Updownarrow\\
&\overline{u_i'u_j'}\left(P^k - \varepsilon\right) = P_{ij} - c_1\frac{\varepsilon}{k}\left(\overline{u_i'u_j'} - \frac{2}{3}k\delta_{ij}\right)\\
&\qquad - c_2\left(P_{ij} - \frac{2}{3}P^k\delta_{ij}\right) + \Pi_{ij}^{1w} + \Pi_{ij}^{2w} - \frac{2}{3}\varepsilon\delta_{ij}
\end{aligned} \tag{8.175}$$

Multiplying both sides of Eq. (8.175) by k/ε results in:

$$\begin{aligned}
&\frac{\overline{u_i'u_j'}}{\varepsilon}\left(P^k - \varepsilon\right) = \frac{k}{\varepsilon}P_{ij} - c_1\left(\overline{u_i'u_j'} - \frac{2}{3}k\delta_{ij}\right) - c_2\frac{k}{\varepsilon}\left(P_{ij} - \frac{2}{3}P^k\delta_{ij}\right)\\
&\quad + \frac{k}{\varepsilon}\left(\Pi_{ij}^{1w} + \Pi_{ij}^{2w}\right) - \frac{2}{3}k\delta_{ij} \Leftrightarrow \overline{u_i'u_j'}\left\{\frac{1}{\varepsilon}\left(P^k - \varepsilon\right) + c_1\right\}\\
&= \frac{k}{\varepsilon}P_{ij} + c_1\frac{2}{3}k\delta_{ij} - c_2\frac{k}{\varepsilon}\left(P_{ij} - \frac{2}{3}P^k\delta_{ij}\right) + \frac{k}{\varepsilon}\left(\Pi_{ij}^{1w} + \Pi_{ij}^{2w}\right)\\
&\quad - \frac{2}{3}k\delta_{ij} \Leftrightarrow \overline{u_i'u_j'}\left(c_1 + \frac{P^k}{\varepsilon} - 1\right) = c_1\frac{2}{3}k\delta_{ij} + \frac{2}{3}\frac{k}{\varepsilon}P^k\delta_{ij} - \frac{2}{3}\frac{k}{\varepsilon}P^k\delta_{ij}\\
&\quad - \frac{2}{3}k\delta_{ij} + \frac{k}{\varepsilon}P_{ij} - c_2\frac{k}{\varepsilon}P_{ij} + c_2\frac{2}{3}\frac{k}{\varepsilon}P^k\delta_{ij} + \frac{k}{\varepsilon}\left(\Pi_{ij}^{1w} + \Pi_{ij}^{2w}\right).
\end{aligned} \tag{8.176}$$

In the third line of Eq. (8.176), for algebraic convenience, the term $(2k/3\varepsilon)P^k\delta_{ij}$ was added and subtracted. Isolating $\overline{u_i'u_j'}$, follows:

$$\overline{u_i'u_j'} = \frac{2}{3}k\delta_{ij} + \frac{k}{\varepsilon}\frac{(1-c_2)\left(P_{ij} - \frac{2}{3}P^k\delta_{ij}\right) + \left(\Pi_{ij}^{1w} + \Pi_{ij}^{2w}\right)}{\left(c_1 + \frac{P^k}{\varepsilon} - 1\right)}. \tag{8.177}$$

The constants are specified here again for ease of reading and use: $c_1 = 1.5$ and $c_2 = 0.6$. It is observed that Eq. (8.177) is better arranged, but an implicit algebraic equation for $\overline{u_i'u_j'}$ remains. In fact, unknowns appear on both sides of this equation. This equation continues to be dependent on k and ε. The turbulent kinetic energy that is part of the unknowns of this equation is calculated with the diagonal of the tensor $\overline{u_i'u_j'}$. The specific power of viscous transformation, ε, must be calculated via the solution of its balance equation. Therefore, this model, known as an algebraic model, requires the solution of a PDE for ε, which makes it quasi-algebraic. In fact, this model is an algebraic-differential turbulence closure model.

It should be noted that the derivation presented here was made based on Davidson [7]. To verify this deduction, the model given by Eq. (8.177) can be particularized for the problem of a Turbulent Plane Poiseuille Flow, for which the Boussinesq–Reynolds Tensor is reduced to $\overline{u_i'u_j'} = \overline{u_x'u_y'}$, being the coordinate x in the flow direction and the coordinate y in the direction normal to the walls. Thus, Eq. (8.177) is reduced to:

$$\begin{aligned}
\overline{u_x'u_y'} &= 0 + \frac{k}{\varepsilon}\frac{(1-c_2)\left(P_{xy} - 0\right) + \left(\Pi_{xy}^{1w} + \Pi_{xy}^{2w}\right)}{\left(c_1 + \frac{P^k}{\varepsilon} - 1\right)^2} \\
&= \frac{2}{3}(1-c_2)\frac{\left(c_1 - 1 + c_2\frac{P^k}{\varepsilon}\right)}{\left(c_1 - 1 + \frac{P^k}{\varepsilon}\right)^2}\frac{k^2}{\varepsilon}\frac{\partial \overline{u}}{\partial y} \\
&= c_\mu \frac{k^2}{\varepsilon}\frac{\partial \overline{u}}{\partial y}.
\end{aligned} \tag{8.178}$$

It can be seen from Eq. (8.178), that it was possible to reduce the algebraic model, given by Eq. (8.163), to the model based on the concept of turbulent viscosity, where the classical constant c_μ is no longer a constant and starts to be evaluated dynamically, as a function of P^k/ε, i.e., $c_\mu = \frac{2}{3}(1-c_2)\left(c_1 - 1 + c_2\frac{P^k}{\varepsilon}\right)/\left(c_1 - 1 + \frac{P^k}{\varepsilon}\right)^2$.

The solution of the implicit model, given by Eq. (8.177), it is laborious and requires algorithms whose execution is highly computationally expensive. It may be advantageous to develop similar models, but of an explicit nature, that is, for the calculation of the components of the Boussinesq–Reynolds Tensor to be straightforward.

8.5.2 Explicit Algebraic Stress Model-EASM

As explained in the previous section, there are motivations for proposing an explicit algebraic model. Pope [38] proposed an EASM version, based on a general function of the type:

$$\frac{\overline{u_i' u_j'}}{k} - \delta_{ij} = b_{ij} = B_{ij}\big(\overline{S}_{ij}, \overline{\Omega}_{ij}\big), \tag{8.179}$$

where $\overline{S}_{ij} = (1/2)\big(\partial\overline{u}_i/\partial x_j + \partial\overline{u}_j/\partial x_i\big)$ is the strain rate tensor and $\overline{\Omega}_{ij} = (1/2)\big(\partial\overline{u}_i/\partial x_j - \partial\overline{u}_j/\partial x_i\big)$ is the vorticity tensor. The proponent of the model asserts that, based on the Cayley–Hamilton theorem, the most general way to establish this function is as follows:

$$B_{ij}\big(\overline{S}_{ij}, \overline{\Omega}_{ij}\big) = \sum_{n=1}^{10} G^{(n)} \mathfrak{I}_{ij}^{n}. \tag{8.180}$$

The ten tensors $\mathfrak{I}_{ij}^{k}$, $k = 1, \ldots, 10$, are built with a combination of products between $\overline{S}_{ij}$ and $\overline{\Omega}_{ij}$, as shown below:

$$\begin{aligned}
&\mathfrak{I}_{ij}^{1} = \overline{S}_{ij};\ \mathfrak{I}_{ij}^{2} = \overline{S}_{i\ell}\overline{\Omega}_{\ell j} - \overline{S}_{j\ell}\overline{\Omega}_{\ell i};\ \mathfrak{I}_{ij}^{3} = \overline{S}_{i\ell}\overline{S}_{\ell j} - \frac{1}{3}\delta_{ij}\overline{S}_{mn}\overline{S}_{nm};\\
&\mathfrak{I}_{ij}^{4} = \overline{\Omega}_{i\ell}\overline{\Omega}_{\ell j} - \frac{1}{3}\delta_{ij}\overline{\Omega}_{mn}\overline{\Omega}_{nm};\ \mathfrak{I}_{ij}^{5} = \overline{\Omega}_{i\ell}\overline{S}_{\ell m}\overline{S}_{mj} - \overline{S}_{im}\overline{S}_{m\ell}\overline{\Omega}_{\ell j};\\
&\mathfrak{I}_{ij}^{6} = \overline{\Omega}_{im}\overline{\Omega}_{m\ell}\overline{S}_{\ell j} + \overline{S}_{i\ell}\overline{\Omega}_{\ell m}\overline{\Omega}_{mj} - \frac{2}{3}\delta_{ij}\overline{\Omega}_{pm}\overline{\Omega}_{m\ell}\overline{S}_{\ell p};\\
&\mathfrak{I}_{ij}^{7} = \overline{\Omega}_{im}\overline{S}_{m\ell}\overline{\Omega}_{\ell n}\overline{\Omega}_{nj} - \overline{\Omega}_{im}\overline{\Omega}_{m\ell}\overline{S}_{\ell n}\overline{\Omega}_{nj};\\
&\mathfrak{I}_{ij}^{8} = \overline{S}_{im}\overline{\Omega}_{m\ell}\overline{S}_{\ell n}\overline{S}_{nj} - \overline{S}_{im}\overline{S}_{m\ell}\overline{\Omega}_{\ell n}\overline{S}_{nj};\\
&\mathfrak{I}_{ij}^{9} = \overline{\Omega}_{im}\overline{\Omega}_{m\ell}\overline{S}_{\ell n}\overline{S}_{nj} - \overline{S}_{im}\overline{S}_{m\ell}\overline{\Omega}_{\ell n}\overline{\Omega}_{nj} - \frac{2}{3}\delta_{ij}\overline{\Omega}_{pm}\overline{\Omega}_{m\ell}\overline{S}_{\ell n}\overline{S}_{np};\\
&\mathfrak{I}_{ij}^{10} = \overline{\Omega}_{im}\overline{S}_{m\ell}\overline{S}_{\ell n}\overline{\Omega}_{np}\overline{\Omega}_{pj} - \overline{\Omega}_{im}\overline{\Omega}_{m\ell}\overline{S}_{\ell n}\overline{S}_{np}\overline{\Omega}_{pj}.
\end{aligned} \tag{8.181}$$

These tensors form the basis used in Eq. (8.180). These tensors are symmetric and deviatoric. Any other possible combination of $\overline{S}_{ij}$ and $\overline{\Omega}_{ij}$ would result in a linear combination of part of these ten tensors. This statement can be demonstrated with the help of the Cayley–Hamilton Theorem, as stated by Pope [38]. The coefficients $G^{(n)}$ of Eq. (8.180) can be determined as a function of five invariants presented below:

$$\frac{k^2}{\varepsilon^2}\overline{S}_{ij}\overline{S}_{ij};\ \frac{k^2}{\varepsilon^2}\overline{\Omega}_{ij}\overline{\Omega}_{ij};\ \frac{k^3}{\varepsilon^3}\overline{S}_{ij}\overline{S}_{j\ell}\overline{S}_{\ell i};\ \frac{k^3}{\varepsilon^3}\overline{\Omega}_{ij}\overline{\Omega}_{j\ell}\overline{S}_{\ell i};\ \frac{k^4}{\varepsilon^4}\overline{\Omega}_{ij}\overline{\Omega}_{j\ell}\overline{S}_{\ell m}\overline{S}_{mi}. \tag{8.182}$$

The procedure for determining coefficients $G^{(n)}$ is laborious. The model will be individualized, for verification purposes, through its reduction to simpler models. Making up $G^{(1)} = -\,C_\mu\left(k^2/\varepsilon\right)$ and $G^{(n)} = 0$ for $n \geq 2$, then Eq. (8.180) is simplified to:

$$b_{ij} = B_{ij} = \overline{u_i'u_j'} - \frac{2}{3}k\delta_{ij} = -\,C_\mu\frac{k^2}{\varepsilon}\left(\frac{\partial\overline{u}_i}{\partial x_j} + \frac{\partial\overline{u}_j}{\partial x_i}\right). \tag{8.183}$$

This is the $k-\varepsilon$ type model based on the concept of turbulent viscosity, where $\nu_t = C_\mu k^2/\varepsilon$ and $\overline{u_i'u_j'} - (2/3)k\delta_{ij} = 2\nu_t\overline{S}_{ij}$. As established by Pope [39], the determination of $G^{(n)}$ para $n \geq 2$ is non-trivial and would result in an explicit nonlinear model for the components of the Boussinesq–Reynolds Tensor.

For two-dimensional flows, the tensors $\Im_{ij}^{(n)}$, $n = 1, 2, 3$, form the necessary basis for the model given through Eq. (8.180). To determine the coefficients $G^{(n)}$ for $n = 1, 2, 3$ the following two invariants are enough $\overline{S}_{ij}^2$ and $\overline{\Omega}_{ij}^2$, in accordance with Pope [38] and Gatski and Speziale [9, 10]. Thus, the coefficients $G^{(n)}$ with $n = 4, 5, \ldots, 10$ are considered null. A simplification was proposed by Pope [39], considering that the divergence of the tensor $\Im_{ij}^{(3)}$ can be incorporated into the pressure gradient term, thus modifying the pressure that would be resolved. By making this option, the value of the coefficient $G^{(3)}$ will not affect the velocity fields and can therefore be assumed to be null. However, this coefficient $G^{(3)}$ can be determined without the need for pressure manipulation, which is convenient as allows the pressure to be directly determined. Therefore, the non-linear model for the closure of Boussinesq–Reynolds Tensions for two-dimensional flows is established as follows:

$$b_{ij} = B_{ij} = \overline{u_i'u_j'} - \frac{2}{3}k\delta_{ij} = G^{(1)}\Im_{ij}^{(1)} + G^{(2)}\Im_{ij}^{(2)} + G^{(3)}\Im_{ij}^{(3)}. \tag{8.184}$$

Pope [38] presents the models for the coefficients $G^{(1)}$, $G^{(2)}$ and $G^{(3)}$, the results of which are transcribed below:

$$G^{(1)} = -\,C_\mu\frac{k^2}{\varepsilon};$$

$$G^{(2)} = -\,C_\mu\frac{k^3}{\varepsilon^2}\left(\frac{1-c_2}{c_1 - 1 + \frac{P}{\varepsilon}}\right);$$

$$G^{(3)} = 2C_\mu\frac{k^4}{\varepsilon^3}\left(\frac{1-c_2}{c_1 - 1 + \frac{P}{\varepsilon}}\right);$$

$$C_\mu = \frac{\frac{2}{3}\left(\frac{1-c_2}{c_1-1+P/\varepsilon}\right)}{1 - \left(\frac{1-c_2}{c_1-1+P/\varepsilon}\right)^2\left[\frac{2}{3}\left(\frac{k^2}{\varepsilon^2}\overline{S}_{ij}\overline{S}_{ij}\right)^2 + 2\left(\frac{k^2}{\varepsilon^2}\overline{\Omega}_{ij}\overline{\Omega}_{ij}\right)^2\right]}. \tag{8.185}$$

It is observed that the non-linear algebraic model defined with Eqs. (8.184) and (8.185) still retains its implicit nature, since the P/ε which appears in Eq. (8.185) is modeled with the following equation:

$$\frac{P}{\varepsilon} = -2\left(\overline{u_i'u_j'} - \delta_{ij}k\right)\frac{\overline{S}_{ij}}{\varepsilon}. \tag{8.186}$$

From a numerical point of view, what can be done is to make this term explicit. Girimaji [12] presents a completely explicit version of this model. Gatski and Speziale [10] presented a version for three-dimensional flows, for which all coefficients $G^{(n)}$, $n = 1, \ldots, 10$, were considered non-zero.

8.6 Quality Analysis of URANS Closure Models

This analysis begins by remembering that a representative model of turbulent flows cannot be confused with a turbulence closure model. A turbulence model must be understood as a complete set of algebraic-differential equations, which makes up a model with a unique solution (the same number of equations and unknowns). The continuity equation, the Navier–Stokes equations, the energy equation and the balance equations for scalars and/or chemical components, all of them filtered or averaged, and the algebraic closure equations must be included. In this chapter, a review was made of all the closure models that, historically, have been proposed, implemented and tested. This analysis aims to provide a joint view of functionalities and deficiencies related to the existing models presented here.

8.6.1 Classification of URANS/RANS Closure Models

The Prandtl Mixture Length Model was developed during the first generation of closure models. Such a model is based on the concept of turbulent viscosity. It is purely algebraic and is called a zero-balance equation model.

The model $k-\ell$, based on the concept of turbulent viscosity, belongs to the second generation. It comprises one balance equation for k and an algebraic relation for the characteristic length ℓ.

The third generation of closure models includes the $k-\varepsilon$ and $k-\omega$, both based on the concept of turbulent viscosity and two transport equations, one for k and the other for ε or to ω. The model based on a transport equation for turbulent viscosity, presented by Spalart–Allmaras, also belongs to the third generation.

The fourth generation of closure models is based on balance equations for the components of the Boussinesq–Reynolds tensor (RSM). Contrary to what is stated in the literature, this model is not independent of the concept of turbulent viscosity, ν_t. This function continues to be used to close third-order moments. Another category of closure models, which also belongs to the fourth generation, are nonlinear algebraic models, involving the components of the Boussinesq–Reynolds Tensor.

The fifth generation of closure models are the sub-grid models, used for the methodology Large Turbulence Structures Simulation, which will be presented in Chap. 10.

8.6.2 Analysis and Highlights of Closure Models

The Prandtl Mixing Length Model and the balance equation model for k ($k-\ell$ model) are incomplete models, because they require the evaluation of the spatio-temporal distribution of length and/or speed, algebraically. The Prandtl mixing length model is not universal, but it is simple to implement and, when adjusted for a given non-complex flow, for example, in a flat channel or circular pipe, displays a high level of accuracy.

The second and third generation of models, $k-\varepsilon$ and $k-\omega$, and the Spalart–Allmaras (SA) Model are said to be complete, as the characteristic quantities of turbulence or the turbulent viscosity itself are obtained directly from the balance equations. The drawback of this generation of closure models is the assumption that the Boussinesq–Reynolds tensor can be modeled using local turbulence information. However, such models are easy to implement and easy to use. They provide plausible results, with low computational cost, when used in conjunction with wall proximity functions, known as wall law or wall functions. The $k-\omega$ model was developed to solve flows close to walls, without the use of wall laws. However, this requires high refinement in this flow region. The $k-\omega-$SST model, derived from the $k-\omega$ model, represents a significant advance in terms of greater mathematical representation of the physical behavior of turbulent flows; however, the first cell in the vicinity of a wall must be placed in $y^+ \approx 1$, which is computationally expensive.

The Boussinesq–Reynolds Stress Model (RSM) presents the possibility of treating the anisotropy of turbulent flows. This is one of its positive characteristics. As disadvantages, we can highlight the fact that the representative turbulence model itself is composed of 11 differential balance equations, including a continuity equation, three average equations related to Newton's Second Law, six equations for the components of the Boussinesq–Reynolds tensor, $\overline{u_i'u_j'}$, and an equation for the specific power of viscous transformation, ε, for isothermal flows and without chemical reactions. If there are thermal effects, the average thermal energy equation must be added, which gives rise to three more unknowns, in the form of the turbulent flux vector, $q_{t,i} = \overline{u_i'T'}$. This term can be closed using the turbulent Prandtl number hypothesis, $\mathrm{Pr_t}$. In this case, there will be an additional equation to the previous ones, yielding 12 coupled partial differential equations. The determination of $\mathrm{Pr_t}$ is a limitation of this modeling. To avoid its use, three additional equations for turbulent flux are derived, $q_{t,i} = \overline{u_i'T'}$, which triggers the need to derive additional equations for the temperature variance, $\overline{T'T'}$ and for the viscous transformation $\varepsilon_{\overline{T'T'}}$. As a result, the complete turbulence model is composed of 17 coupled partial differential equations. Naturally, its implementation is laborious, and convergence is very difficult to achieve. This additional difficulty in convergence is due, in part, to the fact that the term $\partial\left[\nu_t\left(\partial u_i/\partial x_j + \partial u_j/\partial x_i\right)\right]/\partial x_j$ that appears in the $k-\varepsilon$ model is replaced by the term $-\partial\left(\overline{u_i'u_j'}\right)/\partial x_j$, directly, and the Boussinesq–Reynolds Tensor, $\overline{u_i'u_j'}$, is brought from the solution of additional balance equations. An application for the RSM Model, including gravitational effects, was made by Zeman and Lumley [57] for studying an atmospheric mixed layer.

Explicit nonlinear algebraic models were proposed as an alternative to the convergence difficulties and computational cost of the RSM model. However, the algebraic models resulted in neither greater savings nor greater stability, as expected. They are composed of coupled algebraic equations, whose solutions are multiple roots, which require great computational effort and sophisticated algorithms to achieve numerical stability.

The use of these models leads to important findings for users new to turbulence modeling. The use of the $k-\varepsilon$ model ensures good results for free shear flows, which do not have wall interference. For flows with walls, its concomitant use with wall laws allows good results for some part of the flow. It is an inappropriate modeling for flows that present large curvatures in the streamlines, flows with strong adverse pressure gradients and flows with strong anisotropies, for example, rotating flows. In flows over airfoils, which are of the boundary layer type with adverse pressure gradients, the $k-\varepsilon$ model fails severely. The $k-\omega$ and *SA* models are more appropriate, if the mesh refinement requirements are respected. For complex flows, such as flows over a sudden expansion, collision of jets against a wall, interaction of jets with transverse flows, the use of modeling $k-\varepsilon$ leads to non-negligible errors. Features have been incorporated to the $k-\omega-$SST model, derived from a combination between the Models $k-\varepsilon$ and $k-\omega$, to address the effects of adverse pressure gradients, but these still require a high level of refinement in the vicinity of the walls, $y^+ \leq 1$.

The anisotropies and other characteristics of turbulent flows, already mentioned, are not considered with the URANS modeling described above. Boussinesq–Reynolds Stress modeling, RSM, was proposed to cover the effects of anisotropy, strong curvatures in streamlines, adverse pressure gradients, strong flow rotation effects and secondary flows, as in rectangular ducts. It really is, now, the most complete modeling. With this modeling, it becomes possible to obtain results in highly complex turbulent flows. However, it is difficult to implement and difficult to converge and is very sensitive to boundary conditions, which requires considerable user experience. Another shortfall is the higher computational cost.

Algebraic nonlinear modeling has been proposed to cover these negative points of RSM modeling. Because this model is composed of a system of coupled algebraic equations, its solution is equally demanding in terms of stability control and also computationally expensive.

The task to be tackled with turbulence closure modeling associated with the URANS Methodology is of great scope, since only average information must be calculated. This requires that the entire process of nonlinear interactions between the average information and the turbulence itself be modeled with a given type of closure model.

Smagorinsky's Proposal [47] consists of calculating more and modeling less. This means that this proposal seeks to resolve as much of the spectrum of turbulent structures that make up turbulent flow as possible, as well as their nonlinear interactions.

The objective of the closure models is to model the transfer of information between the band of resolved turbulent structures and the band of unresolved structures. As a result of the essence of this proposal, closure modeling for this methodology is known as sub-mesh modeling. This methodology is called Large Eddy Simulation (LES-Large Eddy Simulation). Now, understanding that turbulence is much richer than a simple composition of vortices, it is more appropriate to use the turbulent structure, or even vortex structure, which represents a spatio-temporal composition of fluid dynamic instabilities that make up turbulent flows. This methodology and the associated sub-grid models are presented and discussed in the next chapter.

8.7 Exercises

1. What differentiates a turbulence model from a closure model? Discuss and comment on the topic.
2. Describe and exemplify two closure models that depend on the concept of turbulent viscosity and two closure models that do not depend on the concept of turbulent viscosity.
3. Describe and classify turbulence closure models according to the number of balance equations that compose them.
4. Discuss and describe the order of statistical moments that must be closed in the balance equations that make up the closure model.
5. Choose one of the nomenclatures: "transport equations" or "balance equations" or "conservation equations". Justify your choices.
6. What is the first gradient hypothesis used to propose turbulence closure models? Who first established this proposal?
7. Write a balance equation for generic *information.* Formally filter this equation. Define the turbulent information flux vector. Physically interpret each term that makes up this balance equation. Use the concept of turbulent "Schmidt Number" and the first gradient hypothesis to propose a closure model for this filtered balance equation.
8. Seeking to improve the closure model, return to exercise 7. Instead of using the concept of Turbulent Schmidt Number, formally derive a balance equation for the turbulent flux of information $\phi(\vec{x}, t)$, i.e., $\overline{\phi' u_j'}$. The resulting equation is an open equation, right? What is the order of the statistical moment(s) that must be closed? Propose the closure(s) using the first gradient hypothesis.

9. Seeking to further improve the proposal of exercise 8, reflect on the possibility of, and, if possible, derive one or more balance equation(s) for the open statistical moments of order $n + 1$, where n is the order of the statistical moment, subject of the balance equation(s), deduced in exercise 8. Once the balance equations for the statistical moments of order $n + 1$ have been deduced, propose a closure for the statistical moments of order $n + 2$. Did you understand why, normally, closing models are completed at the level of second-order moments?
10. Is the Spalart–Allmaras model a turbulence model or a closure model? Why? Is it a dynamic model? Explain what you understand by the dynamic closure model.
11. Use dimensional analysis to derive Eqs. (8.50), (8.51), (8.52) and (8.69).
12. The model $k-\omega-$SST, represented a significant improvement of the family of closure modeling to two transport equations. What are the main physical aspects that were not well modeled and that were incorporated into this new modeling proposal? What physical aspect was the reason for the *SST* acronym? What is the biggest limitation of using this closure model?
13. The closing model *RSM* is one of the most complete proposals ever made. What are the statistical moments and what are their orders, for which closures are proposed? What is the most important physical characteristic of turbulence that is covered by this closure modeling proposal? Despite how much more complete and more representative this closure model of turbulence is, it is little used by the scientific community and even by engineers. Why?
14. What is an algebraic closure model? What differentiates this class of modeling from differential closure models?
15. Differentiate implicit and explicit algebraic models.
16. Why is the $k-\varepsilon$ model the standard suitable for free flows and unsuitable for flows with walls? What can be done to make its application viable for flows with walls?
17. Why does the $k-\omega$ model, which was developed for parietal flows, not work well even for this class of turbulent flows?
18. Why does the merge of $k-\varepsilon$ and $k-\omega$ closure models represent a promise of generalization to complex flows? What is the characteristic of this modeling that continued to be a limitation?

References

1. Baldwin, B.S., Barth, T.J.: A one-equation turbulence transport model for high Reynolds number wall-bounded flows. NASATM-102847 or AIAA paper 91-0610 (1990)
2. Bobyleva, I.M., Zilitinkevich, S.S., Laykhtman, D.L.: Turbulent exchange in thermally stratified atmospheric planetary boundary layer. In: Atmospheric Turbulence and Radio Wave Propagation, pp. 179–190. Nauka Press, Moscow (1967)
3. Chou, P.Y.: On the velocity correlations and the solution of the equation of turbulent fluctuation. Quart. Appl. Math. **3**, 38 (1945)

4. Chung, M.K., Sung, H.J.: Four equation turbulence model for prediction of turbulent boundary layer affected by buoyancy force over a flat plate. Int. J. Heat Mass Transf. **27**, 2387–2395 (1984)
5. Coder, J.G.: A Scale-Adaptive Variant of the Spalart–Allmaras Eddy-Viscosity Model, pp. 2015–2462. AIAA (2015)
6. Davidov, B.I.: On the statistical dynamics of an incompressible fluid. Dokl. Akad. Nauk SSSR **136**, 47 (1961)
7. Davidson, L.: Fluid Mechanics, Turbulent Flows and Turbulence Modeling, Course Material. Division of Fluid Dynamics, Department of Applied Mechanics, Chalmers University of Technology, Goteborg, Sweden (2014)
8. Emmons, H.W.: Shear flow turbulence. In: Proceedings of the 2nd U.S. Congress of Applied Mechanics. ASME (1954)
9. Gatski, T.B., Speziale, C.G.: On explicit algebraic stress models for complex turbulent flows. ICASE Report No. 92-58 (1992)
10. Gatski, T.B., Speziale, C.G.: On explicit algebraic stress models for complex turbulent flows. J. Fluid Mech. **254**, 59–78 (1993)
11. Gibson, M.M., Launder, B.E.: Ground effects on pressure fluctuations in the atmospheric boundary layer. J. Fluid Mech. **86**, 491–511 (1978)
12. Girimaji, S.S.: Fully explicit and serf-consistent algebraic Reynolds stress model. Theor. Comput. Dyn. **8**, 387–402 (1996)
13. Glushko, G.: Turbulent boundary layer on a flat plate in an incompressible fluid. Izv. Akad. Nauk SSSR Mekh. **4**, 13 (1965)
14. Harlow, F.H., Nakayama, P.I.: Transport of turbulence energy decay rate. Report LA-3854. Los Alamos Sci. Lab., University of California (1968)
15. Johnson, D.A., King, L.S.: Mathematically simple turbulent closure model for attached and separated turbulent boundary layers. AIAA J. **23**(11), 1684–1692 (1985)
16. Jones, W.P., Launder, B.E.: The prediction of laminarization with a two-equation model of turbulence. Int. J. Heat Mass Transf. **15**, 301–314 (1972)
17. Kolmogorov, A.N.: The equation of turbulent motion in an incompressible fluid. Izv. Acad. Sci. USSR Phys. 56–58 (1942)
18. Launder, B.E., Sharma, B.I.: Application of energy dissipation model of turbulence to the calculation of flow near a spinning disc. Lett. Heat Mass Transf. **1**(2), 131–138 (1974)
19. Launder, B.E., Spalding, D.B.: Mathematical Models of Turbulence. Academic Press, London (1972)
20. Launder, B.E., Spalding, D.B.: The numerical computation of turbulent flows. Comput. Methods Appl. Mech. Eng. **3**, 269 (1974)
21. Launder, B.E.: On the effects of a gravitational field on the turbulent transport of heat and momentum. J. Fluid Mech. **67**, 569–581 (1975)
22. Launder, B.E., Reece, G.J., Rodi, W.: Progress in the development of a Reynolds-stress turbulence closure. J. Fluid Mech. **68**, 537–566 (1975)
23. Launder, B.E.: Phenomenological modeling: present and future. In: Lumlay, J.L. (ed.) Lecture Notes in Physics, vol. 357, p. 439. Springer Verlag (1990)
24. Lemos, M.J.S., Sesonske, A.: Turbulence in combined convection in mercury pipe flow. Int. J. Heat Mass Transf. **28**(6), 1067–1088 (1985)
25. Lumley, J.L.: Computational modeling of turbulent flows. Adv. Appl. Mech. **18**, 123 (1978)
26. Mansur, N.N., Kim, J., Moin, P.: Reynolds stress and dissipation rate budgets in turbulent channel flow. J. Fluid Mech. **194**, 15–44 (1988)
27. Menter, F.R.: Two-equation eddy-viscosity turbulence models for engineering applications. AIAA J. **32**, 1598–1605 (1994)
28. Menter, F.R., Kuntz, M., Bender, R.: A scale-adaptive simulation model for turbulent flow predictions. AIAA paper 2003-0767 (2003)
29. Menter, F.R., Egorov, Y.: Formulation of the scale-adaptive simulation (SAS) model during the DESIDER project. In: Haase, W., Braza, M., Revell, A. (eds.) DESIDER—European Effort on Hybrid RANS—LES Modelling. Springer (2009)

30. Menter, F.R., Egorov, Y.: The scale-adaptive simulation method for unsteady turbulent flow predictions. Part 1: theory and model description. Flow Turbul. Combust. **85**, 113–138 (2010)
31. Monpean, G.M.: Modélisation des écoulments turbulents avec transferts thermiques en convection mixte. Thèse de Doctorat, L'École Centrale de Lyon (1989)
32. Monin, A.S.: On symmetry properties of turbulence in the surface layer of air. Ocean. Phys. **1**, 45 (1965)
33. Nee, V.W., Kovasznay, L.S.G.: The calculation of the incompressible turbulent boundary layer by simple theory. Phys. Fluids **12**, 473 (1968)
34. Newman, G.R., Launder, B.E., Lumley, J.L.: Modeling the behavior of homogeneous scalar turbulence. J. Fluid Mech. **111**, 217–232 (1981)
35. Owen, R.G.: Ph.D. thesis, Mechanical Engineering Department, Pennsylvania State University (1973)
36. Peng, S.H., Davidson, L., Holmberg, S.: A modified low Reynolds number k-ω models for recirculating flows. J. Fluids Eng. **119**, 867–875 (1997)
37. Pimont, V.: Etude et modélisation d'écoulements turbulents de métaux liquides. Thèse de Dr. Ingénieur, Ecole Centrale des Arts et Manufactures (1983)
38. Pope, S.B.: A more general effective viscosity hypothesis. J. Fluid Mech. **72**, 331 (1975)
39. Pope, S.B.: Turbulent Flows. Cambridge University Press (2000)
40. Prandtl, Z.A.: Über ein neues Formelsystem für die ausgebildete Turbulenz. Nachr. Akad. Wiss. Göttingen Math-Phys. **K1**, 6–19 (1945)
41. Ribeiro, F.J.O., Silveira-Neto: Theoretical analyze of turbulent channel flow with thermal effects—the influence of the turbulent Prandtl number. In: 17th Brazilian Congress of Thermal Sciences and Engineering, Águas de Lindoia, SP-Brazil, 25–28 Nov 2018
42. Rodi, W., Scheurer, G.: Scrutinizing the k-ε turbulence model under adverse pressure gradient conditions. J. Fluids Eng. **108**, 174–179 (1986)
43. Rotta, J.C.: Statistische theorie nichthomogener turbulenz. Z. Phys. **129**, 547 (1951)
44. Rotta, J.C.: Über eine methode zur berechnung turbulenter scherströmungen. Aerodyn. Versuch. Göttingen Rep. **69**, A14 (1968)
45. Saffman, P.G.: A model for inhomogeneous turbulent flows. Proc. R. Soc. Lond. A **317**, 417–433 (1970)
46. Sekundov, A.N.: Application of the differential equation for turbulent analysis viscosity of plane non-self-similar flows. Akad. Nauk SSSR Izv. Mekh. Zhidkosti Gaza 114–127 (1971)
47. Smagorinsky, J.: General circulation experiments with the primitive equations. Mon. Weath. Rev. **91**(3), 99–164 (1963)
48. Spalart, P.R., Allmaras, S.R.: A one-equation turbulence model for aerodynamic flows. AIAA paper 92-0439 (1992)
49. Spalding, D.B.: Concentration fluctuation in a round turbulent free jet. Chem. Eng. Sci. **26**, 95–107 (1971)
50. Speciziale, C.G., Abid, R., Anderson, E.C.: A critical evaluation of two equation models for near wall turbulence. AIAA paper 90-1481, Seattle, WA (1990)
51. Speziale, C.G., Sarkar, S., Gatski, T.B.: Modeling the pressure-strain correlation of turbulence: an invariant dynamical systems approach. J. Fluid Mech. 227, 245 (1991)
52. Taylor, G.I.: Statistical theory if turbulence. Proc. R. Soc. Lond. A **151**, 421–450 (1935)
53. Wilcox, C.W.: Reassessment of the scale determining equation for advanced turbulence models. AIAA J. **26**(11), 1299–1310 (1988)
54. Wilcox, C.W.: Turbulence Modeling for CFD. DCW Industries, La Cañada (1998)
55. Wyngaard, J.C., Cote, O.R.: Bound.-Layer Meteorol. **7**, 289 (1974)
56. Yakhot, V., Orzag, S.A.: Renormalization group analysis of turbulence: 1. Basic theory. J. Sci. Comput. **1**, 3–51 (1986)
57. Zeman, O., Lumlay, J.L.: Modeling buoyancy driven mixed layers. J. Atmos. Sci. **33**, 1974–1988 (1976)

Chapter 9
Sub-grid Closure Models for LES Methodology

In this chapter, the sub-grid closure models for the LES methodology are presented. In Sect. 9.1, an introduction to this topic is presented. In Sect. 9.2, the general outline of this family of closure models is presented. The Smagorinsky sub-grid model is presented in Sect. 9.3. In Sect. 9.4, the relationship between the Smagorinsky model and the Prandtl mixing length model is presented. The sub-grid model of the second-order structure function of velocity is described in Sect. 9.5. Germano's dynamic sub-grid modeling is presented in Sect. 9.6. Sub-grid modeling for irregular or nonuniform meshes is presented. In Sect. 9.8, considerations regarding post-processing of results are presented. In Sect. 9.9, considerations are presented regarding the cutting wave number. Considerations of numerical methods for LES are presented in Sect. 9.10. In Sect. 9.11, an analysis of mesh refinement for LES is presented. Illustrative results for LES are presented in Sect. 9.12. An example of computational experimentation in fluid dynamics is presented in Sect. 9.13. In Sect. 9.14, the fundamentals for hybrid turbulence closure models are presented. This chapter ends with a list of exercises.

9.1 Introduction

In Chap. 7, justifications and arguments were presented based on which the model composed of the continuity, Navier–Stokes, and energy equations was established as sufficient for the adequate solution of laminar, transitional and turbulent flows, for Newtonian fluids and any values of the Reynolds Number. This is the best turbulence model, as it only requires the Stokes and Fourier Closure Model for molecular effects, that is, viscous stresses and pressure and diffusive thermal flux. The limiting factor is the fact that the higher the Reynolds Number, the wider the spectrum of turbulent structures associated with the flow becomes. To directly resolve all the turbulent structures that make up the turbulent flow, it is necessary to use sufficiently refined meshes, which implies high or even prohibitive computational costs. Therefore, the

A. da Silveira Neto, *Turbulent Flows*, https://doi.org/10.1007/978-3-031-89222-6_9

practice of Direct Numerical Simulation—DNS is limited to flows characterized by low values of the Reynolds Number.

To solve this problem, the decomposition of the spectrum of turbulent structures into two bands was proposed and presented in Chap. 5, using the concept of filtering. This leads to the derivation of the filtered equations, but with an additional cost associated with the famous turbulence closure problem. Equations (5.44) and (5.46) resulted from this filtering process. In fact, in these equations, called turbulence equations, derived from the filtering process of the original equations: continuity, Navier–Stokes and thermal energy. The Global Boussinesq-Reynolds Tensor appears, $\tau_{ij} = \overline{u_i u_j} - \overline{u}_i \overline{u}_j$, as well as the global turbulent flux of thermal energy, $q_j = \overline{u_j T} - \overline{u}_j \overline{T}$. They are called global because they incorporate all the tensors and vectors that appeared separately, that is, the sub-grid Reynolds tensor, $\overline{u_i' u_j'}$, the crossed tensor, $C_{ij} = \overline{u_i' \overline{u}_j} + \overline{\overline{u}_i u_j'}$ the Leonard tensor, $L_{ij} = \overline{\overline{u}_i \overline{u}_j} - \overline{u}_i \overline{u}_j$ and sub-grid turbulent flux, $\overline{u_j' T'}$, the cross flux $C_{qj} = \overline{u_j' \overline{T}} + \overline{\overline{u}_j T'}$ and Leonard's flux, $L_{qj} = \overline{\overline{u}_j \overline{T}} - \overline{u}_j \overline{T}$. It is noticed that $\tau_{ij} = \overline{u_i u_j} - \overline{u}_i \overline{u}_j = \left(\overline{u_i' u_j'} + C_{ij} + L_{ij}\right)$ and that $q_j = \overline{u_j T} - \overline{u}_j \overline{T} = \left(\overline{u_j' T'} + C_{qj} + L_{qj}\right)$. The τ_{ij} tensor is symmetric and, therefore, along with the vector q_j, represent nine more unknowns in the filtered equations. One of the ways to close this open problem is to write this tensor and this vector as functions of the gradients of the filtered velocity and temperature fields. These closure proposals were presented through Eqs. (6.1) and (6.2), which are rewritten here for ease of reading:

$$\tau_{ij} = f\left(\overline{u}_i, \overline{u}_j\right), \tag{9.1}$$

$$q_j = g\left(\overline{T}\right) \tag{9.2}$$

Boussinesq proposed closing the system of filtered equations of fluid dynamics by modeling the Boussinesq-Reynolds tensor with the concept of turbulent viscosity, $\nu_t(\vec{x}, t)$. An analogous procedure was proposed to close the filtered energy equation, using the concept of turbulent thermal diffusivity, $\alpha_t(\vec{x}, t)$. Other approaches can be used, for which these concepts of turbulent viscosity and thermal diffusivity are not directly necessary. Boussinesq's proposal was addressed at the beginning of Chap. 6.

In the previous chapter, closure was dealt with from the perspective of the URANS methodology, which aims to determine the average behavior of turbulent flows. In this chapter, closure is handled from the perspective of the methodology Large Structures Simulation—LES. Closing is done in the same way as established in Chap. 6, i.e., via Eqs. (6.3) and (6.6), which will be rewritten here to facilitate reading:

$$\tau_{ij} = -\nu_t\left(\frac{\partial \overline{u}_i}{\partial x_j} + \frac{\partial \overline{u}_j}{\partial x_i}\right) + \frac{2}{3}k\delta_{ij}, \tag{9.3}$$

$$q_j = -\alpha_t \frac{\partial \overline{T}}{\partial x_j} \tag{9.4}$$

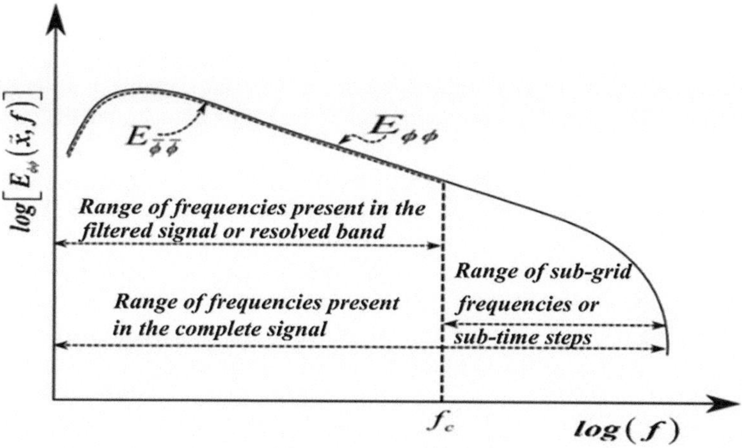

Fig. 9.1 Illustration of the characteristic bands of the energy spectral density associated with the temporal distribution of information $\phi(x_i, t)$

In Chap. 8, several closure models were presented, with which it is possible to calculate the coefficients ν_t and α_t. The closure models for calculating these quantities, associated with the LES Methodology, should preferably be simple algebraic models, with low computational cost. Indeed, the main characteristic of the LES methodology is to calculate more and model less. This requires that the frequency or cutoff wavenumber be in the inertial zone of the turbulent kinetic energy spectrum, as illustrated in Fig. 9.1. The cutoff frequency is determined by the time step used to solve the PDE, $f_c = \pi/\Delta t$, and the cutoff wave number is determined using the size of the cells of the grid, $k_c = \pi/\Delta x$. The greater the refinement of the mesh and time step, the greater the computational cost. Therefore, we seek to develop sub-grid turbulence closure models that are simple and computationally cheap. This is also because the larger the resolved part of the spectrum, the smaller the sub-grid part and the smaller the number of nonlinear interactions to be modeled. In the limit, where the entire spectrum is resolved (DNS), closure modeling must be null.

9.2 Turbulence Closure Sub-grid Modeling

The turbulence models presented here are based on the use of the Boussinesq hypothesis [4]. Taking inspiration from the proposal of the illustrious scholar in fluid mechanics, it was proposed to express the sub-grid Boussinesq-Reynolds tensor as a function of the deformation rate associated with the filtered velocity field and the turbulent kinetic energy, as presented in Eq. (9.3). The turbulent viscosity ν_t can be calculated using different models.

The sub-grid turbulent kinetic energy can be incorporated into the static pressure. The cross and Leonard tensors can be modeled, following Clark et al. [8] idea, who suggested expressing the sum of these tensors as a Taylor Expansion of the filtered velocity field. Based on this, Findikakis and Street [15] demonstrated that:

$$L_{ij} + C_{ij} \cong \frac{\Delta_k}{12} \frac{\partial \overline{u}_i}{\partial x_k} \frac{\partial \overline{u}_j}{\partial x_k} \tag{9.5}$$

where $k = 1, 2, 3$. With this expression, these tensors are explicitly calculated, depending on the filtered velocity field. Shaanan et al. [48] estimated that, when a numerical scheme of up to second order is used to treat advective transport, specific calculations of the Leonard and cross tensors are not necessary. On the other hand, when using higher order schemes or spectral methods, these tensors must be modeled separately from the tensor $\overline{u_i' u_j'}$. Silveira-Neto et al. [50] used computational experiments on a sudden expansion and confirmed, by using second-order schemes, that the influences of the cross and Leonard tensors can be neglected compared to the tensor $\overline{u_i' u_j'}$. So, these researchers defined three diffusion terms, as follows:

$$D_R = \overline{\vec{\nabla}.\left(\overline{u_i' u_j'}\right)} \tag{9.6}$$

$$D_{\mathrm{L}} = \overline{\vec{\nabla}.(L_{ij} + C_{ij})} \tag{9.7}$$

$$D_{\mathrm{M}} = \overline{\vec{\nabla}.(2\nu \overline{S}_{ij})}, \tag{9.8}$$

where D_{R}, D_{L} and D_{M} are, respectively, the norms of the diffusive liquid flux associated with the Sub-Grid Boussinesq-Reynolds tensor, the Leonard and cross tensors, and the molecular viscous tensor, respectively. These effects were calculated using Large Structure Simulations, where the turbulent viscosity was evaluated with a Smagorinsky [51] Model, to be described in Sect. 9.3.

The relative effects of these three types of diffusion are shown in 0. The maximum value was plotted, calculated on each plane perpendicular to the channel, as a function of the longitudinal coordinate. In regions further away from the step (high values of x/H), the effects of D_{R} are on the order of 40 times the effects of D_{L} and D_{M}. This result agrees with the results of Antonopoulos-Domis [1], resolved with a similar discretization method. Therefore, the inclusion or not of these tensors depends on the convergence rate of the advective transport scheme.

The problem is linked to the calculation of turbulent viscosity ν_{t}, still remains unsolved and will be the subject of different types of models to be described in the following items (Fig. 9.2).

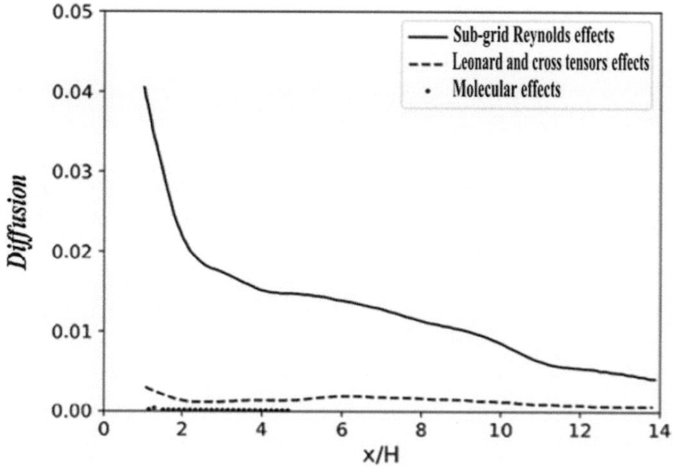

Fig. 9.2 Comparison of molecular diffusion effects, Leonard diffusion effects, cross tensors and sub-grid Reynolds effects [48]

9.3 Smagorinsky Sub-grid Model

This model was proposed by Smagorinsky [51], based on the hypothesis of local equilibrium for sub-grid turbulent structures. For this hypothesis to hold, it is required that the transfer P^k of turbulent kinetic energy, k, from the resolved band to the unresolved band of the turbulent kinetic energy spectrum, is equal to the specific power of viscous transformation ε^k of turbulent kinetic energy k into thermal energy, i.e.:

$$P^k = \varepsilon^k \tag{9.9}$$

where P^k can be written as a function of the shear rate of the filtered velocity field:

$$\begin{aligned} \wp &= -\overline{u_i' u_j'}\frac{\partial \bar{u}_i}{\partial x_j} = \left[\nu_t\left(\frac{\partial \bar{u}_i}{\partial x_j} + \frac{\partial \bar{u}_j}{\partial x_i}\right) - \frac{2}{3}k\delta_{ij}\right]\frac{\partial \bar{u}_i}{\partial x_j} \\ &= \nu_t\left(\frac{\partial \bar{u}_i}{\partial x_j} + \frac{\partial \bar{u}_j}{\partial x_i}\right)\frac{\partial \bar{u}_i}{\partial x_j} - \frac{2}{3}k\boxed{\delta_{ij}\frac{\partial \bar{u}_i}{\partial x_j}} = 2\nu_t \bar{S}_{ij}\bar{S}_{ij} \end{aligned} \tag{9.10}$$

In Equation (9.10), the following equality was used:

$$\begin{aligned} \bar{S}_{ij}\bar{S}_{ij} &= \frac{1}{2}\left(\frac{\partial \bar{u}_i}{\partial x_j} + \frac{\partial \bar{u}_j}{\partial x_i}\right)\cdot\frac{1}{2}\left(\frac{\partial \bar{u}_i}{\partial x_j} + \frac{\partial \bar{u}_j}{\partial x_i}\right) = \frac{1}{4}\left[\left(\frac{\partial \bar{u}_i}{\partial x_j} + \frac{\partial \bar{u}_j}{\partial x_i}\right)\left(\frac{\partial \bar{u}_i}{\partial x_j} + \frac{\partial \bar{u}_j}{\partial x_i}\right)\right] \\ &= \frac{1}{4}\left[\frac{\partial \bar{u}_i}{\partial x_j}\frac{\partial \bar{u}_i}{\partial x_j} + \frac{\partial \bar{u}_i}{\partial x_j}\frac{\partial \bar{u}_j}{\partial x_i} + \frac{\partial \bar{u}_j}{\partial x_i}\frac{\partial \bar{u}_i}{\partial x_j} + \frac{\partial \bar{u}_j}{\partial x_i}\frac{\partial \bar{u}_j}{\partial x_i}\right] \end{aligned}$$

$$= \frac{1}{2}\left[\frac{\partial \bar{u}_i}{\partial x_j}\frac{\partial \bar{u}_i}{\partial x_j} + \frac{\partial \bar{u}_j}{\partial x_i}\frac{\partial \bar{u}_i}{\partial x_j}\right] = \frac{1}{2}\left(\frac{\partial \bar{u}_i}{\partial x_j} + \frac{\partial \bar{u}_j}{\partial x_i}\right)\frac{\partial \bar{u}_i}{\partial x_j}$$

$$\Rightarrow \left(\frac{\partial \bar{u}_i}{\partial x_j} + \frac{\partial \bar{u}_j}{\partial x_i}\right)\frac{\partial \bar{u}_i}{\partial x_j} = 2\bar{S}_{ij}\bar{S}_{ij} \tag{9.11}$$

The specific power of viscous transformation of turbulent kinetic energy, ε^k, can be written as a function of the sub-grid characteristic speed and length:

$$\varepsilon = c_1\left(\overline{u_i'u_j'}\right)^{3/2}/\ell. \tag{9.12}$$

In Eq. (9.12), $(\overline{u_i'u_j'})^{1/2}$ and ℓ are the sub-grid characteristic velocity and length, respectively. It is also assumed that the sub-grid turbulent viscosity is proportional to these two characteristic quantities, resulting in the following equation:

$$\nu_t = c_1\ell\left(\overline{u_i'u_j'}\right)^{1/2} \tag{9.13}$$

By manipulating these equations, we arrive at:

$$\wp^k = \varepsilon^k \Rightarrow 2\nu_t\overline{S}_{ij}\overline{S}_{ij} = c_1\left(\overline{u_i'u_j'}\right)^{3/2}/\ell = \frac{c_1}{\ell}\left(\frac{\nu_t}{c_1\ell}\right)^3 = \frac{1}{c_1^2\ell^4}\nu_t^3 \Rightarrow 2\overline{S}_{ij}\overline{S}_{ij}$$

$$= \frac{1}{c_1^2\ell^4}\nu_t^2 \Rightarrow \nu_t = c_1\ell^2\sqrt{2\overline{S}_{ij}\overline{S}_{ij}} = (C_S\Delta)^2\sqrt{2\overline{S}_{ij}\overline{S}_{ij}} \tag{9.14}$$

From Eq. (9.14) we state the model proposed by Smagorinsky [51]:

$$\nu_t = (C_S\Delta)^2\sqrt{2\overline{S}_{ij}\overline{S}_{ij}} \tag{9.15}$$

The characteristic length ℓ can be modeled depending on the characteristic size Δ of the discretization mesh. The Smagorinsky Constant, $C_S = 0.18$, was determined analytically by Lilly [35] for homogeneous and isotropic turbulence. However, the value of this constant must be adjusted depending on the type of calculation code used and the type of flow under analysis. Despite this, the first sub-grid model has been widely used, which allowed the beginning of one of the most promising topics of research in computational simulation of turbulent flows, that is, the LES Methodology. Considerable advances have been achieved in the field of sub-grid modeling, leading to new concepts, such as dynamic models, which do not require the use of this ad-hoc constant. In this new modeling concept, this constant is replaced by a function, which is dynamically evaluated during the simulation.

With the Smagorinsky model, turbulent viscosity is calculated as a function of the norm of the strain rate tensor, $\sqrt{2\overline{S}_{ij}\overline{S}_{ij}}$. This model, as presented, is not suitable for solving flows with walls. In flow layers very close to a wall, turbulence is dampened by molecular viscous effects. Thus, the turbulent viscosity should also cancel out in the layer of fluid close to the wall. However, as the derivatives of the velocity components are large at the wall, high values of the turbulent viscosity arise in the fluid layers in that region. The palliative solution to this problem is to use a damping function, such as van Driest's proposal [54]. This function is presented as follows:

$$f_\mu(y^+) = 1 - \exp\left(-\frac{y^+}{26}\right) \tag{9.16}$$

In this equation, $y^+ = \rho u_\tau y/\mu$, where $u_\tau = \tau_w/\rho^{1/2}$ and τ_w is the viscous shear stress in the wall. The second weakness of this proposal is the fact that the proportionality constant does not assume a universal value. Its value varies depending on the problem, on the numerical method, and on the refinement of the grid. Its value varies from $C_S = 0.065$, to $C_S = 0.25$, as presented by Moin and Kim [39] and Jones and Launder [23]. These problems motivated researchers to develop dynamic sub-grid modeling, for which a dynamic function replaces this constant. This model will be presented in Sect. 9.6.

The modeling of the turbulent flux of thermal energy can be done using Eq. (9.4). The calculation of thermal diffusivity is done using the concept of the turbulent Prandtl Number:

$$\alpha_t = \nu_t/\mathrm{Pr}_t \tag{9.17}$$

This parameter, Pr_t, varies in the range $0.7 \le \mathrm{Pr}_t \le 0.91$.

9.4 Relationship Between Smagorinsky Sub-grid Model and Prandtl Mixing Length Model

The Prandtl mixing length model was presented in Chap. 7, Hinze [22] and Schlichting [46]. The interest here is to show that this model is a particularity of the Smagorinsky [51] sub-grid model. In fact:

$$\nu_t = (C_S\Delta)^2\sqrt{2\overline{S}_{ij}\overline{S}_{ij}} = \ell^2\sqrt{\left(\frac{\partial \overline{u}_i}{\partial x_j} + \frac{\partial \overline{u}_j}{\partial x_i}\right)\frac{\partial \overline{u}_i}{\partial x_j}} \tag{9.18}$$

The second equality in Eq. (9.18) was demonstrated and presented through Eq. (9.11). The equality $\ell = C_s\Delta$ was also used. For flows where $\overline{u}(y)$ and $\overline{\vartheta} = \overline{w} = 0$, Eq. (9.18) can be simplified to $\nu_t = \ell^2\left[(\mathrm{d}u/\mathrm{d}y)^2\right]^{1/2}$, which results in:

$$\nu_t = \ell^2\frac{\mathrm{d}\overline{u}}{\mathrm{d}y} \tag{9.19}$$

Equation (9.19) is the Prandtl mixing length model for turbulence closure. This model is applicable to turbulent flows that, on average, are one-dimensional.

9.5 Sub-grid Model of the Type "Second-Order Velocity Structure Function"

In a simulation of large structures, turbulent flux of linear momentum and thermal energy are modeled using the concepts of turbulent diffusion of linear momentum and thermal energy. Chollet and Lesieur [7] presented the formalism for calculating $\nu_t(\mathrm{m}^2/\mathrm{s})$ (turbulent kinematic viscosity) and $\alpha_t(\mathrm{m}^2/\mathrm{s})$ (turbulent thermal diffusivity) in Fourier space. These quantities serve to model the interactions of large structures that correspond to the smallest wavenumbers, $k < k_c$, or at lower frequencies, $f < f_c$, with the sub-grid structures relative to the highest wave numbers, $k \geq k_c$, or higher frequencies, $f \geq f_c$, where $k_c = \pi/\Delta$ is the cutoff wave number and $f_c = \pi/\Delta t$ is the cutoff frequency, illustrated in Fig. 9.1, determined with the spatial grid and the time step used in the discretization of PDEs.

Based on the formalism developed by Kraichnan [25], Cholet and Lesieur [7] used the theory EDQNM, Lesieur [33], to close the balance equations. They obtained the following equation for turbulent viscosity:

$$\nu_t(k_c, t) = \nu_t^+ \sqrt{\frac{E(k_c, t)}{k_c}} \tag{9.20}$$

The constant ν_t^+ is determined by making an energy balance, as follows:

$$\int_0^{k_c} 2\nu_t k^2 E(k, t)\mathrm{d}k = \varepsilon(t), \tag{9.21}$$

where $\varepsilon(t)$ is the rate of viscous transformation of turbulent kinetic energy. The turbulent kinetic energy spectral density is used, as proposed by Kolmogorov:

$$E(k, t) = C_\mathrm{K}\varepsilon^{2/3}k^{-5/3}, \tag{9.22}$$

which allows us to obtain:

$$\nu_t^+ = (2/3)C_\mathrm{K}^{-3/2}, \tag{9.23}$$

where $C_\mathrm{K} = 1.4$ is the Kolmogorov Constant. The calculation of ν_t involves calculating the turbulent kinetic energy spectrum in Fourier Space. The transformation of ν_t from Fourier Space to physical space is a difficult task. One way to solve this problem is to use the concept of second-order velocity structure function, which gives rise to a new type of sub-grid modeling, proposed by Métais and Lesieur [37]. A summary of this modeling is presented here.

The second-order velocity structure function is a statistical *information*, defined as follows:

$$F_2(\vec{x}, r, t) = \overline{\vec{u}(\vec{x} + \vec{r}, t) - \vec{u}(\vec{x}, t)^2}, \tag{9.24}$$

where the operator $\overline{(\cdot)}$ denotes the spatial filtering process around the point $\vec{r}$ and inside a sphere of radius $r = \vec{r}$. Batchelor [3] showed a dualism between the structural function $F_2(\vec{r}, t)$, defined in physical space, and the turbulent kinetic energy spectrum $E(k, t)$ defined in Fourier Space, valid for homogeneous and isotropic turbulence. Under these conditions, an energy spectrum is used that presents an inertial zone in $k^{-5/3}$ obtaining the desired relationship:

$$E(\vec{x}, k_c, t) = 0.03\ \Delta\ F(\vec{x}, \Delta, t) \tag{9.25}$$

where Δ is the characteristic size of the mesh and $k_c = \pi/\Delta$ is the cutoff wave number. After substituting Eq. (9.25) in Eq. (9.20), the following result is obtained:

$$\nu_t(\vec{x}, \Delta, t) = 0.067 C_K^{-3/2} \Delta \sqrt{F_2(\vec{x}, \Delta, t)} \tag{9.26}$$

Thus, the problem of calculating ν_t is resolved, based on the hypotheses used. However, $F_2(\vec{x}, \Delta, t)$ is the structure function, relative to the complete energy spectrum, with $k \in [0 - \infty]$. On the other hand, in Large Structures Simulation, this calculation must be carried out using the resolved part of the filtered velocity field, averaging over a sphere of radius $0\ r \geq \Delta$. Consequently, part of the sub-grid structures is missing. Metals and Lesieur [35] proposed a correction, assuming an extension of the inertial zone of the energy spectrum to $k > k_c$. A truncated structure function is then defined, as follows:

$$\bar{F}_2(\vec{x}, r, t) = \overline{\left\|\vec{\bar{u}}(\vec{x} + \vec{r}, t) - \vec{\bar{u}}(\vec{x}, t)\right\|^2} \quad \text{para} \quad \vec{r} = \Delta \tag{9.27}$$

The simplest way to calculate the spatial average is over the six points closest to the point $\vec{x}$. In Fig. 9.3, the distribution of the points used is shown, in the specific case of two dimensions.

The following equation is used to calculation of ν_t, considering this correction:

$$\nu_t(\vec{x}, \Delta, t) = 0.104\ C_K^{-3/2} \Delta \sqrt{\bar{F}_2(\vec{x}, \Delta, t)} \tag{9.28}$$

The turbulent viscosity is calculated in physical space, for each position $\vec{x}$ and for every moment t, using the filtered velocity field. It is observed that the passage of a turbulent viscosity $\nu_t(k_c, t)$ in spectral space to a turbulent viscosity $\nu_t\left(\vec{x}, \Delta, t\right)$ in physical space is justified by the hypothesis of homogeneity and isotropy in small

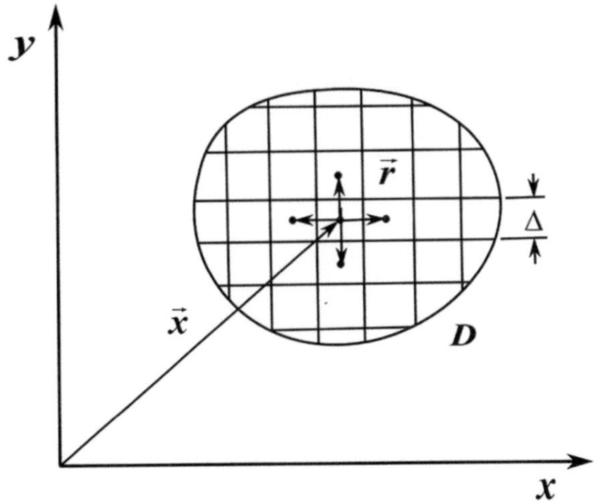

Fig. 9.3 Distribution of points used to calculate the truncated structure function $\overline{F}_2(\vec{x}, \Delta, t)$ over a domain D

turbulent structures, which are considered via sub-grid modeling. The inhomogeneity of large structures must be considered, explicitly, via numerical-computational solution of the filtered and modeled balance equations.

The modeling of the filtered thermal energy balance equation is carried out with the concept of turbulent thermal diffusivity α_t, calculated via the turbulent Prandtl Number concept, $Pr_t = \nu_t/\alpha_t = 0.95$, according to Ribeiro and Silveira-Neto [44].

In this kind of model, an ad-hoc constant, determined analytically, is also present, which renders this model similar to the Smagorinsky model, whose application is limited to fully turbulent flows. These models are not suitable for simulating transitional flows and flows in the vicinity of walls.

Modifications to this basic version of the structure–function model were proposed to make it applicable to all flow regimes, including transition. David [10] proposed the selective structure–function model, in which the turbulent viscosity "is turned off" if the flow does not present the appropriate level of three-dimensionalization. Fallon [14] applied this model to simulate the transition over a sudden expansion, obtaining more coherent results than with the use of the base model. Ducros [12] proposed the filtered structure–function model, in which low frequencies are eliminated via a high-pass filter. Tests were carried out for the boundary layer transition. However, there remains a gap in the generality of these new versions, as well as the possibility of correctly simulating what happens next to walls where, normally, the use of specific models or extremely fine meshes is necessary. Modeling is also needed for transition problems to turbulence, where the hypotheses of homogeneity and isotropy are no longer appropriate, and the Kolmogorov theory is not valid. With regards to the Smagorinsky model, new proposals have been made, among which the so-called dynamic modeling stands out, which is presented in the next item.

9.6 Dynamic Sub-grid Modeling

9.6.1 Fundamental Aspects

Conventional sub-grid modeling involves an ad-hoc, imposed proportionality constant. Despite the limitations arising from this fact, important advances have been achieved in recent years in computational simulation of turbulent flows. The results that can be obtained in fully developed turbulence, and outside the parietal regions, place the LES methodology as a parallel tool for material experimentation [5, 19].

One of the main limitations concerns the analysis of flows in transition and in the vicinity of walls, because of the proportionality of ad-hoc constant imposition. The dynamic determination of a proportionality function, for the calculation of turbulent viscosity, can represent important advances. Germano et al. [17] proposed a dynamic sub-grid closure model.

9.6.2 Dynamic Sub-grid Modeling

In this type of model, the proportionality coefficient is not constant but a function that adjusts to the flow, in time and space. The basis of this modeling is the use of two filters with different characteristic lengths: for the first, the dimensions of the mesh are used to calculate its characteristic length. This is called a mesh-level filter; for the second, a multiple of the mesh dimensions is used to calculate the characteristic length. It is called a test filter.

Based on the use of two levels of characteristic lengths (above the mesh), it is concluded that, in dynamic modeling, information on the energy level contained in the smallest resolved structures, located between the characteristic sizes of the two filters, is used. This way, the energy transfer between resolved structures and unresolved structures is modeled, as illustrated in Fig. 9.4.

It is important to realize that the function to be determined varies over time and space. It should be canceled in regions of laminar flows and close to solid boundaries, implying better asymptotic behavior.

It is also expected to be possible to simulate the effect of inverse transfer of turbulent kinetic energy from sub-grid structures to resolved turbulent structures (back-scattering), a phenomenon that takes place, for example, in boundary layer regions.

It is necessary to emphasize that the main ingredient for establishing the dynamic model is the identity between the sub-grid turbulent stress and the spectrum range defined using two filters of different widths.

The mathematical basis of the dynamic model for incompressible flows of Newtonian fluids is the Navier–Stokes equations:

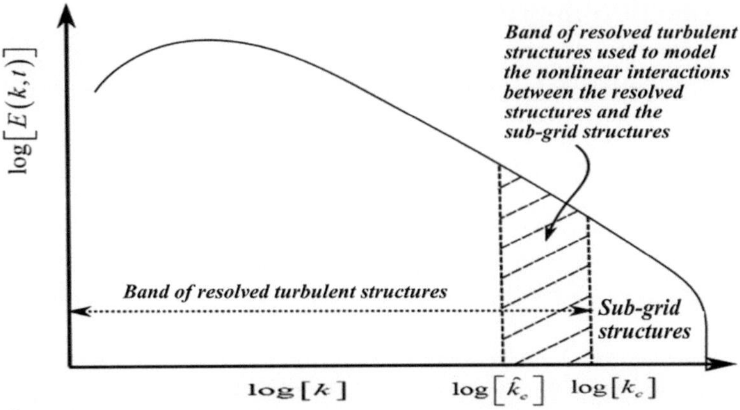

Fig. 9.4 Energy spectrum, illustrating the two quantities that define the two filters used and the wavenumber band that serves to model the process of nonlinear interactions between the resolved structures and the sub-grid structures

$$\frac{\partial\, u_i}{\partial\, t} + \frac{\partial}{\partial\, x_j}\left(u_i u_j\right) = -\frac{1}{\rho}\frac{\partial\, p}{\partial\, x_i} + \frac{\partial}{\partial\, x_j}\left[\nu\left(\frac{\partial\, u_i}{\partial x_j} + \frac{\partial u_j}{\partial x_i}\right)\right] \tag{9.29}$$

After applying a filter $\overline{G}$ of characteristic length $\overline{\Delta}$, calculated using the mesh, the following filtered equation is obtained:

$$\frac{\partial\, \overline{u}_i}{\partial\, t} + \frac{\partial}{\partial\, x_j}\left(\overline{u_i u_i}\right) = -\frac{1}{\rho}\frac{\partial\, \overline{p}}{\partial\, x_i} + \frac{\partial}{\partial\, x_j}\left[\nu\left(\frac{\partial\, \overline{u}_i}{\partial x_j} + \frac{\partial \overline{u}_j}{\partial x_i}\right)\right] \tag{9.30}$$

The Boussinesq-Reynolds sub-grid global tensor was defined in Chap. 6, Eq. (6.42), which is presented here:

$$\tau_{ij} \equiv \overline{u_i u_j} - \overline{u}_i \overline{u}_j \tag{9.31}$$

Then the following filtered equation is obtained:

$$\frac{\partial\, \overline{u}_i}{\partial\, t} + \frac{\partial}{\partial\, x_j}\left(\overline{u}_i \overline{u}_j\right) = -\frac{1}{\rho}\frac{\partial\, \overline{p}}{\partial\, x_i} + \frac{\partial}{\partial\, x_j}\left[\nu\left(\frac{\partial\, \overline{u}_i}{\partial x_j} + \frac{\partial \overline{u}_j}{\partial x_i}\right) - \tau_{ij}\right] \tag{9.32}$$

Applying a new filter $\hat{G}$ with width $\hat{\Delta} > \overline{\Delta}$ about Eq. (9.30) yields

$$\frac{\partial \hat{\overline{u}}_i}{\partial t} + \frac{\partial}{\partial x_j}\widehat{\overline{u_i u_j}} = -\frac{1}{\rho}\frac{\partial \hat{\overline{p}}}{\partial x_i} + \frac{\partial}{\partial x_j}\left[\nu\left(\frac{\partial \hat{\overline{u}}_i}{\partial x_j} + \frac{\partial \hat{\overline{u}}_j}{\partial x_i}\right)\right], \tag{9.33}$$

where the choice $\hat{\Delta} = 2\overline{\Delta}$ has been used by the scientific community because of numerical empiricism. The stress tensor, relative to the second filter, also called sub-test tensor, is defined as:

$$T_{ij} \equiv \widehat{\overline{u}_i\overline{u}_j} - \widehat{\overline{u}}_i\widehat{\overline{u}}_j \tag{9.34}$$

Thus, Eq. (9.33) can be rewritten:

$$\frac{\partial\,\widehat{\overline{u}}_i}{\partial\,t} + \frac{\partial}{\partial\,x_j}\left(\widehat{\overline{u}}_i\widehat{\overline{u}}_j\right) = -\frac{1}{\rho}\frac{\partial\,\widehat{\overline{p}}}{\partial\,x_i} + \frac{\partial}{\partial\,x_j}\left[\nu\left(\frac{\partial\,\widehat{\overline{u}}_i}{\partial x_j} + \frac{\partial\widehat{\overline{u}}_j}{\partial x_i}\right) - T_{ij}\right] \tag{9.35}$$

Filtering Eq. (9.40), using the filter $\hat{G}$, results in:

$$\frac{\partial\widehat{\overline{u}}_i}{\partial t} + \frac{\partial}{\partial x_j}\left(\widehat{\overline{u}_i\overline{u}_j}\right) = -\frac{1}{\rho}\frac{\partial\widehat{\overline{p}}}{\partial x_i} + \frac{\partial}{\partial\,x_j}\left[\nu\left(\frac{\partial\widehat{\overline{u}}_i}{\partial x_j} + \frac{\partial\widehat{\overline{u}}_j}{\partial x_i}\right) - \hat{\tau}_{ij}\right] \tag{9.36}$$

Subtracting Equation (9.35) from Eq. (9.36), it follows that:

$$\frac{\partial}{\partial x_j}\left(\widehat{\overline{u}_i\overline{u}_j} - \widehat{\overline{u}}_i\widehat{\overline{u}}_j\right) = \frac{\partial}{\partial x_j}\left(T_{ij} - \hat{\tau}_{ij}\right) \tag{9.37}$$

The global Leonard tensor is defined in Eq. (9.37):

$$L_{ij} = \widehat{\overline{u}_i\overline{u}_j} - \widehat{\overline{u}}_i\widehat{\overline{u}}_j = T_{ij} - \hat{\tau}_{ij} \tag{9.38}$$

Equation (9.38) can be rewritten to relate the deviatoric parts of the tensors involved:

$$L_{ij} - \frac{1}{3}\delta_{ij}L_{ij} = \left(T_{ij} - \frac{1}{3}\delta_{ij}T_{ij}\right) - \left(\hat{\tau}_{ij} - \frac{1}{3}\delta_{ij}\hat{\tau}_{ij}\right) \tag{9.39}$$

This equality is known as German Identity. It is used to determine the coefficient function $c(\vec{x}, t)$ which appears in turbulence closure models. The deviatoric part of the global sub-grid tensor can be modeled with the Boussinesq hypothesis:

$$\tau_{ij} - \frac{\delta_{ij}}{3}\tau_{ij} = -\nu_t\left(\frac{\partial\overline{u}_i}{\partial x_j} + \frac{\partial\overline{u}_j}{\partial x_i}\right) = -2\nu_t\bar{S}_{ij} = -2c(\vec{x}, t)\bar{\Delta}^2\left\|\bar{S}_{ij}\right\|\bar{S}_{ij}, \tag{9.40}$$

where $\overline{S}_{ij} = (1/2)\left(\partial\overline{u}_i/\partial x_j + \partial\overline{u}_j/\partial x_i\right)$ and $\left\|\bar{S}_{ij}\right\| \equiv \sqrt{2\bar{S}_{ij}\bar{S}_{ij}}$. Modeling the sub-test turbulent stresses T_{ij} in a similar way, we arrive at:

$$T_{ij} - \frac{\delta_{ij}}{3}T_{ij} = -2c(\vec{x}, t)(\hat{\Delta})^2\left\|\widehat{\bar{S}}_{ij}\right\|\widehat{\bar{S}}_{ij}, \tag{9.41}$$

where

$$\hat{\bar{S}}_{ij} = \frac{1}{2}\left(\frac{\partial \hat{\bar{u}}_i}{\partial x_j} + \frac{\partial \hat{\bar{u}}_j}{\partial x_i}\right) \quad \text{and} \quad \left\|\hat{\bar{S}}_{ij}\right\| = \sqrt{2\hat{\bar{S}}_{ij}\hat{\bar{S}}_{ij}} \tag{9.42}$$

Filtering Eq. (9.40), results in:

$$\hat{\tau}_{ij} - \frac{\delta_{ij}}{3}\hat{\tau}_{ij} = -2\nu_t \hat{\bar{S}}_{ij} = -2c(\vec{x}, t)(\bar{\Delta})^2 \widehat{\left\|\bar{S}_{ij}\right\|\bar{S}_{ij}} \tag{9.43}$$

Substituting Eqs. (9.41) and (9.43) in Eq. (9.39) results in:

$$\begin{aligned} L_{ij} - \frac{1}{3}\delta_{ij}L_{ij} &= \left(T_{ij} - \frac{1}{3}\delta_{ij}T_{ij}\right) - \left(\hat{\tau}_{ij} - \frac{1}{3}\delta_{ij}\hat{\tau}_{ij}\right) \\ &= -2c(\vec{x}, t)(\hat{\Delta})^2\left\|\hat{\bar{S}}_{ij}\right\|\hat{\bar{S}}_{ij} + 2c(\vec{x}, t)(\bar{\Delta})^2\widehat{\left\|\bar{S}_{ij}\right\|\bar{S}_{ij}} \\ &= -2c(\vec{x}, t)\left[(\hat{\Delta})^2\left\|\hat{\bar{S}}_{ij}\right\|\hat{\bar{S}}_{ij} - (\bar{\Delta})^2\widehat{\left\|\bar{S}_{ij}\right\|\bar{S}_{ij}}\right] \end{aligned} \tag{9.44}$$

Equation (9.44) is a tensor equation, composed of symmetric tensors and with zero traces. This results in five algebraic equations with a single unknown, that is, the scalar function $c(\vec{x}, t)$. Lily [33] proposed to determine the function $c(\vec{x}, t)$ using the least squares method, which makes it possible to determine it in such a way as to minimize the error resulting from the adjustment. Thus, a squared error function is defined:

$$E = \left[L_{ij} - \frac{1}{3}\delta_{ij}L_{ij} + 2c(\vec{x}, t)M_{ij}\right]^2, \tag{9.45}$$

where

$$M_{ij} = (\hat{\Delta})^2\left\|\hat{\bar{S}}_{ij}\right\|\hat{\bar{S}}_{ij} - (\bar{\Delta})^2\widehat{\left\|\bar{S}_{ij}\right\|\bar{S}_{ij}}. \tag{9.46}$$

This error function will have a minimum or a maximum when

$$\frac{\partial E}{\partial c} = 4M_{ij}\left[L_{ij} - \frac{1}{3}\delta_{ij}L_{ij} + 2c(\vec{x}, t)M_{ij}\right] = 0 \tag{9.47}$$

The second derivative of the error function results in:

$$\frac{\partial^2 E}{\partial c^2} = 8M_{ij}M_{ij}, \tag{9.48}$$

which is greater than or equal to zero. Thus, Eq. (9.45) will always have a minimum point and, knowing that $M_{ij}\delta_{ij} = 0$, results in:

$$c(\vec{x}, t) = -\frac{1}{2}\frac{L_{ij}M_{ij}}{M_{ij}M_{ij}} \tag{9.49}$$

The calculation of the dynamic coefficient only depends on resolved quantities and an explicit filtering process. However, this coefficient varies over time and space and can assume negative values. Considering numerical stability, the function $c(\vec{x}, t)$ and its average are calculated in the direction of flow homogeneity, when it exists. It is also necessary that the effective viscosity $\nu_{\text{ef}} = \nu + \nu_t \geq 0$. This condition allows us to obtain numerical stability but eliminates the possibility of modeling the physical effect of backscattering. In industrial three-dimensional flows, there is no direction of homogeneity, allowing for the possibility of calculating a local average, that is, a spatial filtering of the function $c(\vec{x}, t)$.

Some complementary publications are cited below: Germano [16], Germano et al. [17], Moin et al., Lilly [35], Germano [18], Piomeli [40], Carati et al. [6], Ghosal et al. [20] and Spyropoulos and Blaisdel [52].

The mesh-level filtering is actually an implicit filtering of the PDE discretization process. The test filtering must be an explicit filtering to be applied to the fields of the variables of interest. This filtering was explained and illustrated in Chap. 6, Fig. 6.8, for a one-dimensional flow situation.

9.7 Sub-grid Modeling in Irregular Meshes

Sub-grid models may or may not be based on the turbulent viscosity concept. In this book, sub-grid models were discussed only based on this concept. Two ways of calculating turbulent viscosity were presented: the first requires the calculation of deformation rates, which characterizes a family of Smagorinsky-type Models [51]; the second involves calculating velocity differences, which characterizes a family of models based on the concept of second-order velocity structure function, Lesieur and Métais [34]. The application of these models to regular meshes does not present difficulties, since the characteristic lengths of the filters are calculated directly from the regular dimensions of the mesh. In cases of irregular meshes, corrections must be made to the models originally developed for regular meshes.

9.7.1 Correction in Smagorinsky Sub-grid Modeling for Irregular Meshes

Smagorinsky's original model can be rewritten as follows:

$$\tau_{ij} = -2\left[L(\overline{\Delta})\right]^2 \left[2\overline{S}_{mn}\overline{S}_{mn}\right]^{1/2}\overline{S}_{ij}, \tag{9.50}$$

where $L(\overline{\Delta})$ depends on the degree of anisotropy or irregularities of the mesh used. If the mesh is irregular and rectangular, as illustrated in Fig. 9.5:

$$L(\overline{\Delta}) = C_s \Delta, \tag{9.51}$$

where $\Delta = \Delta x_1 = \Delta x_2 = \Delta x_3$ and C_S is the Smagorinsky constant, already presented. For weak anisotropy, Deardorff [11] proposed:

$$L(\overline{\Delta}) = C_s \Delta_{eq}, \tag{9.52}$$

$$\Delta_{eq} = (\Delta_1 \Delta_2 \Delta_3)^{1/3} \tag{9.53}$$

For strongly anisotropic meshes, a more rigorous development was proposed by Scotti and Meneveau [47], which can be summarized in the following equations:

$$L(\overline{\Delta}) = C_s \Delta_{eq} f(a_1, a_2), \tag{9.54}$$

where

$$f(a_1, a_2) = \cos h \sqrt{\frac{4}{27}\left[(\ln a_1)^2 - \ln a_1 \ln a_2 + (\ln a_2)^2\right]}, \tag{9.55}$$

and a_1 and a_1 are the aspect factors, defined as follows:

$$a_1 = \frac{\Delta_1}{\Delta_{max}} \quad \text{and} \quad a_2 = \frac{\Delta_2}{\Delta_{max}}, \tag{9.56}$$

where

$$\Delta_{max} = \Delta_3 = \max\{a_1, a_2, a_3\} \tag{9.57}$$

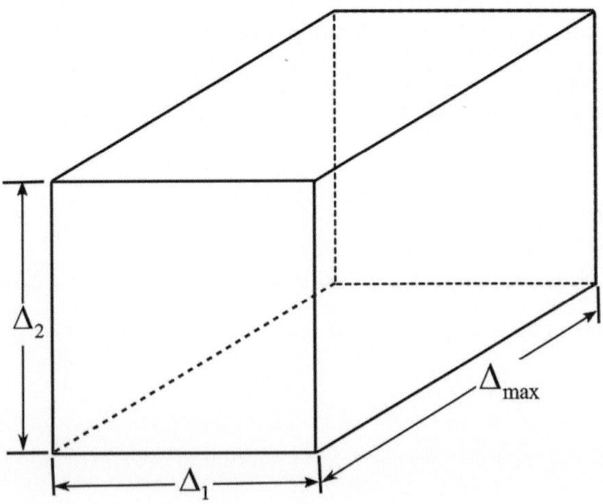

Fig. 9.5 Cell of an irregular mesh

This type of correction can be adapted to dynamic modeling as well, as presented by Scotti and Meneveau [47].

9.7.2 Correction in Sub-grid Modeling Type Second-Order Structure Function of Velocity in Irregular Meshes

In this type of model, the turbulent viscosity is calculated with an algebraic equation of the following type:

$$\nu_t(\vec{x}, \Delta x, t) = 0.104\, C_K^{-3/2} \Delta x \left[F_2(\vec{x}, \Delta x, t)\right]^{1/2}, \tag{9.58}$$

where

$$F_2(\vec{x}, \Delta x) = < \|\overline{u}(\vec{x}, t) - \overline{u}(\vec{x} + \vec{r}, t)\|^2 >_{\|\vec{r}\|=\Delta x} \tag{9.59}$$

When the mesh is structured and regular, Δx is obtained from the mesh itself. When the mesh is structured and irregular, correction is necessary. Lesieur and Métais [34] suggested the following:

$$F_2(\vec{x}, \Delta c, t) = \frac{1}{6}\sum_{i=1}^{3}[\|\overline{u}(\vec{x}, t) - \overline{u}(\vec{x} + \Delta x_i\vec{e}_i, t)\|^2 + \|\overline{u}(\vec{x}, t) - \overline{u}(\vec{x} - \Delta x_i\vec{e}_i, t)\|^2]\left(\frac{\Delta c}{\Delta x_i}\right)^{(2/3)} \tag{9.60}$$

where

$$\Delta c = (\Delta x_1 \Delta x_2 \Delta x_3)^{1/3} \tag{9.61}$$

Kolmogorov's [24] Law was taken into account, with which it was established that the second-order structure function is proportional to $(\varepsilon r)^{2/3}$, where r is the distance between the two considered points. The above turbulent viscosity is then recalculated by replacing Δx by Δc, which results in Eq. (9.60) for the six-point formulation. In this equation, $\vec{e}_i$ is the vector in x_i direction. To better understand, this expression is rewritten in an extended form:

$$F_2(\vec{x}, \Delta c, t) = \frac{1}{6}\left\{\begin{array}{c} [\|\overline{u}(\vec{x}, t) - \overline{u}(\vec{x} + \Delta x_1\vec{e}_1, t)\|^2 + \|\overline{u}(\vec{x}, t) - \overline{u}(\vec{x} - \Delta x_1\vec{e}_1, t)\|^2]\left(\frac{\Delta c}{\Delta x_1}\right)^{(2/3)} + \\ [\|\overline{u}(\vec{x}, t) - \overline{u}(\vec{x} + \Delta x_2\vec{e}_2, t)\|^2 + \|\overline{u}(\vec{x}, t) - \overline{u}(\vec{x} - \Delta x_2\vec{e}_2, t)\|^2]\left(\frac{\Delta c}{\Delta x_2}\right)^{(2/3)} + \\ [\|\overline{u}(\vec{x}, t) - \overline{u}(\vec{x} + \Delta x_3\vec{e}_3, t)\|^2 + \|\overline{u}(\vec{x}, t) - \overline{u}(\vec{x} - \Delta x_3\vec{e}_3, t)\|^2]\left(\frac{\Delta c}{\Delta x_3}\right)^{(2/3)} \end{array}\right\} \tag{9.62}$$

Imagine strong anisotropy in the mesh, for example, $\Delta x_1 \gg \Delta x_2 \approx \Delta x_3$. In this case,

$$\frac{\Delta c}{\Delta x_1} \ll \frac{\Delta c}{\Delta x_2} \approx \frac{\Delta c}{\Delta x_3} \tag{9.63}$$

and anisotropy causes fluctuations in the direction Δx_1 and has less influence on the calculation of turbulent viscosity than in other directions.

When the mesh is unstructured, it is suggested to interpolate the velocity over a circle of radius $r = \|\vec{r}\|$. With the velocities located at equidistant positions, the turbulent viscosity is calculated at a point, in the center of this circle.

David [10] proposed the selective function-structure model, in which the turbulent viscosity "is turned off" if the flow does not present the appropriate level of three-dimensionalization. Fallon [14] applied this model to simulate the transition over a sudden expansion, obtaining more coherent results than with the use of the base model.

9.8 Considerations About Statistical Post-processing of Results

The basis for the LES methodology is the separation of the spectrum of turbulent structures into one part relating to large structures and the other part relating to sub-grid structures. The large structures are solved when the filtered equations are solved. The frequency or cutoff wavenumber must be in the inertial zone of the spectrum. The sub-grid structures are not resolved, but the non-linear interactions between the resolved structures and the sub-grid structures are modeled via closure models. The sub-grid structures are not modeled, but rather the nonlinear interactions, between the resolved band and the sub-grid band of the turbulence spectrum, are modeled via a turbulence closure model. The results obtained with the LES methodology are always transient: $\overline{u}_i(\vec{x}, t)$, $\overline{p}(\vec{x}, t)$, $\overline{T}(\vec{x}, t)$ etc. Once the temporal and spatial samples of the quantities of interest have been obtained, the temporal average can be calculated, for example: $\overline{u}_i(\vec{x})$, $\overline{p}(\vec{x})$ and $\overline{T}(\vec{x})$. These functions provide the time-averaged spatial fields. Once the temporal averages are determined, the fluctuations relative to large turbulence structures are also determined: $\overline{u}_i'(\vec{x}, t) = \overline{u}_i(\vec{x}, t) - \overline{u}_i(\vec{x})$, $\overline{p}'(\vec{x}, t) = \overline{p}(\vec{x}, t) - \overline{p}(\vec{x})$ and $\overline{T}'(\vec{x}, t) = \overline{T}(\vec{x}, t) - \overline{T}(\vec{x})$. As fluctuations are known, second-order statistical moments can be determined, such as $\overline{u}_i'\overline{u}_j'$, $\overline{T}'\overline{T}'$, $\overline{u}_i'\overline{T}'$, $\overline{u}_i'\overline{p}'$ etc. Note that these statistics do not include the sub-grid contribution of information. However, it is a good approximation, since large structures carry the largest amount of information transported.

9.9 Cutoff Wavenumber Considerations

It is important to know how to evaluate the cutoff wave number relative to a given signal acquired or calculated in experiments, whether they are material experiments (using a bench for material experimentation) or computational experiments (using a computational platform). Let us consider any signal relating to information $\phi(x, t)$, as illustrated in Fig. 9.6a and b. In Fig. 9.6a, a wavelength $\lambda = 2\pi$ was divided into two cells of size Δx_1. Certainly, the resolution of the problem will be very poor, with this number of points per wavelength. But, following the definition of cutoff wavenumber, i.e.:

$$k_{c1} = \frac{2\pi}{\lambda} = \frac{2\pi}{2\Delta x_1} = \frac{\pi}{\Delta x_1} \tag{9.64}$$

To better describe the physics of the problem, the number of points for this same wavelength was doubled. At Fig. 9.6b, the wavelength $\lambda = 2\pi$ was divided into four cells of size Δx_2. The cutoff wave number will then be:

$$k_{c2} = \frac{2\pi}{\lambda} = \frac{2\pi}{4\Delta x_2} = \frac{\pi}{2\Delta x_2} = \frac{\pi}{\Delta x_1} = k_{c1} \tag{9.65}$$

The conclusion reached is that if the wavelength is fixed ($\lambda = 2\pi$), the greater the number of nodes used to describe a given wavelength, the more detailed the capture of the physics of the problem will be, but the number of cutting waves will not change. Observing the two figures in the case of Fig. 9.6a, with only three discretization nodes, the curve would be poorly calculated. In Fig. 9.6b, the curve would be better described.

In Fig. 9.7, two different wavelengths are illustrated. If four nodes per wavelength are required for your solution, then only the longest wavelength represented in this figure can be solved. To solve the wavelength represented by the dashed line, it will be required to double the number of nodes.

The question is: what is the number of nodes per wavelength necessary to solve a turbulent flow problem? With numerical empiricism, it is shown that the answer to this question depends on the convergence rate of the numerical method used. For methods up to second order, it is recommended to use sixteen nodes per wavelength. For methods above third order, at least eight nodes per wavelength are recommended. For example, if one wishes to perform DNS of a turbulent flow, characterized by a given Reynolds Number, it is first necessary to calculate the Kolmogorov characteristic

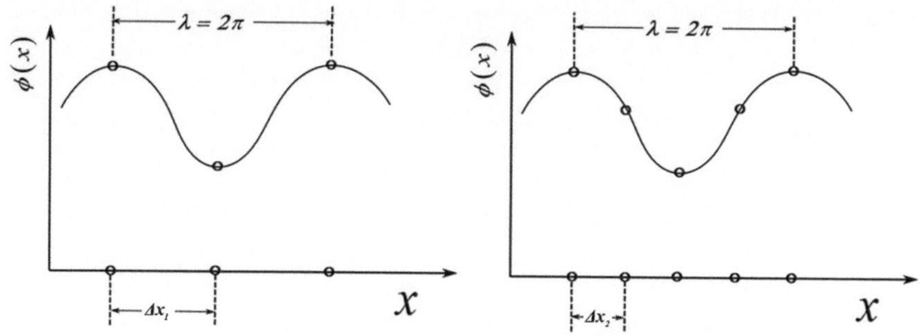

Fig. 9.6 Illustration of the number of nodes per wavelength of an acquired or calculated signal

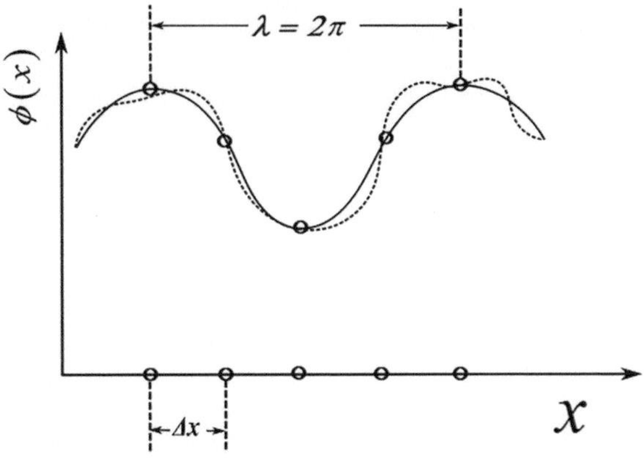

Fig. 9.7 Illustration of two different wavelengths to be resolved

quantities of time and length. The way to make this calculation has already been presented in Chap. 5. Having this information and the characteristics of the numerical method to be used, the characteristics of the mesh to be used are defined. On the other hand, if one wants to make use of the LES methodology, the available computational capacity is known and knowing that the cutoff wavenumber or cutoff frequency must be in the inertial zone of the spectrum, the size of the finest cells that will be used to solve the problem can be found using the rule above.

9.10 Considerations About the Numerical Method for LES

9.10.1 Introduction

The turbulence closure sub-grid models serve to mathematically represent the transfer of information in question (linear momentum, energy, etc.) between the resolved turbulence structures and the sub-grid structures, which are not resolved. It is not correct to say that sub-grid models serve to model these structures, but rather that they serve to model the nonlinear interactions between the band of resolved structures and the band of unresolved structures, or the sub-grid structures. The effects of numerical diffusion and dispersion are harmful to the closure modeling process. The ideal numerical scheme would be free of these two bad characteristics. The centered scheme is free from numerical diffusion, but may present dispersion, depending on the value of the Reynolds Number and the level of refinement used. When using an UPWIND scheme, the high numerical diffusion of this scheme masks turbulent diffusion, which must be modeled with a mathematically consistent and well-structured model. Numerical diffusion appears as a function of local velocity and mesh characteristics. Thus, it will appear, even in regions where the flow is laminar or even potential. To better understand the effects of numerical diffusion and dispersion, a brief study on this topic will be presented.

Closure modeling, including the effects of numerical diffusion, can be expressed by the following equation:

$$\left\{ \begin{array}{c} \text{Diffusive liquid flux of linear} \\ \text{momentum over a differential} \\ \text{fluid particle} \end{array} \right\}_i = \frac{\partial}{\partial x_j} \left\{ (\nu + \nu_t + \nu_{\text{num}}) \frac{\partial u_i}{\partial x_j} \right\}, \tag{9.66}$$

where ν is the fluid molecular viscosity, ν_t is the sub-grid turbulent viscosity and ν_{num} is the numerical viscosity (numerical diffusion coefficient).

If refinement is not sufficient and if turbulence closure modeling is not used, then there should not be numerical convergence. The simulation should diverge! As a rule, numerical methods are characterized by the presence of numerical diffusion, which models the closure of turbulence. It is common to hear researchers say: "I am simulating a turbulent flow without closure modeling". This is not possible! If there is no mesh for DNS, then what is being practiced is LES, at the expense of numerical diffusion implicit in the discretization process. However, this type of "modeling" should not be of interest, since it has no relationship with the physical behavior of the problem. It only depends on the numerical method and the size of the discretization cells. Numerical diffusion will appear throughout the analysis domain, even in regions of laminar flow! Therefore, a numerical method for LES Methodology should display zero numerical diffusion, which only happens with centered schemes. On the other hand, these schemes incorporate the dispersive effect. To illustrate these two characteristics, numerical diffusion and dispersion, the numerical analysis of a problem of a purely mathematical nature will be presented, but one which, due to its simplicity, makes it possible to develop a numerical analysis.

9.10.2 Concepts of Numerical Diffusion and Dispersion

In the literature, there are several methods for numerical-computational solution of differential models. The most used are finite differences, finite volumes, finite elements and finite volume-elements. With discretization, a differential equation is transformed into a set of algebraic equations. In this exercise, a pure advection model will be used, composed of a single partial differential equation:

$$\frac{\partial \phi}{\partial t} + c\frac{\partial \phi}{\partial x} = 0 \tag{9.67}$$

The finite difference method will be used for discretization. The variable $\phi(x, t)$ represents any scalar information. The constant $c(\text{m/s})$ is a velocity of advective transport of information $\phi(x, t)$. To illustrate the effects of numerical diffusion and dispersion, two discretization schemes will be used: first-order upwind and Lax-Wendroff. Performing Taylor expansions in time and space, of the function $\phi(x, t)$, starting with the first order upwind scheme, we have:

$$\phi(t+\Delta t, x) = \phi(x,t) + \sum_{n=1}^{\infty} \frac{\Delta t^{(2n-1)}}{(2n-1)!} \frac{\partial^{(2n-1)}\phi(x,t)}{\partial t^{(2n-1)}} + \sum_{n=1}^{\infty} \frac{\Delta t^{(2n)}}{(2n)!} \frac{\partial^{(2n)}\phi(x,t)}{\partial t^{(2n)}}, \tag{9.68}$$

$$\phi(t, x-\Delta x) = \phi(x,t) + \sum_{n=1}^{\infty} \frac{\Delta x^{(2n-1)}}{(2n-1)!} \frac{\partial^{(2n-1)}\phi(x,t)}{\partial x^{(2n-1)}} + \sum_{n=1}^{\infty} \frac{\Delta x^{(2n)}}{(2n)!} \frac{\partial^{(2n)}\phi(x,t)}{\partial x^{(2n)}}. \tag{9.69}$$

From these two equations, the discrete forms of the derivatives are isolated, plus the errors that are made when truncating the discretization, in the first term of the expansion:

$$\frac{\partial \phi(x,t)}{\partial t} = \frac{\phi(t+\Delta t, x) - \phi(x,t)}{\Delta t} - \sum_{n=1}^{\infty} \frac{\Delta t^{(2n)}}{(2n+1)!} \frac{\partial^{(2n+1)}\phi(x,t)}{\partial t^{(2n+1)}} - \sum_{n=1}^{\infty} \frac{\Delta t^{(2n-1)}}{(2n)!} \frac{\partial^{(2n)}\phi(x,t)}{\partial t^{(2n)}}, \tag{9.70}$$

$$\frac{\partial \phi(x,t)}{\partial t} = \frac{\phi(x,t) - \phi(t, x-\Delta x)}{\Delta x} - \sum_{n=1}^{\infty} \frac{\Delta x^{(2n)}}{(2n+1)!} \frac{\partial^{(2n+1)}\phi(x,t)}{\partial x^{(2n+1)}} + \sum_{n=1}^{\infty} \frac{\Delta x^{(2n-1)}}{(2n)!} \frac{\partial^{(2n)}\phi(x,t)}{\partial x^{(2n)}} \tag{9.71}$$

The approximate form of these two equations is obtained by substituting the series in terms of first order:

$$\frac{\partial \phi(x,t)}{\partial t} \approx \frac{\phi(t+\Delta t, x) - \phi(x,t)}{\Delta t}, \tag{9.72}$$

$$\frac{\partial \phi(x,t)}{\partial x} \approx \frac{\phi(x,t) - \phi(t, x-\Delta x)}{\Delta x} \tag{9.73}$$

Substituting Eqs. (9.72) and (9.73) in Eq. (9.67), its approximate shape is obtained using the UPWIND scheme:

$$\frac{\phi(t+\Delta t, x) - \phi(x,t)}{\Delta t} + c\frac{\phi(x,t) - \phi(t, x-\Delta x)}{\Delta x} \approx 0 \tag{9.74}$$

In order to determine an approximate computational solution for $\phi(x, t)$, the inequality (9.74) will be rewritten in a discrete form for the discrete spatial positions $x_i = i\Delta x$ and for the discrete times $t^n = n\Delta t$, i and n being integers. The equivalent discretized equation is:

$$\frac{\phi_i^{n+1} - \phi_i^n}{\Delta t} + c\frac{\phi_i^n - \phi_{i-1}^n}{\Delta x} = 0 \tag{9.75}$$

The solution ϕ_i^n of the discrete Eq. (9.75) is an approximation for $\phi(x, t)$. Looking for an equation for a continuous function $\psi(x, t)$ that is equal to ϕ_i^n at discrete points i in time t. In summary, $\phi(x, t)$ is the exact solution, ϕ_i^n is the approximate discrete solution and $\psi(x, t)$ is the continuous solution that equals ϕ_i^n at discrete points and times, but which differs from the exact solution $\phi(x, t)$. As the function $\psi(x, t)$ is continuous, it can be expanded into Taylor Series. Substituting the result into Eq. (9.67) results in:

$$\begin{aligned}\frac{\partial \psi}{\partial t} + c\frac{\partial \psi}{\partial x} = &-\sum_{n=1}^{\infty} \frac{\Delta t^{(2n)}}{(2n+1)!}\frac{\partial^{(2n+1)}\psi(x,t)}{\partial t^{(2n+1)}} - \sum_{n=1}^{\infty}\frac{\Delta t^{(2n-1)}}{(2n)!}\frac{\partial^{(2n)}\psi(x,t)}{\partial t^{(2n)}}\\ &- c\sum_{n=1}^{\infty}\frac{\Delta x^{(2n)}}{(2n+1)!}\frac{\partial^{(2n+1)}\psi(x,t)}{\partial x^{(2n+1)}} + c\sum_{n=1}^{\infty}\frac{\Delta x^{(2n-1)}}{(2n)!}\frac{\partial^{(2n)}\psi(x,t)}{\partial x^{(2n)}}\end{aligned} \tag{9.76}$$

By comparing Eq. (9.76) with Eq. (9.67), it can be seen that the right side of Eq. (9.76) is the error of the first-order UPWIND approximation. After manipulations carried out on Eq. (9.76), details of which can be found in Warming and Hyett [55] and Ávila et al. [2], the following equation is obtained:

$$\frac{\partial \psi}{\partial t} + c\frac{\partial \psi}{\partial x} = -\frac{c\Delta x}{2}(\mathrm{CFL}-1)\frac{\partial^2 \psi}{\partial x^2} - \frac{c\Delta x^2}{6}\left(2\mathrm{CFL}^2 - 3\mathrm{CFL} + 1\right)\frac{\partial^3 \psi}{\partial x^3} + \ldots, \tag{9.77}$$

where CFL was defined by Courant et al. [9], such that $\Delta t^n = \mathrm{CFL}\Delta x^n / c^n$.

Repeating the entire procedure for the Lax-Wendroff spatial discretization scheme, which is second order, we obtain:

$$\frac{\partial \psi}{\partial t} + c\frac{\partial \psi}{\partial x} = -\frac{c\Delta x^2}{6}(1-\mathrm{CFL}^2)\frac{\partial^3 \psi}{\partial x^3} - \frac{c\mathrm{CFL}\Delta x^3}{6}\left(1-\mathrm{CFL}^2\right)\frac{\partial^4 \psi}{\partial x^4} + \cdots \tag{9.78}$$

It is observed that the right sides of Eqs. (9.77) and (9.78) are the errors of the first-order UPWIND and second-order Lax-Wendroff discretization. The error in the UPWIND scheme is in the order of Δx, while the Lax-Wendroff Scheme error is in the order of Δx^2, as can be seen from the first terms that appear on the right sides of the two equations. When the errors for both methods are zero, that is, at discrete points and times, the computational solution is equal to the exact solution.

The errors given by the right sides of Eqs. (9.77) and (9.78) can be divided into numerical diffusion type error and numerical dispersion type error. To characterize what these errors are, the equations will be transformed into Fourier Space. The exact solution of Eq. (9.67), using the Spatial Fourier Transform, is given by:

$$\phi(x,t) = \frac{1}{\sqrt{2\pi}} \int_{-\infty}^{\infty} \widehat{\phi_0}(\xi) \mathrm{e}^{(x-ct)i\xi}\, \mathrm{d}\xi \tag{9.79}$$

Note that this integral is equal to $\phi_0(x-ct)$ which means that the solution $\phi(x,t)$, for any time, can be obtained from the initial condition, calculated at the new position $x-ct$. This translates exactly what is expected from the transport process of an initial wave $\phi(x)$ knowing that during transport nothing will change either its width or its shape. It will simply be transported exactly as it started. In other words, the exact solution $\phi(x,t)$ does not incorporate any numerical diffusion and/or dispersion effects.

Let's now see what will happen with the approximate solution $\psi(x,t)$, which carries the effects of numerical discretization methods. Let's start with the approximate solution $\psi(x,t)$, for the Lax-Wendroff scheme. Starting from Eq. (9.77) and truncating it from the third-order term into Δx, we obtain:

$$\frac{\partial \psi}{\partial t} = -c\frac{\partial \psi}{\partial x} - \frac{c\Delta x^2}{6}\left(1-\mathrm{CFL}^2\right)\frac{\partial^3 \psi}{\partial x^3} \tag{9.80}$$

Applying the Fourier transform to Eq. (9.80) results in:

$$\frac{\partial \hat{\psi}}{\partial t} = \left[\xi^2\frac{c\Delta x^2}{6}\left(1-\mathrm{CFL}^2\right) - c\right] i\xi\hat{\psi}(\xi,t) \tag{9.81}$$

The initial condition $\psi_0(x) = \phi_0(x)$ has a Fourier transform equal to $\hat{\psi}_0(\xi)$. Thus, the exact solution of Eq. (9.81) is:

$$\hat{\psi}(\xi,t) = \mathrm{e}^{\left[\xi^2\frac{c\Delta x^2}{6}\left(1-\mathrm{CFL}^2\right)-c\right]i\xi t}\widehat{\psi_0}(\xi) \tag{9.82}$$

Performing the inverse transform of Eq. (9.82), the approximate solution is obtained $\hat{\psi}(\xi,t)$:

$$\psi(x,t) = \frac{1}{\sqrt{2\pi}} \int_{-\infty}^{\infty} \widehat{\psi_0}(\xi)\mathrm{e}^{\left[\xi^2\frac{c\Delta x^2}{6}\left(1-\mathrm{CFL}^2\right)-c\right]i\xi t}\,\mathrm{d}\xi \tag{9.83}$$

By comparing Eq. (9.83) with Eq. (9.80), it can be seen that the numerical error related to the Lax-Wendroff numerical method modifies the wave propagation speed. If this error is zero, the solution by the numerical method coincides, at the discretization nodes, with the exact solution of the wave equation, $\phi(x, t)$, since it was said that $\psi_0(x) = \phi_0(x)$. This change in wave transport speed depends on the wave number ξ. Each wave number has its transport velocity. This is not physical since the wave transport velocity is unique and equal to c. This results in oscillations in the solution and is therefore called the numerical dispersion effect. If $\left(1 - \text{CFL}^2\right) > 0$ the wave will lose velocity. If $\left(1 - \text{CFL}^2\right) < 0$ the wave will gain velocity. This gain or loss in wave velocity will also depend on the wave number in question. Although this method presents a dispersion error, the wave amplitude is not affected by the truncation proposed in the present numerical analysis.

We now present an analysis of the modified equation for the UPWIND scheme, making a truncation from the second-order term in Δx:

$$\frac{\partial \psi}{\partial t} = -c\frac{\partial \psi}{\partial x} - \frac{c\Delta x}{2}(\text{CFL} - 1)\frac{\partial^2 \psi}{\partial x^2} \tag{9.84}$$

Performing the Fourier transform of Eq. (9.84), we have:

$$\frac{\partial \hat{\psi}}{\partial t} = \left[\frac{c\Delta x}{2}(\text{CFL} - 1)\xi^2 - ci\xi\right]\hat{\psi}(\xi, t) \tag{9.85}$$

The solution of Eq. (9.85) is:

$$\hat{\psi}(\xi, t) = \mathrm{e}^{\left[\frac{c\Delta x^2}{2}(\text{CFL}-1)\xi^2 - ci\xi\right]t}\,\widehat{\psi_0}(\xi), \tag{9.86}$$

whose inverse transform provides the approximate solution $\psi(x, t)$:

$$\psi(x, t) = \frac{1}{\sqrt{2\pi}}\int_{-\infty}^{\infty} \widehat{\psi_0}(\xi)\mathrm{e}^{\left[\frac{c\Delta x^2}{2}(CFL-1)\xi^2 - ci\xi\right]t}\mathrm{d}\xi \tag{9.87}$$

With the imaginary part already analyzed, the transport velocity of each wavelength is modeled. In Eq. (9.87), the numerical error does not affect the imaginary part. Therefore, it does not affect the wave transport velocity. The wave will be transported at the exact velocity. However, if the second-order term in Δx is preserved, then the imaginary part of the solution will be affected by the error, and the dispersive effect will also appear in the UPWIND scheme. The real part of the exponent has not yet been analyzed. If this part is zero, then the amplitude and phase of the wave will be preserved. If this part is negative, the amplitude will decay exponentially. If it is positive, the amplitude will grow exponentially. The higher the wave number, the higher the wave decay or amplification rates.

These effects are of a numerical nature and are undesirable. Growth in amplitude will lead to numerical instability of the method, while decay will lead to the effect of numerical diffusion of the information being transported. Both effects are unphysical and appear due to discretization error. The amplification errors imply problems with method stability. If there is a numerical diffusion error, which will dampen physical effects can be undesirable to computational simulation. It should be noted that, if terms involving derivatives of higher even order are maintained, the Lax-Wendroff Method will also display diffusion or wave amplification errors. Depending on the terms retained in the discrete model, both Lax-Wendroff and UPWIND methods will present all the numerical problems described, that is, dispersion, damping and amplification of the information being transported. However, for the Lax-Wendroff Scheme, the dispersive error is predominant and for the UPWIND scheme, amplification and diffusion errors are predominant.

9.10.3 Case Study

The example presented in this item can be found, for more details, in Queiroz [43]. The domain of interest is given by $x \in [0, 2]$. The boundary conditions are periodic, and the initial condition is given by:

$$\phi(x,0) = \begin{cases} e^{\log\left[50\left(\frac{x-0.15}{0.05}\right)^2\right]}, & x \in [0, 0.2] \\ 1, & x \in [0.3, 0.4] \\ 20x - 10, & x \in [0.5, 0.55] \\ -20x + 12, & x \in [0.55, 0.66] \\ \sqrt{1 - \left(\frac{x-0.75}{0.05}\right)^2} & x \in [0.7, 0.8] \\ 0 & x \in [0.2, 0.3] \cup [0.4, 0.5] \cup [0.66, 0.7] \cup [0.8, 0.2] \end{cases} \tag{9.88}$$

The domain discretization was done with 1000 nodes. Different values were assigned to the parameter *CFL*, with a view to analyzing the effects of errors arising from the discretization process. Starting with the Lax-Wendroff Scheme and making CFL $= 0.5$, the simulation was developed over time. In Fig. 9.8, the results are shown for the time equal to the first-time step, $t = \Delta t$, Fig. 9.8a, and for the last simulated time, Fig. 9.8b. From a very early time, $t = \Delta t$, oscillation already appears in the computational solution, which deviates from the continuous solution. As time evolves, the values of function $u(x, t)$, calculated in other wave numbers, are amplified, resulting in numerical oscillations of a dispersive nature, Fig. 9.8b.

In Fig. 9.9, the results are shown for the Lax-Wendroff Scheme (CFL $= 1.0$). Even for a long time, for which the wave has already been transported over a long distance, the computational method reproduces the exact theoretical solution with

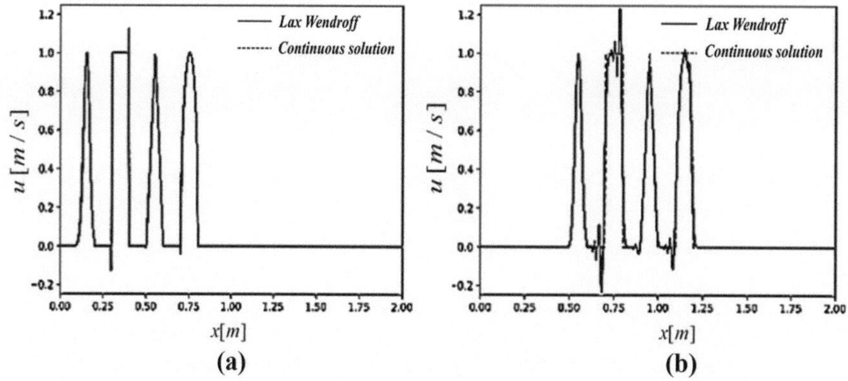

Fig. 9.8 Transport of a wave, numerical discretization using the Lax-Wendroff scheme, $CFL = 0.5$; **a** $t = \Delta t$ and **b** $t > \Delta t$

zero error. The method does not display any numerical dispersion. The solutions are coincidental.

As already pointed out with the numerical analysis, the first-order UPWIND scheme is diffusive for CFL = 1.0. In Fig. 9.10, the result obtained for wave transport is illustrated using this numerical scheme. It can be seen that the effect of diffusion intensifies with simulation time, causing very significant errors between the computational solution and the exact solution.

As expected, the diffusion error of the UPWIND scheme disappears when CFL = 1.0, as illustrated in Fig. 9.11.

The errors associated with the UPWIND and Lax-Wendroff schemes, used for the discretization of the one-dimensional wave equation, were calculated. Numerical dispersion and diffusion errors were identified and quantified. The analysis showed that these errors significantly affect the computational solution, which was verified

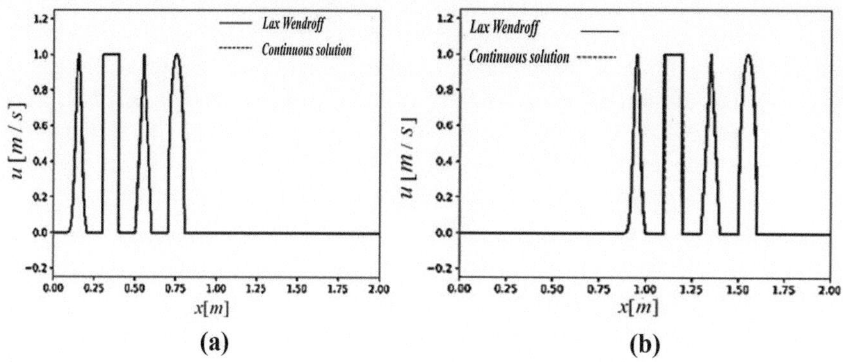

Fig. 9.9 Transport of a wave, numerical discretization using the Lax-Wendroff scheme, CFL = 1.0; **a** $t = \Delta t$ and **b** $t > \Delta t$

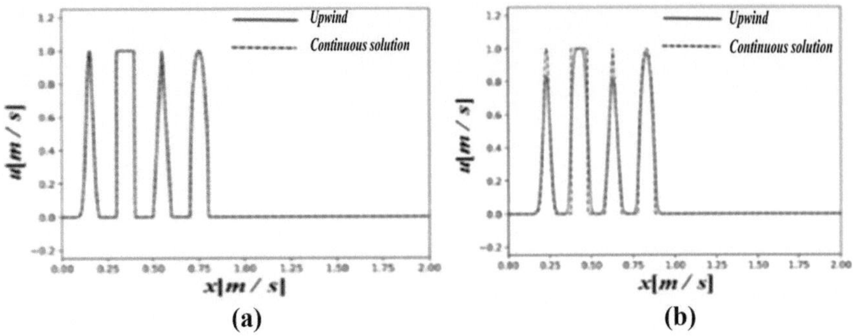

Fig. 9.10 Transport of a wave, numerical discretization using the UPWIND scheme, $CFL = 0.1$; **a** $t = \Delta t$ and **b** $t > \Delta t$

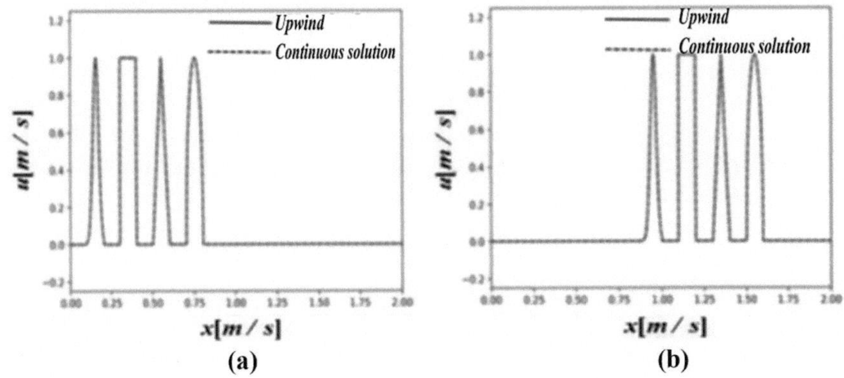

Fig. 9.11 Transport of a wave, numerical discretization using the UPWIND scheme, CFL $= 1.0$; **a** $t = \Delta t$ and **b** $t > \Delta t$

by comparing the numerical results with the results obtained via the exact solution of this equation. The extension of these conclusions to the solution of more complex models composed of a set of partial differential equations is complex. Models for fluid dynamics are too complex for the development of numerical analysis. Thus, what has been done by the global scientific community is the use of various numerical schemes, with the results of which the behavior of the solution is verified, regarding the presence or absence of these numerical diffusion and dispersion effects.

9.11 Mesh Refinement Requirement for LES Use

The application of LES requires appropriate mesh refinement, according to the physics of the problem to be solved. Problems such as free shear flows, flows over blunt bodies, such as automobiles, piping for thermic energy exchangers, flows under the effects of gravitational forces have been the object of daily practice in the use of LES.

When the problem involves flows in the vicinity of walls, typically boundary layers, the requirement for mesh refinement reaches the level of unity, in wall coordinates, for the size of the cells neighboring the walls. This requirement is like what occurs with the URANS Methodology, adapted for flows at low Reynolds numbers. This resolution makes it possible to solve the fluid dynamic instabilities characteristic of a boundary layer, that is, it makes it possible to solve the so-called internal region, composed of the viscous sublayer, the buffer region and the logarithmic region. For engineering applications, this level of mesh refinement can rarely be achieved.

The application of LES to boundary layer problems is still limited. However, for flows whose boundary layer solution is not a quality limiter, applied simulations of flows characterized by a Reynolds Number in the order of $10^6 - 10^7$, due to the availability of computational resources and, mainly, the development of efficient, highly scalable algorithms. The ability to solve problems depends much more on the efficiency of the algorithms than on increasing the capacity of the hardware.

9.12 Illustrative Results for the LES Methodology

9.12.1 Computer Simulation of an Incompressible Single-Phase Jet

In this item, results of simulations carried out using the pseudo-spectral Fourier methodology are shown. Germano's dynamic sub-grid closure modeling was used. A jet in space development was simulated using three levels of mesh refinement. The value of the Reynolds Number that characterizes this flow is $\mathrm{Re}_d = 1,050$. The refinement levels used were: $64 \times 64 \times 256$; $96 \times 96 \times 384$ and $128 \times 128 \times 512$. The criterion $Q = \left(\Omega_{ij}\Omega_{ij} - S_{ij}S_{ij}\right)/2 = 0,1\left(\mathrm{s}^{-2}\right)$. was used to produce the images presented.

The temporal evolution of the jet is shown in Fig. 9.12, relative to the most refined mesh ($128 \times 128 \times 512$), for four different times, reported in the legend of this figure. Initially, Kelvin–Helmholtz Toroidal Instabilities were developed, after which the transition to turbulence occurred and, finally, full turbulence was established. Some transition mechanism to turbulence is illustrated in Fig. 9.12, in which the spatial sequence of the formation mechanism of a new mode of instabilities is presented, for which the toroids oscillate, moving in relation to the axial axis. This displacement results in a reconnection mechanism between the toroids. In fact, the part of the toroid that comes closer to the center of the jet is transported more quickly thanhe part that moves away from the center. Thus, the reconnection results, which can be seen in Fig. 9.12b. These reconnections can also be viewed in Fig. 9.12d.

The distribution of the average of the axial component of the velocity is presented in 0. Three distributions corresponding to the three refinement levels are presented, compared with Todde et al. [53] experiment. The importance of refining the mesh can be seen for a better computational representation of the physical problem. For the more refined mesh, an excellent agreement can be observed between the result obtained through computational simulation and the result obtained through the material experiment of Todde et al. [53].

The computational modeling of a jet is highly complex because this flow transitions over a very small distance in relation to the fluid injection nozzle. On the other hand, the dynamic of this flow is highly dependent on the level of turbulence at the entrance to the domain. Modeling this effect when using LES is complex, as it requires detailed information from material experiments. Taking these observations into account, the high quality of the results obtained with computational experimentation stands out here (Figs. 9.13 and 9.14).

9.12.2 Modeling and Computer Simulation of a Crossed Two-Phase Jet

In this item, an industrial application of Large Structures Simulation (LES) will be presented. A liquid jet, crossing a gas flow, is shown in Fig. 9.15. The simulated problem was characterized under three fundamental aspects: geometric, physical and operational. Information relating to these aspects is presented in Table 9.1.

This simulation was carried out at the Fluid Mechanics Laboratory—MFLab-FEMEC-UFU, using the MFSim Computational Modeling and Simulation Platform (Multiphysics Simulator), entirely developed at MFLab. The gas flows from left to right on a flat channel, as shown in Fig. 9.15. The liquid is injected crosswise into the gas flow through an orifice in the lower wall of the flat channel. The objective sought with this flow configuration is to break the liquid jet into drops, with a view to a given industrial application. These drops must evaporate, promoting a mixture of vapor and gas. This gas contains a corrosive substance; the vapor from the liquid jet will react with the corrosive agent. Thus, the ultimate objective is to mitigate the corrosive effect of gas flow in the channel.

This image shows the result of a computer simulation, considered to be extremely complex. Soon, it is believed that this task will become trivial! Mathematical and computational modeling must be adequate to deal with several concomitant complex physical effects: turbulent regime, two-phase flow, breakage of a liquid jet into drops, transport and evaporation of thousands of drops. To this end, the model used is composed of differential models for incompressible flows (continuity equation,

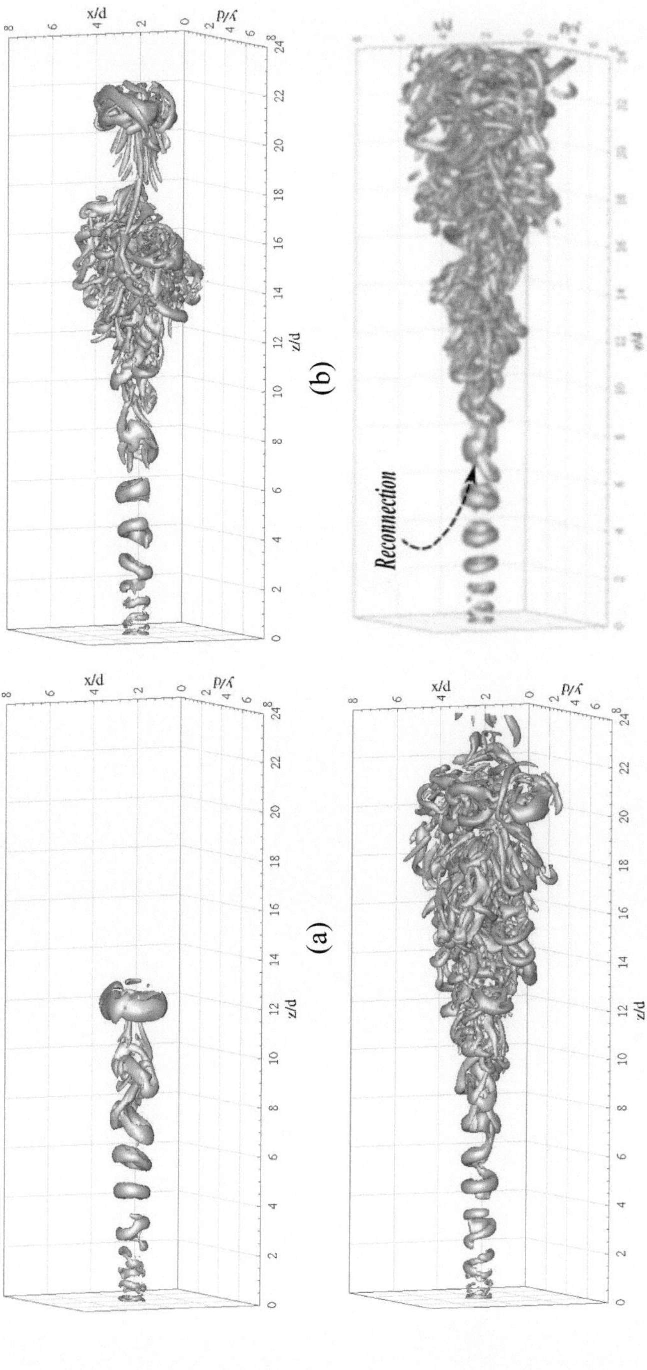

Fig. 9.12 Temporal evolution of a circular jet in spatial development, characterized by $\mathrm{Re}_d = 1,050$: **a** $t = 10$ (s); **b** $t = 20$ (s); **c** $t = 30$ (s) and **d** $t = 40$ (s). Figure taken from personal collection, with thanks to Dr. Leonardo Queiroz [43]

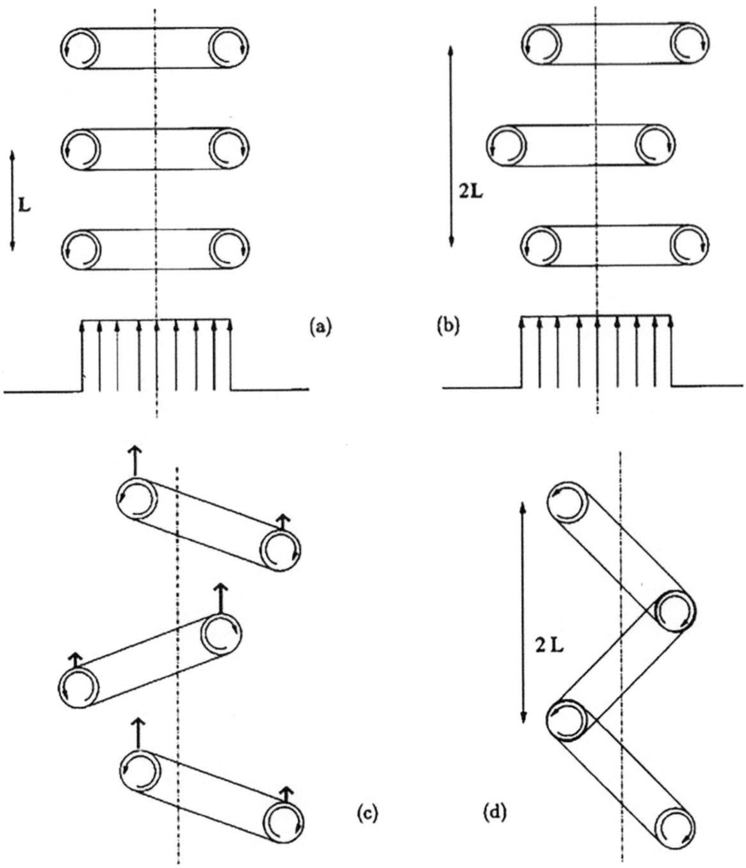

Fig. 9.13 Illustrative scheme of toroid arrangements, leading to reconnection between toroids

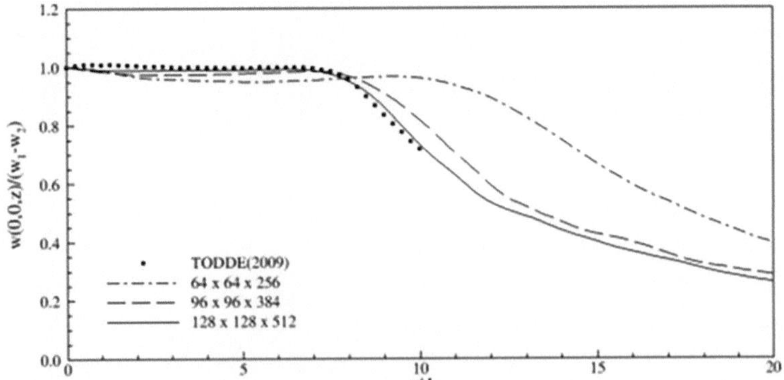

Fig. 9.14 Spatial distribution of the axial component of the average velocity at the center of the jet. Personal collection, Queiroz and Silveira-Neto [39, 43]

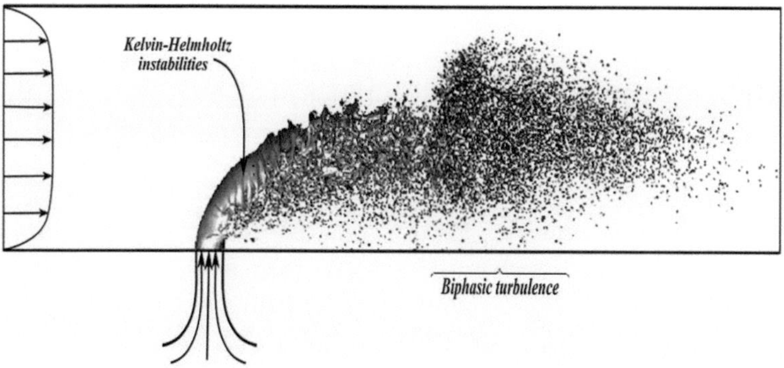

Fig. 9.15 LES applied to the simulation of a two-phase evaporative cross jet. The physical, geometric and operational characteristics of the system are provided in Table 9.1. This result is part of the MFLab scientific collection. Thanks to the researcher Dr. Bernardo Alan

Table 9.1 Characterization table of a two-phase cross jet

Characteristics	Values	Units
Liquid jet diameter	0.012	(m)
Specific mass of the gas	1.225	(kg/m^3)
Specific mass of the liquid jet	12.25	(kg/m^3)
Average gas velocity	120.4	(m/s)
Liquid jet velocity	97.84	(m/s)
Gas dynamic viscosity	1.82×10^{-5}	(kg/ms)
Dynamic viscosity of liquid jet	1.11×10^{-5}	(kg/ms)
Interfacial tension coefficient of the fluid pair	0.07	(N/m)
Ratio of linear momentum flows	6.6	–
Weber Number value for gas	330	–
Weber Number value for liquid	2178	–
Reynolds Number value for gas	5.7×10^5	-
Reynolds Number value for the liquid	1.4×10^4	–

Navier–Stokes equations, thermal energy equation, all of them filtered), state equations and droplet evaporation equation, VOF method (Volume of Fluid) to model two-phase flow, Eulerian and Lagrangian transport, conversion model from Eulerian to Lagrangian and vice versa. The LES Methodology was used with sub-grid modeling of turbulence closure, proposed by Smagorinsky [51]. The Smagorinsky constant was used with the value $C_S = 0.1$. The result of the computer simulation of the physical problem can be viewed in Fig. 9.16, for which an isovalue of the VOF function equal to 0.5 was used, which makes it possible to visualize the interface between the two fluids.

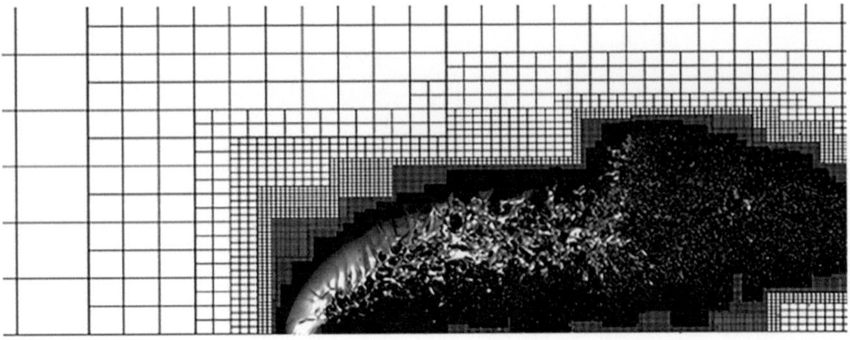

Fig. 9.16 Illustration of mesh refinement adaptivity. The physical, geometric and operational characteristics of the system are provided in Table 9.1. This result is part of the MFLab scientific collection. Thanks to researcher Dr. Bernardo Alan

The formation of ripples on the back of the curved jet is observed, which can be classified as Kelvin–Helmholtz instabilities. It is also possible to observe the primary breakdown of the liquid jet into thousands of drops, which are transported by the gas. These drops will evaporate and disappear. It is worth highlighting some of the most important physical aspects that were captured through computer simulation. Among them, the bending of the jet to the right stands out, resulting from the interaction of the gas flow with the jet injected transversely.

The importance of this type of computational experimentation is to enable the optimization of the liquid evaporation system, up to a given distance from the jet injection point. This will also optimize the process of protection against corrosion, promoted by the presence of chemical substances in the main flow.

This simulation was possible thanks to the computational resources available in MFLab, but the modeling methods used are of greater power in solving problems. Among them, the adaptivity mesh refinement feature stands out. This feature is illustrated in Fig. 9.16, where several levels of refinement can be seen, with the finest levels being found above the jet, where the vorticity gradients are high. The reduction in computational cost, with the use of adaptive refinement, can reach more than 50%, depending on the physical nature of the problem.

As already explained, the views presented in Figs. 9.15 and 9.16 were made with an iso value of the VOF function equal to 0.5, which made it possible to draw the interface between the fluid's phases. To explore other physical aspects that can be revealed through computational experimentation, the combination of an iso value of the VOF function and an iso value of the criterion $Q = \left(\Omega_{ij}\Omega_{ij} - S_{ij}S_{ij}\right)/2 = 0.1\left(\mathrm{s}^{-2}\right)$ was also visualized. This view is presented in Fig. 9.17. Observe, in more detail, the Kelvin–Helmholtz instabilities at the rear of the jet. It is also possible to observe the strong three-dimensional effects that characterize the transition process from the laminar regime to the turbulent regime.

Fig. 9.17 Visualization of the crossed jet using two simultaneous criteria: criterion (blue color) and VOF function (gray color). The physical, geometric and operational characteristics of the system are provided in Table 9.1. This result is part of the MFLab scientific collection. Thanks to researcher Dr. Bernardo Alan

9.12.3 Modeling and Simulating a Dense Spray

A second industrial problem presented here, subject to computer modeling and simulation, was that of a dense spray of drops, as illustrated in Fig. 9.18. To obtain a dense spray, an efficient operating mechanism was used: the jet rotation effect. The non-rotating spray typically develops as a cloud of droplets, leaving an empty core. With the rotation effect, the spray becomes dense; that is, a cone full of drops is obtained, as can be seen in Fig. 9.18.

The geometric, physical and operational information imposed for the simulation carried out is presented in Table 9.2.

It is noteworthy that the tangential component of the velocity vector in the jet injection is different from zero, which models its rotation. Liquid filament-like instabilities are evident in the development of the spray, thanks to the rotation effect of the jet. The core of the spray is dense with droplets of different sizes.

9.12.4 Modeling and Simulation of Complex Turbulent Flows

A sudden expansion composed of a channel with a descending step is a simple geometry, in which, however, very complex turbulent flows occur. This is the fourth

Fig. 9.18 Breaking of a circular liquid jet into the form of dense spray with rotation. This result is part of the MFLab scientific collection. Thanks to researcher Dr. Bernardo Alan

Table 9.2 Characterization of a dense spray, with a rotational effect

Characteristics	Values	Units
Liquid jet diameter	0.05	m
Specific mass of the gas	1.0	kg/m^3
Specific mass of the liquid jet	100.0	kg/m^3
Axial component of jet velocity	10.0	m/s
Tangential component of jet velocity	1.0	m/s
Dynamic viscosity of gas	1.0×10^{-4}	kg/ms
Dynamic viscosity of liquid jet	1.0×10^{-3}	kg/ms
Interfacial tension coefficient of the fluid pair	0.1	N/m
Reynolds Number value for the liquid jet	1000	–

case to be presented, with the aim of illustrating the application of the theory exposed in this chapter. A backward-facing step was used. The simulations were carried out using the first two sub-grid models proposed by the scientific community, namely the Smagorinsky Sub-Grid model [51] and the sub-grid model of the order 2 velocity structure function type, proposed by Métais and Lesieur [37]. The coordinate system and stations where the velocity and turbulent kinetic energy profiles will be compared are presented.

The results to be presented here relate to the first application of the Large Structure Simulation (LES) for the computational modeling of complex turbulent flows, the details of which can be found in Silveira-Neto et al. [50]. Until then, this methodology had only been applied to homogeneous and isotropic flows and to flat channels. The simulated flow is characterized by the geometry illustrated in Fig. 9.19. The Reynolds Number was specified as $Re_H = 38,000$. The height of the channel at the outlet and its width were specified as 2.5H. The mesh used was $200 \times 30 \times 30$, which amounted to, at the time, the fantastic amount of 180,000 cells.

A high-performance computing system was used, consisting of a CRAY2-type data processing center, which had 256 processors and vectorization resources. This system was installed at CEA-Grenoble-France. The mass of data that resulted from the simulation was stored in a tower of magnetic tapes, which were accessed by a robot. The CEA-G laboratory did not yet have graphic terminals, which required this data to be transported to the University of Grenoble, linked to the INPG- National Polytechnic Institute of Grenoble. There was also no internet connection between CEA-G and the University. Thus, the data was loaded onto large magnetic tapes and downloaded to LEGI, the Geophysical and Industrial Flow Laboratory, after which it was post-processed on a graphic terminal.

In Fig. 9.20a, the speed distributions are presented, $\overline{u}/U_0$ and in Fig. 9.20b, turbulent kinetic energy is presented, k/U_0^2. The symbols ϕ and $\overline{\phi}$ mean, respectively, the temporal average and filtering of any function ϕ. These averaging and filtering operators are effectively associated with the discretization mesh of the differential equations. Two sub-grid closure models were used: the Smagorinsky Model and the second-order velocity structure function model. A simulation was also carried out without explicit closure modeling. In the figure captions, the nomenclature "Dir. Simulate" appears. This cannot be understood by Direct Computational Simulation,

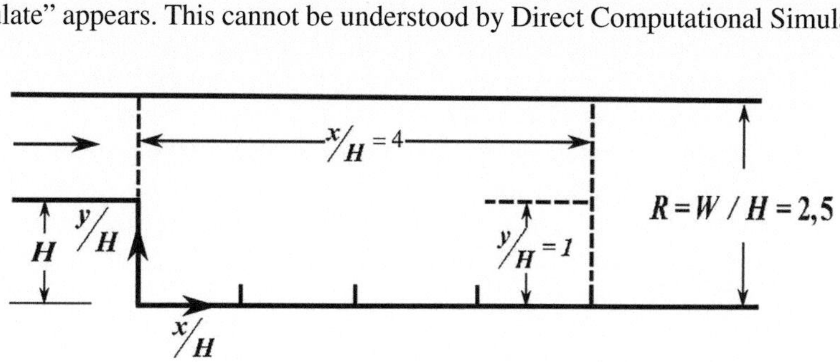

Fig. 9.19 Descending step. Illustration of the axes system and the location of the stations at which the profiles will be presented

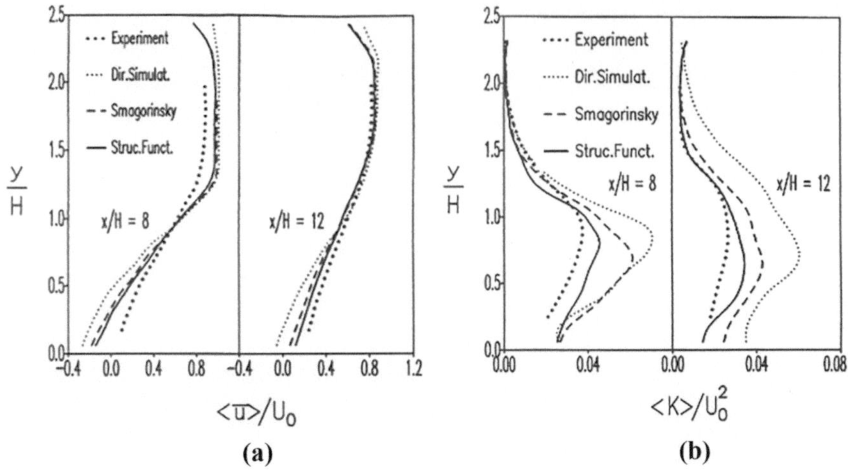

Fig. 9.20 Turbulent flow over a descending step. Distribution of average velocity (**a**) and turbulent kinetic energy (**b**)

since the mesh refinement used would not be enough for a good use of this methodology. It is understood that sub-grid modeling was not explicitly used. In fact, sub-grid modeling associated with numerical diffusion still remains. Otherwise, that is, if numerical diffusion were null, the simulation could not have converged.

With the comparison made with experimental data from Eaton and Johnston [13], there is a small difference between the results obtained with the two sub-grid modeling, that is: the Smagorinsky Sub-Grid Model and the second-order velocity structure function. The quality of the results obtained with the simulation without explicit modeling of turbulence closure is inferior to the results obtained with the two sub-grid modeling. The same findings can occur when observing Fig. 9.20b, for turbulent kinetic energy.

In Fig. 9.21, the spectral densities $E_{uu}(S)$ are presented, relating to temporal fluctuation $u'(\vec{x}, t) = u(\vec{x}, t) - u(\vec{x})$ of velocity. This fluctuation was sampled over time and at a position $x = \{(x/H) = 12, (y/H) = 12, (z/H) = 0.5\}$. The parameter $S = f/(U/H)$ is the Strouhal Number, or the dimensionless frequency of formation of turbulent structures. The presence of a band of more energized structures around $S = 0.03$. An inertial band is also observed, with a slope close to -5/3, in accordance with Kolmogorov's law. As expected, the simulation without explicit sub-grid modeling resulted in energy accumulation at the cutoff frequency, approximately $S_c = 0.35$. Effectively, there is an accumulation of turbulent kinetic energy in the vicinity of the cutoff frequency, when explicit sub-grid modeling was not used, that is, only numerical diffusion was used as sub-grid modeling. This shows that modeling with only numerical viscosity is not sufficient for correct energy transfer between the resolved band and the sub-grid band.

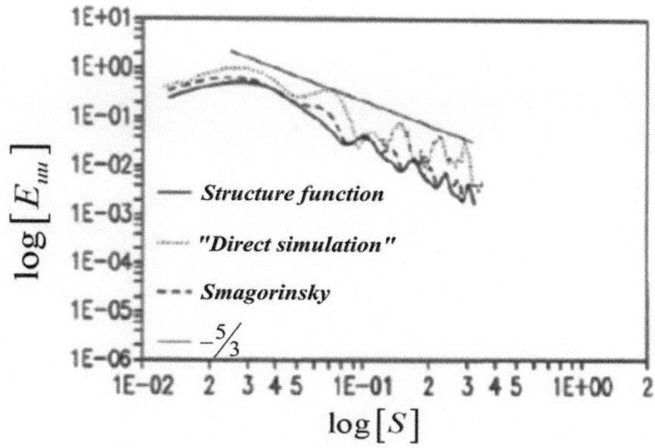

Fig. 9.21 Turbulent flow over a backward-facing step. Spectral density $E_{uu}(S)$, for the flow over the sudden expansion already described

9.13 An Example of Computational Experimentation in Fluid Dynamics

As an illustration of what is called computational experimentation in this book, results and analyses of a classic problem of the natural convection on a horizontal cylinder heated in relation to the environment will be presented. The models will be presented: physical, algebraic-differential mathematics and computational. Illustrative results of the experiment will be presented.

9.13.1 Physical Models

Two problems were studied. The first is a heated cylinder inside a circular cavity, with non-zero eccentricities. This problem is illustrated in Fig. 9.22. The dimensions of the cavity are the same as those used by Kuehn and Goldstein [26]. The domain of physical interest is the annular cylindrical cavity. For the numerical solution, using the immersed boundary method, it is necessary to specify a rectangular external domain of dimensions ($L = 7.5D$; $H = 7.5D$; $W = 1.25D$), where L and H are the dimensions transverse to the axis of the cylinder and W is the length in the axial direction of the cylinder, D is the diameter of the outer cylinder. The diameter of the inner cylinder, d, is such that $(D/d) = 2.6$. The physical properties were chosen such that the Rayleigh number, $\text{Ra} = \left[g\beta(T - T_r)L^3\right]/(\alpha\mu/\rho) = 4.8 \times 10^4$, was used to characterize the simulations. Four different eccentricities were used: $\varepsilon_\vartheta/\ell = 0$; $\varepsilon_\vartheta/\ell = 2/3$; $\varepsilon_\vartheta/\ell = -2/3$ and $\varepsilon_h/\ell = 2/3$, where represents the eccentricity relative to the space between the cylinders, $\ell = D - d$. Indices (ϑ) and (h) refer to the vertical and horizontal directions, respectively. More details about this physical model can be found in Romão et al. [45].

In the second problem, a heated cylinder is placed inside a rectangular cavity, as illustrated in Fig. 9.23, where the main dimensions are given. The diameter of the

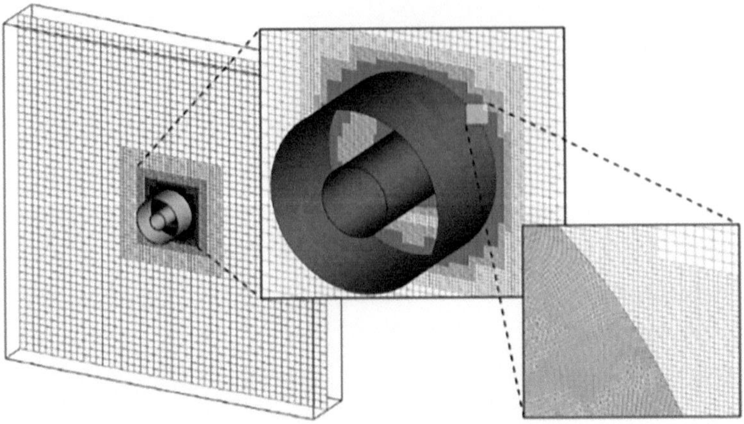

Fig. 9.22 Physical model of the natural convection problem around a heated horizontal cylinder placed inside a circular cavity

Fig. 9.23 Physical model of the natural convection problem in a heated cylinder placed inside a rectangular cavity

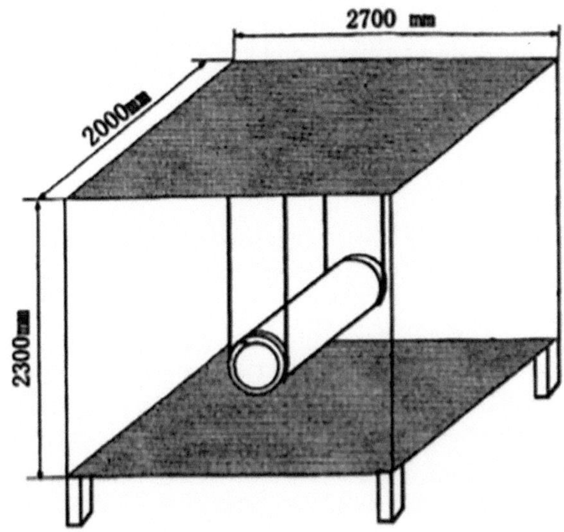

cylinder is $d = 0.4$ (m). The sides and top plane of the domain are open, while the bottom is closed. The physical properties and operating conditions were specified, such as the Rayleigh number, $\mathrm{Ra} = \left(g\beta q'' d^4\right)/(\nu\alpha k)$, for which some values are assigned, $\mathrm{Ra} = 1.97 \times 10^9$; 1.5×10^{10} and 1.12×10^{11}.

9.13.2 Algebraic-Differential Mathematical Model

These are incompressible flows, but their specific mass varies with temperature. The mathematical model for this class of problems, natural convection, was established in Chap. 2, using the so-called Oberbeck-Boussinesq approximation. This model will be rewritten here for educational reasons:

$$\vec{\nabla} \cdot \overrightarrow{\overline{V}} = 0, \tag{9.89}$$

$$\frac{\partial \overrightarrow{\overline{V}}}{\partial t} + \vec{\nabla}.\left(\overrightarrow{\overline{V}}\,\overrightarrow{\overline{V}}\right) = -\frac{1}{\rho_o}\vec{\nabla}\overline{p} - \beta(\overline{T} - T_o)\vec{g} + \vec{\nabla} \cdot \left(\nu_{ef}\left(\nabla \overrightarrow{\overline{V}} + \nabla \overrightarrow{\overline{V}}^T\right)\right) + \vec{f}, \tag{9.90}$$

$$\frac{\partial \overline{T}}{\partial t} + \vec{\nabla}.\left(\overrightarrow{\overline{V}}\,\overline{T}\right) = \vec{\nabla} \cdot \left(\alpha_{ef}\vec{\nabla}\overline{T}\right) + f_T/\rho C_p. \tag{9.91}$$

The term $\beta(T - T_o)\vec{g}$, which appears in the Navier–Stokes equations, represents the combined effect of the weight force and the buoyancy force acting on a fluid particle. A reference state of the problem is represented by the properties ρ_o and T_o. They can be, for example, the average values of the specific mass and temperature within the domain of interest, the object of the problem under analysis. Properties ν_{ef} and α_{ef} are the effective diffusivities of linear momentum and thermal energy, respectively. The specific mass of the fluid is represented by ρ and the thermal capacity is represented by C_p. These equations are solved in Cartesian domains, but non-Cartesian bodies placed inside are modeled dynamically and thermally with the immersed boundary method. Details of this modeling can be found in Romão et al. [45]. Turbulence modeling was done using Germano's [18] dynamic sub-grid closure model, presented in Sect. 6.2.

9.13.3 Numerical Mathematical Model

The differential equations were discretized using the finite volume method for space and a second-order method for discretization in time, as presented by Romão et al. [45]. To define the numerical mesh, consistent with LES modeling, the number of degrees of freedom of the turbulent flow was evaluated, relative to $\mathrm{Ra} = 1.97 \times 10^{11}$. The value of the Reynolds number for this flow was estimated. To this end, the following characteristic quantities were determined: $U = \sqrt{gd} = \sqrt{9.8\left(\mathrm{m/s^2}\right) \times 0.8\,(\mathrm{m})} = 2.8\,(\mathrm{m/s})$; $d = 0.8\,(\mathrm{m})$ and $\nu = 4.5 \times 10^{-7}\left(\mathrm{m^2/s}\right)$. This resulted in $\mathrm{Re}_d = 5 \times 10^6$. Using Eq. (6.2), $(\ell_I/\eta)^3 \approx \mathrm{Re}_{\ell_I}^{9/4}$, getting $ngl = 1.2 \times 10^{15}$ (number of degrees of freedom) per unit volume (d^3). From this equation, assuming $\ell_I = d = 0.8\,(\mathrm{m})$, the characteristic Kolmorov length was determined, $\eta = 7.6 \times 10^{-6}$ (m). The finest level of the composite mesh used corresponds to $\Delta x = 6 \times 10^{-3}$ (m). With this mesh, approximately 50% of the spectrum of turbulent structures can be resolved. This ensures that the frequency or cutoff wave number is in the inertial zone of the spectrum. As presented by Romão et al. [45], almost all the energy contained in this spectrum of turbulent structures was calculated. The sub-grid energy is very small compared to the energy contained in the resolved spectrum. This result corroborates the nomenclature associated with the integral turbulent structures that make up the spectrum of structures. They are also referred to as the structures that carry turbulent kinetic energy.

9.13.4 Computational Model

The MFSim computational modeling platform, developed at MFLab-FEMEC-UFU, was used, which has as its main characteristic the dynamic adaptivity of mesh refinement. This platform offers possibilities for modeling problems with complex non-Cartesian geometries, using the dynamic and thermal immersed boundary methodology. To characterize the computational model related to the problems presented in Sect. 13.1, the described boundary and initial conditions were considered, which were translated into input data for the computational model. For the annular cavity problem, wall temperatures were specified. For the horizontal cylinder problem, a thermal flow was imposed at the interface between the cylinder and the fluid $q''(\mathrm{W/m^2})$. The initial conditions were in equilibrium with the environment. The results of the computational experiments developed are presented below.

9.13.5 Illustrative Results

9.13.5.1 Natural Convection in Annular Cavities

In Fig. 9.24, images relating to natural convection flows are presented for the four specified eccentricities. In Fig. 9.24a, there are the isotherms related to the computational experiment developed in MFLab for the Rayleigh already specified and for the four eccentricities also already specified. The flow regime for this *Ra* is laminar. In Fig. 9.24b, we have the corresponding results, obtained with the material experiment, carried out by Kuehn and Goldstein [26]. From the qualitative comparison between the experiments, a great similarity between them was obtained.

Quantitative comparisons are shown in Fig. 9.25a–c, for three different eccentricities and for three levels of refinements. In Fig. 9.25a, the dependence of the results on refinement is shown. The results are consistent and the difference between theoretical modeling and material modeling decreases as refinement increases. The agreement

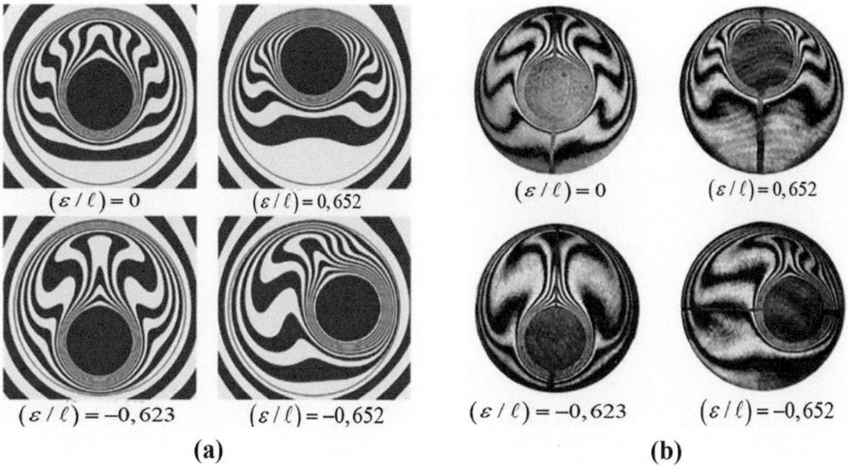

Fig. 9.24 Natural convection in an eccentric annular cavity. Comparison between computational (**a**) and material (**b**) experiments, Kuehn and Goldstein [26]. Isotherms are visualized

between the results for different eccentricities is very good. The enormous similarity between the images leads to the notion of what is called computational experimentation in this book. However, it is necessary to quantitatively prove the agreement between the results obtained with the two experiments, which is presented in Fig. 9.25a–c. Three levels of mesh refinement were used, and the importance of good computational resolution is shown. The quantitative agreement between the results of the experiments is also of excellent quality, which can be seen in Fig. 9.25a–c.

9.13.5.2 Natural Convection Around a Heated Horizontal Cylinder

The Q criterion, calculated with Eq. (2.126), is presented in Fig. 9.26. The exuberance of turbulence in a natural convection flow over a horizontal cylinder can be visualized in Fig. 9.26a. In Fig. 9.26b, the adaptive mesh for turbulent flow is visualized. Refinement is carried out as necessary. The vorticity level was adopted as an adaptive refinement criterion.

The temperature fields promoted by the natural convection over the heated horizontal cylinder can be visualized in Fig. 9.27. The isotherms obtained through computational experimentation are shown in Fig. 9.27a, while in Fig. 9.27b, the temperature field resulting from material experimentation, carried out by Missumi and Kitamura [38], is visualized. The qualitative similarity between the two fields is notable. It is difficult to say which of the two fields comes from one or the other type of experimentation.

Computational probes were distributed over the turbulent wake to sample information of interest. In particular, time series of the velocity components were sampled, $\vec{V} = u\vec{i} + \vartheta\vec{j} + w\vec{k}$, at the points $(z/d) = 1.0$, $(z/d) = 2.5$, $(z/d) = 4.0$ and

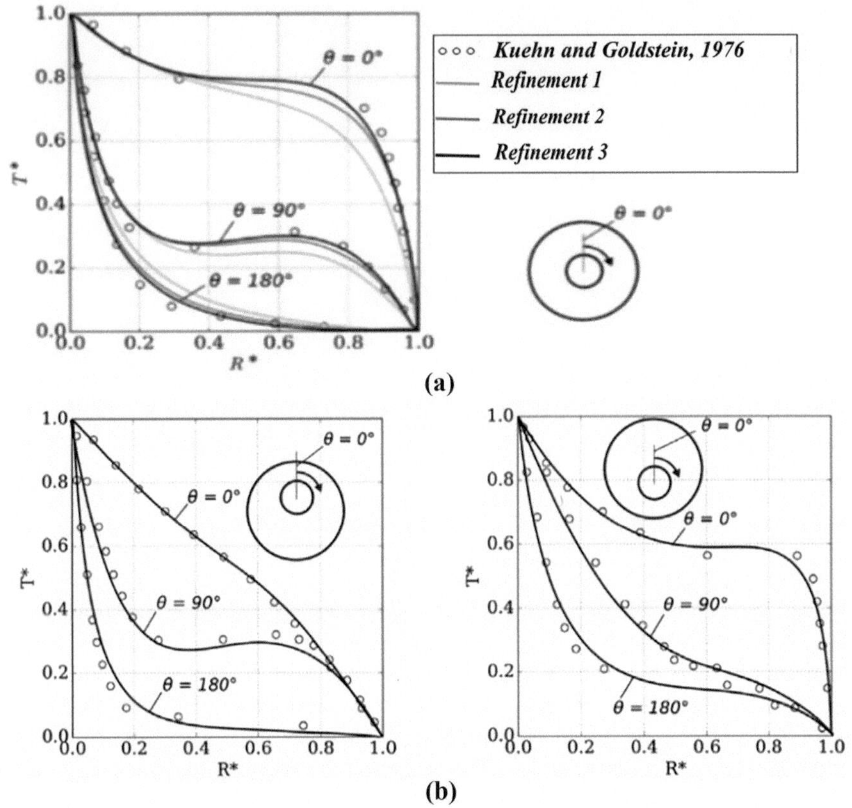

Fig. 9.25 Natural convection in an eccentric annular cavity. Comparison between computational and material experiments, Kuehn and Goldstein [26]; **a** zero eccentricity, $\varepsilon_\vartheta/\ell = 0$, and effect of mesh refinement, **b** eccentricity $\varepsilon_\vartheta/\ell = \frac{2}{3}$ and **c** $\varepsilon_\vartheta/\ell = -2/3$

$(z/d) = 4.6$. These probes were placed on a vertical line passing through the center of the cylinder, as illustrated in Fig. 9.27a. Variable d is the diameter of the cylinder. Once computationally sampled, the temporal distributions of the velocity components were transformed to Fourier space, and the turbulent specific kinetic energy spectral densities were determined. The treatment presented in Chap. 4, Eqs. (4.4) and (4.14). The spectral distributions are presented in Fig. 9.28, where it is noted that they are continuous and wide, characterizing a turbulent regime. These spectra were truncated using the cutoff frequency, determined with the time step used in the computational simulations.

The Nusselt Number was calculated for different values of the Rayleigh Number. The results obtained with the computational experiment were compared with the results of Morgan's material experiment [38]. Comparisons are presented in Table 9.3. Differences in comparisons with Morgan's work [38] were within the error range of the material experiment.

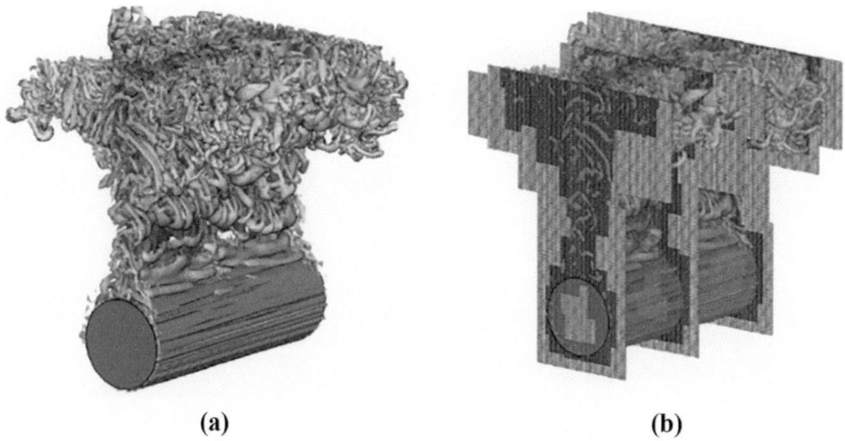

Fig. 9.26 Computational experimentation for analysis of natural convection over a horizontal cylinder. The criteria are displayed Q in (**a**) and the adaptive mesh in (**b**)

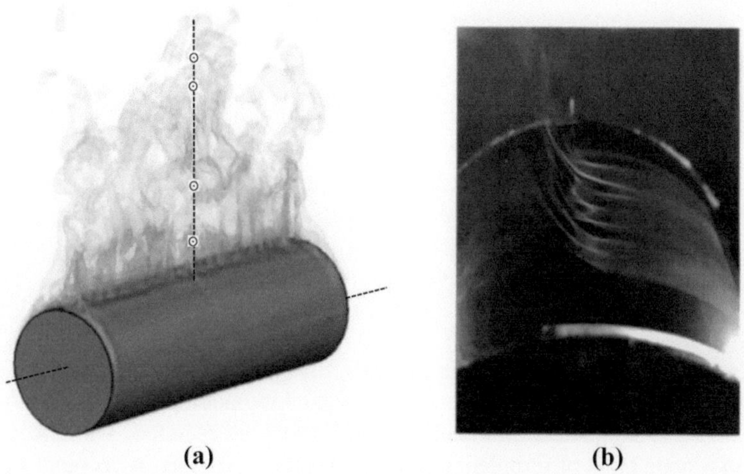

Fig. 9.27 Natural convection over a heated horizontal cylinder. The temperature from the computational experiment is visualized in (**a**) and the temperature from the material experiment, Missumi and Kitamura [38] in (**b**)

9.14 Hybrid Turbulence Closure Models

As explained in the previous items, the characteristics, advantages and disadvantages of the turbulence modeling methodologies were highlighted: LES, URANS and RANS. When using DNS, the entire spectrum of turbulent structures is resolved. Flows in the vicinity of walls are resolved accurately. However, enormous computational power is required, with large storage capacity, as well as sophisticated methods

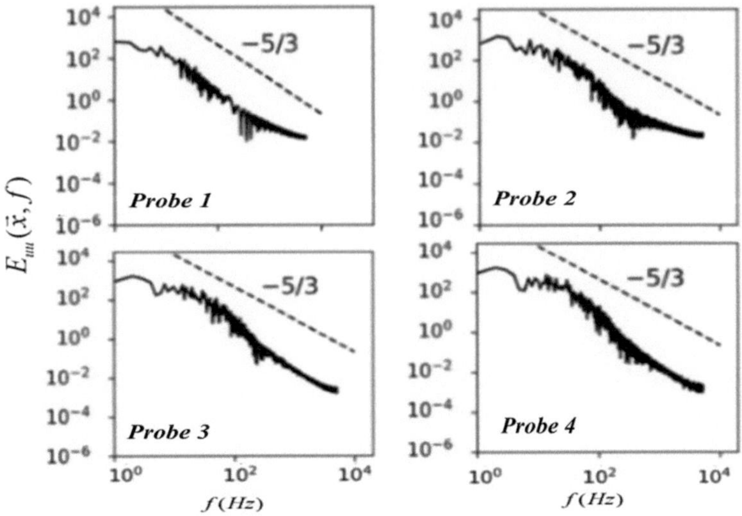

Fig. 9.28 Spectral density of turbulent specific kinetic energy. Frequency distributions relative to the four probes placed in the natural wake behind a heated cylinder

Table 9.3 Mean Nusselt Number for different values of the Rayleigh Number, compared with references

	Ra = 1.97×10^9	Ra = 1.5×10^{10}	Ra = 1.97×10^{11}
Morgan [38]	44.24	73.49	121.49
Present work	40.01	55.74	115.34
Relative difference (Morgan) (%)	9.56	2.36	5.06

for solving PDEs. DNS applications are, therefore, limited to flows characterized by low Reynolds Number values. To use the LES Methodology, less computational power and less sophistication in numerical methods are required. The direct calculation of a wide band of turbulent structures becomes possible. It becomes possible to solve flows at high values of the Reynolds Number. However, to solve flows with walls, a very high spatial resolution—mesh refinement—is necessary, which makes simulations very expensive. The URANS Methodology is used to solve problems at high Reynolds Number values, as well as modeling flows close to walls. However, the solutions obtained using URANS are only for the steady state or for the lowest frequencies in the spectrum band of turbulent structures. Given this framework, the scientific community has been working on the hybridization process of the LES and URANS methodologies, which will be the subject of the next chapter.

9.15 Exercises

1. It was seen in Chap. 8 that a turbulence closure model, in the context of the URANS methodology, has the purpose of modeling the process of transferring a given *information* between the entire spectrum of turbulent structures and the average behavior of the turbulent flow. What can be said in this regard for the LES methodology? Is this transfer process reversible or irreversible?
2. Demonstrate that the Prandtl mixing length model is a particularity of the Smagorinsky model.
3. The Smagorinsky Sub-Grid model and the second-order structure function model of velocity are both algebraic and are similar. However, they have a fundamental difference. Identify and discuss this difference. Can it have useful implications to be explored regarding numerical-computational implementation facilities?
4. What does the turbulence closure dynamic modeling add more to, when compared to the Original Smagorinsky modeling?
5. Review the deduction of the dynamic sub-grid model and point out what allows us to conclude that this is the only model, until the publication of this book, that, conceptually, can be used in transition to turbulence.
6. In dynamic sub-grid modeling, at what point is the first filtering, known as mesh filtering, applied? And at what point is the so-called sub-test filtering applied?
7. Why are Smagorinsky's original sub-grid models not suitable for non-uniform meshes? How to fix it? Is it different for the model F_2?
8. Use any computational platform and simulate a temporal wake using the three sub-grid models presented. Compare the results.
9. Differentiate the concepts of numerical dispersion and diffusion. For LES, is it recommended to eliminate these effects? Why? For URANS, these effects have little effect on the results. Explain?
10. Can a URANS closure model be used as a sub-grid model? Discuss and argue.
11. How to specify a fluid dynamic boundary condition at the entrance of a domain, considering the use of the LES methodology, and considering the turbulent regime upstream of the domain?
12. Consider the modeling presented in Sect. 9.3. Deduce the model for statistical treatment for the second-order moment $\langle u_i' T' \rangle$ and for the third-order moment $\langle u_i' p' \rangle$.
13. Repeat exercise 12 for the third-order moment $\left\langle u_i' u_i' u_j' \right\rangle$.

References

1. Antonopoulos, D.: Large-Eddy simulation of a passive scalar in isotropic turbulence. J. of Fluid Mech. **104**, 55–79 (1981)
2. Ávila, Í.A.M., Ribeiro Neto, H., Damasceno, M.M.R., Silveira-Neto, A.: XI EPTT—Escola Brasileira de Transição e Turbulência – Uberlândia-MG, Anais (2018)
3. Batchelor, G.K.: The Theory of Homogeneous Turbulence. Cambridge University Press (1953)
4. Boussinesq, J.: Essai sur la théorie des eaux courantes. Mém. Acad. Sci. Paris, **22**, 46 (1877)
5. Bradshaw, P., Launder, B.E., Lumlay, J.L.: Collaborative testing of turbulence models. J. Fluid Eng. Trans. ASME **118**, 243–247 (1996)
6. Carati, D., Ghosal, S., Moin, P.: On the representation of backscatter in dynamic localization models. Phys. Fluids **7**(3), 606–626 (1995)
7. Chollet, J.P., Lesieur, M.: Parametrization of small scales of three-dimensional isotropic turbulence utilizing spectral closures. J. Atms. Sci. **38**, 2747–2757 (1982)
8. Clark, R., Ferziger, J., Reynolds, W.: Evaluation of sub grid-scale models using an accurately simulated turbulent flow. J. Fluid Mech. **91**, 1–16 (1979)
9. Courant, R., Friedrichs, K., Lewy, H.: On the partial difference equations of mathematical physics. IBM J. Res. Dev. **11**(2), 215–234 (1967). ISSN 0018-8646
10. David, E.: Modélisation des écoulments compresibles et hypersonics: une approache instationnaire, Thèse, Institut National Polytechnique de Grenoble (1993)
11. Deardorff, J.W.: AS numerical study of three-dimensional turbulent channel flow at Large Reynolds Number. J. Fluid Mech. **14**, 453–480 (1970)
12. Ducros, F.: Simulations numériques directes et des grandes échelles de couches limites compressibles, Thèse, Insitut National Polytechnique de Grenoble (1995)
13. Eaton, J.K., Johnston, J.P.: Turbulent flow reattachment: an experimental study of the flow and structure behind a backward-facing step. Stanford University, Re. MD-39 (1980)
14. Fallon, B., Lesieur, M., Grand, D.E Delcaycre, F.: Large-Eddy simulations of stable-stratification effects upon a backstep flow, Report LEGI-IMG/CEA-Grenoble (1996)
15. Findikakis, A., Street, R.L.: An algebraic model for sub grid-scale turbulence in stratified flows. J. Atmos. Sci. **36**, 1934–1949 (1979)
16. Germano, M.: A proposal for a redefinition of the turbulent stresses in Filtered Navier-Stokes Equations. Phys. Fluids **29**(7), 2323–2324 (1986)
17. Germano, M., Piomelli, U, Moin, P., Cabot, W.H.: A dynamic sub grid-scale eddy viscosity model. Phys. Fluids A **3**(7), 1760–3128 (1991)
18. Germano, M.: Turbulence, the filtering approach. J. Fluid Mech. **238**, 325–336 (1991)
19. Gharib, M.: Perspective: the experimentalist and the problem of turbulence in the age of supercomputers. J. Fluids Eng. **118**(2), 233–242 (1996)
20. Ghosal, S., Lund, T.S., Moin, P., Vol., K.A.: A dynamic localization models for large-eddy simulation of turbulent flows. J. Fluid Mech. **286**, 229–255 (1995)
21. Harlow, F.H., Nakayama, P.I.: Transport of turbulence energy decay rate, Los Alamos Sci. Lab., University of California Report LA-3854 (1968)
22. Hinze, J.O.: Turbulence, 2nd edn. McGraw-Hill, New York (1975)
23. Jones, W.P., Launder, B.E.: The prediction of laminarization with a two-equation model of turbulence. Int. J. Heat Mass Transf. 15, 301–314 (1972)
24. Kolmogorov, A.N.: The local structure of turbulence in incompressible viscous fluid for very Large Reynolds Number. Doklady Akademiya Nauk SSSR **30**, 299–303 (1941)
25. Kraichnan, R.H.: Eddy viscosity in two and three-dimensions. J. Atmos. Sci. **33**, 1521–1536 (1976)
26. Kuehn, T.H., Goldstein, R.J.: An experimental study OD natural convection heat transfer in concentric and eccentric horizontal cylindrical annuli. J. Heat Transfer **100**, 635–640 (1976)
27. Launder, B.E., Sharma, B.I.: Application of energy dissipation model of turbulence to the calculation of flow near a spinning disc. Lett. Heat Mass Transf. **1**(2), 131–138 (1974)
28. Launder, B.E., Spalding, D.B.: Mathematical Models of Turbulence. Academic Press, London (1972)

29. Launder, B.E., Spalding, D.B.: The numerical computation of turbulent flows. Comput. Methods Appl. Mech. Eng. **3**, 269 (1974)
30. Launder, B.E.: On the effects of a gravitational field on the turbulent transport of heat and momentum. J. Fluid Mech. **67**, 569–581 (1975)
31. Launder, B. E., 1975b, Progress in modeling of turbulent transport, Lecture series 76, Von Kàrmàn Institute.
32. Launder, B.E.: Phenomenological modeling: present and future. In: Lumley, J.L. (Ed.) Lecture Notes in Physics, vol. 357, p. 439. Springer Verlag (1990)
33. Lesieur, M.: Turbulence in Fluids, revised edn. Kluwer Publishers (1990)
34. Lesieur, M., Métais, O.: New trends in Large-Eddy simulation of turbulence. Ann. Rev. Fluid Mech. **28**, 45–82 (1996)
35. Lilly, D.K.: A proposed modification of the Germano sub grid-scale closure method. Phys. Fluids A **4**(3), 633–635 (1992)
36. Menter, F.R.: Influence of freestream values on $k - \omega$ turbulence model predictions. AIAA J. **30**, 1657–1659 (1992)
37. Métais, O. e Lesieur, M.: Spectral large-eddy simulation of isotropic and stably stratified turbulence. J. Fluid Mech. **239**, 157–194 (1992)
38. Missumi, T., Kitamura, K.: Fluid Flow and heat transfer of natural convection around large horizontal cylinder: experiments with air. Heat Transf. Asian Res. **32**(4), 293–305 (2003)
39. Moin, P., Kim, J.: J. Fluid Mech. **118**, 341–377 (1982)
40. Monpean, G. M., 1989, Modélisation des écoulments turbulents avec transferts thermiques en convection mixte, Thèse de Doctorat, L´École Centrale de Lyon.
41. Moreira, L.Q., Silveira-Neto: Modelagem Matemática de jatos em desenvolvimento espacial usando a Metodologia Pseudo Espectral de Fourier, Tese de doutorado, PPG Engenharia Mecânica da UFU (2011)
42. Piomelli, U.: High Reynolds Number calculations using the dynamic sub grid-scale stress model. Phys. Fluids A **5**(6), 1484–1490 (1993)
43. Queiroz, R.A.B.: Desenvolvimento e teste de esquemas "upwind" de alta resolução e suas aplicações em escoamentos incompressíveis com superfícies livres. Dissertação de mestrado em ciências de computação e matemática computacional, Instituto de Ciências Matemáticas e de Computação, Universidade de São Paulo. (2011). https://doi.org/10.11606/D.55.2009.tde-27052009-141016
44. Ribeiro, F.J.O., Silveira-Neto: Theoretical analyze of turbulent channel flow with thermal effects—the influence of the turbulent Prandtl number. In: 17th Brazilian Congress of Thermal Sciences and Engineering, November 25th-28th, Águas de Lindoia, SP-Brazil (2018)
45. Romão, R.S.M., Kinoshita, D., Villar, M.M., Serfaty, R., Silveira-Neto, A.: Numerical experiment of turbulent transfer of thermal energy using the immersed boundary method and adaptative mesh refinement. Int. J. Therm. Sci. **151**, 1–11 (2020)
46. Schlichting, H.: Boundary-Layer Theory. McGraw-Hill Inc., New York (1979)
47. Scotti, A., Meneveau, C., Fatica, M.: Dynamic Smagorinsky model on anisotropic grids. Phys. Fluids **9**, 1856–1858 (1996)
48. Shaanan, S., Ferziger, J.H., Reynolds, W.C.: Numerical simulation of turbulence in the presence of shear. Dept. Mech. Engng, Stanford University, Rep. TF-6 (1975)
49. Silva, C.B., Métais, O.: Vortex control in bifurcation jets: a numerical study. Phys. Fluids **14**(11), 3798–3819 (2002)
50. Silveira-Neto, A., Grand, D., Lesieur, M., Metais, O.: A numerical investigation of coherent vortices in turbulence behind a backward-facing step. J. Fluid Mech. **256**, 1–25 (1993)
51. Smagorinsky, J.: General circulation experiments with the primitive equations. Mon. Weath. Rev. **91**(3), 99–164 (1963)

52. Spyropoulos, E.T., Blaisdel, G.A.: Evaluation of the dynamic models for simulation of compressible decaying isotropic turbulence. AIAA J. **34**(5), 990–998 (1996)
53. Todde, V., Spazzini, P.G.: Experimental analyses of low Reynolds number free jets evolution along the jet centerline and Reynolds Number Effects. Exp. Fluids **3**(47), 279–294 (2009)
54. van Driest, E.R.: On turbulent flow near a wall. J. Aerospace Sci. **23**, 1007–1011 (1956)
55. Warming, R., Hyett, B.: The modified equation approach to the stability and accuracy analysis of finite-difference methods. J. Comput. Phys. **14**(2), 159–179 (1974)

Chapter 10
URANS-URANS and LES-URANS Hybrid Closure Models

In this chapter, hybrid turbulence closure models are discussed. In Sect. 10.1, an introduction to the topic is presented. In Sect. 10.2, hybrid turbulence closure models of the URANS–URANS type are presented. In Sect. 10.3, comments are presented on the transient nature of the URANS Methodology. In Sect. 10.4, illustrative results of the URANS–URANS hybrid models are shown. Hybrid URANS–LES turbulence closure models are presented in Sect. 10.5. In Sect. 10.6, a list of exercises is proposed. Finally, in the references are presented.

10.1 Hybrid Turbulence Closure Models

The main characteristics highlighted will be summarized here, as well as the advantages and disadvantages of the turbulence modeling methodologies: DNS, LES, URANS and RANS. With the DNS methodology, it is possible to solve the entire spectrum of turbulent structures that make up a turbulent flow. It is possible to resolve flows near to walls, accurately. To achieve this, enormous computational power and large storage capacity must be available, as well as sophisticated methods for solving PDEs. These requirements increase non-linearly, as the Reynolds Number is increased. Applications with DNS are, therefore, limited to flows characterized by low Reynolds Number.

Solving flows using the LES methodology requires less computational power and less sophistication in numerical methods, when compared to DNS. The direct calculation of a wide band of turbulent structures became possible, as well as the solution of flows at high Reynolds Number. However, to solve flows with the presence of walls, very high spatial resolution and high mesh refinement are necessary, which makes simulations very expensive. Despite the high costs, they are possible. When flows are intrinsically transient, currently, there is no alternative methodology. For

A. da Silveira Neto, *Turbulent Flows*, https://doi.org/10.1007/978-3-031-89222-6_10

example, a fluid–structure interaction problem can be solved mainly with the LES methodology. In this case, the cost becomes less limited compared to DNS.

Using URANS models, it is possible to solve problems at high Reynolds Number, which has brought significant advances in the quality of modeling flows close to walls. It was also possible to develop URANS models for flows without walls. The model $k-\varepsilon$ pattern is an example. The use of wall laws is a palliative to solve the problem of turbulent flows in the vicinity of walls. The Spalart and Almaras model (S–A) [16], for instance, was proposed to simulate boundary layer flows. The $k-\omega$ model, was proposed by Wilcox [21] to simulate flows in the vicinity of walls, but it is inappropriate for regions of the flow where the Reynolds Number assumes high values, such as the turbulent core of a flow. Another limiting characteristic of the URANS Models is that they are only viable for solving the lowest frequencies that compose the band of turbulent structures.

Given these arguments, the scientific community has been working on the process of hybridization of methodologies, which occurs involving URANS Models, such as, $k-\omega-$SST, proposed by Menter [7]. In this approach, the author hybridized the $k-\varepsilon$ standard closure model, adapted for flows at high Reynolds Number, with the model $k-\omega$, developed for flows close to walls. This hybridization resulted in a model with which both regions, close to the walls and the turbulent core, are well modeled. However, with this hybrid closure modeling, one remains limited to calculating only the lowest frequencies of the spectrum of turbulent structures. An alternative that has been the subject of developments is URANS–LES hybridization. These two hybridization lines will be presented below, starting with the URANS–URANS hybridization.

10.2 URANS–URANS Hybrid Turbulence Closure Models

10.2.1 Preliminary

A family of hybrid models has been presented in scientific literature. In this text, some of them are presented to facilitate the learning of the beginner on the topic of turbulence modeling. They will be briefly commented on in this chapter, after which they will be illustrated with results from computer simulations.

10.2.2 k–ω–SST Closure Model

The model $k-\omega-$SST, presented in Sect. 9.3.5, will again be presented here, in summary, for didactic reasons. This model is composed of an equation for turbulent kinetic energy, k, and an equation for the Kolmogorov frequency, ω. The model $k-\omega-$SST resulted from the hybridization of $k-\varepsilon$ closure models, suitable for flows far from walls, and the $k-\omega$ model, suitable for areas close to walls:

$$\frac{\partial k}{\partial t}+\frac{\partial}{\partial x_j}(\overline{u}_jk)=-\overline{u_i'u_j'}\frac{\partial\overline{u}_i}{\partial x_j}+\frac{\partial}{\partial x_j}\left((\nu+\sigma^k\nu_t)\frac{\partial k}{\partial x_j}\right)-\beta^k k\omega+g_i\beta\alpha_t\frac{\partial\overline{T}}{\partial x_i} \tag{10.1}$$

$$\frac{\partial\omega}{\partial t}+\frac{\partial}{\partial x_j}(\overline{u}_j\omega)=-\frac{\alpha^\omega}{\nu_t}\overline{u_i'u_j'}\frac{\partial\overline{u}_i}{\partial x_j}-\beta^\omega\omega^2+\frac{\partial}{\partial x_j}\left[(\nu+\sigma^\omega\nu_t)\frac{\partial\omega}{\partial x_j}\right]$$
$$+2(1-F_1)\frac{1}{k}(\nu+\sigma_2^\omega\nu_t)\frac{\partial\omega}{\partial x_j}\frac{\partial k}{\partial x_j}+\gamma^\omega g_i\beta\alpha_t\frac{\partial\overline{T}}{\partial x_i} \tag{10.2}$$

As explained in Chap. 9, function F_1 depends on the distance between a considered point and the nearest wall. With this function, hybridization is carried out between the two models mentioned above. All details of this model, including the constant values, are available in Chap. 9.

10.2.3 SST–SAS Closure Model

One of the greatest difficulties encountered in modeling turbulence closure is determining its characteristic quantities (length, time, etc.). It is impossible in a turbulent flow to determine unique values of these quantities, with which to represent the entire richness of its physical characteristics, as a function of space and time. SAS (Scale Adaptive Simulation) models resulted from the pursuit of a way to obtain adaptivity of these quantities, with the local physical characteristics of the flows. This type of model was presented in Chap. 9, Eqs. (9.97) and (9.102):

$$\frac{\partial k}{\partial t}+\frac{\partial}{\partial x_j}(\overline{u}_jk)=-\overline{u_i'u_j'}\frac{\partial\overline{u}_i}{\partial x_j}+\frac{\partial}{\partial x_j}\left((\nu+\sigma^k\nu_t)\frac{\partial k}{\partial x_j}\right)$$
$$-\beta^k k\omega+g_i\beta\alpha_t\frac{\partial\overline{T}}{\partial x_i} \tag{10.3}$$

$$\frac{\partial\omega}{\partial t}+\frac{\partial}{\partial x_j}(\overline{u}_j\omega)=-\frac{\alpha^\omega}{\nu_t}\overline{u_i'u_j'}\frac{\partial\overline{u}_i}{\partial x_j}-\beta^\omega\omega^2+\frac{\partial}{\partial x_j}\left[(\nu+\nu_t\sigma^\omega)\frac{\partial\omega}{\partial x_j}\right]$$
$$+2(1-F_1)\frac{1}{k}(\nu+\sigma_2^\omega\nu_t)\frac{\partial\omega}{\partial x_j}\frac{\partial k}{\partial x_j}+\gamma^\omega g_i\beta\alpha_t\frac{\partial\overline{T}}{\partial x_i} \tag{10.4}$$

where

$$q_{\mathrm{SAS}}=\max\left[\rho\eta_2\kappa S^2\left(\frac{L}{L_{\vartheta K}}\right)^2-C_{\mathrm{SAS}}\frac{2\rho k}{\sigma_\phi}\max\left(\frac{1}{\omega^2}\frac{\partial\omega}{\partial x_j}\frac{\partial\omega}{\partial x_j},\frac{1}{k^2}\frac{\partial k}{\partial x_j}\frac{\partial k}{\partial x_j}\right),0\right] \tag{10.5}$$

The characteristic lengths are given by:

$$L = \frac{\sqrt{k}}{c_\mu^{1/4}\,\omega}, \tag{10.6}$$

$$L_{\vartheta K} = \max\left[\kappa\left(\frac{U'}{U''}\right), C_s\sqrt{\frac{\kappa\eta_2}{(\beta/c_\mu)-\alpha}}\Delta\right], \tag{10.7}$$

$\Delta = (\Delta x \Delta y \Delta z)^{(1/3)}$ and Δx, Δy and Δz, the sizes of the discretization cells. The other variables are specified in Chap. 9.

10.2.4 SA−SAS Closure Model

This is a variant of the Spalart–Allmaras Model [15], proposed by Coder [1], with the same purpose of incorporating the adaptivity of characteristic quantities into this model. It was presented in Sect. 9.3.8 and will be presented again here:

$$\frac{\partial\tilde{\nu}}{\partial t} + U_j\frac{\partial\tilde{\nu}}{\partial x_j} = c_{b1}\tilde{S}\tilde{\nu} - c_{w1}f_w\left(\frac{\tilde{\nu}}{d}\right)^2 + \frac{1}{\sigma}\frac{\partial}{x_k}\left[(\nu+\tilde{\nu})\frac{\partial\tilde{\nu}}{\partial x_k}\right] + \frac{c_{b2}}{\sigma}\frac{\partial\tilde{\nu}}{\partial x_k}\frac{\partial\tilde{\nu}}{\partial x_k} - q_{\mathrm{SAS}}, \tag{10.8}$$

where

$$q_{\mathrm{SAS}} = c_{s1}\max\left\{\left(\frac{\kappa^2}{L^2} - \frac{c_{s2}}{d^2}\right), 0\right\}\tilde{\nu}, \tag{10.9}$$

$$L = \max(L_{\vartheta K}, \kappa C_{\mathrm{SA-SAS}}\Delta) \quad \text{and} \quad L_{\vartheta K} = \kappa S/\sqrt{\frac{\partial^2 u_i}{\partial x_n \partial x_n}\frac{\partial^2 u_i}{\partial x_n \partial x_n}}. \tag{10.10}$$

The constants are the following: $c_{s1} = 4.9$, $c_{s2} = 2.0$ and $C_{\mathrm{SA-SAS}} = C_{\mathrm{DES}} = 0.65$. Again, it can be noted that the size of the mesh cells, Δ, was used as a limit for the von Kármán characteristic length. In a similar way to that presented in the Model SST–SAS, the von Kármán characteristic length was introduced independently from the Original Spalart–Allmaras [15] Model, through the transformation term q_{SAS}. The resulting model tends to be the behavior of the classics URANS models in the vicinity of walls, while, far from the walls, it works as a SAS model.

10.3 Comments on the Transient Nature of the URANS Methodology

As mentioned in this chapter, the LES Methodology represents a great potential for solving turbulent flows, allowing the resolution of a band of the spectrum of turbulent structures. Therefore, this methodology is still the most appropriate for solving flows dominated by coherent turbulent structures with long wavelengths. For example, wake-type flows, left by bodies immersed in flows.

An alternative to LES in engineering applications is the Transient RANS Methodology, whose acronym in English is URANS (Unsteady Reynolds Averaged Navier–Stokes Equations). As already mentioned in Chap. 6, Eqs. (6.45) and (6.46) were derived and they were called global equations for turbulence. It was demonstrated that the filtered equations are the same for both LES and URANS Methodologies. What differs between both are the objectives of the simulation, the closure models, the numerical methods used and the level of refinement of the meshes.

When the objective of the simulation is to obtain the entire spectrum of turbulent structures, we use DNS. When the objective is to obtain the band of large turbulent structures, one can choose one of the two methodologies, LES or URANS. With LES, the spectrum bandwidth to be obtained depends on the level of mesh refinement used. With URANS, only the largest structures can be calculated, due to the highly diffusive nature of this methodology. In reality, the transient nature of the URANS methodology originates from the numerical convergence procedure, when compared to a RANS methodology, which is purely steady state in nature. However, it was possible to realize that it is possible to capture low frequencies or small wavelengths relative to the complete spectrum. In Fig. 10.1, time distributions relating to velocity are illustrated, $u_i(\vec{x}, t)$, obtained with DNS, which can be decomposed into its filtered value $\overline{u}_i(\vec{x}, t)$ (LES or URANS) and its average value $u_i(\vec{x})$.

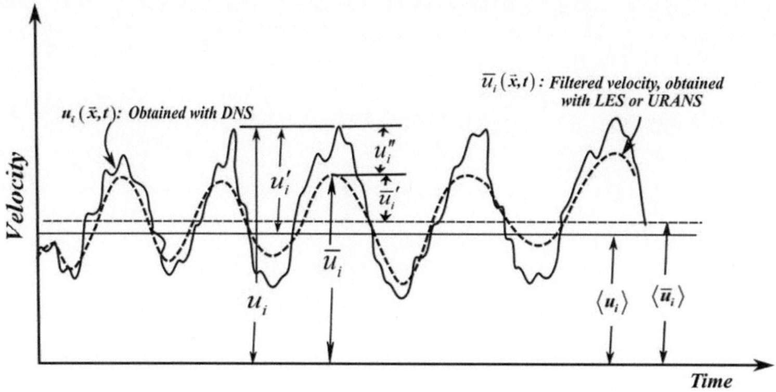

Fig. 10.1 Schematic illustration of velocity distributions obtained with DNS, $u_i(\vec{x}, t)$, LES and URANS, $\overline{u}_i(\vec{x}, t)$.

As can be seen qualitatively, the fluctuations in the signal obtained with DNS are characterized by a wide band of frequencies. With the filtering inherent to LES and URANS, the high-frequency band is eliminated. The signal of LES and URANS, $\overline{u}_i(\vec{x}, t)$, can also be decomposed into an average part, $\overline{u}_i(\vec{x})$ and a fluctuating part, $\overline{u}'_i(\vec{x}, t)$. The URANS signal is not necessarily a constant and may remain transient. Temporal or spatial averages, $\overline{u}_i(\vec{x})$ or $\overline{u}_i(t)$, representative of signals obtained with LES or URANS were drawn as being different from the averages $u_i(\vec{x})$ or $u_i(t)$, representative of DNS. Only one dashed curve was used for both methodologies, LES and URANS, but solely for the purpose of not overloading the figure. With the use of these methodologies, the results obtained will not necessarily be the same (Fig. 10.2).

In Figs. 10.3, 10.4 and 10.5, what is expected from the turbulent kinetic energy spectra, representative of results that can be obtained using DNS, LES, URANS and RANS, is schematically presented. In Fig. 10.3 representative spectra of DNS and LES are presented, where it is observed that, with DNS, a full and wide spectrum is obtained, with characteristic lengths and times that vary between the integral characteristic quantities and the Kolmogorov characteristic quantities. With LES, what is expected is to obtain a spectrum truncated in frequency or cutoff wavenumber, f_c or k_c. The turbulent kinetic energy level obtained with LES is expected to have a similar magnitude to that obtained via DNS, as illustrated with the solid and dotted curves. In this case, the integral of the LES spectrum does not provide all the turbulent kinetic energy obtained with the integral of the DNS spectrum. Thus, it will be modeled with Eqs. (10.11) and (10.12), which serve to represent the turbulent kinetic energy as a function of spatial position and function of time, respectively.

$$k(\vec{x}) = \underbrace{\int_0^{f_c} E(\vec{x}, f)\mathrm{d}f}_{\text{Calculated or simulated part}} + \underbrace{<k>(\vec{x})}_{\text{Sub-grid part (temporal average)}} , \qquad (10.11)$$

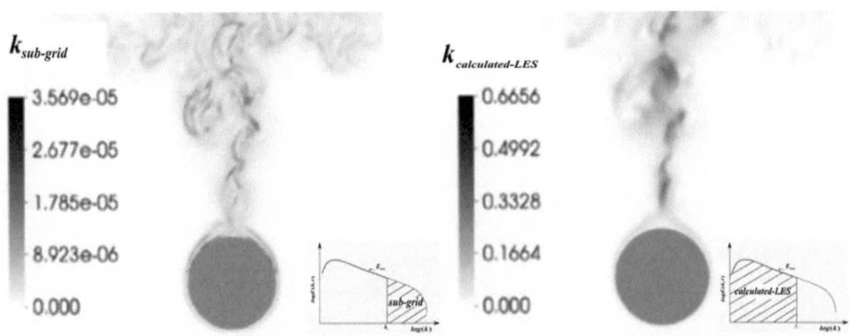

Fig. 10.2 Illustration of sub-grid turbulent kinetic energy and calculated turbulent kinetic energy. Natural convection over a horizontal heated cylinder

$$k(t) = \underbrace{\int_{0}^{k_c} E\left(\vec{k}, t\right) \mathrm{d}\vec{k}}_{\substack{\text{Calculated part, or} \\ \text{simulated part}}} + \underbrace{<k>(t)}_{\substack{\text{Sub-grid part} \\ \text{(spacial average)}}} \tag{10.12}$$

The integrations are carried out in the calculated part of the spectrum, $0 < f \leq f_c$ and $0 < k \leq k_c$, which makes it possible to obtain part of the turbulent kinetic energy. Corrections relating to the sub-grid band of the spectrum are applied to these quantities. These corrections, in the case of LES, must be modeled. However, it should be noted that if these corrections are smaller, the more refined the mesh is and the shorter the time step used to carry out the simulations. So, normally, they are neglected.

In Fig. 10.2, relating to a computational experiment on natural convection over a horizontal and heated cylinder, presented in Sect. 9.13.5.2, the sub-grid and total turbulent viscosities are shown. In Fig. 10.2a, sub-grid turbulent kinetic energy is presented, $k_{\text{sub-grid}}$, obtained through a sub-grid model, reported by Romão et al. [10], $k_{\text{sub-grid}} = (\nu_t / C_K \Delta)^2$, where ν_t is calculated with the dynamic sub-grid model, Germano et al. [1]. This portion of the total turbulent kinetic energy, obtained with a closure model, is equivalent to the integrals:

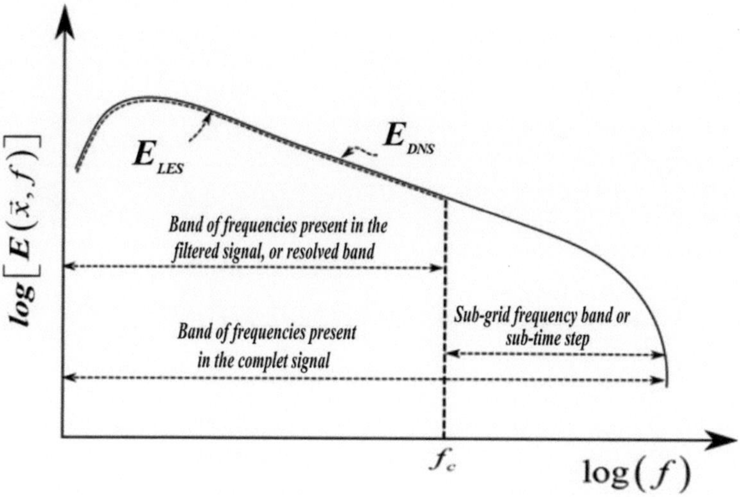

Fig. 10.3 Schematic illustration of energy spectra representative of DNS modeling, E_{DNS}, and LES, E_{LES}

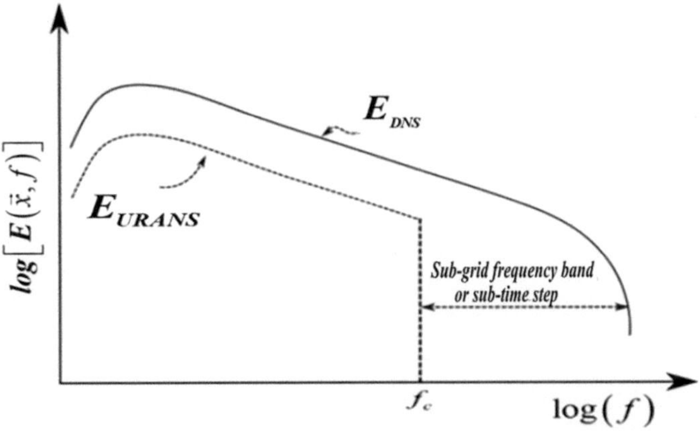

Fig. 10.4 Schematic illustration of turbulent kinetic energy spectra to DNS model E_{DNS} and E_{URANS}

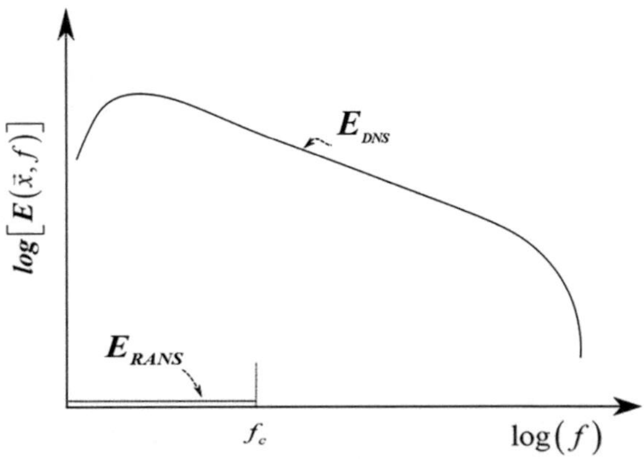

Fig. 10.5 Illustration of turbulent kinetic energy spectra relative to DNS models, E_{DNS} and RANS, E_{RANS}

$$k(t) = \int_{k_c}^{k_K} E\left(\vec{k}, t\right) \mathrm{d}\vec{k}\; k(\vec{x}) = \int_{f_c}^{f_K} E(\vec{x}, f) \mathrm{d}f \tag{10.13}$$

These integrals go through the range of the band indicated in Fig. 10.2a i.e., from the cutoff frequency, f_c, up to the Kolmogorov frequency, f_κ, or the equivalent in wavenumber. The calculated turbulent kinetic energy, via post processing, which appears in Eqs. (10.11) and (10.12), is illustrated in Fig. 10.2b with the hatched area. The turbulent kinetic energy levels are shown with the scales to the left of the figures. The sub-grid portion of the turbulent kinetic energy is much smaller than the total turbulent kinetic energy, approximately five orders of magnitude smaller.

In Fig. 10.4, the expected spectrum using URANS is represented. This spectrum is also characterized by a cutoff frequency or cutoff wave number that depend on the size of the cells that make up the mesh and the calculation time step. However, the expected energy level is lower than that obtained with DNS, due to the high level of turbulent viscosity inherent to URANS. Thus, again, corrections are necessary on the calculation of turbulent kinetic energy, which can be obtained through Eqs. (10.14) and (10.15):

$$k(\vec{x}) = \underbrace{\int_0^{f_c} E(\vec{x}, f)\mathrm{d}f}_{\substack{\text{Calculated part}\Rightarrow \\ \text{comes from the simulated part}}} + \underbrace{k(\vec{x})}_{\substack{\text{Modeled part with a balance} \\ \text{equation for } k\text{-time average}}}, \tag{10.14}$$

$$k(t) = \underbrace{\int_0^{k_c} E\left(\vec{k}, t\right)\mathrm{d}\vec{k}}_{\substack{\text{Calculated part}\Rightarrow \\ \text{comes from the simulated part}}} + \underbrace{k(t)}_{\substack{\text{Modeled part with a balance} \\ \text{equation for } k\text{-spatial average}}} \tag{10.15}$$

In this case, the correction is obtained from the turbulence closure model itself, by solving the balance equation for turbulent kinetic energy, k. The integrals are the area under the curve for URANS and the correction corresponds to the area between the two curves, DNS and URANS.

Figure 10.5 represents what is expected in terms of the turbulent kinetic energy spectrum when using RANS modeling. This modeling serves to directly obtain average *information*, in time and in space, through the solution of the differential model, composed of balance equations. The energy spectrum, in this case, is zero. All energy must be calculated via the balance equation for turbulent kinetic energy. In this case, all turbulent kinetic energy is determined by solving the balance equation for this *information.* Thus,

$$k(\vec{x}) = \underbrace{k_{\text{balance}}(\vec{x})}_{\substack{\text{Fully modeled with a} \\ \text{balance equation for } k}} \tag{10.16}$$

10.4 Illustrative Results of the URANS–URANS Hybrid Models

For illustrating the functionality of the URANS–URANS hybrid models, flow simulations were carried out on a backward facing step, using the MFSim computational code. The closure models type standard $k{-}\varepsilon$ and $k{-}\omega$ were used to develop hybridization, which resulted in the Model $k{-}\omega{-}\text{SST}$.

The backward facing step is illustrated in Fig. 10.6, presented below, where the main geometric and physical characteristics of this problem are represented. The settings for the computer simulation were based on the Jovic and Driver experiment [2]. The value of the Reynolds Number, based on the height of the step, is $\mathrm{Re}_h = 5000$. In this material experiment, two wires of diameter $d = 0.1667h$ were used upstream of the step, to force transition of the boundary layer, which reaches the vicinity of the step, with a thickness of $\delta = 1.15h$.

The domain chosen to carry out the computational experiments is characterized by $h = 0.5m$; $L = 16m$; $H = 3m$; $W = 2m$ and $X_i = 1.5m$. The dimension L was thus chosen, as it provides a relationship of $L = 32H$, thus avoiding the effects of the boundary condition imposed on the channel outlet [14]. An advective boundary condition was used at the channel exit. The length of the entrance channel was chosen as $X_i = 3h$, since sufficient experimental data is provided at this position to model the boundary conditions at the inlet. The height $H = 6h$ was prescribed, where a superior horizontal plane of symmetry was defined. $W = 4h$ was chosen based on other works [14, 17].

At the entrance to the domain, a boundary condition of the first kind was used, for which the velocity is imposed, representing the material experiment; on the lower wall, the non-slip condition was imposed, which also corresponds to the boundary condition of the first kind; at the upper limit, a symmetry or second kind condition

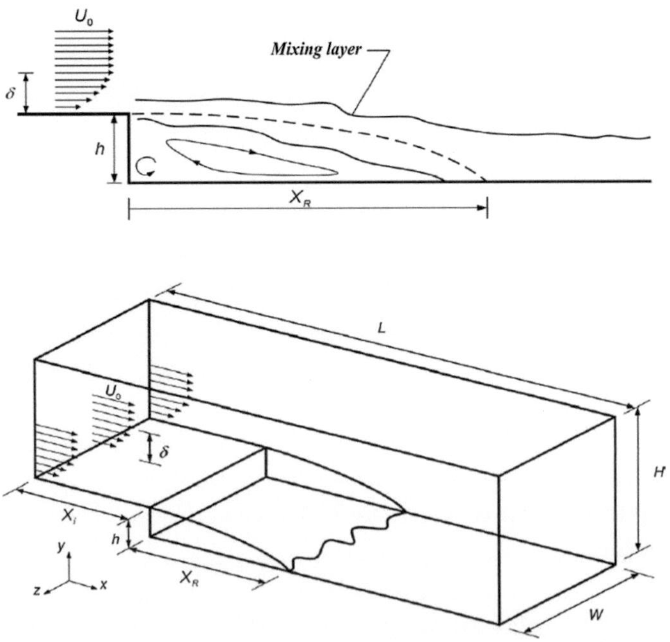

Fig. 10.6 Illustration of turbulent flows over a descending step. The main geometric and physical characteristics of the problem are presented

was imposed; Likewise, in the lateral vertical planes, conditions of symmetry or of second kind were imposed. At the exit, the third type or advective type boundary condition was used. The boundary condition for the turbulent kinetic energy at the channel entrance, $k_\infty = 3(IU_\infty)^2/2$, was used, where U_∞ and I are the velocity and turbulent intensity of the free stream. The evaluation of the Kolmogorov frequency at the inlet can be made using experimental evidence for turbulent viscosity, Nichols [8]. For the present work, the intensity $I = 5\%$ and turbulent viscosity $\nu_t = 5\nu$ where ν is the molecular kinematic viscosity, were employed. Once k_∞ and $\nu_{t\infty}$ have been prescribed, ω_∞ can be calculated.

In Fig. 10.7, the distributions of the horizontal component of the average velocity are presented, in different vertical sections, in the channel downstream of the step. These sections are represented by different x/h. Close to the lower wall, there are negative velocities behind the step, indicating the existence of fluid recirculation. Both models, $k-\varepsilon$ and $k-\omega$, seriously fail to compare with the results of Jovic and Driver's [2] material experiment. This was already expected, since the model was developed for flows without the presence of a wall, and the $k-\omega$ model was developed only for flows close to walls, but it fails for flows at high Reynolds Numbers, as it moves away from the walls. Both models fail close to the wall. The hybrid model $k-\omega-$SST, which is the fusion of the two others, $k-\varepsilon$ and $k-\omega$, reproduces the flow throughout the channel with superiority.

Downstream the channel, a boundary layer develops on the lower wall, which detaches at the upper corner of the sudden expansion, characterized by the line with coordinates as shown in Fig. 10.6. As a result of the detachment, a recirculation region is formed behind the step. The boundary layer upstream of the step and the flow recirculation region are also illustrated in Fig. 10.6. It also illustrates the length of the recirculation region, also known as boundary layer reattachment length, denoted by X_R. This is statistical *information* that is difficult to predict using mathematical models. It is a rigorous test for the purposes of validating and evaluating theoretical methodologies. In Fig. 10.8, the distributions of the horizontal component of velocity are presented in the first layer of computational cells adjacent to the lower horizontal wall. Inside the recirculation, this velocity component must be negative, while, after the recirculation region, it must be positive. Therefore, the average reattachment

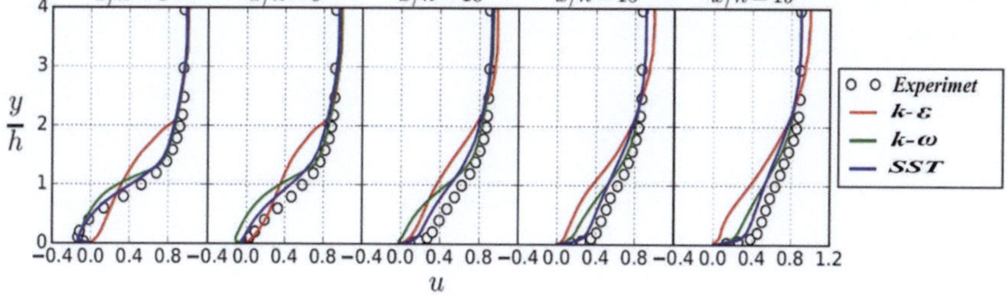

Fig. 10.7 Velocity distributions in different vertical sections in the channel downstream of the step. Personal collection, with thanks to Rafael Romão

point, $x = X_R$, will be found when this velocity component is equal to zero. This can be better seen in Fig. 10.8, in which the three velocity distributions are presented, relative to the three closure models in question. Thus, where the curves cross the elevation $u = 0$ from the graph, the reattachment points are determined, relative to the three closure models. It is noticed that the values of x/h, where $u = 0$ takes place, are very different for the three closure models. In Table 10.1, the values of the boundary layer reattachment lengths at the lower wall are presented, calculated using the three closure models in question. It is observed that the value of X_R/h was under-predicted using the $k{-}\varepsilon$ model; the value of this *information* was over-predicted using the $k{-}\omega$ model, while the value obtained for X_R/h, using the Hybrid model $k{-}\omega{-}$SST, was practically within the error range of Jovic and Driver's [2] material experiment.

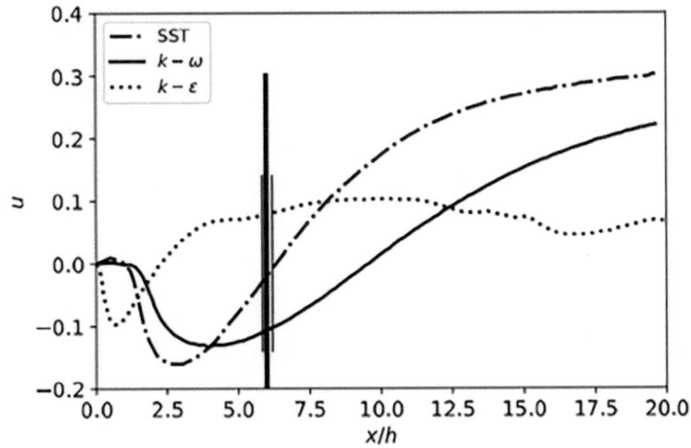

Fig. 10.8 Distribution of the horizontal component of velocity, on the lower wall of the channel, downstream of the step, for $\mathrm{Re}_h = 5,000$

Table 10.1 Boundary layer reattachment length, X_R/h. Personal collection, with thanks to Rafael Romão

Closure turbulence models and material experiment	Reattachment length of the boundary layer: X_R/h
$k{-}\varepsilon$	2.23
$k{-}\omega$	9.47
$k{-}\omega{-}$SST	6.24
Material experiment: Jovic and Driver [2]	6.0 ± 0.15

10.5 URANS–LES Hybrid Turbulence Closure Models

10.5.1 S–A Closure Model

Spalart and Almaras [15] proposed the model S–A, which is characterized by a single balance equation, presented in Chap. 7 and again presented here for didactic reasons:

$$\nu_t = \tilde{\nu} f_{\vartheta 1} \tag{10.17}$$

The function $\tilde{\nu}$ is determined by solving the balance equation proposed by the authors of this closure model:

$$\frac{\partial \tilde{\nu}}{\partial t} + U_j \frac{\partial \tilde{\nu}}{\partial x_j} = c_{b1} \tilde{S} \tilde{\nu} - c_{w1} f_w \left(\frac{\tilde{\nu}}{d} \right)^2 + \frac{1}{\sigma} \frac{\partial}{x_k} \left[(\nu + \tilde{\nu}) \frac{\partial \tilde{\nu}}{\partial x_k} \right] + \frac{c_{b2}}{\sigma} \frac{\partial \tilde{\nu}}{\partial x_k} \frac{\partial \tilde{\nu}}{\partial x_k} \tag{10.18}$$

where

$$c_{w1} = \frac{c_{b1}}{\kappa^2} + \frac{(1 + c_{b2})}{\sigma}, \tag{10.19}$$

$$f_{\vartheta 1} = \frac{\chi^3}{\chi^3 + c_{\vartheta 1}^3}, \tag{10.20}$$

$$f_{\vartheta 2} = 1 - \frac{\chi}{1 + \chi f_{\vartheta 1}}, \tag{10.21}$$

$$f_w = g \left[\frac{1 + c_{w3}^6}{g^6 + c_{w3}^6} \right]^{1/6}, \tag{10.22}$$

$$\chi = \frac{\tilde{\nu}}{\nu}, \tag{10.23}$$

$$g = r + c_{w2} \left(r^6 - r \right), \tag{10.24}$$

$$r = \frac{\tilde{\nu}}{\tilde{S} \kappa^2 d^2}, \tag{10.25}$$

$$\tilde{S} = S + \frac{\tilde{\nu}}{\kappa^2 d^2} f_{\vartheta 2}, \tag{10.26}$$

$$S = \sqrt{2 S_{ij} S_{ij}}, \tag{10.27}$$

$$\Omega_{ij} = \frac{1}{2}\left(\frac{\partial U_i}{\partial x_j} - \frac{\partial U_j}{\partial x_i}\right) \tag{10.28}$$

The constants are: $c_{b1} = 0.1355$, $c_{b2} = 0.622$, $c_{\vartheta 1} = 7.1$, $\sigma = 2/3$, $c_{w2} = 0.3$, $c_{w3} = 2.0$, $\kappa = 0.41$, Ω_{ij} is the rotation rate tensor and d_w is the distance from a cell to the nearest surface. The proposal presented above is URANS in nature. The S–A model was transformed into a hybrid LES–URANS model. Therefore, in this closure modeling, d_w, the distance to the nearest wall was replaced by the variable $\tilde{d}$, which is now evaluated with the following equation:

$$\tilde{d} = \min(d_w, C_{\text{DES}}\Delta), \tag{10.29}$$

where $\Delta = \max(\Delta x, \Delta y, \Delta z)$. The length $\tilde{d} = d_w$ is supplied close to walls, according to the S–A model. Thus, the hybrid model acts as S–A in the vicinity of walls and, as it moves away from a wall, $\tilde{d} = C_{\text{DES}}\Delta$ becomes a characteristic length calculated with the mesh and automatically, the model starts to behave like a sub-grid model for LES. When the transformation terms $c_{b1}\tilde{S}\tilde{\nu}$ and $(c_{b2}/\sigma)(\partial\tilde{\nu}/\partial x_k)(\partial\tilde{\nu}/\partial x_k)$ balance each other, the S–A model assumes a similar behavior to the Smagorinsky sub-grid model, for which $\nu_t \propto \Delta^2 S$, where S is the norm of the strain rate tensor. The value of C_{DES} can be adjusted. For example, Shur et al. [13] used $C_{\text{DES}} = 0.6$ for homogeneous turbulence. Oliveira and Silveira-Neto [9] used the same constant to model and simulate flows over blunt bodies and airfoils.

10.5.2 Flows Over a Circular Cylinder

Illustrative and comparative results of applications of the three turbulence modeling methodologies will be presented: URANS, LES and Hybrid LES–URANS. Such results were developed by Oliveira and Silveira-Neto [9]. A computational, finite-volume code was developed, with second-order numerical schemes in time and space. The Immersed Boundary Method, MFV (Virtual Physical Model), developed by Lima and Silva et al. [5], was used. The hypotheses of incompressible flows of Newtonian fluids were considered. The URANS, LES and hybrid LES–URANS methodologies, described above, were used. Flows over a circular cylinder were simulated for various Reynolds Numbers.

In Fig. 10.9, streamlines are shown to visualize the flow and comparison with the image obtained through material experiment, Schilichting [12]. It can be seen how well the MFV method represents the presence of the cylinder, in a virtual way. The captured recirculations are very similar to what was observed in the material experiment. The simulation was carried out by Oliveira and Silveira-Neto [9] for $\text{Re}_\text{D} = 1 \times 10^4$.

In Fig. 10.10, the effective viscosity normalized by the molecular kinematic viscosity, ν_{ef}/ν, is shown. It was obtained using three turbulence models: LES, URANS and DES. As can be seen, the intensity of turbulent viscosity increases

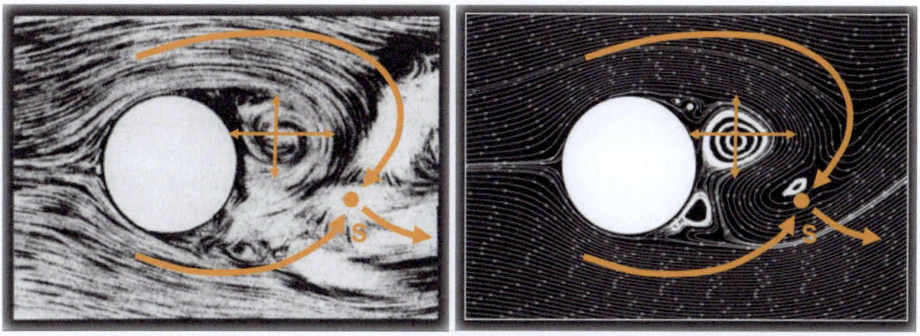

Fig. 10.9 Visualization of flow over a circular cylinder. **a** material experiment by Schlichting [12] and **b** computer simulation with the LES Methodology by Oliveira and Silveira-Neto [9], for $\mathrm{Re_D} = 1 \times 10^4$

when starting from LES, passing through DES and arriving at URANS. The ratio between the effective viscosity for URANS and LES reaches three orders of magnitude. The turbulent viscosity for the DES is at an intermediate level between the previous two, which is compatible with its hybrid nature. For this reason, among others, the LES methodology provides the highest level of physical detail as results, while URANS provides the physical effects related to low frequencies. However, for any of the three methodologies, statistical results can be obtained.

In Tables 10.2 and 10.4, part of the statistical results are presented, which were extracted from the masses of instantaneous *information* stored during processing or simulations. In Table 10.2, the drag coefficient resulting from the flow over the cylinder is presented. In Table 10.3, the Strouhal Number is presented, relating to the coherent turbulence structures that form downstream of the cylinder. In Table 10.4, the detachment angles of the boundary layer that forms on the cylinder are shown. The comparison between mathematical and computational modeling methodologies and material experimentation shows that they are similar. The fact that the drag crisis was not captured using any of the three methodologies stands out.

10.5.3 Flows Over a Backward Facing Step

10.5.3.1 Preliminary

Turbulent flows over a backward facing step were simulated using the presented turbulence models. This kind of flow is interesting, since the geometry is simple, including that it is Cartesian. This problem is illustrated in Fig. 10.6, where the main geometric and physical characteristics are presented.

The boundary layer that forms upstream of the step detaches exactly at the corner, which keeps the detachment line fixed, regardless of the operating conditions. This

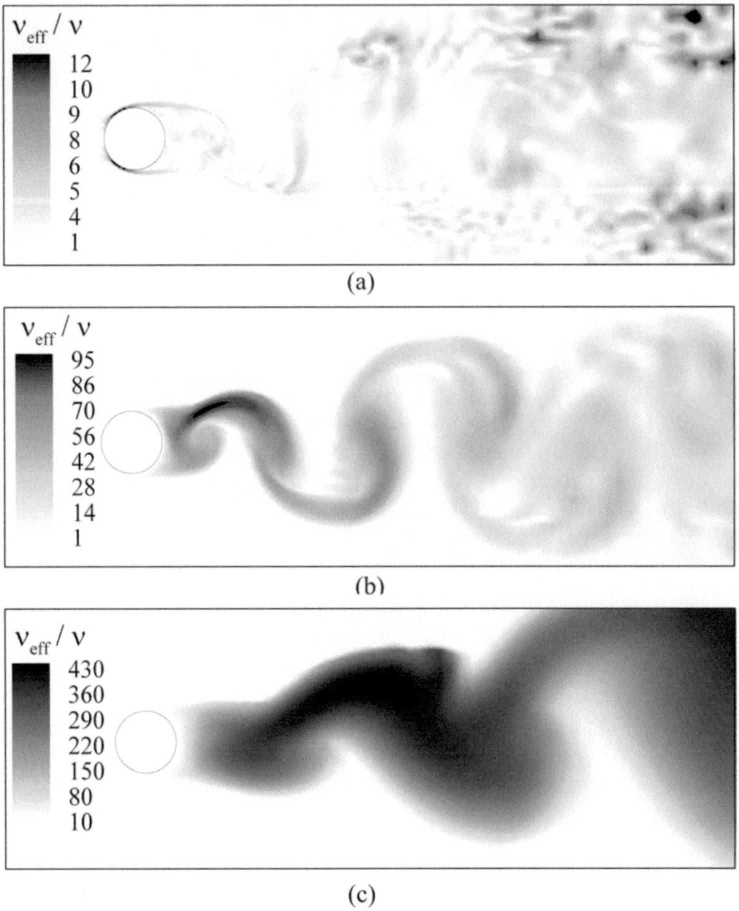

Fig. 10.10 Flow over a circular cylinder, using the normalized effective turbulent viscosity ν_{ef}/ν, using the methodologies: **a** LES; **b** DES and **c** URANS; $Re_D = 10{,}000$. Data taken from Oliveira and Silveira-Neto [9]

fact is important to ensure control over this operational parameter. Due to the recirculation that forms after the step, an inflectional velocity field appears, determining the existence of a shear layer. As a result of the expansion, the pressure increases towards the channel exit. This characterizes an adverse pressure gradient, which acts on the shear layer, bending it towards the lower wall. As illustrated in Fig. 10.6, the meeting point of the dashed line, central to the shear layer, with the lower wall, determines the reattachment length of the boundary layer, X_R. This parameter is difficult to predict theoretically and makes the case interesting for testing the functionality of a modeling proposal.

The flow will be characterized by the Reynolds Number, based on the height of the step:

Table 10.2 Average drag coefficient, C_D, comparing the results obtained with the three methodologies with results obtained with material experiments

Re_D	Average drag coefficient—C_D				
	Present work [9]			Sucker and Brauer [18], White [19]	Wieselsberger [20]
	URANS	DES	LES	Fitted curve	Experimental results
1×10^4	0.8782	0.9840	1.2203	1.091	1.139
5×10^4	0.8958	0.9332	1.1930	1.166	1.209
2×10^5	0.8746	0.9422	1.1900	1.178	1.138
5×10^5	0.8751	0.9394	1.2184	–	0.295
1×10^6	0.8787	0.9434	1.2082	–	0.350

Table 10.3 Average Strouhal number, St, comparing the results obtained by computational simulations with the three methodologies with material experiments results

Re_D	Average Strouhal number—St			
	Present work [9]			Roshko's [11] experiment
	URANS	DES	LES	
1×10^4	0.1877	0.2125	0.2507	0.2117
5×10^4	0.1752	0.2125	0.2569	0.2119
2×10^5	0.1877	0.2125	0.2451	0.2120
5×10^5	0.1877	0.2125	0.2463	0.2120
1×10^6	0.1857	0.2125	0.2403	0.2120

Table 10.4 Detachment angle, ($\theta_{desc.}(°)$), comparing the results obtained with the three turbulence modeling methodologies

Re_D	Boundary layer detachment angle, ($\theta_{desc.}(°)$)		
	URANS	DES	LES
1×10^4	88.8	88.1	91.7
5×10^4	95.4	92.0	95.8
2×10^5	95.8	95.8	95.1
5×10^5	96.0	93.1	96.7
1×10^6	96.7	94.5	94.4

$$\mathrm{Re}_h = \frac{U_o h}{\nu}. \tag{10.30}$$

The following geometric parameters are important to characterize a configuration that will affect the physical nature of the flow: step elongation, $\mathrm{AL} = W/h$; expansion ratio, $\mathrm{ER} = H/(H - h)$ and the thickness of the boundary layer over the step. These parameters are illustrated in Fig. 10.6. In addition to these geometric factors, the flow regime (laminar or turbulent), the intensity of the turbulence at the domain inlet and the value of the Reynolds Number are important factors for the flow dynamics and, mainly, for the value of the attachment length, X_R/h.

As the Reynolds Number increases and, considering that the velocity field after the step is inflectional, Kelvin–Helmholtz Instability develops, resulting from the amplification of disturbances injected into the flow. The reattachment length depends on the kinetic energy associated with the disturbances present in the flow upstream of the step. A detailed study of this flow was carried out by Spode and Silveira-Neto [17], using turbulence modeling of the following types: LES, LESVD (LES with Van Driest Damping Function), S–A and DES (S–A + LES).

The physical nature of the flow that develops downstream of the step is very dependent on the intensity of the turbulence at the domain inlet. Therefore, to simulate a turbulent flow and compare the computational results with the results of material experiments, it is necessary that the flow at the entrance of the domain is well specified. The velocity distribution must be imposed so that the thickness of the boundary layer corresponds to that of the material experiment. Furthermore, the distribution of turbulence intensity must also be specified at the domain inlet. This *information* is more delicate to impose, especially when the turbulence is being modeled with LES and LES–URANS approaches.

The easiest way to model a turbulent boundary condition is to superimpose random fluctuations on the average velocity profile imposed at the entrance of the domain. But for low-order spatial and temporal numerical methods, these small fluctuations are quickly diffused, and their effect is annihilated in the first cells adjacent to the domain entry. Furthermore, this type of fluctuation does not have any turbulent character, energy is applied only at high frequencies, which justifies its rapid diffusion. Therefore, it is necessary that the vector of boundary conditions is a function of time and space, satisfying at least the first- and second-order statistical moments, as well as the energy spectrum.

Keating's et al. [3] proposal is based on simulating, in detail, a channel, with periodicity in the longitudinal and transverse directions of the channel. They used dynamic sub-grid modeling. From this simulation, *information* is taken from several planes normally to the walls to be used as boundary conditions at the channel entrance upstream of the step.

A more versatile proposal, from a theoretical and implementation point of view, was published by Lund, Wu and Squires [6]. The method consists of extracting a velocity plane downstream of the domain entrance and reconditioning the inner and outer portions of the boundary layer, separately, considering the different similarity laws in these regions. The corrected velocity is then reintroduced into the input. This modeling for input boundary condition was used by Spode and Silveira-Neto [17]. The basic idea associated with the method is to estimate the velocity profile at the entrance of the domain, based on a plane downstream in the flow, which is resized and reimposed at the entrance, see Fig. 10.11. Details of the method can be found in Spode and Silveira-Neto [17].

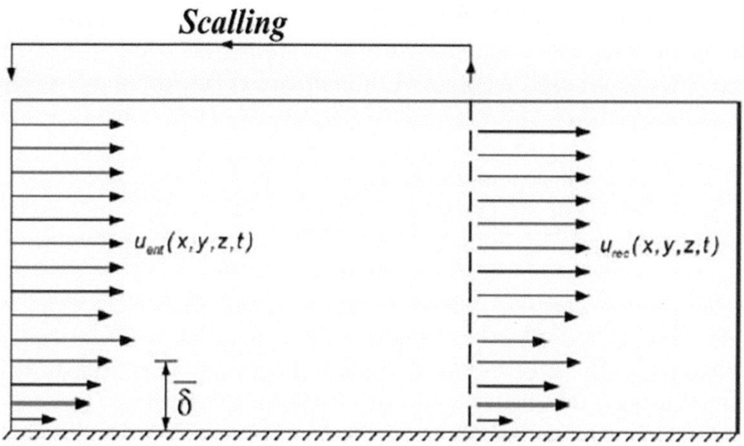

Fig. 10.11 Rescaling process to obtain the turbulent boundary condition at the domain entrance

10.5.3.2 Flow at the Inlet Considered Laminar

Flows in backward facing step can be classified as: laminar flow during detachment and laminar flow during reattachment; laminar flow during detachment and turbulent flow during reattachment; turbulent flow during detachment and turbulent flow during reattachment. The results that will be presented below refer to the third group. The turbulence modeling methodologies used are described in Table 10.5.

The results will initially be presented for inlet laminar boundary conditions. Subsequently, turbulent inlet conditions will be used, demonstrating the importance of modeling turbulence in the boundary condition at the channel entrance. Details for the analysis of domain size independence, as well as mesh refinement, can be found in Spode and Silveira-Neto [17].

The turbulent kinematic viscosity levels, normalized by the molecular kinematic viscosity, can be visualized in Fig. 10.12 for the four turbulence modeling methodologies used. The levels and spatial distributions are completely different between the methodologies. The URANS methodology has the highest kinematic viscosity, between 15 and 30 times higher than the others. This is natural and expected, since,

Table 10.5 Description of the turbulence modeling methodologies used

Acronym	Description of the methodologies used
LES	Simulation of Large Structures (LES) using the Smagorinsky model ($C_S = 0.1$)
LESVD	Large Structure Simulation (LES), Smagorinsky Model ($C_S = 0.1$)with addition of the Van Driest damping function
S–A	Average transient Reynolds equations (URANS), using the Spalart–Allmaras model
DES	Hybrid, using the Spalart–Allmaras model modified to act as URANS close to walls and as LES on remainder of the domain

in the case of URANS (S-A), the role of the closure model is to model the interaction between all the turbulence and the mean flow fields. For the other three, the role of closure models is to model the transfer of *information* between the resolved band and the unresolved band, of the spectrum in frequencies or wavenumbers.

URANS models are characterized by the fact that turbulent viscosity does not directly depend on the size of the mesh cells. The turbulent viscosity depends directly on the square of the cell size in LES and DES models. To illustrate these characteristics, simulations relating to Fig. 10.12 were done with a mesh twice as fine.

The turbulent viscosities, related to the S–A and DES methodologies, are presented in Fig. 10.13. As can be seen, the viscosity field in the S–A case is virtually identical to that of Fig. 10.12, differing only in scale, with an increase of about 10% in maximum viscosity. The DES case proved to be much more sensitive to variations in mesh refinement, presenting a 500% increase in maximum viscosity. Similar results are observed for the LESVD case, where the increase in maximum viscosity was 100%. These results were already expected, as sub-grid modeling is characterized by a decrease in turbulent viscosity as the mesh is refined. In the limit, with the mesh very refined, reaching the Kolmogorov characteristic length, the turbulent viscosity tends to zero.

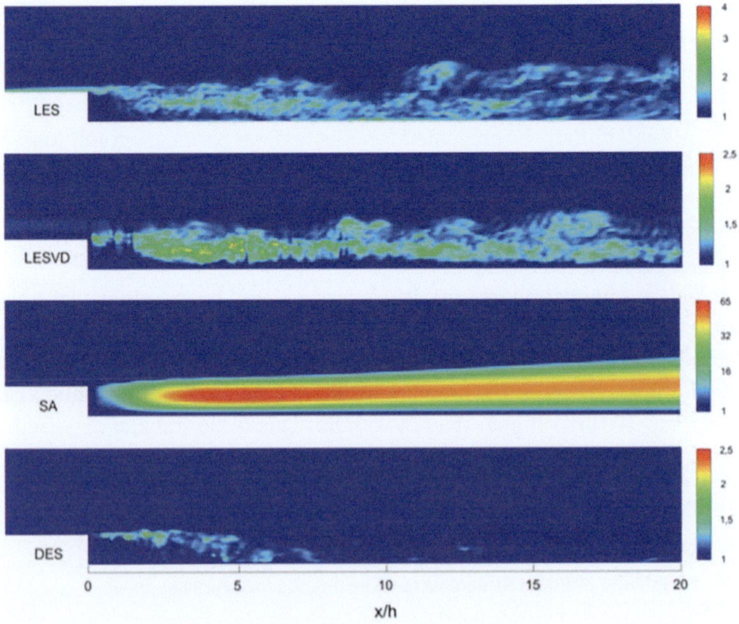

Fig. 10.12 Normalized effective turbulent kinematic viscosity fields, $(\mu + \mu_{\text{turb}})/\mu$, relating to the four turbulence modeling methodologies used. Plan $(x, y, z = W/2)$ and $t = 2000h/U_o$

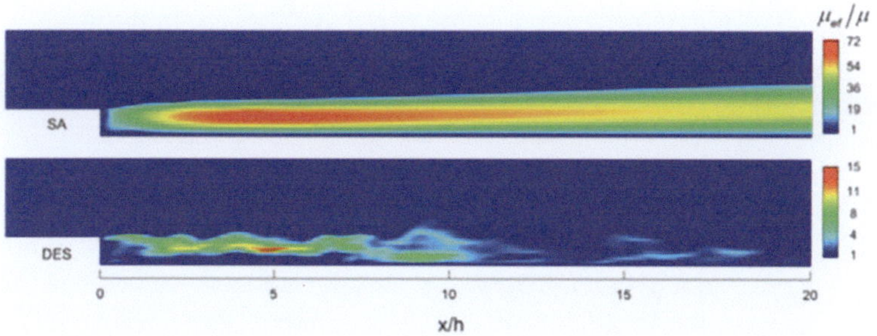

Fig. 10.13 Instantaneous viscosity fields. Mesh is twice as thick as compared to the mesh of Fig. 10.12

The boundary layer reattachment length is illustrated in Fig. 10.6. This is the most important *information* to be evaluated in this problem. All the influences and iterations of different flow zones resonate in the reattachment length. Average velocity profiles, fluctuations and pressure coefficient change, according to the reattachment length, which proved to be sensitive to the modeling of turbulence and, mainly, to the disturbance of the inlet boundary layer.

Two methods were used to measure the reattachment length X_R/h: the first consists of locating the point of zero average horizontal velocity in the first volume of the mesh, at the coordinate z, normal to the lower wall. The second is the method of the p.d.f. (probability distribution function), where the reattachment length is determined by the fraction $\Gamma(\%)$, which measures the percentage of time in which the flow acts in the positive direction, indicated with the axis system.

The reattachment point is the location where this fraction reaches 50%. The mean velocity distributions U and fraction $\Gamma(\%)$ for cases LES, LESVD, S–A and DES, depending on the coordinate x/h, are presented in Fig. 10.14 a and b, respectively. With these methods, very close values of reattachment length are obtained, with the reattachment length being determined using the $\Gamma(\%)$ function method, slightly greater than the length calculated using average velocity, presenting an average increase of 2.34%, as presented in Table 10.6. Similar results were found by Le et al. [4]. The velocity profiles used as boundary conditions at the domain entrance are presented in Fig. 10.15.

10.5.3.3 Flow at the Inlet Considered Turbulent

To model turbulence at the domain inlet, the Lund, Wu and Squires method was used [6], as previously commented. More details about this modeling can be obtained from Spode and Silveira-Neto [17]. The following notation will be used: LES–LWS, LESVD–LWS and DES–LWS to represent the use of the turbulent boundary condition, where LWS refers to Lund, Wu and Squires. Component $\sqrt{\overline{u'^2}}/U_0$, $\sqrt{\overline{\vartheta'^2}}/U_0$ and $\sqrt{\overline{u'\vartheta'}}/U_0$ distributions of the Boussinesq–Reynolds tensor are presented in

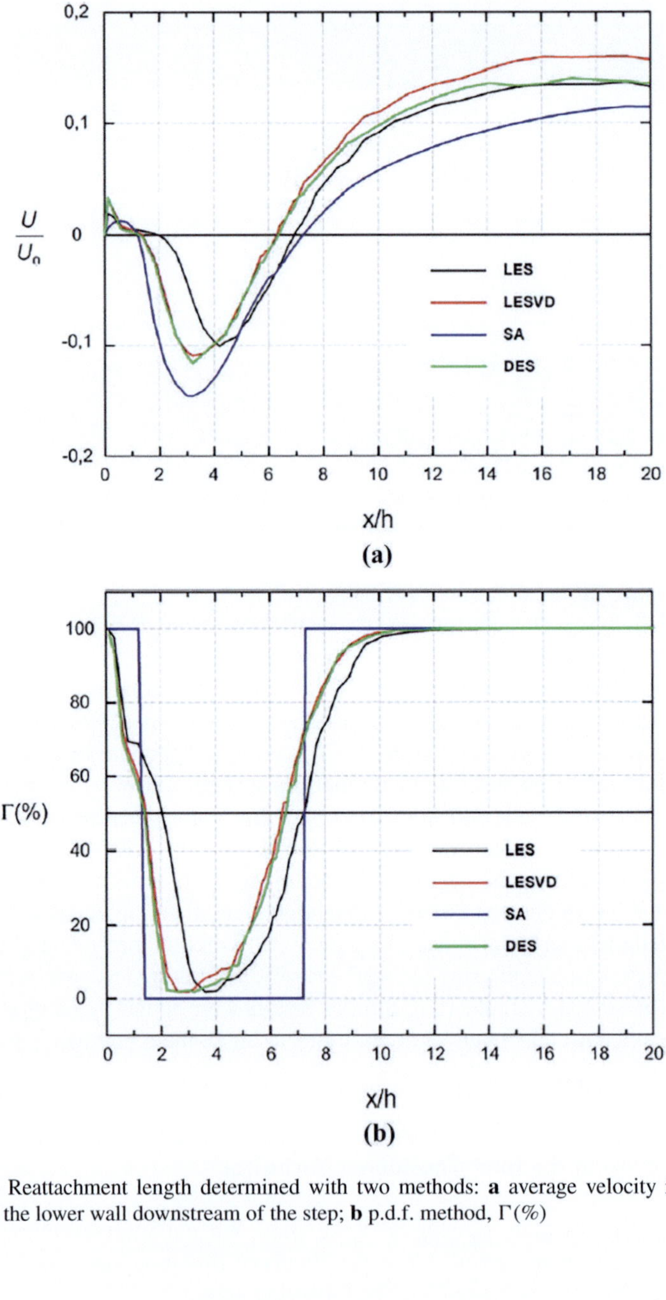

Fig. 10.14 Reattachment length determined with two methods: **a** average velocity in the cells adjacent to the lower wall downstream of the step; **b** p.d.f. method, $\Gamma(\%)$

Table 10.6 Reattachment length (X_R/h), compared with material experimentation and Direct Computational Simulation (DNS)

Cases	Reattachment length (X_R/h) method			Error (%)
	Average velocity	Function $\Gamma(\%)$	Average value between the two methods	
Jovic and Driver [2]—Experiment	–	–	6.0 ± 0.15	–
Le et al. [4]—DNS	–	–	6.28	4.6
LES	6.94	7.23	7.09	18.2
LESVD	6.31	6.43	6.37	6.1
S–A	7.26	7.26	7.26	21.0
DES	6.36	6.56	6.46	7.6

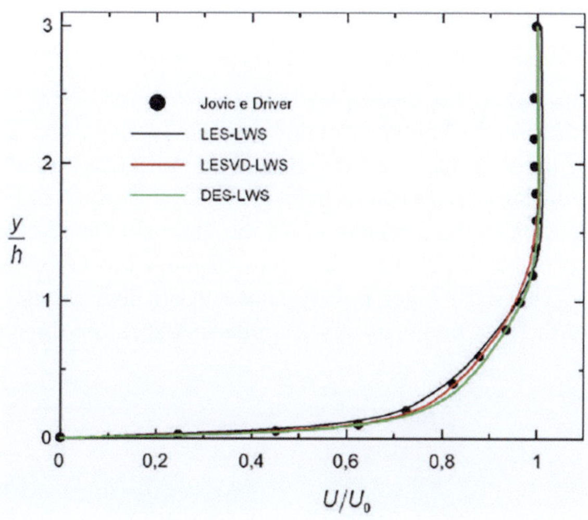

Fig. 10.15 Comparison between the average velocity profiles at the channel entrance, obtained with the Lund, Wu and Squires method [6] and the Experimental Data of Jovic and Driver [2]

Fig. 10.16, including comparisons with experimental data. These are the profiles used as domain entry conditions. There is excellent agreement with the data from the Jovic and Driver material experiment [2].

With the results presented, we point to the use of the Lund, Wu and Squires methodology [6] as a suitable way to model the real boundary conditions of a turbulent boundary layer. The data is now used as an input condition for the flow over the step, thus demonstrating the great influence that the boundary condition has on the flow modeling.

LES was selected to show this influence. The influence of the position of the plane upstream of the step, on which the boundary condition will be imposed, will also be tested. Three cases will be presented, SLE–A, SLE–B, SLE–C, as illustrated in Fig. 10.17. The laminar boundary condition was imposed on the step, in Case

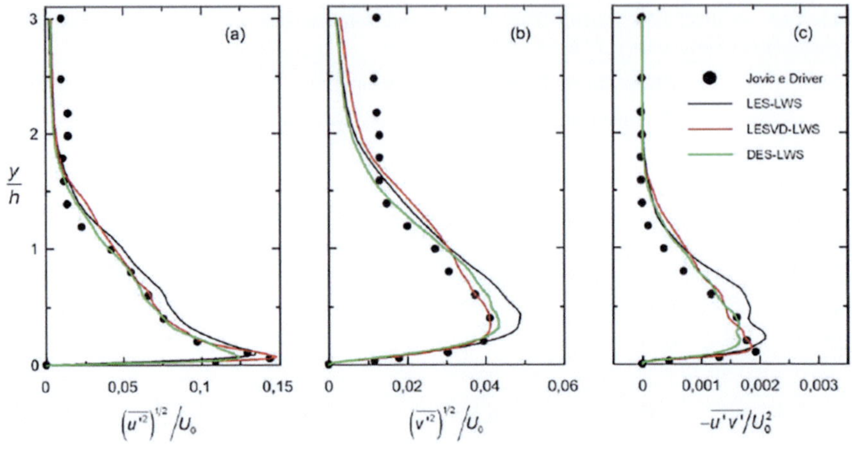

Fig. 10.16 Distributions of RMS $\sqrt{\overline{u'^2}}/U_0$, $\sqrt{\overline{\vartheta'^2}}/U_0$ and the Boussinesq–Reynolds tensor $\sqrt{\overline{u'\vartheta'}}/U_0$

LES–A; the boundary condition, also laminar, was applied at a distance upstream of the step, in Case LES–B, and the turbulent boundary condition was imposed at a distance, upstream of the step, in Case LES–C. In Fig. 10.17, $\overline{u}(\vec{x}, t)$ is presented, relative to the direction x, for the three simulated cases presented above. The plan $(x, y, z = W/2)$ was selected to visualize this component of the velocity vector.

The shear layer is very active in all three cases, with regions of development of the new boundary layer similar between the three cases, for $x/h > 10$. The most

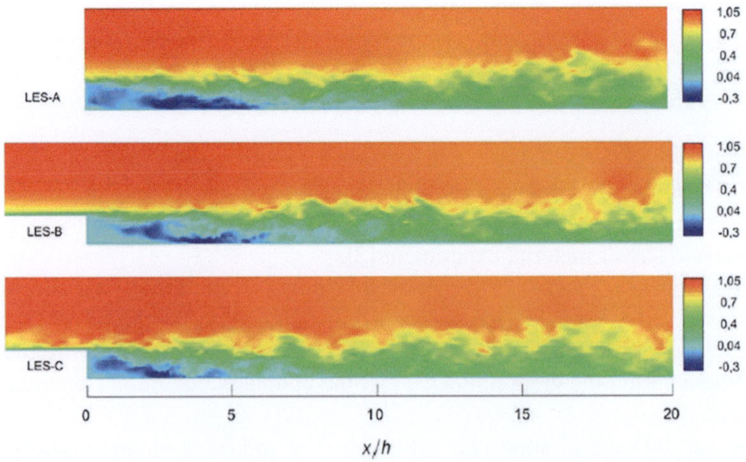

Fig. 10.17 Fields of the component $\overline{u}(\vec{x}, t)$ of the filtered velocity vector, at the plane $z = W/2$ and for time $t = 1.200h/U_0$

important difference occurs in the reattachment length, which, visually, is smaller in the LES–C case, for which the level of fluctuations in the inlet boundary layer is not zero. Qualitatively, the reattachment length can be identified by the sign of the velocity component in the main flow direction, which can be visualized in the color palette to the right of Fig. 10.17.

To quantify the effect of the boundary condition, the boundary layer reattachment length was determined for the three cases. The system presented by Silveira-Neto et al. [14] was adopted, which consists of placing velocity probes adjacent to the lower wall of the channel and sampling the percentage of time for which the component $\overline{u}(x/h)$ remains positive. This percentage is represented by the function $\Gamma(x/h)(\%)$. The reattachment length was also calculated using the average field of $\overline{u}(x/h)$ looking at the distribution into positive and negative values, near the lower wall.

The function $\Gamma(x/h)(\%)$ is presented, in Fig. 10.18, for the three configurations under test. The reattachment length for each case is determined by the crossing point with the line of $\Gamma = 50\%$. The Case LES–C presents the smallest value for this length. At Table 10.7, the values of the reattachment length obtained for the three geometric configurations, and the two forms of boundary conditions are presented: LES–A, LES–B and LES–C, and using two ways to determine reattachment length: using the average velocity field and the method $\Gamma(\%)$.

There is a difference between the results, with a reduction of 15% in the reattachment length for the case LES–C, compared to cases LES–A and LES–B, reaching $X_R/h = 6.07$. The value of X_R/h for the method $\Gamma(\%)$, Fig. 10.18, is slightly higher than that determined using the average velocity. The average reattachment length obtained with the case LES–C is very close to the experimental result, as can be seen

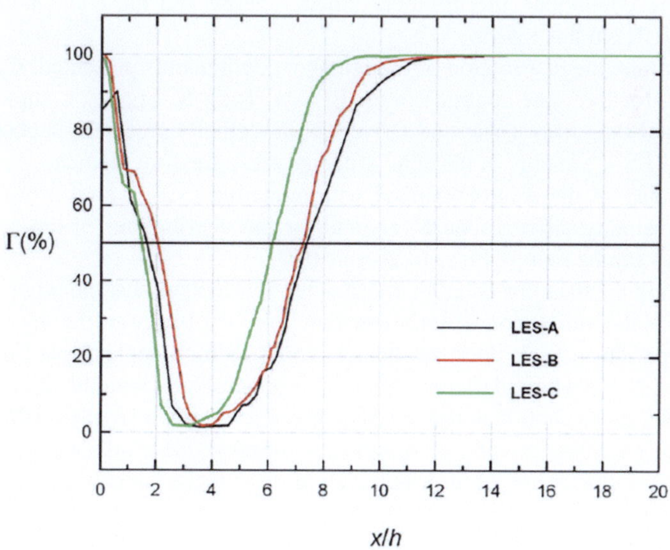

Fig. 10.18 Boundary layer reattachment length using the $\Gamma(\%)$ method

Table 10.7 Boundary layer reattachment length using the three Methodologies LES–A, LES–B, LES–C, and the two ways to evaluate this parameter

Cases	Reattachment length, X_R/h			Error (%)
	Average velocity field	Método $\Gamma(\%)$	Average of both forms	
Jovic and Driver [2]—Material experiment	–	–	6.0 ± 0.15	–
Le et al. [4]—Simulation with DNS	–	–	6.28	4.60
LES–A	7.20	7.36	7.28	21.3
LES–B	6.94	7.23	7.09	18.2
LES–C	6.02	6.12	6.07	1.0

in Table 10.7. The error is smaller than the uncertainty of the experiment. This makes it clear that, as important as modeling turbulence is the application of boundary conditions with which reality is adequately modeled, which is of great importance when the objective is to compare the results with a material experiment already carried out. It is important to highlight that when performing a material experiment, it is necessary to rigorously measure the data that will be used as boundary conditions at the entrance of the domain.

A good prediction of the reattachment length leads to better results for other statistical quantities associated with the flow. For example: the velocity components, the pressure coefficient, the turbulent kinetic energy and the components of the Boussinesq–Reynolds tensor.

To illustrate, the distribution of the pressure coefficient is presented, $C_p(x/h)$, in Fig. 10.19, for the three configurations, LES–A, LES–B, LES–C, comparing with experimental data from Jovic and Driver [2]. Excellent agreement is observed for modeling LES–C, mainly in the zone of the greatest pressure gradient.

The quality of the results obtained with modeling LES–C is also observed for the mean velocity distributions U/U_0 and for the distributions of the component $\overline{u'\vartheta'}/U_o^2$, as can be seen in Figs. 10.20 and 10.21.

Analyzing the data presented in this chapter, it is evident that the input boundary condition in the simulation of turbulent flow over the backward facing step plays a very important role. It does not matter if it is using a sophisticated turbulence closure model if the boundary conditions are not consistent with the reference data, for example, those from material experiments. Even using a simple model, such as Smagorinsky's, it was possible to significantly improve the results when a boundary condition was used at the input, which well represents the material experiment.

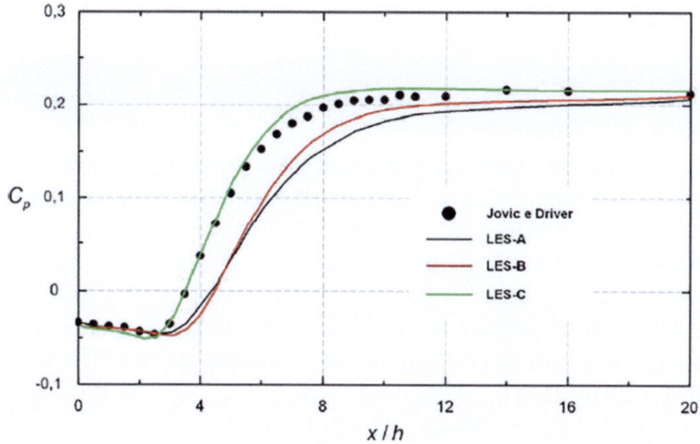

Fig. 10.19 Distribution of the pressure coefficient obtained using the modeling types: LES–A, LES–B and LES–C

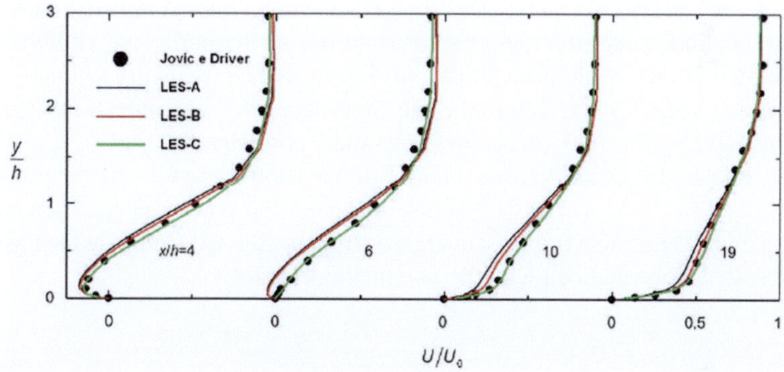

Fig. 10.20 Average velocity distributions, $\overline{u}/U_0$

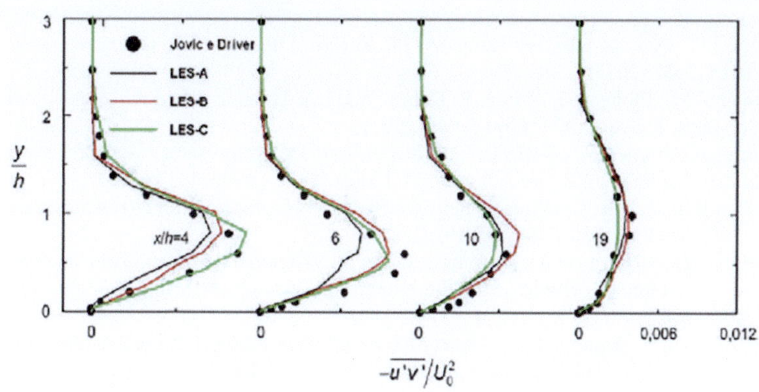

Fig. 10.21 Distributions of the component $\sqrt{\overline{u'\vartheta'}}/U_0$ of the Boussinesq–Reynolds tensor

10.6 Exercises

1. In Sect. 10.2.5, turbulence induction was presented with a wire glued upstream of the step, with a diameter $d = 0.1667h$. Develop a routine to mathematically model the disturbance injected by the wire glued to the lower wall of the channel upstream of the step. Available noise modeling routines can be used.
2. Propose an experimental method (material or computational) to determine the reattachment length of the boundary layer on the lower wall of the channel, after the step, presented in Sect. 10.2.5.
3. All models are based on the concept of turbulent viscosity, ν_t, are proposed using two characteristic quantities of turbulence. List at least three of them. Discuss and justify this statement. Is there any model based on ν_t for which these requirements does not need to be met? Which and why?
4. What is a permanent regime? And a statistically permanent regime?
5. Consider the flow over a cylinder, illustrated in Fig. 10.4. Simulate this flow using the LES and URANS methodologies. For LES, the simulation must be in 3D, while for URANS, it can be 2D. Why? Place computational probes downstream of the cylinder. Plot the temporal distributions of the horizontal component of velocity obtained with these probes. (a) Compare the distributions obtained with LES and URANS; (b) determine the time necessary to reach the statistically permanent regime; (c) Using these samples, eliminate the part of the statistically transient regime. Perform a DFFT of the samples, and, using the procedure described in Chap. 4, Eqs. 4.3 and 4.14, obtain the power spectral densities in frequency. Determine the most energized frequencies; (d) Compare the turbulent viscosity levels obtained with the two methodologies.

References

1. Coder, J.G.: A scale-adaptive variant of the Spalart-Allmaras eddy-viscosity model. AIAA pp. 2015–2462 (2015)
2. Germano, M., Piomelli, U, Moin, P., Cabot, W.H.: A dynamic sub grid-scale eddy viscosity model. Phys. Fluids A **3**(7), 1760–3128 (1991)
3. Keating, A., Piomelli, U., Ballaras, E., Kaltenbach, H.J.: A priori and a posteriori tests of inflow conditions for large-eddy simulation. Phys. Fluids **16**(12), 4696–4712 (2004)
4. Le, H., Moin, P., Kim, J.: Direct numerical simulation of turbulent flow over a backward facing step. J. Fluid Mech. **330**, 349–374 (1997)
5. Silva, A.L.E., Silveira-Neto, A., Damasceno, J.J.R.: Numerical simulation of two-dimensional flows over a circular cylinder using the immersed boundary method. J. Comput. Phys. **189**, 351–370 (2003)
6. Lund, T., Wu, X., Squires, K.D.: Generation of turbulent inflow data for spatially- developing boundary layer simulations. J. Comput. Phys. **140**, 233–258 (1998)
7. Menter, F.R.: Two-equation eddy-viscosity turbulence models for engineering applications. AIAA J. **32**, 1598–1605 (1994)
8. Nichols, R.: Turbulence models and their application to complex flows (2010)

9. Oliveira, J.E.S., Silveira-Neto, A.: Método da fronteira imersa aplicado À modelagem matemática e simulação computacional de escoamentos turbulentos sobre geometrias móveis e deformáveis, Tese de doutorado, FEMEC-UFU (2006)
10. Romão, R.S.M., Kinoshita, D., Villar, M.M., Serfaty, R., Silveira-Neto, A.: Numerical experiment of turbulent transfer of thermal energy using the immersed boundary method and adaptative mesh refinement. Int. J. Therm. Sci. **151**, 1–11 (2020)
11. Roshko, A.: Experiments on the flow past a circular cylinder at very High Reynolds Number. J. Fluid Mech. **10**, 345–356 (1967)
12. Schlichting, H.: Boundary-Layer Theory. McGraw-Hill Inc., New York (1979)
13. Shur, M., Spalart, P., Strelets, M., Travin, A.: Detached-Eddy simulation of an airfoil at high angle of attack. In: 4th International Symposium on Engineering Turbulence Modelling and Measurements, pp. 669–678. Elsevier, Amsterdam (1999)
14. Silveira-Neto, A., Grand, D., Lesieur, M., Métais, O.: A numerical investigation of coherent vortices in turbulence behind a backward-facing step. J. Fluid Mech. **256**, 1–25 (1993)
15. Spalart, P.R., Allmaras, S.R.: A One-Equation Turbulence Model for Aerodynamic Flows, AIAA Paper 92–0439 (1992)
16. Spalart, P.R., Allmaras, S.R.: A one-equation turbulence model for aerodynamic flows. La Recherche Aerospatiale **1**, 5–21 (1994)
17. Spode, C., Silveira-Neto, A.: Simulação de grandes escalas e simulação híbrida RANS/LES do escoamento sobre um degrau com condições de contorno turbulentas, dissertação de mestrado apresentada ao programa de pós-graduação em engenharia mecânica da Universidade Federal de Uberlândia (2006)
18. Sucker, D., Brauer, H.: Investigation of the flow around transverse cylinders. Waerme und Stoffuebertragung **8**(3), 149–158 (1975)
19. White, F.M.: Mecânica dos Fluidos. McGraw Hill, Rio de Janeiro, Brasil (1991)
20. Wieselberger, C.: Der Luftwiderstand von Kugeln. Z. Flugtech. Motor **5**, 140–144 (1914)
21. Wilcox, C.W.: Turbulence Modeling for CFD, La Cañada, Ca, DCW Industries (1998)

Glossary

"Before establishing a dialogue, it is necessary to make the meaning of words common knowledge."
René Descartes
"If you wish to converse with me define your terms."
Voltaire

Advection Advection is a process of transporting a given information (i.e., a physical information, a fluid information, a chemical information, or energy information) due to the movement of matter. A fluid medium in motion may transport various information. Commonly occurring processes are the *advective transport of thermal energy*; *the advective transport of linear momentum*; and/or *the advective transport of the concentration of a chemical species* by fluid motion.

Balance A balance of a given information consists of observing whether the amount of the physical information contained in a control volume is changing in time. The rate of change of the amount of that information contained in the control volume must be accounted for by only two processes: the information is transported and/or transformed into another form. The control volume can experience the flow of this information across its boundary. Inside the control volume, transformations of this information into other forms of correlated informations can occur. In this interpretation of balance, it is assumed that processes of "generation" and "destruction/dissipation" of an information are not considered. For example, assume that the information is "energy". In classical physics, it is not possible to generate or to destroy energy. All that can be done, by an appropriate process, is to accumulate, transport or transform energy.

Balance Equations Any physical information can be the subject of a balance, with which the so-called balance equations are obtained. For example, mass balance, linear momentum balance, angular momentum balance, energy balances, scalars, etc. The term "transport equations" is not consistent, since in a balance process, other types of occurrences are found, in addition to transport. A balance is composed of storage processes (accumulation), transport processes (advective, diffusive and others), and transformation of an information. "Conservation equations" is not an appropriate terminology for balance equations, as rarely does a balance imply conservation. Nor can these equations be called "governing equations", since no physical process can be governed by equations. Physical processes are governed by natural, universal and immutable laws. In this book, these equations are called "balance equations".

A. da Silveira Neto, *Turbulent Flows*, https://doi.org/10.1007/978-3-031-89222-6

Compressible Flow A characteristic associated with fluid flow, compressibility is related to the Mach number, $M = U/c$. The higher the value of Mach number, the more compressible the flow behavior becomes, and the more sensitive the density of a fluid particle is with pressure variations. For $M \leq 0.34$, the flow can be considered incompressible, since the pressure fluctuations due to the flow are negligibly small. The fluid density is a function of pressure, so negligibly small pressure fluctuations allow the flow to be considered incompressible. We must not confuse or interchange the concept of "a compressible fluid" with "a compressible flow". Both are valid concepts, but with different meanings. The first refers to the fluid which can be conveniently measured with the fluid at rest. The second concept only makes sense when the fluid is in motion.

Compressible Fluids Air is a compressible fluid. Water is a slightly compressible fluid. The reduction a given volume of liquid water by 1% requires a pressure of tens of atmospheres. However, a cyclist in motion experiences an incompressible flow of air since $M \leq 0.34$. On the other hand, a missile moving in water at $M > 0.34$ experiences a compressible flow around it. Thus, care must be taken with the use of the expressions "compressible flow" and "compressible fluid". They have very different meanings.

Conservation An information is said to be conserved, during a given process, when the total amount of this information present in the universe does not vary with time. For example, thermal energy is not conserved, since other types of energy can be transformed into thermal energy and vice versa. Kinetic energy, potential energy, etc., are not conservatives quantities. The total energy contained in the universe is conserved. However, the total energy contained in a system is not necessarily conserved, since, normally, there are exchanges with the environment that surrounds it, implying an accumulation of the total energy in the system.

Convection Combined process of transporting a given information. Convection is the sum of advection and diffusion.

Density Versus Specific Mass The concepts of density and specific mass, both in [kg/m^3], are assigned different physical meanings. It is important to rigorously and appropriately use these nomenclatures. The concept of density is used for substances and the concept of specific mass for bodies composed by more than one substance or for multi-substance material media. For example, density of water, density of air, density of steel etc. On the other hand, we speak of the specific mass of a ship. The specific mass of the ship, made of steel and other materials, must be less than the density of water, for flotation to occur.

Dissipation This term is used for the diffusion or dispersion of energy. It appears as "energy dissipation" which suggests that the energy has been expended with no benefit. Dissipated energy leads to the understanding that energy has diffused or "disappeared" (even though it is not a visible information), or that it has gone to some unknown place. In this book, these terminologies, "dissipate" and "dissipation", were avoided, and an attempt was made to use a more appropriate terminology in line with the concept of transformation: in nature, a physical information is "neither created nor destroyed, everything is transformed" according to Lavoisier. The words "everything" and "nothing," used by Lavoisier, should be interpreted with care. For example, one can destroy a seed, but one cannot destroy all the informations stored in it. The word *information* was adopted here as something very general that can be used in place of Lavoisier's "everything" or "nothing". Thus, it is admitted that, in nature, one cannot create or destroy any physical information. In analyzing a thermodynamic process, for example, various forms of energy, entropy, as well as mass, chemical components, linear momentum, angular momentum, etc., can play a role in the process. Thus, each of these contributions will be called an information in this book. Whenever something is identified as an information, it is not created or destroyed but it can be transformed. It can be accumulated, and it can be transported. When the word transformation is used, it is clear that an information has been changed into another related information. In this sense, one is even interested in knowing the destination of the product of this transformative process. It is noticed that the good use of terminology can arouse investigative interest in a given Information.

Divergent Form Versus Non-Divergent Form The advective terms of the *balance equations* can be written in two ways, mathematically equivalent, when they compose these *balance equations*. These are the so-called divergent and non-divergent forms, written below, for the advective transport of any information ϕ, $\vec{\nabla} \cdot \left(\phi \vec{V}\right)$ and $\vec{V} \cdot \vec{\nabla}\phi$, respectively. These forms are named, in parts of the literature, as the *conservative form* and the *non-conservative form*, respectively. The author of this book and some authors of other texts prefer the nomenclature *divergent* and *non-divergent* forms, considering that this term does not represent any type of conservation. It represents the net advective flux of any information ϕ. This term does not even compose conservation equations. It composes balance equations instead.

Experimentation Once a phenomenon is modeled, the model can be used for experiments. Thus, there can be theoretical and/or material experiments. A model of an aircraft, developed with resin and installed in a wind tunnel, constitutes a material model of an aircraft in flight in the Earth's atmosphere. In this case, material experiments are carried out, which the global scientific community usually calls "physical experiments". But this same problem can be modeled from a theoretical point of view, whether mathematical or computational, with which computational simulations will be developed. This type of modeling cannot be devoid of physical meaning. Thus, both modeling processes—the material and the theoretical—must incorporate as much of the physical sense as possible in the modeled problem. Both are physical modeling. Experimentation involves studying a given problem, a given physical effect, a given phenomenon. Albert Einstein said that the most powerful way to experiment with natural phenomena is through thought. To study the problem, material and theoretical experimentation are used. Thus, in this book, it is common to talk about material experiments and theoretical experiments.

Fluid Particle Versus Molecule The concept of a fluid particle is linked to the continuum hypothesis. It is a cluster of molecules or atoms. This cluster must have a volume large enough so that there is independence of statistical moments as to the size of the cluster. On the other hand, the size of this cluster must be small enough so that the property of "limit tending to zero" of differential calculus is valid, when we make the size of the molecule cluster tend to "zero". This "zero" cannot be absolute, for the continuum hypothesis to be valid.

Heat and Work Heat and work, according to established thermodynamics concepts, there are only two ways of exchanging energy between a closed system and its surroundings. To be more precise, heat and work are two ways of exchanging energy. Thus, heat and work are not substances that can be stored, transferred, or transformed. Therefore, the term "heat transfer" is considered inappropriate. In fact, one might ask, by analogy, why not talk about "work transfer"? It is argued that historically, the word heat has also been attributed to the meaning of thermal energy. At the very least, the use of this nomenclature can cause confusion in the students understanding, and it is not necessary to use this nomenclature! Realizing that heat is used instead of thermal energy, it is enough to talk about thermal energy transfer or simply thermal transfer. By extension, one starts talking about thermal exchanger instead of heat exchanger. This is not more work, but it is more correct and more elegant in the context of scientific communication.

Instabilities Once disturbances are injected into the flows, they can be amplified in a given frequency band, giving rise to instabilities. As an example, we can mention Kelvin–Helmholtz instabilities, Rayleigh–Bénard instabilities, etc. A cluster of instabilities, of different frequencies, that coexist in space and time, interacting non-linearly with each other, results in one or more turbulent structures.

Modeling Modeling represents a given *physical phenomenon*, which can be done in the form of a *physical model*, in the form of a *theoretical model* or in the form of a *material model*. Once the phenomenon is modeled, from a theoretical or material point of view, the model can be used to perform simulations. Thus, we can have a theoretical simulation and or a material simulation. A model of an aircraft, developed with resin, installed in a wind tunnel, constitutes a material model of an aircraft in flight in the earth's atmosphere. A mathematical model of this same problem can be extended to a computational model, with which computational simulations will be developed. The *material experiments* are referred to, by the global scientific community, as "physical experiment". But this same problem, can be modeled from the *theoretical point of view*, be it mathematical, be it computational, with which computational simulations will be developed. This type of modeling cannot be devoid of physical sense. Thus, both models, *material* and *theoretical*, must carry all the physical sense of the modeled problem. Both are physical models. As a result, we use the terms "theoretical modeling" and "material modeling". Both are physical models!

Modeling: Closure Models Closure modeling is composed of complementary algebraic-differential equations to those that have their origins in the fundamental balances of some information: mass, linear momentum, energies etc. For example, the balance equations for the information k and ε make up the closure model. All equations together make up the turbulence model.

Modeling: Exact, Continuous and Analytical A mathematical model serves to represent a physical problem mathematically. For a model to be exact, it would have to represent exactly the physical problem, which will hardly happen. We will always have to make simplifying modeling hypotheses. A model will be continuous, when the resulting model, after the integration of the differential model, be continuous. A modeling will be analytical when the model resulting from the integration of the differential equations is continuous, including the derivatives, of any order, of the obtained solution.

Modeling: Turbulence Models Turbulence models are composed of all balance equations, plus the equations for the closure of turbulence. The closure modeling is composed of algebraic-differential equations complementary to those that have their origins in the fundamental balances of physical quantities: mass, linear momentum, energies, etc. The balance equations for the quantities, k and ε, for example, make up a closure model. All equations together make up the turbulence model.

Molecular Diffusion Molecular diffusion is the process of transporting a given physical information into a material medium due to the existence of potential gradients associated with the physical information being transported. The diffusive flux of a given physical information is proportional to the gradient of the potential, and the proportionality function is a diffusion coefficient of the physical information in the material medium in which the diffusion is occurring. Some authors confuse the word "molecular" with the word "laminar". It is common to find mention of "laminar diffusion", which is devoid of physical meaning. The diffusion process is purely molecular in nature. The turbulent flow regime only affects the intensity of the molecular effect, since it increases the intensity of the local gradients, as well as the gradient density, that is, the number of gradients per unit volume.

Oberbeck–Boussinesq Approximation Oberbeck–Boussinesq established an approximation for modeling incompressible flows with variable density. Using this approximation, the differential mathematical model, for this group of flows, assumes that the fluid's density is constant for all terms of the model, except for the term that represents the weight of the fluid particle. It is observed that the equality $(\rho - \rho_o)/\rho_o = \beta(T - T_o)$, normally used for this type of modeling is not part of the approximation in question. This equality is a thermodynamic relation.

Personification In a scientific text, it is undesirable to personify things or objects that take on, in the text, the role of a person. Some examples are presented: "Figure XX shows ...", "This thesis proposes ...", "Equation XX says that ...". In this book, this type of expression was avoided, which, in the author's opinion, makes the text lighter to read and avoids ambiguous interpretations.

Perturbations Perturbations result from external and internal agents to the flows, which are necessary for the flows to become unstable. For example, pressure waves that impinge on the flows and propagate within them. The perturbations can be of mechanical, dynamic, thermal, geometric origin etc. The perturbations can occur at a given frequency or even over a frequency band. Once injected into the flows, they can be amplified over a given frequency band, giving rise to instabilities. A cluster of instabilities, of different frequencies, that coexist in space and time, interacting non-linearly with each other, results in one or more turbulent structures.

Scale There are several definitions of scale. Among them, two stand out in this book. The first means stopping points of a bus, an aircraft, a ship, etc. The second, the most used, refers to a band of possible variations of a given information or some natural phenomenon. With a scale, one can frame the magnitude of the information and know if it is high, medium or low intensity. For example, there is the Celsius Temperature Scale, the Beaufort Scale to check the intensity of the winds, the Fujita Scale to frame the intensities of tornadoes, the Mohs Scale to frame the hardness of minerals, etc. It is common to find in the literature the incorrect use of the concept of scale in place of the intensity of the information itself. For example, in turbulence studies, there are references to "length scale", instead of "characteristic length". There is reference in the literature to the "integral length scale", where one should simply say "integral length", "Kolmogorov Scale", for what should be referred to as "Kolmogorov Characteristic Length". There is also "velocity scale", instead of "characteristic velocity".

Simulation To simulate is to reproduce a given physical phenomenon under controlled conditions. To simulate, it is necessary to model the phenomenon, which can be done theoretically or materially. Once the phenomenon is modeled, the model can be used to carry out its simulation. Thus, we can have a theoretical simulation and or a material simulation. A model of an aircraft, developed with resin, installed in a wind tunnel, constitutes a material model of an aircraft in flight in the earth's atmosphere. A mathematical model of this same problem can be extended to a computational model, with which computational simulations will be developed.

Temperature The temperature scale is used to measure the level of thermal energy of matter. With it, we measure the degree of molecular agitation of matter. Thus, we can identify different states of matter, with different temperature values. We can then talk about high temperature or low temperature. We can never talk about "hot temperature" or "cold temperature". "Hot" or "cold" are states of matter and not of temperature.

Turbulent Diffusion Turbulent diffusion is a modelof the non-linear advective process of transporting an information in space and time. By analogy to the process of molecular diffusion, portions of non-linear interactions are modeled as a turbulent diffusive process, using the concepts of turbulent viscosity and filtered potential gradients. It should be noted that this is just a modeling method. Diffusion only occurs at the molecular level. The turbulent flow regime, physically, accelerates and intensifies molecular diffusion.

Turbulent Structures Once disturbances are injected into the flows, they can be amplified in each frequency, giving rise to instabilities. As an example, in free shear flows, we find Kelvin–Helmholtz instabilities, which can organize and form a turbulent structure with multiple frequencies, or multiple wave numbers. A cluster of instabilities, of different frequencies, coexisting in space and time, interacting non-linearly with each other, results in one or more turbulent structures, which, in turbulent flows, can be identified throughout the observation space, with greater or lesser concentration and intensity, in accordance with the problem under observation and analysis.

Vortex There have been many attempts to define a vortex. An accepted definition is "a mass of fluid in rotational motion around an axis, in which there is invariance of statistical moments, in all directions". According to this and most definitions of "vortex", this is an entity that will hardly be found in any type of flow, even in the laminar regime. Turbulence is composed of three-dimensional and highly rotational movement, whose topology completely deviates from the definition of "vortex". What is observed in turbulent flows is a complex topology, with concentration of vorticity in regions distributed in space and with large temporal variations. In these regions, complex compositions are observed which are referred to, in part of the literature, as *turbulent structures*. This is the nomenclature used in this book.

MIX
Papier aus verantwortungsvollen Quellen
Paper from responsible sources
FSC® C105338

If you have any concerns about our products,
you can contact us on
ProductSafety@springernature.com

In case Publisher is established outside the EU,
the EU authorized representative is:
Springer Nature Customer Service Center GmbH
Europaplatz 3, 69115 Heidelberg, Germany

Printed by Libri Plureos GmbH
in Hamburg, Germany